Technology for Biofuels

Technology for Biofuels

Editor: Damian Price

CALLISTO REFERENCE

www.callistoreference.com

Callisto Reference,
118-35 Queens Blvd., Suite 400,
Forest Hills, NY 11375, USA

Visit us on the World Wide Web at:
www.callistoreference.com

ISBN: 978-1-64116-081-0 (Hardback)

Printed in China.

Cataloging-in-Publication Data

Technology for biofuels / edited by Damian Price.
p. cm.
Includes bibliographical references and index.
ISBN 978-1-64116-081-0
1. Biomass energy. 2. Biomass energy industries--Technological innovations.
I. Price, Damian.

TP339 .T43 2019
662.88--dc23

Table of Contents

Preface VII

Chapter 1 **A type-I diacylglycerol acyltransferase modulates triacylglycerol biosynthesis and fatty acid composition in the oleaginous microalga, *Nannochloropsis oceanica*** 1
Hehong Wei, Ying Shi, Xiaonian Ma, Yufang Pan, Hanhua Hu, Yantao Li, Ming Luo, Henri Gerken and Jin Liu

Chapter 2 **Electricity-assisted production of caproic acid from grass** 19
Way Cern Khor, Stephen Andersen, Han Vervaeren and Korneel Rabaey

Chapter 3 **Enzymatic degradation of sulfite-pulped softwoods and the role of LPMOs** 30
Piotr Chylenski, Dejan M. Petrović, Gerdt Müller, Marie Dahlström, Oskar Bengtsson, Martin Lersch, Matti Siika-aho, Svein Jarle Horn and Vincent G. H. Eijsink

Chapter 4 **Synthetic CO_2-fixation enzyme cascades immobilized on self-assembled nanostructures that enhance CO_2/O_2 selectivity of RubisCO** 43
Sriram Satagopan, Yuan Sun, Jon R. Parquette and F. Robert Tabita

Chapter 5 **Highly efficient methane generation from untreated microalgae biomass** 57
Viktor Klassen, Olga Blifernez-Klassen, Daniel Wibberg, Anika Winkler, Jörn Kalinowski, Clemens Posten and Olaf Kruse

Chapter 6 **High-throughput analysis of chemical components and theoretical ethanol yield of dedicated bioenergy sorghum using dual-optimized partial least squares calibration models** 69
Meng Li, Jun Wang, Fu Du, Boubacar Diallo and Guang Hui Xie

Chapter 7 **Screening and evolution of a novel protist xylose isomerase from the termite *Reticulitermes speratus* for efficient xylose fermentation in *Saccharomyces cerevisiae*** 85
Satoshi Katahira, Nobuhiko Muramoto, Shigeharu Moriya, Risa Nagura, Nobuki Tada, Noriko Yasutani, Moriya Ohkuma, Toru Onishi and Kenro Tokuhiro

Chapter 8 **Genome sequencing and transcriptome analysis of *Trichoderma reesei* QM9978 strain reveals a distal chromosome translocation to be responsible for loss of *vib1* expression and loss of cellulase induction** 102
Christa Ivanova, Jonas Ramoni, Thiziri Aouam, Alexa Frischmann, Bernhard Seiboth, Scott E. Baker, Stéphane Le Crom, Sophie Lemoine, Antoine Margeot and Frédérique Bidard

Chapter 9 **Coexistence and competition of sulfate-reducing and methanogenic populations in an anaerobic hexadecane-degrading culture** 117
Ting-Ting Ma, Lai-Yan Liu, Jun-Peng Rui, Quan Yuan, Ding-shan Feng, Zheng Zhou, Li-Rong Dai, Wan-Qiu Zeng, Hui Zhang and Lei Cheng

Chapter 10 **In vitro oxidative decarboxylation of free fatty acids to terminal alkenes by two new P450 peroxygenases** 130
Huifang Xu, Linlin Ning, Wenxia Yang, Bo Fang, Cong Wang, Yun Wang, Jian Xu, Severine Collin, Frederic Laeuffer, Laurent Fourage and Shengying Li

Chapter 11 **Bioethanol production from spent mushroom compost derived from chaff of millet and sorghum** 145
Peter Ryden, Maria-Nefeli Efthymiou, Teddy A. M. Tindyebwa, Adam Elliston, David R. Wilson, Keith W. Waldron and Pradeep K. Malakar

Chapter 12 **Enhanced polyhydroxyalkanoate (PHA) production from the organic fraction of municipal solid waste by using mixed microbial culture** 156
Bianca Colombo, Francesca Favini, Barbara Scaglia, Tommy Pepè Sciarria, Giuliana D'Imporzano, Michele Pognani, Anna Alekseeva, Giorgio Eisele, Cesare Cosentino and Fabrizio Adani

Chapter 13 **Metabolic engineering of *Rhodopseudomonas palustris* for the obligate reduction of *n*-butyrate to *n*-butanol** 171
Devin F. R. Doud, Eric C. Holmes, Hanno Richter, Bastian Molitor, Georg Jander and Largus T. Angenent

Chapter 14 **Limitation of cellulose accessibility and unproductive binding of cellulases by pretreated sugarcane bagasse lignin** 182
Germano Siqueira, Valdeir Arantes, Jack N. Saddler, André Ferraz and Adriane M. F. Milagres

Chapter 15 **Sustaining fermentation in high-gravity ethanol production by feeding yeast to a temperature-profiled multifeed simultaneous saccharification and co-fermentation of wheat straw** 194
Johan O. Westman, Ruifei Wang, Vera Novy and Carl Johan Franzén

Chapter 16 **Identification of developmental stage and anatomical fraction contributions to cell wall recalcitrance in switchgrass** 210
Jacob D. Crowe, Nicholas Feringa, Sivakumar Pattathil, Brian Merritt, Cliff Foster, Dayna Dines, Rebecca G. Ong and David B. Hodge

Permissions

List of Contributors

Index

Preface

Biofuels are fuels extracted or produced through biological processes or other anaerobic methods of fuel production. They are renewable sources of energy. Some common biofuels include bioethanol, biodiesel, biogas, etc. One of the many prominent advantages offered by biofuels is that they generate energy by undergoing complete combustion. Biofuels are therefore considered the future of energy. The data collated in this book presents an international perspective on the technological development of biofuels. The chapters compiled herein cover an array of significant topics ranging from fundamental concepts and theories to the modern applications. This book will serve as a valuable source of knowledge for graduates, post graduates, researchers and students in this field.

After months of intensive research and writing, this book is the end result of all who devoted their time and efforts in the initiation and progress of this book. It will surely be a source of reference in enhancing the required knowledge of the new developments in the area. During the course of developing this book, certain measures such as accuracy, authenticity and research focused analytical studies were given preference in order to produce a comprehensive book in the area of study.

This book would not have been possible without the efforts of the authors and the publisher. I extend my sincere thanks to them. Secondly, I express my gratitude to my family and well-wishers. And most importantly, I thank my students for constantly expressing their willingness and curiosity in enhancing their knowledge in the field, which encourages me to take up further research projects for the advancement of the area.

Editor

1

A type-I diacylglycerol acyltransferase modulates triacylglycerol biosynthesis and fatty acid composition in the oleaginous microalga, *Nannochloropsis oceanica*

Hehong Wei[1], Ying Shi[1], Xiaonian Ma[1], Yufang Pan[2], Hanhua Hu[2], Yantao Li[3], Ming Luo[4], Henri Gerken[5] and Jin Liu[1*]

Abstract

Background: Photosynthetic oleaginous microalgae are considered promising feedstocks for biofuels. The marine microalga, *Nannochloropsis oceanica,* has been attracting ever-increasing interest because of its fast growth, high triacylglycerol (TAG) content, and available genome sequence and genetic tools. Diacylglycerol acyltransferase (DGAT) catalyzes the last and committed step of TAG biosynthesis in the acyl-CoA-dependent pathway. Previous studies have identified 13 putative DGAT-encoding genes in the genome of *N. oceanica*, but the functional role of *DGAT* genes, especially type-I *DGAT* (*DGAT1*), remains ambiguous.

Results: *Nannochloropsis oceanica* IMET1 possesses two *DGAT1* genes: *NoDGAT1A* and *NoDGAT1B*. Functional complementation demonstrated the capability of NoDGAT1A rather than NoDGAT1B to restore TAG synthesis in a TAG-deficient yeast strain. In vitro DGAT assays revealed that NoDGAT1A preferred saturated/monounsaturated acyl-CoAs and eukaryotic diacylglycerols (DAGs) for TAG synthesis, while NoDGAT1B had no detectable enzymatic activity. Assisted with green fluorescence protein (GFP) fusion, fluorescence microscopy analysis indicated the localization of NoDGAT1A in the chloroplast endoplasmic reticulum (cER) of *N. oceanica*. *NoDGAT1A* knockdown caused ~25% decline in TAG content upon nitrogen depletion, accompanied by the reduced C16:0, C18:0, and C18:1 in TAG *sn*-1/*sn*-3 positions and C18:1 in the TAG *sn*-2 position. *NoDGAT1A* overexpression, on the other hand, led to ~39% increase in TAG content upon nitrogen depletion, accompanied by the enhanced C16:0 and C18:1 in the TAG *sn*-1/*sn*-3 positions and C18:1 in the TAG *sn*-2 position. Interestingly, *NoDGAT1A* overexpression also promoted TAG accumulation (by ~2.4-fold) under nitrogen-replete conditions without compromising cell growth, and TAG yield of the overexpression line reached 0.49 g L^{-1} at the end of a 10-day batch culture, 47% greater than that of the control line.

Conclusions: Taken together, our work demonstrates the functional role of NoDGAT1A and sheds light on the underlying mechanism for the biosynthesis of various TAG species in *N. oceanica*. NoDGAT1A resides likely in cER and prefers to transfer C16 and C18 saturated/monounsaturated fatty acids to eukaryotic DAGs for TAG assembly. This work also provides insights into the rational genetic engineering of microalgae by manipulating rate-limiting enzymes such as DGAT to modulate TAG biosynthesis and fatty acid composition for biofuel production.

Keywords: Diacylglycerol acyltransferase, Functional characterization, Genetic engineering, Microalga, *Nannochloropsis oceanica*, Triacylglycerol

*Correspondence: gjinliu@pku.edu.cn
[1] Institute for Food and Bioresource Engineering, Department of Energy and Resources Engineering and BIC-ESAT, College of Engineering, Peking University, Beijing 100871, China
Full list of author information is available at the end of the article

Background

Triacylglycerols (TAGs) are energy-rich lipids widely used for human nutrition, chemical industries, and the production of high-energy density fuels [1]. Many oleaginous microalgae produce high levels of TAGs as the storage compounds [2–6], triggering substantial interest in exploring microalgae as biofuel feedstock. Efforts have been made for the further improvement of TAG yields, as TAG yields from the naturally occurring microalgal strains currently used for oil production are far less than the theoretical maximum [7, 8]. The understanding of pathways and regulatory mechanisms involved in lipid metabolism will help benefit the rational genetic manipulation of microalgae for improved oil production [9–11].

In higher plants, TAG biosynthesis has been documented and is believed to be mediated mainly via two pathways, acyl-CoA independent pathway and acyl-CoA-dependent Kennedy pathway [12]. The latter pathway starts from glycerol-3-phosphate with three sequential acylation steps, with the last step being mediated by a diacylglycerol acyltransferase (DGAT), which employs an acyl-CoA as the acyl donor and transfers the acyl moiety to the *sn*-3 position of DAG for TAG assembly [13]. DGAT catalyzes the only committed step in Kennedy pathway and is believed to play critical roles in TAG synthesis and accumulation. The well-characterized DGATs in higher plants include type-I (DGAT1) and type-II (DGAT2), the two structurally distinct groups of membrane-bound acyltransferases [14], and type-III (DGAT3), which is soluble and present in the cytosol [15]. Previous studies in higher plants have revealed distinct roles of different types of DGAT in TAG synthesis in different organisms [16, 17]. Bioinformatic analysis reveals the presence of *DGAT1* and *DGAT2* gene homologs in sequenced algal genomes [18–23]. Most of the current knowledge about the role of DGAT in algal TAG biosynthesis is derived from the model alga *Chlamydomonas reinhardtii*, in which DGAT2 has recently been characterized at the molecular and biochemical levels [24–26]. However, *Chlamydomonas* is generally not considered as an oleaginous organism for lipid production and may differ from oleaginous algae in lipid metabolism, driving the research interest to industrially important algae such as *Chlorella* and *Nannochloropsis* [21, 22, 27, 28].

Nannochloropsis oceanica has been recognized as an emerging model oleaginous alga in the study of TAG metabolism because of its fast growth, high TAG content, available genome sequence, and established genetic tools [20, 23, 29–32]. Thirteen putative DGAT-encoding genes were identified in the genomes of two *N. oceanica* strains: IMET1 [23] and CCMP1779 [20]. In the former strain, two out of the thirteen are annotated as *DGAT1* genes, *NoDGAT1A* and *NoDGAT1B*. RNA-Seq analyses following nitrogen depletion revealed the upregulation of *NoDGAT1A* and five *NoDGAT2*s [33], indicative of their roles in the TAG biosynthesis. In a recent study by Zienkiewicz et al. [28], one of the *DGAT2*s in *N. oceanica* was demonstrated to contribute to TAG biosynthesis. However, the role of *NoDGAT1*s in *N. oceanica* TAG synthesis remains ambiguous. Questions remain as to whether they are functional, how they are involved in TAG biosynthesis, and their potential in manipulating algae for improved lipid production.

By employing the industrially important oleaginous alga *N. oceanica*, we here functionally characterized algal *DGAT1* genes in depth by integrating the in silico, ex vivo, subcellular localization, in vitro and in vivo analyses. NoDGAT1A rather than NoDGAT1B shows acyltransferase activity. NoDGAT1A is likely localized in the chloroplast endoplasmic reticulum (cER), and prefers saturated/monounsaturated acyl-CoAs and eukaryotic DAGs for TAG biosynthesis. Overexpression experiments suggest the engineering potential of *NoDGAT1* in modulating TAG accumulation and fatty acid composition in microalgae. We discuss the role of NoDGAT1A in TAG metabolism and its implications for biotechnological applications in microalgae engineering.

Results

Growth, lipid variation, and *NoDGAT1* s expression upon nitrogen depletion

Nitrogen is essential for algal growth and lipid metabolism. In response to nitrogen depletion, *N. oceanica* showed severely impaired growth, revealed by the considerably lower biomass concentration and cell number (Fig. 1a, b). The cell weight of nitrogen-depleted cells was higher than that of nitrogen-replete ones (Fig. 1c). Consistently, nitrogen-depleted cells accumulated substantial amounts of TAG, up to 230 mg g^{-1} cell dry weight, while nitrogen-replete cells synthesized only a basal level of TAG (<20 mg g^{-1}) (Fig. 1d). DGAT is thought to play an important role in nitrogen-depletion-associated TAG accumulation in microalgae [26]. In *N. oceanica*, although a type-II *DGAT* gene has been recently characterized [28], the role of type-I *DGAT*s in TAG synthesis remains ambiguous. Here, two type-I *DGAT* genes, *NoDGAT1A* and *NoDGAT1B*, were examined at the transcriptional level. Obviously, the two genes were differentially regulated by nitrogen depletion: *NoDGAT1A* was upregulated moderately (up to fourfold increase), while *NoDGAT1B* had a much lower level of transcripts than *NoDGAT1A* and remained relatively stable in response to nitrogen depletion (Fig. 1e), suggesting that *NoDGAT1A* is more involved in the nitrogen-depletion-induced TAG biosynthesis. The induced TAG accumulation was

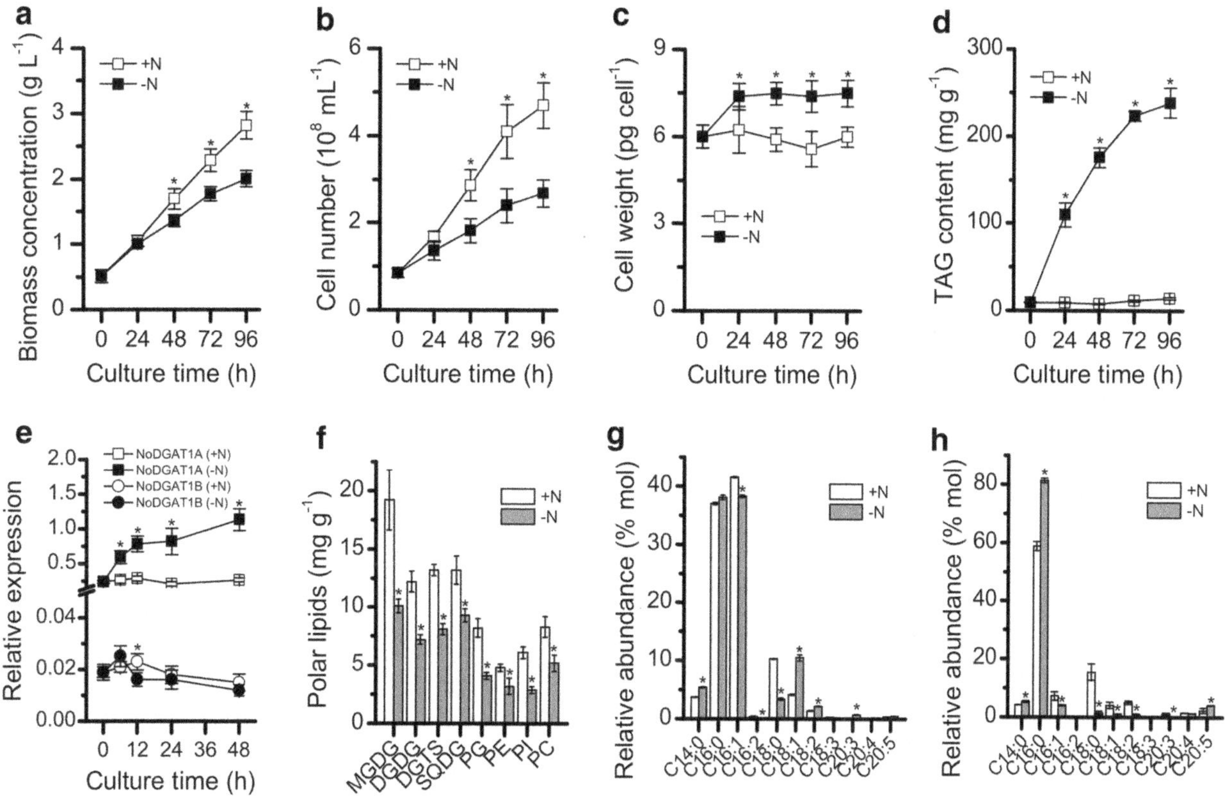

Fig. 1 Growth and lipid profiles of *N. oceanica* in response to nitrogen depletion. **a**–**e** Time course of biomass concentration (**a**), cell number (**b**), cell weight (**c**), TAG content (**d**), and the transcriptional expressions of *NoDGAT1A* and *NoDGAT1B* (**e**). −N and +N represent nitrogen depletion and nitrogen repletion, respectively. **f** Polar lipid contents at 48 h of nitrogen depletion. **g**, **h** Relative abundance of fatty acids in TAG *sn*-1/3 (**g**) and *sn*-2 (**h**) positions at 48 h of nitrogen depletion. C18:1 represents the mixture of C18:1n9 and C18:1n7. Data are expressed as mean ± SD ($n = 3$). *Asterisks* indicate the significant difference compared with +N (*t* test, $P < 0.05$)

accompanied by a decrease in membrane polar lipids including monogalactosyl diacylglycerol (MGDG), digalactosyl diacylglycerol (DGDG), diacylglycerol-N,N,N-trimethylhomoserine (DGTS), sulfoquinovosyldiacylglycerol (SQDG), phosphatidylglycerol (PG), phosphatidylethanolamine (PE), phosphatidylinositol (PI), and phosphatidylcholine (PC) (Fig. 1f), consistent with previous results in *Nannochloropsis* strains [30, 33] and indicative of the turnover of membrane lipids for TAG assembly under nitrogen-depleted conditions. Specifically, MGDG and DGDG, the major lipids of chloroplast, exhibited the most drastic reduction upon nitrogen depletion (Fig. 1f), consistent with the occurrence of shrunken chloroplast and reduced thylakoid membrane (Additional file 1: Figure S1).

The mass spectra of methylated fatty acids from TAG *sn*-1/*sn*-3 and *sn*-2 positions under different culture conditions are shown in Additional file 1: Figure S2. C16:0 and C16:1 represent the major fatty acids of TAG *sn*-1/*sn*-3 positions, which together account for over 75% of the fatty acids regardless of the culture conditions (Fig. 1g). Respective increase in C14:0, C18:1, and C18:2 was observed upon nitrogen depletion, with C18:1 having the largest increase, while C16:1 and C18:0 leveled off (Fig. 1g). In TAG *sn*-2 position, C16:0 is the predominant fatty acid, and it increased from 59% under nitrogen-replete conditions to 81% under nitrogen-depleted conditions, accompanied by decreases in C16:1, C18:0, C18:1, and C18:2 (Fig. 1h), suggesting that *N. oceanica* preferred to utilize the DAGs with *sn*-2 position being C16:0 for TAG synthesis upon nitrogen depletion.

*NoDGAT1*s cloning and functional analysis in yeast

The cDNAs of *NoDGAT1A* and *NoDGAT1B* are 1314 and 2298 bp in length, encoding polypeptides of 427 and 765 amino acids, respectively. To gain insights into the evolutionary relationship between *NoDGAT1* s and other orthologs, a cladogram was reconstructed using MEGA6 [34] based on the multiple sequences from higher plants, animals, fungi, and algae. Phylogenetic analysis reveals that DGAT family proteins are clustered into three major groups: type-I, type-II and type-III (Additional file 1:

Figure S3). NoDGAT1A and NoDGAT1B are closely related to the algal type-I DGAT orthologs. When analyzed on TMHMM Server 2.0 (http://www.cbs.dtu.dk/services/TMHMM/), NoDGAT1A and NoDGAT1B are predicted to have nine and eleven transmembrane domains, respectively (Additional file 1: Figure S4). Protein sequence alignment suggested that NoDGAT1s, NoDGAT1A in particular, contain conserved Motifs (Additional file 1: Figure S5), which are present in most type-I DGAT polypeptides [14].

To test the function of NoDGAT1 s, they were introduced into the TAG-deficient yeast strain H1246 for complementation. The transcriptional expression levels of the two genes were evaluated by quantitative real-time PCR (Additional file 1: Figure S6). The H1246 strain-overexpressing *ScDGA1*, a type-II DGAT from *Saccharomyces cerevisiae*, was used as the positive control. The expression of *NoDGAT1A* restored TAG biosynthesis in H1246 cells, as indicated by the prominent TAG spot on a TLC plate, which fractionated the lipid extracts from the transformed yeast cells (Fig. 2a), and the green fluorescence by BODIPY staining (Fig. 2b). By contrast, *NoDGAT1B* expression in H1246 cells failed to produce any detectable TAG (Fig. 2a, b), indicative of a nonfunctional encoded protein. Nevertheless, compared to *N. oceanica*, *S. cerevisiae* contains no such fatty acids as C14:0, C18:2, C20:4, and C20:5 (Fig. 1g, h), which may lead to the failure of functional complementation for *NoDGAT1B*.

The growth response of *DGAT*-expressing H1246 cells to free fatty acid feeding has been suggested as a feasible way for qualitative evaluation of the enzymatic

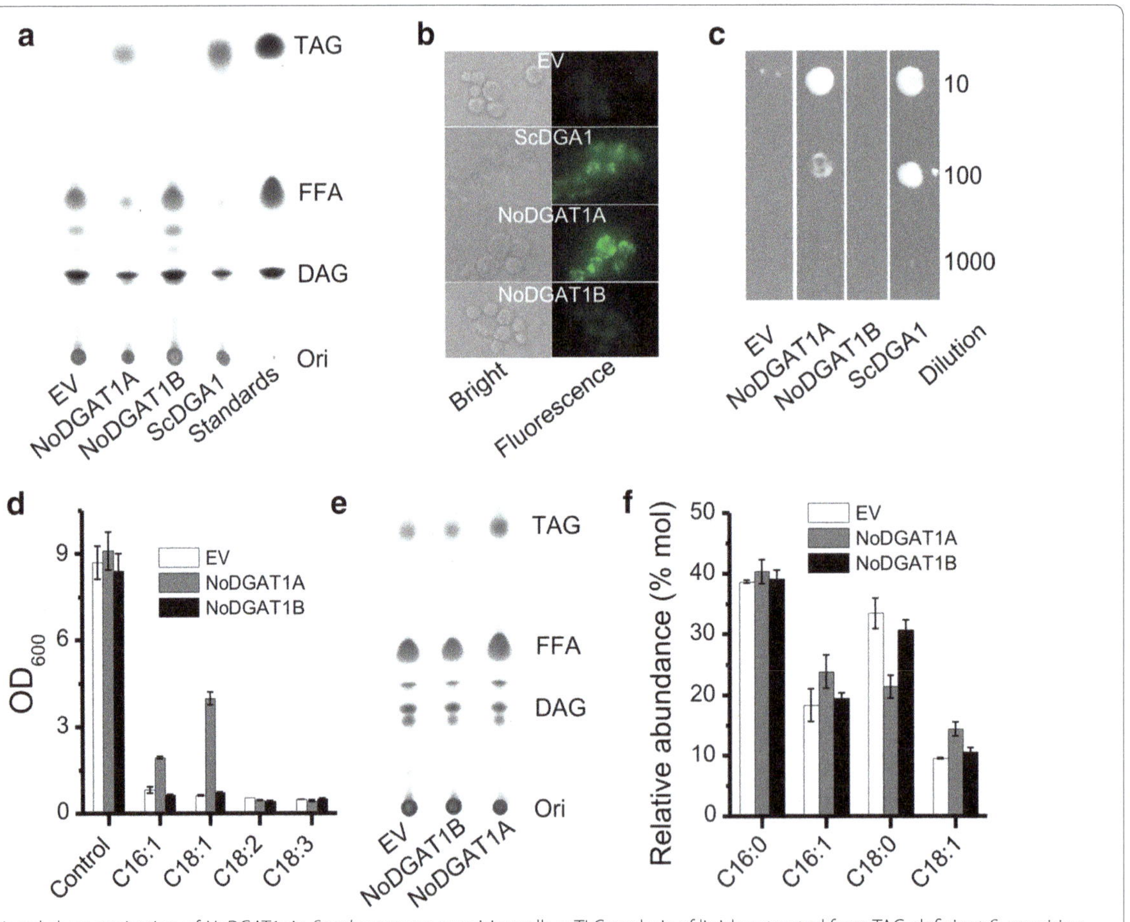

Fig. 2 Functional characterization of *NoDGAT1*s in *Saccharomyces cerevisiae* cells. **a** TLC analysis of lipids extracted from TAG-deficient *S. cerevisiae* H1246 cells transformed with *NoDGAT1*s. Empty vector (EV) and ScDGA1 were used as the negative and positive controls, respectively. **b** BODIPY staining of *NoDGAT1*-expressing H1244 cells. *Left, bright field; right, fluorescent field. Green fluorescence* indicates the BODIPY-bound lipid droplets. **c** Growth of *NoDGAT1*-expressing H1244 cells (OD_{600}, 1) in different dilutions (10- to 1000-fold) plated on 0.25 mM C18:1-containing solid medium for 3 days. 5 μL cells were plated. **d** Growth of *NoDGAT1*-expressing H1244 cells as affected by the feeding of free fatty acids of C16:1, C18:1, C18:2, and C18:3 after a 12-h cultivation. The concentration of FFAs was 125 μM. **e** TLC analysis of lipids extracted from *S. cerevisiae* INVSc1 cells transformed *NoDGAT1*s. **f** Relative abundance of fatty acids in TAG from *S. cerevisiae* INVSc1 cells transformed *NoDGAT1*s. Data in (**d**) and (**f**) are expressed as mean ± SD ($n = 3$)

activity and substrate specificity on acyl-CoAs [35, 36]. H1246 cells lack DGAT activity and may not efficiently metabolize the supplemented free fatty acids, leading to impaired cell growth; by contrast, heterologous expression of an active *DGAT* in H1246 cells enables sequestration of these fatty acids into TAG, and thus can restore cell growth to various extents depending on the enzymatic activity of DGAT. We first investigated the growth of H1246 cells carrying different vectors on agar plates containing 250 μM C18:1. As expected, the dotted H1246 cells carrying empty vector (EV, negative control) showed no growth on the plate, while the H1246 cells carrying *ScDGA1* grew and formed a white colony (Fig. 2c). Similar to the positive control, *NoDGAT1A*-carrying H1246 cells showed growth on the C18:1-containing plates. *NoDGAT1B*-carrying H1246 cells, on the other hand, did not survive, consistent with the functional complementation results (Fig. 2a). Next, four free fatty acids of C16:1, C18:1, C18:2, and C18:3n3 were used to test the growth of yeast cells in liquid cultures. The *NoDGAT1B*-carrying H1246 cells, similar to EV control, did not grow in the presence of any of the four fatty acids, further suggesting its encoded enzyme is nonfunctional toward these corresponding acyl-CoAs (Fig. 2d). By contrast, the *NoDGAT1A*-carrying H1246 cells exhibited differential growth responses: better growth with C18:1 than with C16:1 and no growth with C18:2 or C18:3 (Fig. 2d), indicating that NoDGAT1A has greater activity on monounsaturated acyl-CoAs than that on polyunsaturated ones.

The heterologous expressions of *NoDGAT1A* and *NoDGAT1B* in the TAG-producing *S. cerevisiae* strain INVSc1 were also investigated. The transcriptional expression levels of the two genes were evaluated with quantitative real-time PCR (Additional file 1: Figure S6). In comparison with the EV control, *NoDGAT1A*-carrying INVSc1 cells showed a considerable TAG increase, while *NoDGAT1B*-carrying INVSc1 cells produced TAG at levels similar to the EV control (Fig. 2e), consistent with the functional complementation data in H1246 cells (Fig. 2a). As for the TAG fatty acid profiles, the heterologous expression of *NoDGAT1A* rather than *NoDGAT1B* enabled an increase in the relative contents of both C16:1 and C18:1 at the expense of C18:0 (Fig. 2f).

Substrate specificity of NoDGAT1A

The ability of NoDGAT1A to restore TAG biosynthesis in yeast led us to examine its substrate specificity. We have recently developed a nonradiolabeled DGAT in vitro assay [36], which allows for the measurement of activity and substrate specificity of DGAT and other acyltransferases toward a wide range of acyl-CoAs and DAGs with minimum background signals. Here, we first tested the preference of NoDGAT1A for acyl-CoAs (Fig. 3). Considering that the *sn*-2 position of TAG in *N. oceanica* consists mainly of 16-carbon acyls (Fig. 1h), the prokaryotic C18:1/C16:0-DAG was used as the acyl acceptor for the in vitro assay of NoDGAT1A toward ten various acyl-CoAs. NoDGAT1A preferred the saturated/monounsaturated acyl-CoAs over those polyunsaturated ones for the TAG synthesis (Fig. 3a). Quantitatively, NoDGAT1A had the highest activity on C16:0-CoA, followed by C18:0-, C18:1-, and C16:1-CoAs (Fig. 3b). When the acyl chain length was the same, NoDGAT1A preferred saturated acyl-CoAs; for instance, C16:0 and C18:0 led to a greater TAG level than C16:1 and C18:1, respectively (Fig. 3). These in vitro results suggested that NoDGAT1A contributes to the incorporation of more saturated fatty acids than that of the unsaturated ones into TAG in *N. oceanica*.

The specificity of NoDGAT1A for the other substrate DAG was also investigated (Fig. 4). Eight DAGs were tested, including seven 1,2-DAGs and one 1,3-DAG. Of the seven 1,2-DAGs, three were prokaryotic and four were eukaryotic. As NoDGAT1A showed the highest activity on C16:0-CoA in the above in vitro assay (Fig. 3), C16:0 CoA was first used as the acyl donor to test the

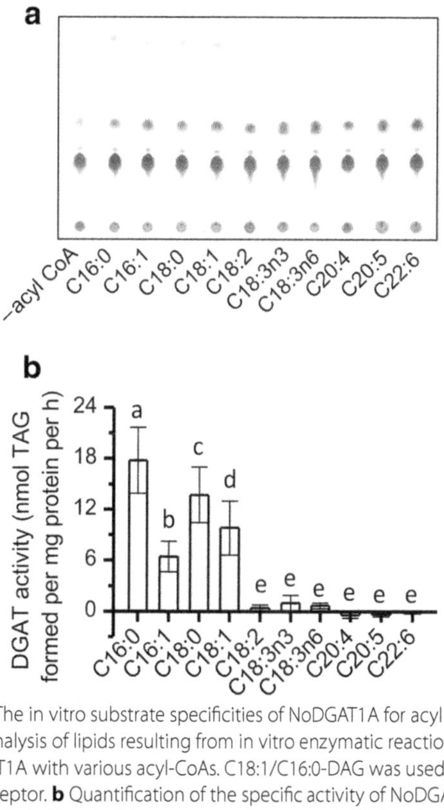

Fig. 3 The in vitro substrate specificities of NoDGAT1A for acyl CoAs. **a** TLC analysis of lipids resulting from in vitro enzymatic reactions of NoDGAT1A with various acyl-CoAs. C18:1/C16:0-DAG was used as the acyl acceptor. **b** Quantification of the specific activity of NoDGAT1A for acyl-CoAs. Data in (**b**) are expressed as mean ± SD ($n = 5$), and those values designated by *different letters* are significantly different ($P < 0.05$)

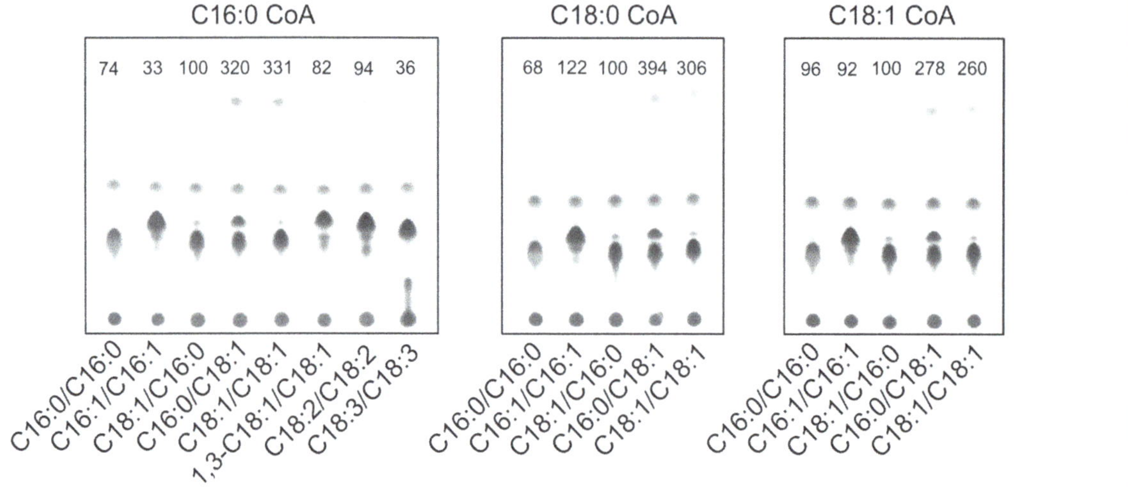

Fig. 4 The in vitro substrate specificities of NoDGAT1A for DAGs. TLC analysis of lipids resulting from in vitro enzymatic reactions of NoDGAT1A with various DAGs. The acyl donor used in each reaction is indicated on the *panel*. The *numbers* above *each panel* indicate the relative TAG values, normalized to those obtained when C18:1/C16:0-DAG was used as the acyl acceptor (set as 100)

DAG preference of NoDGAT1A. Apparently, NoDGAT1A preferred C16:0/C18:1- and C18:1/C18:1-DAGs over other DAGs tested (Fig. 4). When the acyl donor was C18:0- or C18:1-CoA, NoDGAT1A also demonstrated a strong preference for C16:0/C18:1- and C18:1/C18:1-DAGs over the other three tested prokaryotic DAGs. These data indicate that NoDGAT1A can utilize both prokaryotic and eukaryotic DAGs as the acyl acceptor for TAG formation, but prefers eukaryotic DAGs.

Subcellular localization of NoDGAT1A

DGAT proteins are membrane bound and are thought to be ER-localized for TAG assembly in yeast and higher plants [37, 38]. The compartmentalization of DGAT in algae, however, has been rarely discussed earlier and remains ambiguous [26, 39]. We constructed the NoDGAT1A-GFP fusion and introduced it into *N. oceanica* for live cell observation, aiming to examine the subcellular localization of NoDGAT1A experimentally, which will aid in the understanding of its role in TAG biosynthesis. Clearly, the GFP signal was overlaid well with the plastid autofluorescence (PAF, Fig. 5a), indicative of its location in chloroplast rather than in cytosol, mitochondria, or ER [40]. As NoDGAT1A is a membrane-bound protein, it is unlikely to be positioned in stromal or intermembrane space of the chloroplast. To further clarify the compartmentalization, a 3D merger of GFP and PAF signals was built. Interestingly, GFP and PAF signals are not well matched, but instead GFP signal is clustered as an additional irregular layer surrounding PAF (Fig. 5b), suggesting that it is unlikely to be positioned in inner membrane envelope or outer membrane envelope. Considering that *Nannochloropsis* chloroplast evolved via secondary endosymbiosis and is delineated by four distinct membranes, with the outermost membrane being chloroplast ER (cER) connected to cytoplasmic ER [41],

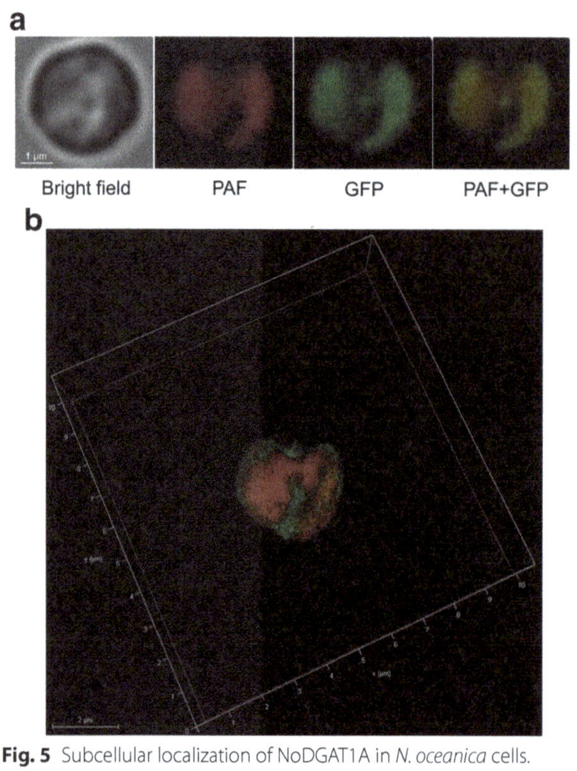

Fig. 5 Subcellular localization of NoDGAT1A in *N. oceanica* cells. **a** Microscopy images of a representative transgenic line expressing NoDGAT1A-GFP fusion. From *right* to *left*: *bright field*, PAF (plastid autofluorescence), GFP, and merger of PAF and GFP (PAF + GFP). **b** 3D microscopy image of PAF + GFP

NoDGAT1A is probably localized in cER. Of course, the co-localization test with a protein known to localize to the cER would provide another layer of evidence for NoDGAT1A's subcellular compartmentalization, which is worth investigating in future when such a marker protein is available for *N. oceanica*.

NoDGAT1A knockdown impairs TAG synthesis and attenuates its relative content of saturated/monounsaturated fatty acids

In order to investigate the function and possible biological role of NoDGAT1A in *N. oceanica*, we first generated *NoDGAT1A* knockdown lines via an RNAi-mediated gene-silencing approach. Through screening over 30 putative transformants (confirmed by genomic PCR) by quantitative real-time PCR, two knockdown lines, NoDGAT1A-i7 and NoDGAT1A-i19, exhibited the maximum reduction in *NoDGAT1A* transcript (~80% lower than the control transformed with the empty vector) under both nitrogen-replete and nitrogen-depleted conditions (Fig. 6a).

To determine whether the knockdown of *NoDGAT1A* could influence TAG synthesis in *N. oceanica*, the TAG content per cell in the empty vector (EV) control and transformants were quantified. No difference in TAG content was observed between EV and *NoDGAT1A* knockdown lines under nitrogen-replete conditions (referred to 0 h of nitrogen depletion in Fig. 6b), although the *NoDGAT1A* expression in knockdown lines was markedly attenuated at the transcriptional level (Fig. 6a), indicating that NoDGAT1A is not involved in the nonstress-associated TAG biosynthesis. Upon stress induction, TAG contents in NoDGAT1A-i7 and NoDGAT1A-i19 decreased by 19 and 25%, respectively, compared with EV (12 h). As nitrogen-depletion stress persisted (24 h), the significant (t test, $P < 0.05$) TAG reduction in the knockdown lines remained prominent. However, no significant difference in TAG content was observed between EV and the knockdown lines after 48 h (Fig. 6b). When calculated as TAG content per dry weight or the ratio of fatty acids in TAG over total fatty acids (TFA), similar results of TAG reduction were

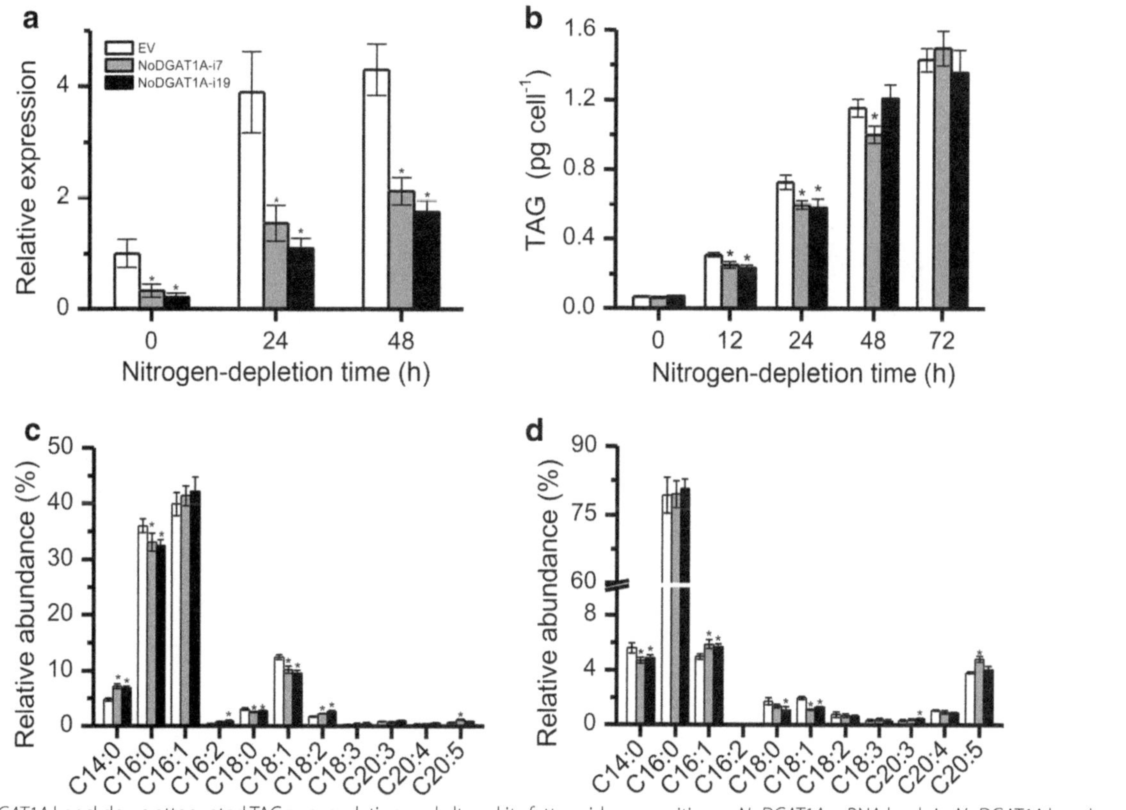

Fig. 6 *NoDGAT1A* knockdown attenuated TAG accumulation and altered its fatty acid composition. **a** *NoDGAT1A* mRNA levels in *NoDGAT1A* knockdown lines and the empty vector (EV) control, as determined by quantitative real-time PCR. **b** TAG contents in *NoDGAT1A* knockdown lines and EV. **c, d** Relative abundance of fatty acids in the *sn*-1/*sn*-3 (**c**) and *sn*-2 (**d**) positions of TAG in *NoDGAT1A* knockdown lines and EV under nitrogen-depleted conditions (24 h). Data in (**a**–**d**) are expressed as mean ± SD ($n = 3$). Algal cells grown in nitrogen-replete medium for 4 days (considered as 0 h of nitrogen depletion) were used for nitrogen-depletion experiment. *Asterisks* indicate the significant difference compared with EV (t-test, $P < 0.05$)

observed for *NoDGAT1A* knockdown lines (Additional file 1: Figure S7).

To study the effect of *NoDGAT1A* knockdown on TAG fatty acid composition, we analyzed the fatty acids in TAG *sn*-1/*sn*-3 and *sn*-2 positions, respectively. Compared to EV, TAG showed a significant decline (*t*-test, $P < 0.05$) in the relative contents of C16:0, C18:0, and C18:1 in TAG *sn*-1/*sn*-3 positions of the knockdown lines under nitrogen-depleted conditions (24 h); by contrast, a significant increase (*t*-test, $P < 0.05$) occurred in the relative contents of C14:0 and C18:2 (Fig. 6c). The relative fatty acid contents in *sn*-2 position of TAG were also changed, with C14:0 and C18:1 being significantly (*t*-test, $P < 0.05$) decreased (Fig. 6d). These in vivo results from *NoDGAT1A* knockdown analysis were in agreement with the in vitro assay data that NoDGAT1A preferred the saturated and monounsaturated acyl-CoAs and eukaryotic DAGs (Figs. 3 and 4).

NoDGAT1A overexpression promotes TAG synthesis and its relative content of saturated/monounsaturated fatty acids

We also generated *NoDGAT1A* overexpression lines to investigate the functional role of NoDGAT1A in *N. oceanica*. Over fifty putative transformants (confirmed by genomic PCR) were screened by quantitative real-time PCR. Two overexpression lines, NoDGAT1A-o5 and NoDGAT1A-o26, which exhibited the maximal increases in *NoDGAT1A* transcripts (by ~10-fold higher than the control) were subsequently chosen for the analysis (Fig. 7a).

Unlike *NoDGAT1A* knockdown (Fig. 6b), *NoDGAT1A* overexpression led to considerable increases (1.9- and 2.4-folds for NoDGAT1A-o5 and NoDGAT1A-o26, respectively) in TAG contents per cell under nitrogen-replete conditions (referred to 0 h of nitrogen depletion in Fig. 7b). The TAG enhancement by *NoDGAT1A* overexpression was also observed at 12–48 h of nitrogen depletion (by 1.25- to 1.44-fold increase); but further nitrogen depletion to 72 h led to no difference (Fig. 7b). TAG enhancement caused by *NoDGAT1A* overexpression was also evidenced when calculated as TAG content per dry weight or the ratio of fatty acids in TAG over TFA (Additional file 1: Figure S8).

NoDGAT1A overexpression affected the fatty acid profiles in TAG *sn*-1/*sn*-3 and *sn*-2 positions (Fig. 7c, d). A

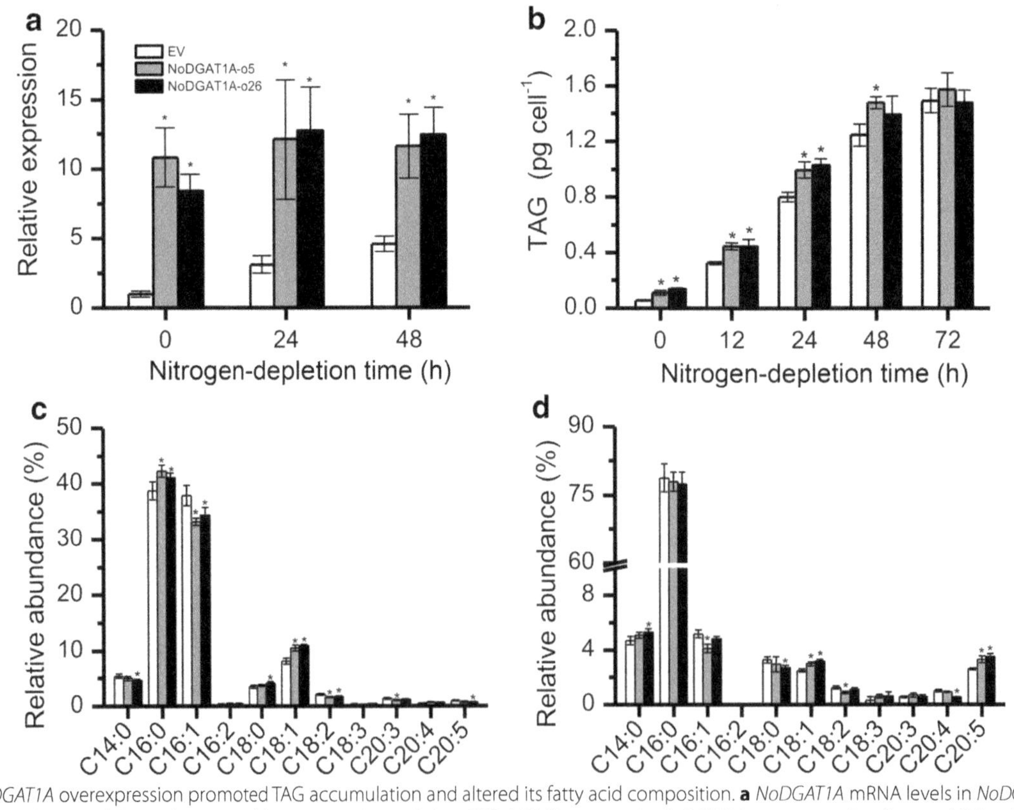

Fig. 7 *NoDGAT1A* overexpression promoted TAG accumulation and altered its fatty acid composition. **a** *NoDGAT1A* mRNA levels in *NoDGAT1A* overexpression lines and EV, as determined by quantitative real-time PCR. **b** TAG contents in *NoDGAT1A* overexpression lines and EV. **c**, **d** Relative abundance of fatty acids in the *sn*-1/*sn*-3 (**c**) and *sn*-2 (**d**) positions of TAG in *NoDGAT1A* overexpression lines and EV under nitrogen-depleted conditions (24 h). Algal cells grown in nitrogen-replete medium for 4 days (considered as 0 h of nitrogen depletion) were used for nitrogen-depletion experiment. Data in (**a**–**d**) are expressed as mean ± SD ($n = 3$). *Asterisks* indicate the significant difference compared with EV (*t*-test, $P < 0.05$)

significant increase (*t*-test, $P < 0.05$) was observed in the relative content of C16:0 and C18:1 in TAG *sn*-1/*sn*-3 positions, accompanied by a significant decrease (*t*-test, $P < 0.05$) in C16:1 and C18:2 (Fig. 7c). As for TAG *sn*-2 position, C18:1 and C20:5 showed a significant increase (*t*-test, $P < 0.05$) while C16:1 dropped significantly (*t*-test, $P < 0.05$) (Fig. 7d). In general, these *NoDGAT1A* overexpression results are in agreement with the in vitro assay data that NoDGAT1A preferred C16:0- and C18:1-CoAs and eukaryotic DAGs (Figs. 3 and 4).

Heterologous expression of *NoDGAT1A* in *Chlamydomonas* enhances TAG accumulation

In order to see if *NoDGAT1A* has the potential to improve TAG synthesis in other algae, we introduced this gene into the model alga *Chlamydomonas reinhardtii*. The mutant strain UVM4, which shows improved transgene expression efficiency [42], was employed in the present study as a host for the heterologous expression of *NoDGAT1A*. More than one hundred putative transformants (confirmed by genomic PCR) were screened, and NoDGAT1A-he33 and NoDGAT1A-he72 showing the highest expression of *NoDGAT1A* (Fig. 8a) were selected for phenotype examination. Unlike in *N. oceanica*, *NoDGAT1A* expression in *C. reinhardtii* UVM4 had no effect on TAG accumulation under nitrogen-replete conditions (referred to 0 h of nitrogen depletion in Fig. 8b), possibly due to the difference in genetic traits of these two organisms (*N. oceanica* belongs to Eustigmatophyceae, while *C. reinhardtii* belongs to Chlorophyceae). By contrast, TAG enhancement was observed under nitrogen-depleted conditions, which is obvious at early stress stages (12–24 h) and TAG contents increased by 34 and 42% at 24 h for NoDGAT1A-he33 and NoDGAT1A-he72, respectively. However, the TAG increase became insignificant during the late stage of nitrogen depletion (48–72 h). The phenotype of TAG enhancement in heterologous expression lines of *Chlamydomonas* was also supported by the data of TAG content per dry weight or the ratio of fatty acids in TAG over TFA (Additional file 1: Figure S9).

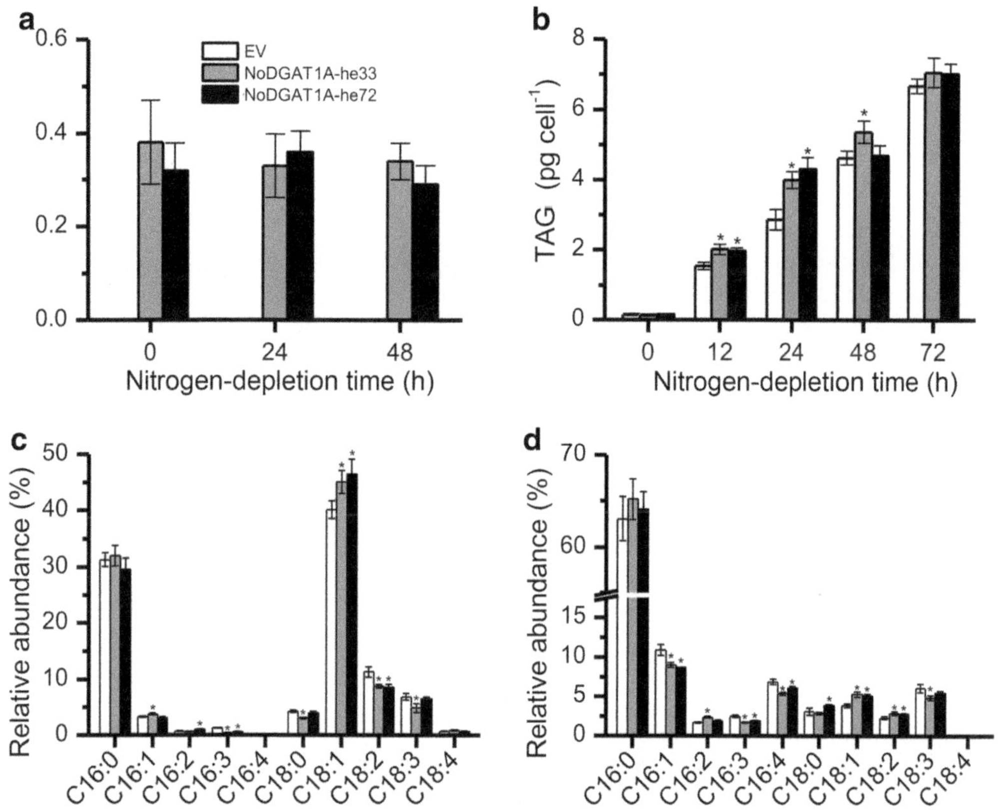

Fig. 8 Heterologous expression of *NoDGAT1A* enhanced TAG synthesis in the *C. reinhardtii* strain UVM4. **a** *NoDGAT1A* mRNA levels (normalized to endogenous β-actin gene) in *NoDGAT1A* heterologous expression lines and EV, as determined by quantitative real-time PCR. **b** TAG contents in *NoDGAT1A* heterologous expression lines and EV. **c**, **d** Relative abundance of fatty acids in the *sn*-1/sn-3 (**c**) and *sn*-2 (**d**) positions of TAG in *NoDGAT1A* heterologous expression lines and EV under nitrogen-depleted conditions (24 h). Algal cells grown in nitrogen-replete medium for 4 days (considered as 0 h of nitrogen depletion) were used for nitrogen-depletion experiment. Data in (**a**–**d**) are expressed as mean ± SD ($n = 3$). *Asterisks* indicate the significant difference compared with EV (*t*-test, $P < 0.05$)

The positional fatty acid profile of TAG from 24-h nitrogen-depleted algal samples was examined for both NoDGAT1A-he33 and EV control (Fig. 8c, d). *NoDGAT1A* heterologous expression led to significant increases (*t*-test, $P < 0.05$) in the relative contents of C16:1 and C18:1 in TAG *sn*-1/*sn*-3 positions, accompanied by the significantly (*t*-test, $P < 0.05$) attenuated C16:3 and C18:2 (Fig. 8c). As for TAG *sn*-2 position, C18:1 and C18:2 showed significant increases (*t*-test, $P < 0.05$) while C16:1, C16:3, and C16:4 exhibited significant decreases (*t*-test, $P < 0.05$) compared with EV control (Fig. 8d).

Comparison of lipid productions between *NoDGAT1A* overexpression lines and EV control

Lipid production by microalgae depends not only on intracellular lipid content but also on biomass concentration. In order to evaluate the lipid-production performance of *NoDGAT1A* overexpression lines, a comparative analysis of biomass concentrations, TAG content and yield, and TFA content and yield was conducted for a 10-day batch culture under nitrogen-replete conditions. As indicated in Fig. 9a, no difference in biomass concentrations was observed between *NoDGAT1A* overexpression lines and EV control during the whole culture period. TAG content, on the other hand, showed a considerable difference: it reached 118.9 mg g^{-1} in the overexpression lines at the end of culture period, 50% higher than that in EV control (Fig. 9b). As regards the TAG yield, the overexpression lines reached 0.49 g L^{-1}, 47% greater than that of EV control (Fig. 9c), indicative of the feasibility of the overexpressing *NoDGAT1A* for achieving improved TAG production. The overexpression of *NoDGAT1A* also led to a significant increase (*t*-test, $P < 0.05$) in the content (Fig. 9d) and yield (Fig. 9e) of TFA, although it was small compared with TAG increase.

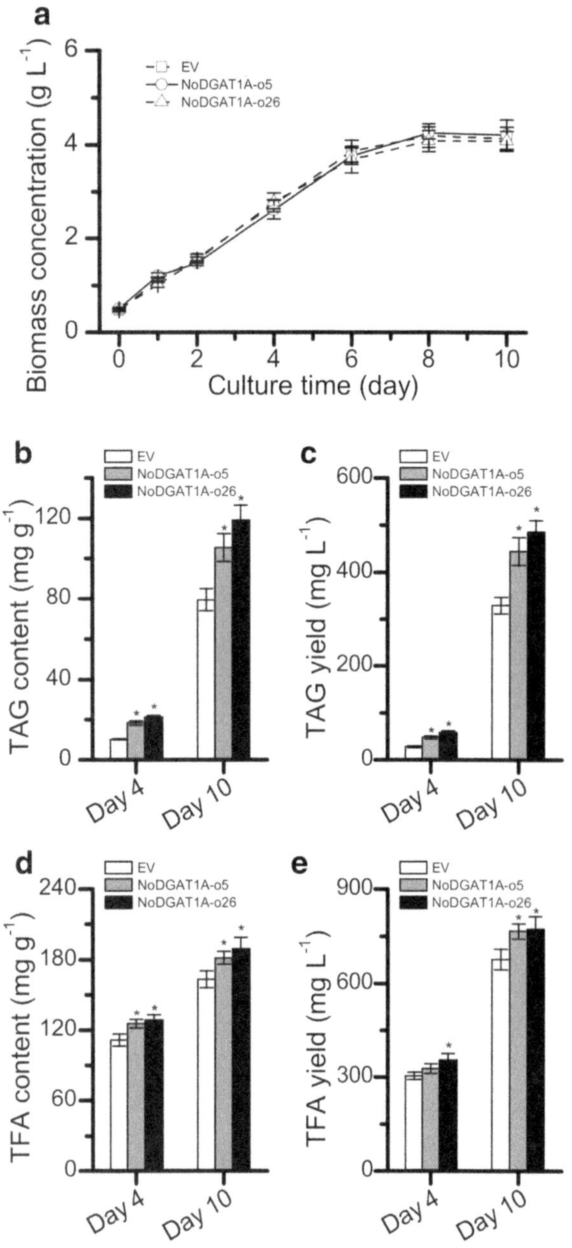

Fig. 9 Comparison of growths (**a**), TAG contents (**b**), and yields (**c**), and TFA contents (**d**), and yield (**e**) between *NoDGAT1A* overexpression lines and EV control in a 10-day batch culture. The initial N concentration for the batch culture was 100 mg L^{-1}. *Asterisks* indicate the significant difference compared to EV (*t*-test, $P < 0.05$)

Discussion

NoDGAT1A is functional and has distinctive substrate preference

Many microalgal strains are able to accumulate TAG in response to stress conditions, particularly nitrogen depletion [2]. The fatty acyls for TAG assembly can be either de novo synthesized or recycled from membrane polar lipids. In *N. oceanica*, the sharp increase in TAG was accompanied by an increase in TFA (Fig. 1d; Additional file 1: Figure S10), suggesting the contribution from the de novo fatty acid biosynthesis. This is further confirmed by the application of a specific fatty acid biosynthesis inhibitor, cerulenin, which led to a severe decrease in intracellular TAG content (Additional file 1: Figure S11). As TAG increases are accompanied by degradation of membrane lipids (Fig. 1f), the recycling of membrane lipids may be involved in TAG accumulation, which has been observed in several algal species [26, 43, 44]. It is generally agreed that glycerolipids with a 16-carbon acyl group in the *sn*-2 position are synthesized in plastid (prokaryotic pathway), while glycerolipids with an 18-carbon acyl group in *sn*-2 position are assembled at ER (eukaryotic pathway) in plants [13]. If this holds true in algae, *N. oceanica* may involve predominantly the prokaryotic pathway for TAG biosynthesis, as the *sn*-2 position of its TAG consists predominantly of 16-carbon acyl group (Fig. 1h).

DGAT catalyzes the committed and the last step in acyl-CoA-dependent TAG biosynthetic pathway and is critical for TAG accumulation in yeast and higher plants [9]. Compared with higher plants, microalgae possess a relatively high dose of annotated *DGAT* genes (5–13; [20, 21, 23, 45, 46]). The presence of high doses of *DGAT* genes suggests distinct yet complicated TAG biosynthesis and regulation mechanisms in *N. oceanica*. In accordance with higher TAG content upon nitrogen depletion, *NoDGAT1A* showed a great increase at the mRNA level, while *NoDGAT1B* had almost no change (Fig. 1d, e), indicating that *NoDGAT1A* contributes to stress-associated TAG accumulation. Functional complementation also indicated that the expression of *NoDGAT1A* but not *NoDGAT1B* restored TAG accumulation in H1246 cells (Fig. 2a). Even when the cells were fed with certain fatty acids (e.g., C18:2, C18:3, 20:4, 20:5) that are present in *N. oceanica* but not in yeast cells, no detectable TAG was observed for *NoDGAT1B*-carrying H1246 cells (Additional file 1: Figure S12). Consistent with the ex vivo data in H1246 (Fig. 2), in vitro DGAT assay demonstrated the enzymatic activity of NoDGAT1A (Fig. 3) rather than NoDGAT1B (Additional file 1: Figure S13). Of course, we cannot exclude the possibility that the functional failure of *NoDGAT1B* may be due to the codon usage preference of yeast. It is also possible that the predicted NoDGAT1B might not be a real DGAT but in fact a monoacylglycerol acyltransferase or other type of acyltransferase, since this distinction is based only on sequence data.

We have recently developed a nonradiolabeled in vitro assay for qualification and quantification of DGAT activity, which is featured by the use of widely available and regular acyl-CoAs and DAGs [36]. By means of this assay, a wide range of acyl-CoAs present in *N. oceanica* were evaluated with NoDGAT1A, revealing its preference for saturated/monounsaturated acyl-CoAs (Fig. 3). Notably, NoDGAT1A showed little in vitro activity on C20:5, although it is relatively abundant in *N. oceanica* [47], which may partially explain the extremely low abundance of C20:5 in TAG (Additional file 1: Figure S14). Interestingly, *C. reinhardtii* is devoid of C20:5, but its DGATs such as CrDGTT1 and CrDGTT2 showed respective high activity on C20:5-CoA in an in vitro assay [26], indicating that these DGATs may have potential in the engineering of *N. oceanica* for C20:5-enriched TAG production. DAG is the other substrate of DGAT, but the preference of DGAT on DAG has been less studied, and only a limited number of DAG species have been evaluated [17, 48]. In the present study, we evaluated the preference of NoDGAT1A for a wide range of DAGs in vitro. Interestingly, NoDGAT1A showed a strong preference for eukaryotic C16:0/C18:1- and C18:1/C18:1-DAGs over prokaryotic (C16:0/C16:0, C16:1/C16:1, and C18:1/C16:0) DAGs (Fig. 4), indicative of its greater contribution to eukaryotic TAG accumulation. Similarly, we have recently reported a type-II DGAT from *C. reinhardtii* (CrDGTT2) with a preference for eukaryotic DAGs [26]. Although *DGAT1* from higher plants has been well characterized [16, 17, 49–51], there are several reports mostly regarding *DGAT1* from algae including *Phaeodactylum tricornutum* [52], *Myrmecia incisa* [53], *Chlorella vulgaris* [54], and *Chlorella ellipsoidea* [55]. However, these studies about algal *DGAT1* gene characterization were mainly restricted to functional complementation in TAG-deficient yeast strains, without investigating the in vitro enzymatic activity or substrate preference. This work to our knowledge represents the first report of characterizing the in vitro activity and substrate preference of algal DGAT1 covering a wide range of acyl-CoAs and DAGs.

NoDGAT1A is critical for stress-associated TAG biosynthesis in *N. oceanica*

There were no in vivo experiments for evaluating algal *DGAT1* function in previous studies [52–55]. In the present study, we employed the RNAi-mediated gene silencing and overexpression methods to generate *NoDGTAT1A*-knockdown and-overexpression lines, respectively, for in vivo functional analysis. It is not surprising that *NoDGAT1A* knockdown had no effect on TAG level under nitrogen-replete conditions, as phospholipid:DAG acyltransferases instead of DGAT are thought to be the major contributors to TAG synthesis under nonstress conditions in algae [56]. The significant decrease (Fig. 6b, 12–24 h) in TAG level caused by *NoDGAT1A* knockdown suggested its important role in nitrogen-depletion-associated TAG synthesis in *N. oceanica*. *NoDGAT1A* overexpression, on the other hand, enabled *N. oceanica* to synthesize more TAG (Fig. 7b), further confirming its in vivo function in *N. oceanica*. It is worth noting that the TAG content was not significantly altered at the late stage of nitrogen-depleted conditions for both *NoDGAT1A* knockdown (Fig. 6b, 48 h and thereafter) and *NoDGAT1A* overexpression (after 48 h), possibly due in part to the decreased *NoDGAT1A* knockdown efficiency (Fig. 6a) and overexpression level (Fig. 7a), respectively. In addition, many other acyltransferases in *N. oceanica* showed upregulation upon nitrogen depletion [33], which may be functional and compromise the phenotype caused by *NoDGAT1A* knockdown or overexpression. Interestingly, *NoDGAT1A* overexpression also led to a remarkable TAG increase under nitrogen-replete conditions (Fig. 7b). Consistent with the enhanced TAG levels, *NoDGAT1A* transcripts exhibited a drastic increase (by ~10-fold) in the overexpression lines under nitrogen-replete conditions (Fig. 7a), even higher than that in the wild type under nitrogen-depleted conditions (Fig. 1e). This may partially explain that *NoDGAT1A* overexpression enabled *N. oceanica* to synthesize more TAG, while its knockdown had no effect on TAG level under nitrogen-replete conditions (Figs. 6b, 7b). It has been recently reported that in *C. reinhardtii,* TAG attenuation caused by *DGAT* knockdown was accompanied by changes in membrane lipids [26]. In the present study, we did not notice a similar effect of *NoDGAT1A* knockdown on *N. oceanica* membrane lipids (Additional file 1: Figure S15).

As DGAT contributes primarily to the acyl groups in the *sn*-3 position of TAG, the TAG positional analysis may be more informative. The fatty acids in *sn*-1 and *sn*-3 positions cannot be distinguished from each other by the currently available methods and are normally grouped as a fraction separated from *sn*-2 fatty acids by the action of *Rhizopus arrhizus* lipase [26, 57, 58]. Our TAG positional analysis indicated that *NoDGAT1A* knockdown and overexpression each led to differential changes in the fatty acid profiles of TAG *sn*-1/*sn*-3 and *sn*-2 positions (Figs. 6c, d, 7c, d), consistent with the in vitro assay results that NoDGAT1A had distinct substrate specificity on acyl-CoAs and DAGs (Figs. 2 and 3). By combining the ex vivo, in vitro, and in vivo data, we inferred that NoDGAT1A is likely to contribute to transferring saturated and monounsaturated acyl groups (C16:0, C16:1, C18:0, and C18:1) to eukaryotic DAGs for TAG synthesis via the Kennedy pathway.

In *C. reinhardtii,* TAG precursors are thought to be synthesized mainly from the prokaryotic pathway in chloroplast [26, 57, 59, 60]. Similarly, *N. oceanica* TAG contained predominantly C16 acyl groups in its *sn*-2 position (Fig. 1h), suggesting a role of prokaryotic pathway for TAG synthesis. Our in vitro and in vivo data revealed that NoDGAT1A accepted both prokaryotic and eukaryotic DAGs for TAG assembly; however, NoDGAT1A had a strong preference for eukaryotic DAGs, C16:0/C18:1, and C18:1/C18:1 in particular (Fig. 3), suggesting its role in eukaryotic TAG biosynthesis. Thus, NoDGAT2s or other unknown acyltransferases may contribute more to prokaryotic TAG biosynthesis, which needs to be further validated experimentally. NoDGAT1A harbors the sequence YYH/AD (Additional file 1: Figure S4), a putative endoplasmic reticulum (ER) retrieval motif [17], indicative of its localization to the ER. To unravel the subcellular localization of NoDGAT1A experimentally, we constructed NoDGAT1A-GFP fusion, and the results indicate that NoDGAT1A is likely to reside in the cER (Fig. 5). cER bridges the chloroplast and ER and may serve as the transiting sites for lipid trafficking between these two organelles, similar to the chloroplast–ER membrane contact sites in higher plants [61]. The subcellular localization of NoDGAT1A in cER may facilitate its access to the substrates of DAGs and acyl-CoAs from both organelles.

Taking all these results together, we proposed a working model for the role of NoDGAT1A in TAG biosynthesis in the oleaginous alga *N. oceanica* (Fig. 10). The fatty acyls de novo synthesized and/or recycled from the turnover of membrane lipids constitute the acyl-CoA pool and enter Kennedy pathway for the synthesis of both prokaryotic and eukaryotic DAGs. NoDGAT1A, which is upregulated upon nitrogen depletion, accumulates and resides in cER and accesses mainly eukaryotic DAGs and C16–C18 saturated/monounsaturated acyl-CoAs to synthesize eukaryotic TAG species packed into lipid droplets (LDs, the TAG sink).

Biotechnological implications

DGAT has gained increasing interest in various organisms including algae, and its engineering potential for oil production in yeast and higher plants has been well documented [62–66]. We here conducted the in-depth characterization of *NoDGAT1A* from *N. oceanica,* an oleaginous and industrially important marine alga. Both the overexpression of *NoDGAT1A* in *N. oceanica* and the heterologous expression of *NoDGAT1A* in the green alga *C. reinhardtii* enhanced the TAG biosynthesis and the relative content of C18:1 in TAG (Figs. 7 and 8), indicative of its potential application in algae engineering for improved oil production. It has been suggested that C18:1 helps benefit the balance between oxidative stability and low-temperature properties and promote the quality of biodiesel [67]. In this context, DGATs with preference for C18:1 such as NoDGAT1A are preferred for genetic engineering of algae for biodiesel production applications. *NoDGAT1A* overexpression in *N. oceanica* promoted TAG synthesis not only under nitrogen-depleted but also under nitrogen-replete conditions without compromising algal growth leading to considerably higher TAG yield (Figs. 7b and 9a). This is particularly important as it may allow using non-stressed algal cells instead of the stressed ones (typically associated with impaired growth and thus compromised oil production) for oil production. Although efforts to overexpress *NoDGAT2* have been made to enhance TAG synthesis in *N. oceanica,* the significance is partly compromised by reduced algal growth [28, 68]. In this context, *NoDGAT1A* may be superior to *NoDGAT2*s in genetic engineering of oleaginous algae for improved TAG synthesis and production. The difference in algal growth caused by overexpression of *NoDGAT1A* and *NoDGAT2*s may be due to the different properties of these two types of *DGAT* gene. It may also result from the use of different expression vectors with different promoters.

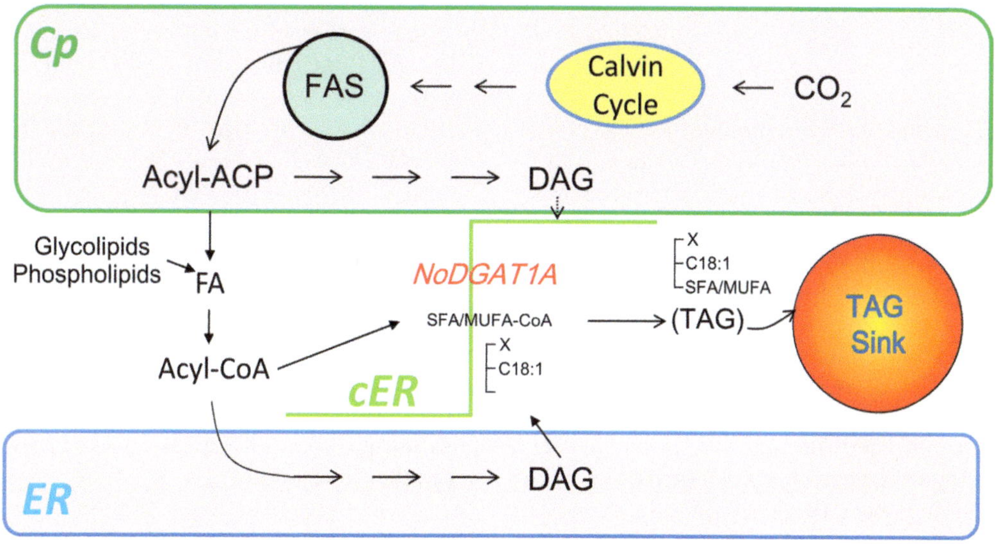

Fig. 10 A hypothesized working model illustrating the role of NoDGAT1A in TAG biosynthesis in *N. oceanica*. For simplicity, a single lipid droplet (TAG sink) is shown bridging the endoplasmic reticulum (ER) and chloroplast (Cp) envelop. The de novo-synthesized fatty acids, together with those released from glycolipids and phospholipids, constitute the acyl-CoA pool. NoDGAT1A, probably located in cER, utilizes eukaryotic DAGs and saturated/monounsaturated (SFA/MUFA)-CoAs for TAG assembly. Not all intermediates or reactions were displayed. For simplicity, cER is expressed as an elbow-like line linking ER and chloroplast envelope. *ER* endoplasmic reticulum, *Cp* chloroplast, and *cER* chloroplast ER

Bioinformatic analysis reveals the presence of multiple conserved motifs in NoDGAT1A ([14]; Additional file 1: Figure S4), which may play an important role in the acyltransferase activity across the DGAT1 family. Previous studies revealed that amino acid mutations occurring in these motifs led to considerably enhanced acyltransferase activity of DGAT1 from *Tropaeolum majus* [50] and *Zea mays* [69]. These findings point to the possible manipulation of *NoDGAT1A* for improved enzymatic activity and bioengineering uses in oil production. In addition, stronger promoters, which have high expression levels under nitrogen depletion such as *LDSP* promoter [20], are needed to drive *DGAT* genes with an aim to maintain TAG enhancement during the late stage of nitrogen-depleted conditions. Future genetic approaches for the further improvement of algal oils may lie in the engineering of multiple genes, e.g., overexpression of *DGAT* (pulling carbon flux toward TAG) together with transcription factors ([63, 65]; upregulating fatty acid/TAG biosynthetic pathway globally to push carbon flux toward TAG) and/or downregulation of TAG lipase genes ([70]; and protecting TAG from degradation).

Conclusions

Several studies have reported the characterization of algal type-I *DGAT* genes, but have been restricted to sequence analysis, functional complementation in yeast and transcriptional expression levels in algae [52–54], and heterologous expression in higher plants [55]. Here, we have solidly demonstrated the function and critical role of *NoDGAT1A* for TAG biosynthesis in the oleaginous alga *N. oceanica* by functional complementation in the TAG-deficient *S. cerevisiae* strain H1246 (Fig. 2), in vitro DGAT assay for substrate specificity (Figs. 3 and 4), knockdown (Fig. 6), and overexpression (Fig. 7). *NoDGAT1A* is likely positioned in cER (Fig. 5), where it prefers to access and utilize saturated/monounsaturated acyl-CoAs and eukaryotic DAGs for TAG biosynthesis. From a biotechnological point of view, *NoDGAT1A* is a useful engineering target for promoting cellular TAG accumulation, which has been evidenced in yeast (Fig. 2), the modal alga *Chlamydomonas* (Fig. 8) and the oleaginous alga *N. oceanica* (Fig. 7). Robust algal TAG production is hampered by the typically required nutrient stress, which is associated with severely compromised biomass production. Unlike *NoDGAT2*s [28, 68], *NoDGAT1A* overexpression in *N. oceanica* enhanced TAG synthesis considerably but without slowing down biomass production under nitrogen-replete conditions (Fig. 9), indicating its potential in industrial uses. More importantly, *NoDGAT1A* expression increased the abundance of C18:1 in TAG (Figs. 7 and 8), which is beneficial to algal lipid-derived biodiesel as C18:1 helps enhance the quality of biodiesel [67]. These results not only shed light on the underlying mechanism for the biosynthesis of various TAG species in *N. oceanica*, but also provide insights into genetic engineering of microalgae by manipulating rate-limiting enzymes such as DGAT to modulate

TAG synthesis and fatty acid composition for biofuel production.

Methods

Nannochloropsis strain and growth conditions

The marine microalga *Nannochloropsis oceanica* IMET1 was from the Institute of Marine and Environmental Technology, the University Systems of Maryland. It was maintained at 16 °C on an agar plate of the modified F/2 medium (100 mg L^{-1} N and 4.5 mg L^{-1} P) containing 20 g L^{-1} sea salt [71]. In brief, 10 mL of modified F/2 liquid medium was inoculated with cells from plates, and the alga was grown aerobically in flasks at 25 °C for 6 days (hand shaking twice per day) illuminated with continuous light of 30 μE m^{-2} s^{-1}. The algal cells were then inoculated at 10% (v/v) into 250-mL columns (3-cm diameter) provided with constant illumination of 70 μE m^{-2} s^{-1} and aeration of 1.5% CO_2 enriched air. Cells grown with nitrogen for 4 days (linear growth phase) were collected, washed three times with nitrogen-free medium, and resuspended in nitrogen-free medium for nitrogen-depletion experiments. For the inhibition of fatty acid biosynthesis, cerulenin (Sigma-Aldrich, St. Louis, MO, USA), which specifically inhibits β-ketoacyl-ACP synthase I (KAS I), was added upon N-depletion at a concentration of 10 μM, according to Liu et al. [72].

Sequence analysis

TargetP 1.1 (http://www.cbs.dtu.dk/services/TargetP/) and SignalP 4.1 (http://www.cbs.dtu.dk/services/SignalP/) were used for the prediction of transit peptides of NoDGAT1A and NoDGAT1B. Sequence alignment of DGAT polypeptides was conducted using ClustalX2.1 (http://www.clustal.org/clustal2/), and the phylogenetic tree was generated using MEGA6 [34]. Proteins used for analysis are listed in Additional file 2: Table S1.

RNA isolation and quantitative real-time PCR

RNA was isolated using the TRI Reagent (Invitrogen, Carlsbad, CA, USA) according to the manufacturer's instructions. The total RNA concentration was determined using NannoDrop 2000c (Thermo Scientific, Wilmington, Delaware, USA) and the quality was checked by electrophoresis. The cDNA synthesis and quantitative real-time PCR were performed as described by Liu et al. [72] using a 7500 Fast Real-Time PCR System (Applied Biosystems, Waltham, MA, USA) with SYBR Green PCR Master Mix (Invitrogen). Primer sequences used for quantitative real-time PCR are listed in Additional file 2: Table S2. The mRNA expression level was normalized using the actin gene as the internal control.

NoDGAT1 gene cloning and expression in yeast

The type-I diacylglycerol acyltransferase genes from *N. oceanica* including *NoDGAT1A* (GenBank Accession No. KY073295) and *NoDGAT1B* (GenBank Accession No. KY073296) were PCR amplified using cDNA as template and cloned into the yeast expression vector pYES2-CT (Invitrogen). PCR primers for cloning are listed in Additional file 2: Table S2. After confirmation by restriction enzyme digestion and sequencing, the recombinant pYES2-NoDGAT plasmids were transformed into the *S. cerevisiae* TAG-producing strain INVSc1 or TAG-deficient quadruple mutant strain H1246 by means of S.c. EasyComp Transformation Kit (Invitrogen). Colony PCR and quantitative real-time PCR were used to verify the presence of the plasmids and the expression of *DGAT1*s, respectively, in the transformants. H1246 cells carrying the empty vector pYES2-CT (EV control) and pYES2-ScDGA1 (containing a type-2 *DGAT* gene from S. *cerevisiae*) were from Liu et al. [26] and Liu et al. [36], respectively. *NoDGAT1*s expression induction by galactose was performed as previously described [36]. When necessary, fatty acids were fed to yeast cultures as described by Siloto et al. [35], with supplementation of (palmitoleic acid) C16:1, oleic acid (C18:1), linoleic acid (C18:2), and α-linolenic acid (C18:3n3) at a concentration of 125 μM upon galactose induction.

Yeast microsome preparation and in vitro DGAT activity assay

The H1244 transformants bearing *NoDGAT1* s were grown in SC-uracil medium with 2% galactose for 12 h at 30 °C. Cells were then harvested and washed twice with ice-cold distilled water. The cell pellets were resuspended in cell lysis buffer (containing 5% glycerol, 20 mM Tris–HCl (pH 8.0), 0.3 M ammonium sulfate, 10 mM $MgCl_2$, 1 mM EDTA, 1 mM DTT, 1 × EDTA-free Protease Inhibitor Cocktail Set X (Calbiochem), 1 mM PMSF) to an OD_{600} of approximately 100 and lysed by passing twice through a French pressure cell (Spectronics Instruments, Rochester, NY, USA) at an internal pressure of 15,000 PSI. Cell debris was removed from the suspension by centrifugation at 10,000×*g* for 10 min at 4 °C, and the supernatant was centrifuged further at 100,000×*g* for 2 h at 4 °C. The resulting microsomal membrane pellets were resuspended in microsome storage buffer (50 mM Tris–HCl, pH 7.5, 10% glycerol) to yield a protein concentration of 10 μg $μL^{-1}$ for immediate use or later use, storing at −80 °C.

The in vitro DGAT activity assay was conducted according to our previously described procedures [26, 36]. The acyl-CoAs tested included palmitoyl-CoA (C16:0-CoA), hexadecenoyl-CoA (C16:1-CoA), stearoyl-CoA (C18:0-CoA), oleoyl-CoA (C18:1-CoA), linoleoyl-CoA (C18:2-CoA), α-linolenoyl-CoA (C18:3n3-CoA), γ-linolenoyl-CoA (C18:3n6-CoA), eicosapentaenoyl-CoA (C20:5-CoA), and docosahexaenoyl-CoA (C22:6-CoA). The DAGs tested were C16:0/C16:0-, C16:1/C16:1-, C18:1/C16:0-, C16:0/C18:1-, C18:1/C18:1-, 1,3-C18:1/C18:1-, C18:2/C18:2-, and C18:3/C18:3-DAGs. C16:1/C16:-, C18:2/C18:2-, and C18:3/C18:3-DAGs were purchased from Larodan Fine Chemicals (Malmo, Sweden), whereas C18:1/C16:0-DAG was prepared by partial digestion of C18:1/C16:0/C18:1-TAG (Larodan Fine Chemicals) with *Rhizopus arrhizus* lipase (Sigma-Aldrich) and recovery of DAG. All other lipid standards were purchased from Avanti Polar Lipids (Alabaster, AL, USA).

Constructs for subcellular localization, knockdown, and overexpression of *NoDGAT1A* in *N. oceanica* and nuclear transformation

For the subcellular localization vector, the coding sequence of *NoDGAT1A* was fused upstream of GFP under the control of violaxanthin/chlorophyll a-binding protein 2 (VCP2) promoter (Additional file 1: Figure S16a), as described by Moog et al. [40]. The RNAi vector construction for *NoDGAT1A* knockdown followed the procedures described by Wei et al. [32], and the resulting vector is depicted in Additional file 1: Figure S16b. For the overexpression vector, *NoDGAT1A* coding sequence was driven by the *Nannochloropsis* ubiquitin extension protein promoter (Additional file 1: Figure S16c). Nuclear transformation of *N. oceanica* was performed by electroporation according to Li et al. [29]. Transformants were selected on modified F/2 plates with 2.5 μg mL^{-1} zeocin (Life Technologies, Carlsbad, CA, USA) and verified by genomic PCR. Quantitative real-time PCR was employed to determine the knockdown efficiency and overexpression of *NoDGAT1A*. The live-cell fluorescence observation for NoDGAT1A-GFP fusion transgenic lines was performed using a Leica TCS SP8 laser scanning confocal microscope (Germany). Fluorescence of GFP and chlorophyll autofluorescence was exited at 488 nm with the emission detected at a bandwidth of 500–525 and 650–750 nm, respectively.

Heterologous expression of NoDGAT1A in *Chlamydomonas*

The coding sequence of *NoDGAT1A* was amplified and cloned into *Nde*I/*Eco*RI sites of pOpt_Clover_Hyg [73], followed by sequencing for verification. The resulting plasmid was linearized by *Xba*I and transformed into the *Chlamydomonas* strain UVM4 [42] via the glass beads method [74]. Transformants were selected on Tris–acetate–phosphate (TAP) plates with 10 μg mL^{-1} hygromycin B (Sigma-Aldrich). The integration of *NoDGAT1A* into *Chlamydomonas* genome was verified by genomic PCR, and its expression was determined by quantitative real-time PCR. To impose nitrogen depletion, UVM4 cells in the stationary growth phase were collected and washed with nitrogen-free TAP medium (TAP-N), resuspended in fresh TAP-N, and cultured at 23 °C under continuous illumination of 40 μE m^{-2} s^{-1}.

Lipid extraction and analysis

Lipid extracts from yeast, *N. oceanica,* and *C. reinhardtii* cells as well as the in vitro DGAT assay mixture were all performed according to our previously described procedures [26, 30].

Neutral lipids were separated on a Silica gel 60 TLC plate (EMD Chemicals, Merck, Darmstadt, Germany) using a mixture of hexane/tert-butylmethyl ether/acetic acid (80/20/2, by volume) as the mobile phase, while polar lipids were separated on a TLC plate using a mixture of chloroform/methanol/acetic acid/water (25/4/0.7/0.3, by volume) as the mobile phase. Lipid detection by charring and lipid quantification by gas chromatography–mass spectrometry (GC–MS) were performed as described previously by Liu et al. [26]. TAG positional analysis was performed using *R. arrhizus* lipase (Sigma-Aldrich) following the procedures of Li et al. [58].

Statistical analysis

All experiments were determined in biological triplicate to ensure the reproducibility. Experimental results were obtained as the mean value ± SD. Statistical analyses were performed using the SPSS statistical package (SPSS Inc., Chicago, IL, USA). Paired-samples *t*-tests were used for two group means, and one-way ANOVA Tukey's HSD test was used for more than two group means. The statistical significances were achieved when $P < 0.05$.

Additional files

Additional file 1: Figure S1. The ultrastructure of *N. oceanica* cells under nitrogen-replete (left) and nitrogen-depleted (right) conditions. **Figure S2.** GC-MS chromatogram of fatty acids in the *sn*-2 and *sn*-1/*sn*-3 positions of TAG from *N. oceanica* cells under nitrogen-replete (+N) and nitrogen-depleted (-N) conditions. **Figure S3.** Cladogram of the DGATs from plants, fungi, algae and animals. The neighbor-joining method was used to reconstruct the cladogram using MEGA6 [34], with the bootstrap value (obtained from 1000 replicates) is shown on each node. The scale bar 0.1 represents 10% divergence, calculated as the estimated number of replacement. Circles, plants; Squares, fungi; Filled circles, algae; triangles, animals. Protein sequences used for the cladogram construction see Additional file 2: Table S1. **Figure S4.** Predicated transmembrane domains for NoDGAT1A NoDGAT1B by TMHMM (V2.0, http://www.cbs.dtu.dk/services/TMHMM/). **Figure S5.** Protein sequence alignment of putative DGAT1s. The alignment was conducted using ClustalX2.1. The sequences used see Additional file 2: Table S1. Red arrows indicate the key amino acid residues identified by previous studies. **Figure S6.** The transcriptional expression levels of *NoDGAT1A and NoDGAT1B* in H1246 (upper) and INVSc1 (lower), as determined by quantitative real-time PCR. The gene expression levels were normalized to the endogenous *ACT1* gene. **Figure S7.** TAG content per dry weight (**a**) and the ratio of fatty acids in TAG over TFA (**b**) in *NoDGAT1A* knockdown lines and EV. Algal cells grown in nitrogen-replete medium for 4 days (considered as 0 h of nitrogen depletion) were used for nitrogen depletion experiment. Data are expressed as mean ± SD (*n*=3). Asterisks indicate the significant difference compared with EV (*t*-test, P<0.05). **Figure S8.** TAG content per dry weight (**a**) and the ratio of fatty acids in TAG over TFA (**b**) in *NoDGAT1A* overexpression lines and EV. Algal cells grown in nitrogen-replete medium for 4 days (considered as 0 h of nitrogen depletion) were used for nitrogen depletion experiment. Data are expressed as mean ± SD (*n*=3). Asterisks indicate the significant difference compared with EV (*t*-test, P<0.05). **Figure S9.** TAG content per dry weight (**a**) and the ratio of fatty acids in TAG over TFA (**b**) in *NoDGAT1A* heterologous expression lines of *Chlamydomonas* and EV. Algal cells grown in nitrogen-replete medium for 4 days (considered as 0 h of nitrogen depletion) were used for nitrogen depletion experiment. Data are expressed as mean ± SD (*n*=3). Asterisks indicate the significant difference compared with EV (*t*-test, P<0.05). **Figure S10.** Time course of total fatty acid (TFA) content of *N. oceanica* in response to nitrogen depletion. **Figure S11.** Effect of cerulenin on TAG content of *N. oceanica* in response to nitrogen depletion. The cerulenin concentration used was 10 µM. **Figure S12.** TLC analysis of lipids extracted from *NoDGAT1B*-carrying H1246 cells without feeding (WF) or fed with free fatty acids of C18:2, C18:3, C20:4, or C20:5 (125 µM). *NoDGAT1A*-carrying H1246 cells were used as the positive control. **Figure S13.** TLC analysis of lipids resulting from *in vitro* enzymatic reactions of NoDGAT1B with various acyl-CoAs. C18:1/C16:0-DAG was used as the acyl acceptor; NoDGAT1A was used as the positive control (C16:0 as the acyl donor). **Figure S14.** Fatty acid composition of TAG in *N. oceanica* upon nitrogen depletion. **Figure S15.** The contents of polar lipids in EV and NoDGAT1A-i19 under nitrogen-depleted conditions (24 h). **Figure S16.** Schematic illustration of constructs for subcellular localization (**a**), knockdown (**b**), and overexpression (**c**) of NoDGAT1A in *N. ocenica* cells.

Additional file 2: Table S1. DGAT protein sequences used for the construction of phylogenetic tree in additional file 1: Figure S3. **Table S2.** Primers used in the present study. Underlined sequences designate the restriction enzyme sites. The sequences in box indicate the linker fragment introduced before GFP coding sequence.

Abbreviations

cER: chloroplast endoplasmic reticulum; CoA: coenzyme A; DAG: diacylglycerol; DGAT: diacylglycerol acyltransferase; DGAT1: type-I DGAT; DGAT2: type-II DGAT; DGAT3: type-III DGAT; DGDG: digalactosyl diacylglycerol; DGTS: diacylglycerol-N,N,N-trimethylhomoserine; ER: endoplasmic reticulum; EV: empty vector; GC-MS: gas chromatography–mass spectrometry; GFP: green fluorescent protein; LD: lipid droplet; MGDG: monogalactosyl diacylglycerol; PAF: plastid autofluorescence; PC: phosphatidylcholine; PCR: polymerase chain reaction; PE: phosphatidylethanolamine; PG: phosphatidylglycerol; PI: phosphatidylinositol; SQDG: sulfoquinovosyldiacylglycerol; TAG: triacylglycerol; TAP: tris-acetate-phosphate; TFA: total fatty acids; TLC: thin-layer chromatography; VCP2: violaxanthin/chlorophyll a-binding protein 2.

Authors' contributions

JL conceived the study and designed the experiments. HW, JL, YS, XM, and YP conducted the experiments. JL, HH, YL, ML, and HG wrote the manuscript. All authors read and approved the final manuscript.

Author details

[1] Institute for Food and Bioresource Engineering, Department of Energy and Resources Engineering and BIC-ESAT, College of Engineering, Peking University, Beijing 100871, China. [2] Key Laboratory of Algal Biology, Institute of Hydrobiology, Chinese Academy of Sciences, Wuhan 430072, China. [3] Institute of Marine and Environmental Technology, University of Maryland Center for Environmental Science and University of Maryland Baltimore County, Baltimore, MA 21202, USA. [4] Guangdong Provincial Key Laboratory of Applied Botany, Key Laboratory of South China Agricultural Plant Molecular Analysis and Genetic Improvement, South China Botanical Garden, Chinese Academy of Sciences, Guangzhou 510650, China. [5] School of Sustainable Engineering and the Built Environment, Arizona State University Polytechnic campus, Mesa, AZ 85212, USA.

Acknowledgements

The authors thank Professor Ralph Bock at Max-Planck-Institut fur Molekulare Pflanzenphysiologie (MPI-MP), Germany for providing the Chlamydomonas strain UVM4, Professor Matthew C. Posewitz at Department of Chemistry and Geochemistry, Colorado School of Mines, USA for providing plasmids TUB and UEP, Professor Uwe-G. Maier at Philipps-Universität Marburgand, Germany for providing the plasmid VCP1-GFP, and Professor Jian Xu at Qingdao Institute of Bioenergy and Bioprocess Technology, Chinese Academy of Sciences for providing the access to *Nannochloropsis* genome database.

Competing interests

The authors declare that they have no competing interests.

Funding

This work is partially supported by grants to J.L. from the National Natural Science Foundation of China (31571807) and the National Youth Thousand Talents Program of China.

References

1. Klok AJ, Lamers PP, Martens DE, Draaisma RB, Wijffels RH. Edible oils from microalgae: insights in TAG accumulation. Trends Biotechnol. 2014;32:521–8.
2. Hu Q, Sommerfeld M, Jarvis E, Ghirardi M, Posewitz M, Seibert M, Darzins A. Microalgal triacylglycerols as feedstocks for biofuel production: perspectives and advances. Plant J. 2008;54:621–39.
3. Zienkiewicz K, Du Z-Y, Ma W, Vollheyde K, Benning C. Stress-induced neutral lipid biosynthesis in microalgae—molecular, cellular and physiological insights. Biochim et Biophys Acta BBA-Mol Cell Biol L. 2016;1861:1269–81.
4. Huerlimann R, de Nys R, Heimann K. Growth, lipid content, productivity, and fatty acid composition of tropical microalgae for scale-Up production. Biotechnol Bioeng. 2010;107:245–57.
5. Ma Y, Wang Z, Yu C, Yin Y, Zhou G. Evaluation of the potential of 9 *Nannochloropsis* strains for biodiesel production. Bioresour Technol. 2014;167:503–9.
6. Sun Z, Zhou ZG, Gerken H, Chen F, Liu J. Screening and characterization of oleaginous *Chlorella* strains and exploration of photoautotrophic *Chlorella protothecoides* for oil production. Bioresour Technol. 2015;184:53–62.
7. Chisti Y. Constraints to commercialization of algal fuels. J Biotechnol. 2013;167:201–14.
8. Mallick N, Bagchi SK, Koley S, Singh AK. Progress and challenges in microalgal biodiesel production. Front Microbiol. 2016;7:1019.
9. Xue J, Niu YF, Huang T, Yang W-D, Liu JS, Li HY. Genetic improvement of the microalga *Phaeodactylum tricornutum* for boosting neutral lipid accumulation. Metab Eng. 2015;27:1–9.
10. Goncalves EC, Wilkie AC, Kirst M, Rathinasabapathi B. Metabolic regulation of triacylglycerol accumulation in the green algae: identification of potential targets for engineering to improve oil yield. Plant Biotechnol J. 2016;14:1649–60.
11. Lenka SK, Carbonaro N, Park R, Miller SM, Thorpe I, Li Y. Current advances in molecular, biochemical, and computational modeling analysis of microalgal triacylglycerol biosynthesis. Biotechnol Adv. 2016;34:1046–63.
12. Bates PD, Browse J. The significance of different diacylgycerol synthesis pathways on plant oil composition and bioengineering. Front Plant Sci. 2012;3:147.
13. Ohlrogge J, Browse J. Lipid biosynthesis. Plant Cell. 1995;7:957–70.
14. Cao H. Structure-function analysis of diacylglycerol acyltransferase sequences from 70 organisms. BMC Res Notes. 2011;4:249.
15. Saha S, Enugutti B, Rajakumari S, Rajasekharan R. Cytosolic triacylglycerol biosynthetic pathway in oilseeds. Molecular cloning and expression of peanut cytosolic diacylglycerol acyltransferase. Plant Physiol. 2006;141:1533–43.
16. Zou J, Wei Y, Jako C, Kumar A, Selvaraj G, Taylor DC. The *Arabidopsis thaliana* TAG1 mutant has a mutation in a diacylglycerol acyltransferase gene. Plant J. 1999;19:645–53.
17. Shockey JM, Gidda SK, Chapital DC, Kuan JC, Dhanoa PK, Bland JM, Rothstein SJ, Mullen RT, Dyer JM. Tung tree DGAT1 and DGAT2 have nonredundant functions in triacylglycerol biosynthesis and are local-

ized to different subdomains of the endoplasmic reticulum. Plant Cell. 2006;18:2294–313.
18. Merchant SS, Prochnik SE, Vallon O, Harris EH, Karpowicz SJ, Witman GB, Terry A, Salamov A, Fritz-Laylin LK, Marechal-Drouard L, et al. The *Chlamydomonas* genome reveals the evolution of key animal and plant functions. Science. 2007;318:245–50.
19. Radakovits R, Jinkerson RE, Fuerstenberg SI, Tae H, Settlage RE, Boore JL, Posewitz MC. Draft genome sequence and genetic transformation of the oleaginous alga *Nannochloropis gaditana*. Nat Commun. 2012;3:686.
20. Vieler A, Wu G, Tsai C-H, Bullard B, Cornish AJ, Harvey C, Reca I-B, Thornburg C, Achawanantakun R, Buehl CJ, et al. Genome, functional gene annotation, and nuclear transformation of the heterokont oleaginous Alga *Nannochloropsis oceanica* CCMP1779. PLoS Genet. 2012;8:e1003064.
21. Fan J, Cui Y, Wan M, Wang W, Li Y. Lipid accumulation and biosynthesis genes response of the oleaginous *Chlorella pyrenoidosa* under three nutrition stressors. Biotechnol Biofuels. 2014;7:17.
22. Gao C, Wang Y, Shen Y, Yan D, He X, Dai J, Wu Q. Oil accumulation mechanisms of the oleaginous microalga *Chlorella protothecoides* revealed through its genome, transcriptomes, and proteomes. BMC Genom. 2014;15:582.
23. Wang D, Ning K, Li J, Hu J, Han D, Wang H, Zeng X, Jing X, Zhou Q, Su X, et al. *Nannochloropsis* genomes reveal evolution of microalgal oleaginous traits. PLoS Genet. 2014;10:e1004094.
24. Hung C-H, Ho M-Y, Kanehara K, Nakamura Y. Functional study of diacylglycerol acyltransferase type 2 family in *Chlamydomonas reinhardtii*. FEBS Lett. 2013;587:2364–70.
25. Miller R, Durrett TP, Kosma DK, Lydic TA, Muthan B, Koo AJK, Bukhman YV, Reid GE, Howe GA, et al. Altered lipid composition and enhanced nutritional value of *Arabidopsis* leaves following introduction of an algal diacylglycerol acyltransferase 2. Plant Cell. 2013;25:677–93.
26. Liu J, Han D, Yoon K, Hu Q, Li Y. Characterization of type 2 diacylglycerol acyltransferases in *Chlamydomonas reinhardtii* reveals their distinct substrate specificities and functions in triacylglycerol biosynthesis. Plant J. 2016;86:3–19.
27. Ma X, Chen T, Yang B, Liu J, Chen F. Lipid Production from Nannochloropsis. Mar Drugs. 2016;14:61.
28. Zienkiewicz K, Zienkiewicz A, Poliner E, Du Z-Y, Vollheyde K, Herrfurth C, Marmon S, Farré EM, Feussner I, Benning C. Nannochloropsis, a rich source of diacylglycerol acyltransferases for engineering of triacylglycerol content in different hosts. Biotechnol Biofuels. 2017;10:8.
29. Li F, Gao D, Hu H. High-efficiency nuclear transformation of the oleaginous marine *Nannochloropsis* species using PCR product. Biosci Biotechnol Biochem. 2014;78:812–7.
30. Ma X, Liu J, Liu B, Chen T, Yang B, Chen F. Physiological and biochemical changes reveal stress-associated photosynthetic carbon partitioning into triacylglycerol in the oleaginous marine alga *Nannochloropsis oculata*. Algal Res. 2016;16:28–35.
31. Wang Q, Lu Y, Xin Y, Wei L, Huang S, Xu J. Genome editing of model oleaginous microalgae *Nannochloropsis* spp. by CRISPR/Cas9. Plant J. 2016;88:1071–81.
32. Wei L, Xin Y, Wang Q, Yang J, Hu H, Xu J. RNAi-based targeted gene-knockdown in the model oleaginous microalgae *Nannochloropsis oceanica*. Plant J. 2017;89:1236–50.
33. Li J, Han D, Wang D, Ning K, Jia J, Wei L, Jing X, Huang S, Chen J, Li Y, Hu Q, Xu J. Choreography of transcriptomes and lipidomes of *Nannochloropsis* reveals the mechanisms of oil synthesis in microalgae. Plant Cell. 2014;26:1645–65.
34. Tamura K, Stecher G, Peterson D, Filipski A, Kumar S. MEGA6: molecular evolutionary genetics analysis version 6.0. Mol Biol Evol. 2013;30:2725–9.
35. Siloto RMP, Truksa M, Brownfield D, Good AG, Weselake RJ. Directed evolution of acyl-CoA:diacylglycerol acyltransferase: development and characterization of *Brassica napus* DGAT1 mutagenized libraries. Plant Physiol Biochem. 2009;47:456–61.
36. Liu J, Lee Y-Y, Mao X, Li Y. A simple and reproducible non-radiolabeled in vitro assay for recombinant acyltransferases involved in triacylglycerol biosynthesis. J Appl Phycol. 2017;29:323–33.
37. Sorger D, Daum G. Synthesis of triacylglycerols by the acyl-Coenzyme A: diacylglycerol acyltransferase Dga1p in lipid particles of the yeast *Saccharomyces cerevisiae*. J Bacteriol. 2002;184:519–24.
38. Chapman KD, Ohlrogge JB. Compartmentation of triacylglycerol accumulation in plants. J Biol Chem. 2012;287:2288–94.
39. Manandhar-Shrestha K, Hildebrand M. Characterization and manipulation of a DGAT2 from the diatom *Thalassiosira pseudonana*: improved TAG accumulation without detriment to growth, and implications for chloroplast TAG accumulation. Algal Res. 2015;12:239–48.
40. Moog D, Stork S, Reislöhner S, Grosche C, Maier UG. In vivo localization studies in the stramenopile alga *Nannochloropsis oceanica*. Protist. 2015;166:161–71.
41. Gibbs SP. The route of entry of cytoplasmically synthesized proteins into chloroplasts of algae possessing chloroplast ER. J Cell Sci. 1979;35:253–66.
42. Neupert J, Karcher D, Bock R. Generation of *Chlamydomonas* strains that efficiently express nuclear transgenes. Plant J. 2009;57:1140–50.
43. Goncalves E, Johnson J, Rathinasabapathi B. Conversion of membrane lipid acyl groups to triacylglycerol and formation of lipid bodies upon nitrogen starvation in biofuel green algae *Chlorella* UTEX29. Planta. 2013;238:895–906.
44. Abida H, Dolch LJ, Mei C, Villanova V, Conte M, Block MA, Finazzi G, Bastien O, Tirichine L, Bowler C, et al. Membrane glycerolipid remodeling triggered by nitrogen and phosphorus starvation in *Phaeodactylum tricornutum*. Plant Physiol. 2015;167:118–36.
45. Miller R, Wu G, Deshpande RR, Vieler A, Gartner K, Li X, Moellering ER, Zauner S, Cornish AJ, Liu B, et al. Changes in transcript abundance in *Chlamydomonas reinhardtii* following nitrogen deprivation predict diversion of metabolism. Plant Physiol. 2010;154:1737–52.
46. Gong Y, Zhang J, Guo X, Wan X, Liang Z, Hu CJ, Jiang M. Identification and characterization of PtDGAT2B, an acyltransferase of the DGAT2 acyl-Coenzyme A: diacylglycerol acyltransferase family in the diatom *Phaeodactylum tricornutum*. FEBS Lett. 2013;587:481–7.
47. Jia J, Han D, Gerken HG, Li Y, Sommerfeld M, Hu Q, Xu J. Molecular mechanisms for photosynthetic carbon partitioning into storage neutral lipids in *Nannochloropsis oceanica* under nitrogen-depletion conditions. Algal Res. 2015;7:66–77.
48. Zhang C, Iskandarov U, Klotz ET, Stevens RL, Cahoon RE, Nazarenus TJ, Pereira SL, Cahoon EB. A thraustochytrid diacylglycerol acyltransferase 2 with broad substrate specificity strongly increases oleic acid content in engineered *Arabidopsis thaliana* seeds. J Exp Bot. 2013;64:3189–200.
49. Chi X, Hu R, Zhang X, Chen M, Chen N, Pan L, Wang T, Wang M, Yang Z, Wang Q, Yu S. Cloning and functional analysis of three diacylglycerol acyltransferase genes from peanut *Arachis hypogaea* L. PLoS ONE. 2014;9:e105834.
50. Xu J, Francis T, Mietkiewska E, Giblin EM, Barton DL, Zhang Y, Zhang M, Taylor DC. Cloning and characterization of an acyl-CoA-dependent diacylglycerol acyltransferase 1 DGAT1 gene from *Tropaeolum majus*, and a study of the functional motifs of the DGAT protein using site-directed mutagenesis to modify enzyme activity and oil content. Plant Biotechnol J. 2008;6:799–818.
51. Zhou X-R, Shrestha P, Yin F, Petrie JR, Singh SP. AtDGAT2 is a functional acyl-CoA:diacylglycerol acyltransferase and displays different acyl-CoA substrate preferences than AtDGAT1. FEBS Lett. 2013;587:2371–6.
52. Guiheneuf F, Leu S, Zarka A, Khozin-Goldberg I, Khalilov I, Boussiba S. Cloning and molecular characterization of a novel acyl-CoA:diacylglycerol acyltransferase 1-like gene PtDGAT1 from the diatom *Phaeodactylum tricornutum*. FEBS J. 2011;278:3651–66.
53. Chen C-X, Sun Z, Cao H-S, Fang F-L, Ouyang L-L, Zhou Z-G. Identification and characterization of three genes encoding acyl-CoA: diacylglycerol acyltransferase DGAT from the microalga *Myrmecia incisa* Reisigl. Algal Res. 2015;12:280–8.
54. Kirchner L, Wirshing A, Kurt L, Reinard T, Glick J, Cram EJ, Jacobsen H-J, Lee-Parsons CWT. Identification, characterization, and expression of diacylgylcerol acyltransferase type-1 from *Chlorella vulgaris*. Algal Res. 2016;13:167–81.
55. Guo X, Fan C, Chen Y, Wang J, Yin W, Wang RRC, Hu Z. Identification and characterization of an efficient acyl-CoA: diacylglycerol acyltransferase 1 (DGAT1) gene from the microalga *Chlorella ellipsoidea*. BMC Plant Biol. 2017;17:48.
56. Yoon K, Han D, Li Y, Sommerfeld M, Hu Q. Phospholipid:diacylglycerol acyltransferase is a multifunctional enzyme involved in membrane lipid turnover and degradation while synthesizing triacylglycerol in the unicellular green microalga *Chlamydomonas reinhardtii*. Plant Cell. 2012;24:3708–24.
57. Fan J, Andre C, Xu C. A chloroplast pathway for the de novo biosynthesis of triacylglycerol in *Chlamydomonas reinhardtii*. FEBS Lett. 2011;585:1985–91.

58. Li X, Moellering ER, Liu B, Johnny C, Fedewa M, Sears BB, Kuo M-H, Benning C. A galactoglycerolipid lipase is required for triacylglycerol accumulation and survival following nitrogen deprivation in *Chlamydomonas reinhardtii*. Plant Cell. 2012;24:4670–86.
59. Goodson C, Roth R, Wang ZT, Goodenough U. Structural correlates of cytoplasmic and chloroplast lipid body synthesis in *Chlamydomonas reinhardtii* and stimulation of lipid body production with acetate boost. Eukaryot Cell. 2011;10:1592–606.
60. Goodenough U, Blaby I, Casero D, Gallaher SD, Goodson C, Johnson S, Lee J-H, Merchant SS, Pellegrini M, Roth R, et al. The path to triacylglyceride obesity in the *sta6* strain of *Chlamydomonas reinhardtii*. Eukaryot Cell. 2014;13:591–613.
61. Block MA, Jouhet J. Lipid trafficking at endoplasmic reticulum-chloroplast membrane contact sites. Curr Opin Cell boil. 2015;35:21–9.
62. Tai M, Stephanopoulos G. Engineering the push and pull of lipid biosynthesis in oleaginous yeast *Yarrowia lipolytica* for biofuel production. Metab Eng. 2013;15:1–9.
63. Vanhercke T, El Tahchy A, Shrestha P, Zhou X-R, Singh SP, Petrie JR. Synergistic effect of WRI1 and DGAT1 coexpression on triacylglycerol biosynthesis in plants. FEBS Lett. 2013;587:364–9.
64. Vanhercke T, El Tahchy A, Liu Q, Zhou XR, Shrestha P, Divi UK, Ral JP, Mansour MP, Nichols PD, James CN, et al. Metabolic engineering of biomass for high energy density: oilseed-like triacylglycerol yields from plant leaves. Plant Biotechnol J. 2014;12:231–9.
65. van Erp H, Kelly AA, Menard G, Eastmond PJ. Multigene engineering of triacylglycerol metabolism boosts seed oil content in Arabidopsis. Plant Physiol. 2014;165:30–6.
66. Greer MS, Truksa M, Deng W, Lung SC, Chen G, Weselake RJ. Engineering increased triacylglycerol accumulation in *Saccharomyces cerevisiae* using a modified type 1 plant diacylglycerol acyltransferase. Appl Microbiol Biotechnol. 2015;99:2243–53.
67. Knothe G. Improving biodiesel fuel properties by modifying fatty ester composition. Energ Environ Sci. 2009;2:759–66.
68. Li DW, Cen SY, Liu YH, Balamurugan S, Zheng X-Y, Alimujiang A, Yang W-D, Liu J-S, Li HY. A type 2 diacylglycerol acyltransferase accelerates the triacylglycerol biosynthesis in heterokont oleaginous microalga *Nannochloropsis oceanica*. J Biotechnol. 2016;229:65–71.
69. Zheng P, Allen WB, Roesler K, Williams ME, Zhang S, Li J, Glassman K, Ranch J, Nubel D, Solawetz W, et al. A phenylalanine in DGAT is a key determinant of oil content and composition in maize. Nat Genet. 2008;40:367–72.
70. Trentacoste EM, Shrestha RP, Smith SR, Glé C, Hartmann AC, Hildebrand M, Gerwick WH. Metabolic engineering of lipid catabolism increases microalgal lipid accumulation without compromising growth. Pro Natl Acad Sci USA. 2013;110:19748–53.
71. Liu J, Sommerfeld M, Hu Q. Screening and characterization of *Isochrysis* strains and optimization of culture conditions for docosahexaenoic acid production. Appl Microbiol Biotechnol. 2013;97:4785–98.
72. Liu J, Sun Z, Zhong Y, Huang J, Hu Q, Chen F. Stearoyl-acyl carrier protein desaturase gene from the oleaginous microalga *Chlorella zofingiensis*: cloning, characterization and transcriptional analysis. Planta. 2012;236:1665–76.
73. Lauersen K, Kruse O, Mussgnug J. Targeted expression of nuclear transgenes in *Chlamydomonas reinhardtii* with a versatile, modular vector toolkit. Appl Microbiol Biotechnol. 2015;99:3491–503.
74. Kindle KL. High-frequency nuclear transformation of *Chlamydomonas reinhardtii*. Pro Natl Acad Sci USA. 1990;87:1228–32.

2

Electricity-assisted production of caproic acid from grass

Way Cern Khor, Stephen Andersen, Han Vervaeren and Korneel Rabaey*

Abstract

Background: Medium chain carboxylic acids, such as caproic acid, are conventionally produced from food materials. Caproic acid can be produced through fermentation by the reverse β-oxidation of lactic acid, generated from low value lignocellulosic biomass. In situ extraction of caproic acid can be achieved by membrane electrolysis coupled to the fermentation process, allowing recovery by phase separation.

Results: Grass was fermented to lactic acid in a leach-bed-type reactor, which was then further converted to caproic acid in a secondary fermenter. The lactic acid concentration was 9.36 ± 0.95 g L^{-1} over a 33-day semi-continuous operation, and converted to caproic acid at pH 5.5–6.2, with a concentration of 4.09 ± 0.54 g L^{-1} during stable production. The caproic acid product stream was extracted in its anionic form, concentrated and converted to caproic acid by membrane electrolysis, resulting in a >70 wt% purity solution. In a parallel test exploring the upper limits of production rate through cell retention, we achieved the highest reported caproic acid production rate to date from a lignocellulosic biomass (grass, via a coupled process), at 0.99 ± 0.02 g L^{-1} h^{-1}. The fermenting microbiome (mainly consisting of *Clostridium* IV and *Lactobacillus*) was capable of producing a maximum caproic acid concentration of 10.92 ± 0.62 g L^{-1} at pH 5.5, at the border of maximum solubility of protonated caproic acid.

Conclusions: Grass can be utilized as a substrate to produce caproic acid. The biological intermediary steps were enhanced by separating the steps to focus on the lactic acid intermediary. Notably, the pipeline was almost completely powered through electrical inputs, and thus could potentially be driven from sustainable energy without need for chemical input.

Keywords: Grass, Lactic acid, Caproic acid, Decane, Fermentation, Chain elongation, Electrolysis

Background

Caproic acid is a medium-chain carboxylic acid, its demand has been growing due to its application as chemical commodity, feed additive and more recently as bio-based fuel precursor. The medium-chain carboxylic acid market (caproic acid, caprylic acid, capric acid, and lauric acid) is predicted to reach USD 1.25 billion globally by 2020 [1]. Generally, these medium chain fatty acids (MCFA) are derived from triglycerides of coconut and palm oil by fractional distillation, ozonolysis or catalytic reduction processes [2]. MCFA can also be microbially synthesized from alcohols and carboxylic acids through fermentation. Zhu et al. [3] recently demonstrated that a solution of lactic acid could enable caproic acid production through the microbial reverse β-oxidation pathway.

Lactic acid is typically formed during the fermentation/ensiling of grass. Grass is a widely available substrate presently underused and often overlooked as a carbon source. In the US, grasslands represent an estimated 2.51 × 10^6 km^2 of available biomass [4], which is either ensiled or lowly used. A grass-based fermentation can lead to the formation of lactic and acetic acid. In an uncontrolled pH, mesophilic fermentation of grass, a concentration of 12.6 g L^{-1} lactic acid can be reached, usually along with 2.0 g L^{-1} of acetic acid [5]. Lactic acid and acetic acid are highly soluble in water, which makes the downstream extraction of lactic acid energy intensive. More hydrophobic products such as caproic acid can

*Correspondence: Korneel.Rabaey@UGent.be
Department of Biochemical and Microbial Technology, Centre for Microbial Ecology and Technology (CMET), Ghent University, Coupure Links 653, 9000 Ghent, Belgium

be produced through the reverse β-oxidation pathway with ethanol as the most common reducing substrate [6] and acetic acid as the electron acceptor. Studies have been carried out with pure cultures such as *Clostridium kluyveri* using ethanol [7, 8] and *Megasphaera elsdenii* using sugars and lactic acid [9]. Caproic acid production from lignocellulosic material [10] and a route using lactic acid was first considered in 1956 [11], with attention returning to this process in recent years, through the reverse β-oxidation pathway pure culture fermentations [12], and mixed culture community fermentations [3, 13–17]. Zhu et al. proposed the stoichiometry of caproic acid formation via lactic acid as 3 $CH_3CHOHCOO^- + 2\ H^+ \rightarrow CH_3(CH_2)_4COO^- + H_2O + 2\ H_2 + 3\ CO_2$, with Gibbs free energy of -123.1 kJ mol^{-1} [3].

The hydrophobicity and low water solubility of caproic acid allow extraction of caproic acid from the fermentation broth by phase separation. If the concentration of caproic acid in its protonated form exceeds the solubility limit (11.0 g L^{-1} at 20 °C), caproic acid will form an immiscible layer and phase separate from the fermentation broth along with other hydrophobic chemical species. Recently, Xu et al. [18] managed to extract n-caproic acid from a bioreactor broth using an in-line membrane electrolysis system, while minimizing caustic and acidic dosing with electrochemistry.

Here, we studied the feasibility of a complete pipeline for caproic acid generation from grass via microbial fermentations and electricity-driven processes alleviating the use of chemicals, and completely avoiding energy intensive dewatering and distillation steps on the carboxylic acid intermediates. Lactic acid fermentation broth from grass elongates to caproic acid by reverse β-oxidation. The caproic acid extracts via membrane electrolysis delivering the acid concentration of caproic acid. As final proof of concept, we also converted caproic acid to decane via Kolbe electrolysis. For each process rate, efficiencies and intermediary concentrations were determined. The overall process is presented in Fig. 1.

Results and discussion

Semi-continuous fermentation of lactic acid from grass

Semi-continuous fermentation of lactic acid using grass as substrate was performed for 33 days, with a solids and liquid retention time of 2 days. The lactic acid concentration stabilized at 9.36 ± 0.95 g L^{-1}, and the acetic acid concentration at 0.90 ± 0.14 g L^{-1} (Fig. 2a). The pH of the fermentation broth was between 4.5 and 5. The lactic acid concentration rapidly reached its final concentration within the first day of fermentation, indicating that the native microorganisms of grass were active as there was no lag time for lactic acid production. In this study, lactic acid and acetic acid were the main products during fermentation due to low pH and short retention time.

The conversion of organic compounds to lactic acid was low relative to our previous study [5] (0.065 g g^{-1} in this study compared to 0.136 g g^{-1} grass volatile solid), while the rate of lactic acid production was 0.197 g L^{-1} h^{-1} in this study. The conversion rate and efficiency of biomass to lactic acid were relatively low in this study, which is likely due to the limited pretreatment and reactor design involving a higher ratio of solution volume relative to mass of grass. Pretreatment (e.g. grinding, steam-explosion, alkaline) can be performed to improve the biomass biodegradability, while taking into consideration the cost-benefit-sustainability nexus. For instance, mild pretreatment methods such as lime pretreatment can be applied without introducing excessive cost, to improve conversion efficiency and production rate, and minimize carbon loss [19]. In earlier studies, it was shown that pretreatment, such as extrusion, can improve the degradability of the grass [19]. At the time of this manuscript, the equipment was not available; however, similar results to Khor et al. [5] can be expected, where a conversion increase of 109% was noted. While pretreatment can enhance biomass biodegradability, it is also necessary to consider the biological impact of the pretreatment. Phenolic compounds and furfural can be released from lignin if the pretreatment is severe (e.g. high temperature, low pH), which can then interfere with fermentative processes and inhibit the growth of microorganisms. Hence, a substrate purification step (e.g. metal catalysts, adsorption or chemical oxidation) to remove inhibitory compounds prior to biological processes could be beneficial to improve product yield. In our study, the protein fraction of grass is assumed to be well retained and not degraded after fermentation due to the short retention time and low pH of the process, and no ammonia gas was detected. In a 2nd generation biorefinery process, complete conversion of lignocellulosic biomass into a single targeted compound is highly unlikely due to the

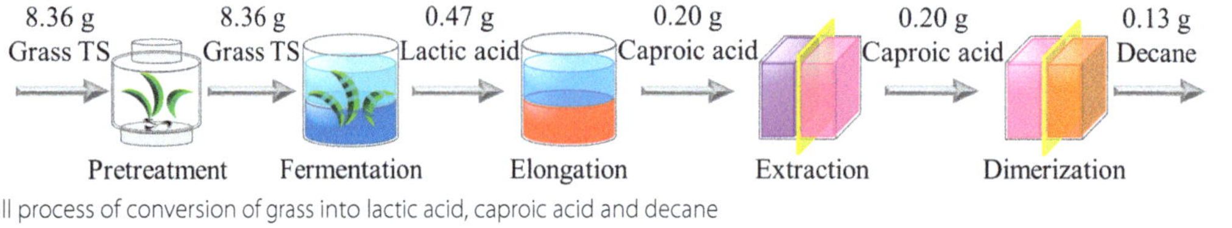

Fig. 1 Overall process of conversion of grass into lactic acid, caproic acid and decane

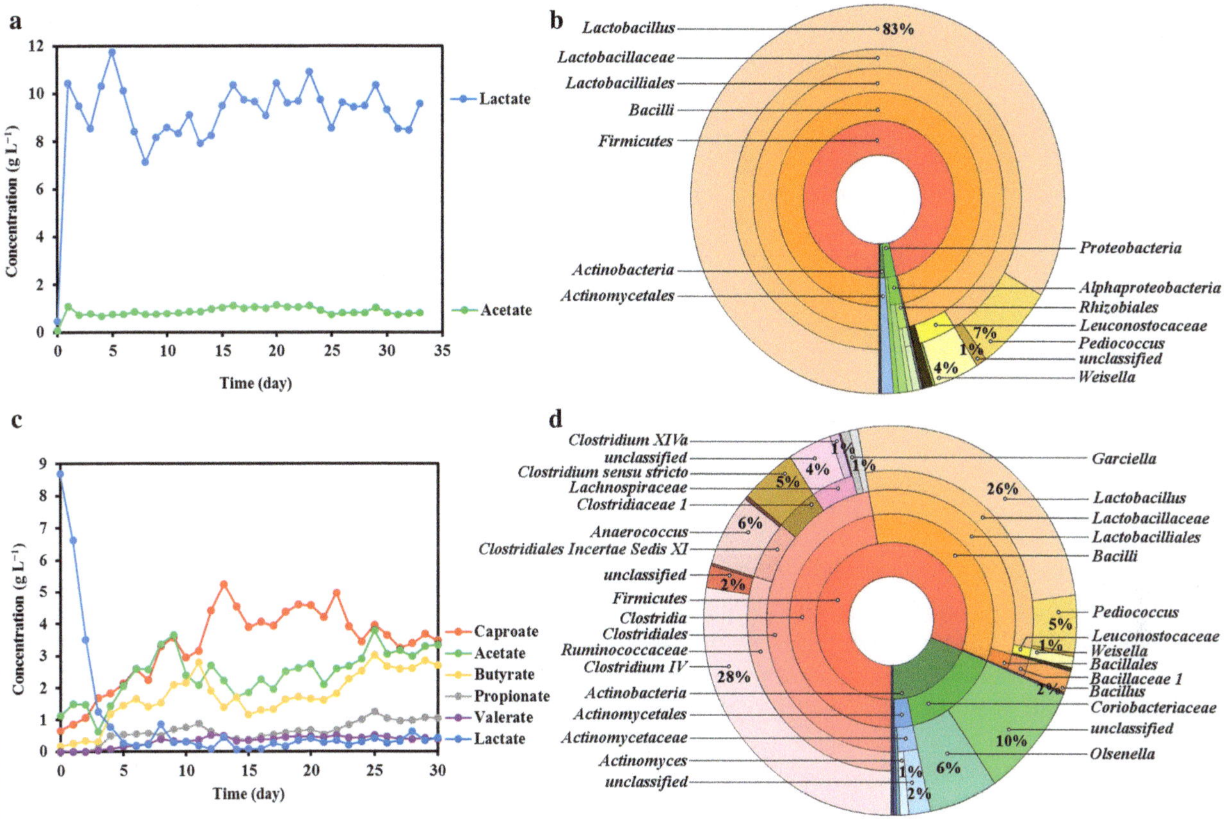

Fig. 2 Carboxylate profile and bacterial community of fermentation and elongation system: **a** carboxylate profile of fermentation system, **b** bacterial community for fermentation system, **c** carboxylate profile of elongation system, **d** bacterial community for elongation system. The concentric circles represent the bacterial taxonomy from phylum (*centre*) to genus (*outermost*). *Each band* indicates the proportion of the bacteria. *One colour* is dedicated to each classification

complexity of the substrate. Instead, available fractions of the biomass can be converted into desired compounds at high rate, with the remainder recovered as by-products. For instance, in this case, one could foresee that the hemicellulose/cellulose fraction can be converted into lactic acid, while the remaining protein/amino acid fraction is recovered via an alternate route, such as the Grassa process [20].

For lactic acid fermentation, community analysis indicated that over time, the population became enriched with lactic acid bacteria such as *Lactobacillus* spp. (83% relative abundance), *Pediococcus* and *Weissella* spp. (Fig. 2b). *Lactobacillus*, *Pediococcus* and *Weissella* species are known for their ability to utilize polysaccharides and monosaccharides with lactic acid as a key outcome, as was observed in our experiments [5, 21].

Semi-continuous elongation of caproic acid through lactic acid

When the grass fermentation effluent was fed into the elongation system, lactic acid was consumed, producing caproic acid (C6), as shown in Fig. 2c, along with butyric acid (C4) and acetic acid (C2). Enanthic acid (C7) and caprylic acid (C8) were not detected, unlike other chain elongation systems [16]. The caproic acid concentration was 4.09 ± 0.54 g L^{-1} (9.03 ± 1.19 g COD L^{-1}), and $49 \pm 9\%$ of products formed was caproic acid during the stable phase (30 day operation, Fig. 2c). The pH of the elongation broth was between 5.5 and 6.2, showing the pH increasing effect of chain elongation.

The stoichiometry of 3 mol lactic acid forming 1 mol caproic acid as proposed by Zhu et al. [3] was employed to assess the caproic acid production in this study. In terms of lactic acid and caproic acid balance, on average 4.68 ± 0.47 g L^{-1} (53 ± 5 mM) of lactic acid was fed and the average caproic acid production was 2.05 ± 0.27 g L^{-1} (18 ± 2 mM). This fits well with the stoichiometry of 3 mol lactate forming 1 mol caproate, or a consumption of 2.32 g lactate per g caproate formed. Lactic acid generated from a low-value substrate can thus act as an effective intermediary for caproic acid production. While caproic acid production can also be achieved via ethanol, this tends to bind one to an integrated biorefinery or other ethanol producing (bio-)processes. Under different circumstances (e.g. substrate availability, market considerations), one can foresee that both pathways (i.e. chain elongation using lactate or ethanol) are interesting.

In the elongation system, a different community developed from the initial fermentation, despite influx via the substrate. *Firmicutes* dominated the population (81% relative abundance), in which *Clostridium* IV- and *Lactobacillus*-related species were most prevalent. *Coriobacteriaceae* and *Anaerococcus* spp. were other highly abundant species (Fig. 2d). *Clostridium* species, including *Clostridium* IV, *Clostridium Sensu Stricto* and *Clostridium* XIVa spp. identified in this study, are the archetype organisms known to perform reverse β-oxidation [22, 23], and also utilize lactic acid. Although being weakly fermentative, *Anaerococcus* spp. is known to be able to metabolize carbohydrate and produce butyric acid and lactic acid as major metabolic end products. Some genus have butyric acid and caproic acid as major end products [23], which fits very well in this context. *Coriobacteriaceae* spp. has been mostly reported to be saccharolytic; however, its genus such as *Olsenella* spp. is able to convert glucose into lactic acid as major product [24]. The family of *Actinomycetaceae* spp. are also able to produce lactic acid and acetic acid under fermentative condition [25]. The bacterial composition of the inoculum and the elongation system were similar (Additional file 1: Fig. S1a; Fig. 2d). This showed that introduction of bacterial community from the fermentation system did not change the bacterial community of the elongation system significantly.

Maximum rate of caproic acid production and maximum concentration of caproic acid test

The highest rate of caproic acid production reached was 0.99 ± 0.02 g L^{-1} h^{-1}, when the cells were retained (Fig. 3—cycle 6). The system was left unattended for 1 week to test the resilience of the elongation microorganisms, and the production rate recovered to the maximum rate after 4 cycles of operation (Fig. 3—cycle 9 to 12). Table 1 compares different feed sources and maximum rates of medium chain fatty acids produced for microbial chain elongation process through a select screening of literature.

The maximum caproic acid production rate achieved in this study is high compared to literature, such as the study of Grootscholten et al. where 4.5 g L^{-1} day^{-1} (0.19 g L^{-1} h^{-1}) was achieved with mixed culture up-flow anaerobic filter system using castor oil, at pH between 6.5 and 7 [29]. Comparable rates as reported here were thus far only obtained with synthetic media, with the highest from acetic acid and ethanol, 57.4 g L^{-1} day^{-1} (2.39 g L^{-1} h^{-1}) [26], albeit with a considerable higher loading rate. Our value considers only the rate of the lactic acid-rich reactor effluent, and biomass hydrolysis is broadly recognized as a rate limiting step in conversion of 'real', complex substrates. Further, it is important to note that to achieve this rate, Grootscholten et al. used a neutral pH between 6.5 and 7.2 [26], thus requiring dosing of base which was not the case in our study.

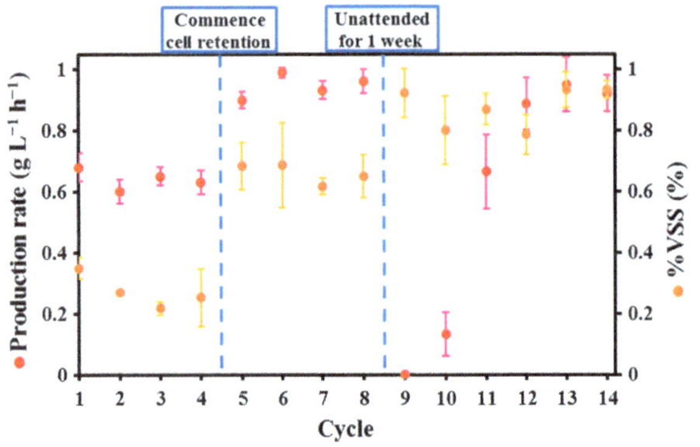

Fig. 3 Production rate of caproic acid (g L^{-1} h^{-1}, *red circles*) and percentage of volatile suspended solids (%VSS, *orange circles*) of elongation broth during maximum rate of caproic acid production test (*error bars* represent the standard deviation of triplicates)

Cell density is often unreported in mixed culture fermentations, and cell retention and maximizing the density in the fermenters can potentially improve conversion dramatically. The production rate achieved here is partially a result of high microbial density, lack of pH controlling salts such as sodium hydroxide and the used substrate (real versus synthetic). *Anaerococcus* spp. (32% relative abundance) became the most dominant species (Additional file 1: Figure S1b), during the maximum rate of caproic acid production test with cell retention, followed by *Lactobacillus* spp., while *Clostridium* IV spp. was merely 6% relative abundance. *Anaerococcus* are typically found in anaerobic conditions and were recently identified under methanogenic conditions [31], and have not been implicated in chain elongation to our knowledge. *Clostridium* IV spp. have been broadly implicated in reverse β-oxidation with lactic acid as an intermediary [3, 6, 17].

In order to determine the maximum concentration of caproic acid the community could sustain without a substrate limitation, an excess of lactic acid was fed to the fermenter. The maximum concentration of caproic acid reached was 10.92 ± 0.62 g L^{-1} (n = 3). By molar concentration, caproic acid consisted of 57 ± 4% of the products formed (Additional file 1: Figure S2), compared to 34 ± 9% during normal elongation operation in this study. While chain elongation is often preferred at

Table 1 Microbial chain elongation operation and maximum rate of medium chain fatty acid production

Feed source	Operation	Culture	pH	Maximum rate of production (g L^{-1} h^{-1})	Reference
Synthetic medium containing acetate, ethanol, and yeast extract, with CO_2 gas flow	Packed bioreactor	Mixed (non-sterilized, mainly *Clostridium kluyveri*)	6.5–7.2	2.39 (medium chain fatty acids)	[26]
Grass fermentation effluent containing lactic acid	Semi-continuous	Mixed (non-sterilized, mainly *Clostridium cluster IV*)	5.5–6.2	0.99 (caproic acid)	This study
Synthetic medium containing galactitol, yeast extract, with in situ extraction	Batch	Pure (*Clostridium* sp. BS-1)	6.5	0.34 (caproic acid)	[12]
Dilute ethanol and acetate, with in-line extraction	Continuous	Mixed (non-sterilized, mainly *Clostridium* spp.)	5.2–5.5	0.33 (caprylic acid)	[27]
Synthetic medium containing acetate and ethanol	Semi-continuous	Mixed (non-sterilized, mainly *Clostridium kluyveri*)	7	0.20 (caproic acid)	[28]
Castor oil	Continuous	Mixed (non-sterilized, mainly *Clostridium kluyveri*)	6.5–7	0.19 (caproic acid)	[29]
Yeast fermentation beer	Continuous	Mixed (non-sterilized, mainly *Clostridium kluyveri*)	5.5	0.14 (caproic acid)	[30]
Diluted yellow water	Semi-continuous	Mixed (non-sterilized, mainly *Clostridium cluster IV*)	5.5–6.5	0.12 (caproic acid)	[3]

neutral pH (6.5–7.2) to avoid toxicity of caproic acid, the highest concentration achieved in this study was under acidic conditions (pH 5.5–6), suggesting some tolerance of the community to the usually anti-microbial caproic acid. During the maximum concentration of caproic acid test in which excess lactic acid was provided, *Clostridium* IV spp. remained the most dominant (Additional file 1: Figure S1c), and other species reduced in terms of abundance. This provides additional evidence that *Clostridium* IV spp. are generating caproic acid through the lactic acid intermediate, and indicates that *Clostridium* IV spp. can tolerate a caproic acid concentration at least up to 10.92 g L^{-1}, at the border of the solubility limit. The mixed population showed functional stability between experiments, while the community members differed.

Electrochemical extraction of caproic acid and decane production through Kolbe electrolysis

Effluent from the elongation system was sent to the cathodic chamber of an electrochemical system for extraction and concentration of caproic acid, with an applied current of 0.4 A (resulting in 25 A m^{-2}). When current was applied, the anion exchange membrane allowed anions such as carboxylates, sulphate, chloride and phosphate to migrate from the cathodic to anodic compartment through the membrane to complete the circuit. The pH dropped at the anodic compartment as protons were generated due to water electrolysis, and carboxylates formed their undissociated counterparts and thus accumulated in the anodic compartment. Over time, the caproic acid phase-separated from the anodic broth as the concentration exceeded its maximum (11 g L^{-1} at 20 °C in water), forming a lower density, hydrophobic layer on top of the solution (Additional file 1: Figure S3), similar to that of Xu et al. [18].

In parallel with the caproic acid extraction, the caproic acid concentrate was also tested for conversion to decane. Energy-dense alkanes are classically obtained from fossil fuels, but can also be generated electrochemically from carboxylic acids. Kolbe electrolysis was performed on the extracted product, introduced in the anolyte again to a concentration of 3.03 g L^{-1} (26 mM), and compared against a synthetic caproic acid electrolysis. The synthetic solution cyclic voltammetry (CV) was shifted earlier compared to the control, which suggested an anodic reaction apart than water electrolysis, while the real solution CV gave a distinctive peak for an unidentified peak for an anodic reaction (Fig. 4b). Increasing the current density beyond the mass transport limit can give rise to oxygen evolution rather than the desired Kolbe process, resulting in loss of efficiency. The electrolysis reaction resulted in decarboxylation and dimerization of caproic acid, leading to the production of decane in the organic phase. Caproic acid was electrolyzed at a rate of 1.21 g L^{-1} h^{-1} (0.202 kg m^{-2} h^{-1}) for the control synthetic solution and 1.05 g L^{-1} h^{-1} (0.174 kg m^{-2} h^{-1}) for the real product effluent solution, only 13% less than the control (Fig. 4a). Compared to literature, conversion rates of valeric acid were 1.9 kg m^{-2} h^{-1} assuming 100% Coulombic efficiency [32]. The coulombic efficiency was relatively low in this study, approximately ten times less, as the un-optimized process also contained other

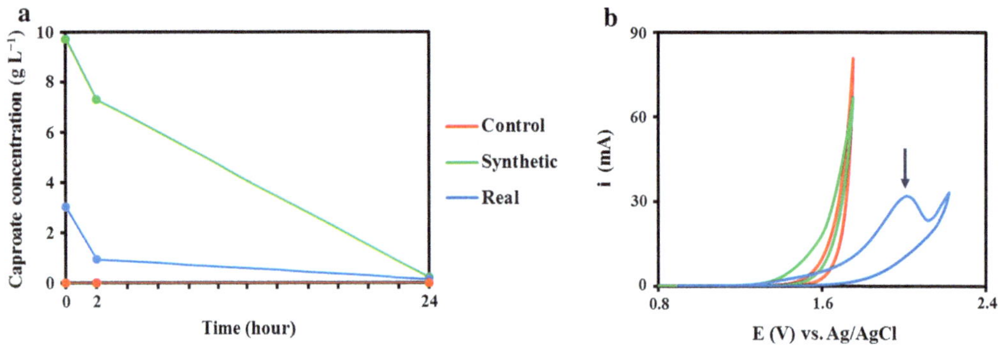

Fig. 4 Caproate concentration and cyclic voltammetry of Kolbe electrolysis process: **a** Caproate concentration over the period of Kolbe electrolysis, and **b** cyclic voltammetry of solution before Kolbe electrolysis. Control (*red line*) contained only 0.5 M Na_2SO_4 in water, synthetic solution (*green line*) contains 0.1 M caproic acid and 0.5 M Na_2SO_4 in water, and real solution (*blue line*) was the anolyte from electrochemical extraction system

carboxylic acids and salts extracted alongside the target product. Furthermore, lower solubility of caproic acid compared to valeric acid also contributed to less reactants in the solution, which can decrease product yield [33–35]. The aqueous phase contains more impurities such as Cl^-, SO_4^{2-}, acetic acid, butyric acid and others, which subsequently led to formation of other impurities and non-Kolbe products which warrants further investigation. NMR analysis reveals that 10% of the product was decane, and indicated that decane was not the sole product. A quantitative analysis was not possible due to small amount that could be sampled throughout the laboratory pipeline. In theory, a range of other Kolbe and non-Kolbe products such as 1-pentene, 2-pentanone, pentan-1-ol, and valeric acid can be formed; however, they could not be accurately identified in this study due to the limited sample amount and complexity of the mixture.

Compared to the literature, Kolbe electrolysis performed at current density of 0.13 A cm^{-2} and temperature of 293 K, using synthetic caproic acid solution and sono-emulsion system in the presence of 190 W cm^{-2} ultrasound, yields 24 $\pm$ 3% of the Kolbe dimer [36]. In this study, the efficiencies of conversion were lower, likely due to the complexity of the fermentation broth and the unoptimized conditions. Increasing current density can potentially result in a higher yield; Wadhawan et al. reported an improvement from 24 to 45% when current density was increased from 0.13 to 0.18 A m^{-2} [36]. Integration of power ultrasound can improve mass transport and help remove the products formed from the electrode surface, resulting in higher conversion efficiency. This study demonstrates that the mixed solution of caproic acid (alongside other carboxylic acids and other salts) can still be converted to alkanes, which is a promising route for carboxylic acid fermentation streams. The tendency for microbial communities to form side products is a challenging issue in biorefinery production from wastes, and effective conversion processes that are unimpeded by side products are valuable to explore.

Grass to chemicals and fuels: preliminary economic assessment

Grass was used as substrate to produce chemicals such as lactic acid, caproic acid and decane. Here, the process was performed in discrete steps, whereas in future phases the caproic acid fermentation and extraction will be directly coupled, and potentially the Kolbe electrolysis can be performed directly in the anode of the extraction cell, pending reactor and electrode design. The system could thus be simplified to two fermenters and a coupled electrochemical system which will likely entail efficiency increases.

An economic estimate was made by considering the operating costs (Table 2). Pretreatment such as extrusion requires around 0.20 kWh kg^{-1} TS of grass [19], and extraction using membrane electrolysis needs approximately 2 kWh kg^{-1} carboxylic acid extracted. Kolbe electrolysis of valeric acid into octane requires around 1.64 kWh kg^{-1} octane, assuming 100% selectivity [32]. Here, 30% conversion of grass TS to lactic acid, and 50% conversion of caproic acid to decane are assumed, and an electricity price of 0.125 USD kWh^{-1}. Production of 1 kg of lactic acid thus costs 0.33 USD. When further converted into caproic acid, the production cost increased to 0.44 USD and conversion to 1 kg decane incurred a cost of 1.29 USD kg^{-1} decane produced. Compared to the market price of the products, lactic acid (1.00 USD kg^{-1}), caproic acid (1.60 USD kg^{-1}) and decane (0.41 USD kg^{-1}, as aviation fuel), one can see that caproic acid has a higher margin for profit; however, one must consider that these prices represent market ready chemicals with a relatively small market. Lactic acid can also

Table 2 Material, energy and costs required to produce 1 kg of lactic acid, caproic acid or decane

	Lactic acid	Caproic acid	Decane
Material (kg TS)			
Grass	*3.33*	*7.69*	*25.00*
Energy (kWh)			
Pretreatment	0.67	1.54	5.00
Extraction	2.00	2.00	2.00
Kolbe electrolysis	–	–	3.28
Total	*2.67*	*3.54*	*10.28*
Cost ($)			
Pretreatment	0.08	0.19	0.63
Extraction	0.25	0.25	0.25
Kolbe electrolysis	–	–	0.41
Total	*0.33*	*0.44*	*1.29*
Selling price ($ kg^{-1})	*1.00*	*1.60*	*0.41*

be a lucrative product, although energy-intense product purification may be needed, generally in the form of distillation. Decane may seem at first glance to be a less attractive end product due to its low economic value at this point of time; however, the liquid fuel market is extensive and well entrenched, and its low-density and low-solubility make it a rather simple product to target in terms of process engineering. If caproic acid can be produced biologically in mass, one can foresee that decane production may become a more attractive end-point due to the much higher demand. Moreover, depending on the final use, the presence of trace compounds in the product can require extensive purification trains. Both energy and unit investments for removing trace impurities, for example removing short chain carboxylic acids including formic and acetic acids, damage the economic argument in producing specialty chemicals like lactic acid from sustainable resources, while this is less of an issue for liquid fuels. In this study, the cost assessment was simplified to give a rough estimate on the economics of each target product. The chemical and electricity prices were also assumed based on literature values and available data, which can differ significantly over time, location and market. Biomass conversion generally varies over a range of 10 to 60% depending on the substrate and efficiency of pretreatment [5, 37]. A consistent supply of biomass is also crucial and biomass storage is often foreseen for continuous biorefinery processes. For a complete and elaborate assessment, other elements including capital and labour costs, and transportation cost for both grass and product need to be taken into account. As in first generation ethanol biorefineries, the cost and sustainability of collection and transport can make or break the argument for the grass to chemicals and fuels process.

In 2005, it was estimated that 2.5×10^{11} kg of municipal solid waste was generated in the United States, in which a significant fraction of the organic material was yard trimmings [38]. Assuming 10% municipal solid waste is made up of yard trimming, this gives an estimation of 1×10^9 kg (1.35×10^9 L) of fuel which can optimistically be produced in a green manner, and contribute to the approximately 78×10^9 L of aviation fuel consumption in year 2012 in the US [39]. Today, the market size and need for energy-dense liquid fuel combined with present non-sustainable fuel production practices will continue to drive development of sustainable fuel production processes outcome.

Conclusions

Here we demonstrated that a lignocellulosic biomass such as grass can be converted without addition of chemicals with electrochemical processes into an energy-dense chemical, through the production and utilization of a lactic acid intermediary towards caproic acid. The system can produce up to a maximum concentration in the fermentation of 10.92 ± 0.62 g L^{-1} caproic acid and achieved a production rate of 0.99 ± 0.02 g L^{-1} h^{-1}, unmatched thus far for real substrates and at pH values relevant for fermentation. Despite a lower amount of end product as compared to lactic acid, caproic acid is simpler to harvest through phase separation, avoiding heat-based extraction. This process also opens an avenue towards the sustainable production of decane, a Kolbe electrolysis derivative of caproic acid.

Methods

Substrate and microorganisms

Farmland grass was harvested from a meadow on January 2016 (East Flanders, Belgium). The grass was air-dried at 28 °C to total solid content of 0.836 ± 0.006 g g^{-1} grass (volatile solid 0.728 ± 0.011 g g^{-1} grass) and stored at 4 °C until used. To make the grass more accessible for fermentation, size reduction was performed with a blender (Philips Daily HR2100/90) before fermentation. The grass particles sizes were between 24 and 453 µm, with a mean of 98 ± 74 µm over 200 measurements. The bacterial culture for lactic acid fermentation was native to the grass itself without extra inoculation. The microbiome for elongation, dominated by *Clostridium*- and *Lactobacillus* spp.-related species, was obtained from a continuous reactor producing caproic acid from thin stillage containing both ethanol and lactate [40]. 25 mL of thin stillage reactor effluent was collected and centrifuged at 5000*g* for 5 min. The supernatant was removed and the pellet obtained was washed 3 times with tap water and dissolved in 10 mL tap water before it was used as inoculum for elongation test.

Fermentation

The semi-continuous fermentation test for grass to lactic acid consisted of a 100-mL reactor filled with 50 mL of tap water and 10 g (wet weight) of grass. The grass was packed in 2 bags made of AISI 316 stainless steel mesh (44 μm mesh size, 33 μm wire thickness), each containing 5 g (wet weight) of grass. The experiment was run under anaerobic condition for 33 days at 32 °C. Steady-state values are reported as the average of 20 samples taken over a stable period (7% variation) of 20 days (i.e. 10 HRTs). The substrate in one of the stainless steel bags was replaced every day (2 days of solid retention time). Half of the liquid phase was removed and refilled with tap water (2 days of hydraulic retention time). The system was sparged with nitrogen each time during substrate and liquid replacement and flushed with nitrogen after substrate replacement to ensure anaerobic condition. Samples were collected every day for carboxylic acids analysis. Samples were collected at day 0 and 33 for bacterial community analysis. pH was between 4.8 and 5.8, and it was not adjusted or controlled throughout the experiment.

Microbial elongation

Semi-continuous operation

The semi-continuous elongation test consisted of a 100-mL reactor filled with 50 mL of elongation broth. Half of the solution was replaced with effluent from the fermentation test every day, resulting in 2 days of hydraulic retention time (HRT). The experiment was run under anaerobic condition for 30 days at 32 °C. Steady-state values are reported as the average of 20 samples taken over a stable period (13% variation) of 20 days (i.e. 10 HRTs). Samples were collected every day for carboxylic acids analysis. The system was sparged with nitrogen each time during substrate replacement and flushed with nitrogen after substrate replacement. Samples were collected at day 0, 9, 16 and 30 for bacterial community analysis. pH was between 5.5 and 6.3, and was not adjusted or controlled throughout the experiment. At the end of experiment, the elongation broth was split into six parts, which served as the inoculum for biologically independent replicates to determine maximum caproic acid concentration ($n = 3$) and maximum rate of caproic acid production ($n = 3$).

Maximum rate of caproic acid production test

This test was performed ($n = 3$) in 50-mL Falcon tubes, each containing 30 mL of elongation broth. Half of the solution was replaced with effluent from the fermentation test every day for 4 days, for each replacement it was expressed as 1 cycle. From the 5th to 8th cycle, cells were retained by centrifugation. Before replacement of solution, the Falcon tubes were centrifuged at 5000*g* for 5 min to retain the cells in the elongation broth. The tubes were then left unattended for 1 week. After that, the operation resumed with cell retention for the 9th to 14th cycle. Samples were collected at 0, 1 and 24 h for carboxylic acids analysis. 1 mL of sample was collected for volatile suspended solid analysis, before the solution was centrifuged and replaced.

Maximum concentration of caproic acid test

The test was performed ($n = 3$) in 50-mL Falcon tubes, each containing 30 mL of elongation broth. 1.135 mL of 50% sodium lactate (VWR) solution was added to the elongation broth to give an extra 20 g L^{-1} lactate in the tubes. In addition to the lactate coming from the lactic acid fermentation, the resulting lactate concentration in the tubes was approximately 25 g L^{-1}. Samples were taken at 0, 2, 4, 6, 8, 12, 24 and 48 h for carboxylic acids analysis.

Electrochemical extraction

The electrochemical cell was constructed from Perspex™, consisting of an anode chamber ($20 \times 5 \times 0.3$ cm^3) and a cathode chamber ($20 \times 5 \times 2.6$ cm^3), separated by an anion exchange membrane (Fujifilm Manufacturing B.V., Netherlands) with a surface area of 0.01 m^2. The cathode was an AISI Type 316L stainless steel felt ($20 \times 5 \times 0.15$ cm^3) with 1-mm wire thickness (LierFilter Ltd., China), and the anode was a mixed metal oxide iridium oxide-coated titanium electrode (IrO_2/TaO_2: 0.65/0.35), 20 cm × 5 cm, with a centrally attached, perpendicular current collector (Magneto Special Anodes BV, The Netherlands). The cathode chamber was filled with effluent from the elongation system, and the anode chamber was filled with tap water. The extraction was performed in batch, a recirculation flow rate of 1.67 mL s^{-1} was maintained for both cathode and anode chambers to ensure mixing. 0.4 A of current was applied using a potentiostat (VSP, Biologic, France) in chronopotentiometry mode to drive the extraction operation, resulting in a current density of 25 A m^{-2} (membrane area). The pH of cathode chamber was controlled at pH 5.5 ± 0.3 by electrochemical water reduction and dosing of 2 M sulphuric acid solution with a pH controller.

Kolbe electrolysis

The Kolbe electrolysis process was performed according to the study of Wadhawan et al. [35]. 50-mL one chamber dimerization reactor consisted of a spiral wire platinum anode (0.5 mm diameter, 3 cm^2 surface area, Bio-Logic, France), a stainless steel plate cathode (12.5 cm^2 surface area, height 5 cm × length 2.5 cm, and width 0.5 cm), and an Ag/AgCl reference electrode (filled with 3 M KCl solution). The chamber was filled with either control

(0.5 M sodium sulphate in distilled water), synthetic solution (0.1 M caproic acid and 0.5 M sodium sulphate) or real solution (electrochemical extraction anolyte, containing 26 mM caproic acid), with pH adjusted to 7. Before electrolysis, cyclic voltammetry (CV, 10 mV s^{-1}), with ohmic drop compensation, was performed on the broth using a potentiostat (SP-50, BioLogic, France), at 25 °C. Kolbe electrolysis experiments were performed at chronopotentiometry mode at a fixed current density of 0.133 A cm^{-2} ($i = 0.4$ A) for 24 h. Liquid phase of reactor was sampled at time 0, 2, and 24 h for analysis.

Bacterial community analysis

DNA extraction

DNA extraction was performed as previously reported [5]. Samples were taken from initial inoculum, initial substrate and end of fermentation broth for analysis. 1.5 mL samples were centrifuged at 11,000*g* for 300 s in a 2-mL Lysing Matrix E tube (Qbiogene, Alexis Biochemicals, Carlsbad, CA). Pelleted cells were re-suspended in 1 mL of lysis buffer containing Tris/HCl (100 mM at pH 8.0), 100 mM EDTA, 100 mM NaCl, 1% (w vol^{-1}) polyvinylpyrrolidone and 2% (w vol^{-1}) sodium dodecyl sulphate. Cells were lysed using 0.2 cm^3 beads of 0.1 mm size in a Fast Prep-96 homogenizer for 40 s at 1600 rpm twice. Samples were centrifuged at 18,000*g* for 60 s at room temperature and washed with phenol/chloroform (1:1) and chloroform. After centrifugation, nucleic acids (supernatant) were precipitated with 1 volume of isopropanol at −20 °C and 1:10 volume of 3 M sodium acetate. After centrifugation and washing with 80% ethanol, the pellet was re-suspended in 20 µL of Milli-Q water. The quality and quantity of DNA samples were analysed using Illumina sequencing primers by polymerase chain reaction (PCR). Amplified sequences were separated by electrophoresis on 1% agarose gels.

DNA sequencing and bioinformatics processing

DNA sequencing was performed as previously reported [5]. The V3–4 region of the bacterial 16S rRNA gene was sequenced with Illumina sequencing Miseq v3 Reagent kit (http://www.illumina.com/products/miseq-reagent-kit-v3.ilmn, by LGC Genomics GmbH, Berlin, Germany) using 2 × 300 bp paired-end reads, and primers 341F (5′-NNNNNNNNTCCTACGGGNGGCWGCAG) and 785R (5′-NNNNNNNNTGACTACHVGGGTATCTAAK CC). Each PCR included DNA extract (~5 ng), forward and reverse primer (~15 pmol for each) and MyTaq buffer (20 µL containing 1.5 units MyTaq DNA polymerase (Bioline) and 2 µL of BioStabII PCR Enhancer). 8-nt barcode sequence was performed for both forward and revers primers of each sample. PCRs were carried out for 96 °C pre-denaturation for 120 s and 30 cycles using the following parameters: 96 °C for 15 s, 50 °C for 30 s and 72 °C for 60 s. DNA concentration of amplicons of interest was determined by gel electrophoresis. Amplicon DNA of each sample (~20 ng) were pooled for up to 48 samples carrying different barcodes. PCRs showing low yields were further amplified for 5 cycles. The amplicon pools were purified with one volume AMPure XP beads (Agencourt) to remove primer dimer and other small mispriming products, followed by an additional purification on MinElute columns (Qiagen). Each purified amplicon pool DNA (~100 ng) was used to construct Illumina libraries using the Ovation Rapid DR Multiplex System 1–96 (NuGEN). Illumina libraries were pooled and size selected by preparative gel electrophoresis. Sequencing was done on an Illumina MiSeq using v3 Chemistry (Illumina).

16S rRNA sequence analysis was performed with mothur community pipeline and clustering into operational taxonomic units (OTUs). The analysis was initiated by clipping 16S rRNA sequences from primers. The fragments after removing the primer sequences were combined into forward and reverse primer orientation sequences. The sequences were then processed to remove wrong size sequence and identify unique sequences. Sequences which were not matched or overhung were removed by aligning with a V3–V4 customized SILVA database v123. Chimera were removed using UCHIME algorithm. Taxonomical classification of sequences and removal of non-bacterial sequences were done using Silva database v123. OTU were picked by clustering at 97% identity level using the cluster split method.

Analytical methods

Determination of fermentation products including organic acids (lactic acid, acetic acid, propionic acid, butyric acid) was performed with Metrohm ion chromatography equipped with Metrosep organic acids column and Metrosep organic acids guard column, and an ion chromatography conductivity detector, using 1 mM H_2SO_4 as eluent at flow rate of 0.0083 mL s^{-1}, oven temperature at 35 °C, and 500 mM LiCl as regenerant for suppressor. Caproic acid in aqueous phase was measured by gas chromatography (GC-2014, Shimadzu®, The Netherlands) with DB-FFAP 123-3232 column (30 m × 0.32 mm × 0.25 µm; Agilent, Belgium) and a flame ionization detector. Total solids (TS), volatile solids (VS) and volatile suspended solids (VSS) were measured via the standard method [41]. The particle size of blended grass was determined using a Zeiss Axioskop microscope and Image-Pro Insight software. Phase-separated caproic acid and decane were measured using nuclear magnetic resonance (NMR) spectroscopy. ^1H NMR and ^{13}C NMR were performed at 400 and 100 MHz, respectively, on a Bruker Avance III Nanobay 400 MHz spectrometer. 400 μL DMSO-d6 was added to 25 mg of caproic acid and decane sample, stirred and transferred to an NMR tube. Quantification was performed relative to caproic acid and decane as standard (0.2 M in DMSO-d6), contained in a NORELL 100 μL capillary insert. ^{1}H NMR experiments were run with 8 scans and 1-s relaxation delay.

Additional file

Additional file 1: Figure S1. Bacterial community for (a) inoculum for elongation system, (b) maximum rate test, (c) maximum concentration test. **Figure S2.** Carboxylate profile of maximum caproic acid concentration test – elongation under excess lactic acid condition (error bars represent the standard deviation of triplicates, and they are consistently less than 6% (n = 3) and obstructed by the medallions). **Figure S3.** Phase separation of caproic acid and aqueous elongation broth.

Authors' contributions

WCK designed and performed the experiment, analysed the data and wrote the manuscript. SA designed the experiment and revised the manuscript. VH and KR revised the manuscript and supervised the project. All authors read and approved the final manuscript.

Acknowledgements

The authors would like to thank Pieter Naert for NMR analysis, Antonin Prévoteau and Xu Zhang for assisting in electrochemistry operation, Greet Van de Velde (CMET, UGent) for the analysis using ion chromatography, Tim Lacoere for illustration of graphical abstract and analysis using gas chromatography, Mike Taghon for his support, and the Department of Biochemical and Microbial Technology, Ghent University.

Competing interests

The authors declare that they have no competing interests.

Funding

This work was funded by Special Research Fund (BOF, Project Number: DEF13/AOF/010) of the University of Ghent (Belgium). The funder has no role in study design, data collection and interpretation, or the decision to submit the work for publication. SA is supported by Ghent University Multidisciplinary Research Partnership (MRP)-Biotechnology for a sustainable economy (01 MRA 510 W). KR is supported by European Research Council Starter Grant ELECTROTALK.

References

1. Zion Market Research. Medium-chain triglycerides (lauric acid, caprylic acid, caproic acid and capric acid) market for food, medical and other applications: global industry perspective, comprehensive analysis, size, share, growth, segment, trends and forecast, 2014–2020. 2016. Accessed 21 July 2016.
2. Larranaga MD, Richard J, Lewis S, Lewis RA. Hawley's condensed chemical dictionary. 16th ed. 2016.
3. Zhu X, Tao Y, Liang C, Li X, Wei N, Zhang W, Zhou Y, Yang Y, Bo T. The synthesis of n-caproate from lactate: a new efficient process for medium-chain carboxylates production. Sci Rep. 2015;5:14360.
4. Food and Agriculture Organization of the United Nations (FAOSTAT). Composition of global agricultural area. http://faostat3.fao.org/. Accessed 7 June 2016.
5. Khor WC, Roume H, Coma M, Vervaeren H, Rabaey K. Acetate accumulation enhances mixed culture fermentation of biomass to lactic acid. Appl Microbiol Biotechnol. 2016;100(19):8337–48.
6. Angenent LT, Richter H, Buckel W, Spirito CM, Steinbusch KJ, Plugge C, Strik DPBTB, Grootscholten TIM, Buisman CJN, Hamelers HVM. Chain elongation with reactor microbiomes: open-culture biotechnology to produce biochemicals. Environ Sci Technol. 2016;50(6):2796–810.
7. Ding HB, Tan GY, Wang JY. Caproate formation in mixed-culture fermentative hydrogen production. Bioresour Technol. 2010;101:9550–9.
8. Barker HA, Kamen MD, Bornstein BT. The synthesis of butyric and caproic acids from ethanol and acetic acid by *Clostridium kluyveri*. Proc Natl Acad Sci USA. 1945;31:373–81.
9. Choi K, Jeon BS, Kim BC, Oh MK, Um Y, Sang BI. In situ biphasic extractive fermentation for hexanoic acid production from sucrose by *Megasphaera elsdenii* NCIMB 702410. Appl Biochem Biotechnol. 2013;171:1094–107.
10. Xiong B, Richard TL, Kumar M. Integrated acidogenic digestion and carboxylic acid separation by nanofiltration membranes for the lignocellulosic carboxylate platform. J Membr Sci. 2015;489:275–83.
11. Elsden SR, Volcani BE, Gilchrist FMC, Lewis D. Properties of a fatty acid forming organism isolated from the rumen of sheep. J Bacteriol. 1956;72:681–9.
12. Jeon BS, Moon C, Kim BC, Kim H, Um Y, Sang B-I. In situ extractive fermentation for the production of hexanoic acid from galactitol by *Clostridium* sp. BS-1. Enzym Microb Technol. 2013;53:143–51.
13. Sträuber H, Schröder M, Kleinsteuber S. Metabolic and microbial community dynamics during the hydrolytic and acidogenic fermentation in a leach-bed process. Energy Sustain Soc. 2012;2(1):13.
14. Sträuber H, Lucas R, Kleinsteuber S. Metabolic and microbial community dynamics during the anaerobic digestion of maize silage in a two-phase process. Appl Microbiol Biotechnol. 2016;100(1):479–91.
15. Spirito CM, Richter H, Rabaey K, Stams AJM, Angenent LT. Chain elongation in anaerobic reactor microbiomes to recover resources from waste. Curr Opin Biotechnol. 2014;27:115–22.
16. Andersen SJ, Candry P, Basadre T, Khor WC, Roume H, Hernandez-Sanabria E, Coma M, Rabaey K. Electrolytic extraction drives volatile fatty acid chain elongation through lactic acid and replaces chemical pH control in thin stillage fermentation. Biotechnol Biofuels. 2015;8:1–14.
17. Kucek LA, Nguyen M, Angenent LT. Conversion of L-lactate into n-caproate by a continuously fed reactor microbiome. Water Res. 2016;93:163–71.
18. Xu J, Guzman JJ, Andersen SJ, Rabaey K, Angenent LT. In-line and selective phase separation of medium-chain carboxylic acids using membrane electrolysis. Chem Commun (Camb). 2015;51:6847–50.
19. Khor WC, Rabaey K, Vervaeren H. Low temperature calcium hydroxide treatment enhances anaerobic methane production from (extruded) biomass. Bioresour Technol. 2015;176:181–8.
20. Grassa—Mini grass refinery process which focuses on protein/amino acid extraction. http://grassa.nl/. Accessed 29 May 2017.
21. Fusco V, Quero GM, Cho G-S, Kabisch J, Meske D, Neve H, Bockelmann W, Franz CMAP. The genus *Weissella*: taxonomy, ecology and biotechnological potential. Front Microbiol. 2015;6:155.
22. Wiegel J, Tanner R, Rainey FA. An introduction to the family *Clostridiaceae*. In: Dworkin M, Falkow S, Rosenberg E, Schleifer KH, Stackebrandt E, editors. The prokaryotes: volume 4: bacteria: *Firmicutes*, *Cyanobacteria*. Springer: New York; 2006. p. 654–78.
23. Schleifer KH. Phylum XIII. *Firmicutes*. Gibbons and Murray 1978, 5 (*Firmicutes* [sic] Gibbons and Murray 1978, 5). In: De Vos P, Garrity GM, Jones D, Krieg NR, Ludwig W, Rainey FA, Schleifer KH, Whitman WB, editors. Bergey's manual® of systematic bacteriology: volume three the *Firmicutes*. New York: Springer; 2009. p. 19–1317.
24. Clavel T, Lepage P, Charrier C. The family *Coriobacteriaceae*. In: Rosenberg E, DeLong EF, Lory S, Stackebrandt E, Thompson F, editors. The prokaryotes: *Actinobacteria*. Berlin: Springer; 2014. p. 201–38.
25. Schaal KP, Yassin AF, Stackebrandt E. The family *Actinomycetaceae*: The genera *Actinomyces*, *Actinobaculum*, *Arcanobacterium*, *Varibaculum*, and *Mobiluncus*. In: Dworkin M, Falkow S, Rosenberg E, Schleifer K-H, Stackebrandt E, editors. The prokaryotes: volume 3: Archaea. Bacteria: *Firmicutes*, *Actinomycetes*. New York: Springer; 2006. p. 430–537.
26. Grootscholten TIM, Steinbusch KJ, Hamelers HV, Buisman CJN. High rate heptanoate production from propionate and ethanol using chain elongation. Bioresour Technol. 2013;136:715–8.
27. Kucek LA, Spirito CM, Angenent LT. High n-caprylate productivities and specficities from dilute ethanol and acetate: chain elongation with microbiomes to upgrade products from syngas fermentation. Energy Environ Sci. 2016;9(11):2482–3494.
28. Steinbusch KJJ, Hamelers HVM, Plugge CM, Buisman CJN. Biological formation of caproate and caprylate from acetate: fuel and chemical production from low grade biomass. Energy Environ Sci. 2011;4:216–24.

29. Grootscholten TIM, Steinbusch KJJ, Hamelers HVM, Buisman CJN. Improving medium chain fatty acid productivity using chain elongation by reducing the hydraulic retention time in an upflow anaerobic filter. Bioresour Technol. 2013;136:735–8.
30. Ge S, Usack JG, Spirito CM, Angenent LT. Long-term n-caproic acid production from yeast-fermentation beer in an anaerobic bioreactor with continuous product extraction. Environ Sci Technol. 2015;49:8012–21.
31. Li A, Chu YN, Wang X, Ren L, Yu J, Liu X, Yan J, Zhang L, Wu S, Li S. A pyrosequencing-based metagenomic study of methane-producing microbial community in solid-state biogas reactor. Biotechnol Biofuels. 2013;6:3.
32. Nilges P, dos Santos TR, Harnisch F, Schroder U. Electrochemistry for biofuel generation: electrochemical conversion of levulinic acid to octane. Environ Sci Technol. 2012;5(1):5231–5.
33. Schäfer HJ. Recent contributions of Kolbe electrolysis to organic synthesis. In: Steckhan E, editor. Electrochemistry IV. Berlin: Springer; 1990. p. 91–151.
34. Stang C, Harnisch F. The dilemma of supporting electrolytes for electroorganic synthesis: a case study on Kolbe electrolysis. Chemsuschem. 2016;9(1):50–60.
35. Levy PF, Sanderson JE, Cheng LK. Kolbe electrolysis of mixtures of aliphatic organic acids. J Electrochem Soc. 1984;131:773–7.
36. Wadhawan JD, Del Campo FJ, Compton RG, Foord JS, Marken F, Bull SD, Davies SD, Walton DJ, Ryley S. Emulsion electrosynthesis in the presence of power ultrasound biphasic Kolbe coupling processes at platinum and boron-doped diamond electrodes. J Electroanal Chem. 2001;507:135–43.
37. Choi HY, Ryu HK, Park KM, Lee EG, Lee H, Kim SW, Choi ES. Direct lactic acid fermentation of Jerusalem artichoke tuber extract using *Lactobacillus paracasei* without acidic or enzymatic inulin hydrolysis. Bioresour Technol. 2012;114:745–7.
38. U. S. Composting Council. Keeping organics out of Landfills. 2005. Accessed on 17 May 2016.
39. Davidson C, Newes E, Schwab A, Vimmerstedt L. An overview of aviation fuel markets for biofuels stakeholders (Technical report). Golden: National Renewable Energy Laboratory (NREL); 2014.
40. Andersen SJ, De Groof V, Khor WC, Roume H, Props R, Coma M, Rabaey K. A *Clostridium* group IV species dominates and suppresses a mixed culture fermentation by tolerance to medium chain fatty acids products. Front Bioeng Biotechnol. 2017;5.
41. Association APHA, AWWA and WEF. Standard methods for the examination of water and wastewater. 19th ed. Washington: American Public Health Association/America Water Works Association/Water Environment Federation; 1995.

3

Enzymatic degradation of sulfite-pulped softwoods and the role of LPMOs

Piotr Chylenski[1], Dejan M. Petrović[1], Gerdt Müller[1], Marie Dahlström[2], Oskar Bengtsson[2], Martin Lersch[2], Matti Siika-aho[3], Svein Jarle Horn[1] and Vincent G. H. Eijsink[1*]

Abstract

Background: Recent advances in the development of enzyme cocktails for degradation of lignocellulosic biomass, especially the discovery of lytic polysaccharide monooxygenases (LPMOs), have opened new perspectives for process design and optimization. Softwood biomass is an abundant resource in many parts of the world, including Scandinavia, but efficient pretreatment and subsequent enzymatic hydrolysis of softwoods are challenging. Sulfite pulping-based pretreatments, such as in the BALI™ process, yield substrates that are relatively easy to degrade. We have assessed how process conditions affect the efficiency of modern cellulase preparations in processing of such substrates.

Results: We show that efficient degradation of sulfite-pulped softwoods with modern, LPMO-containing cellulase preparations requires the use of conditions that promote LPMO activity, notably the presence of molecular oxygen and sufficient reducing power. Under LPMO activity-promoting conditions, glucan conversion after 48-h incubation with Cellic® CTec3 reached 73.7 and 84.3% for Norway spruce and loblolly pine, respectively, at an enzyme loading of 8 mg/g of glucan. The presence of free sulfite ions had a negative effect on hydrolysis efficiency. Lignosulfonates, produced from lignin during sulfite pretreatment, showed a potential to activate LPMOs. Spiking of Celluclast®, a cellulase cocktail with low LPMO activity, with monocomponent cellulases or an LPMO showed that the addition of the LPMO was clearly more beneficial than the addition of any classical cellulase. Addition of the LPMO in reactions with spruce increased the saccharification yield from approximately 60% to the levels obtained with Cellic® CTec3.

Conclusions: In this study, we have demonstrated the importance of LPMOs for efficient enzymatic degradation of sulfite-pulped softwood. We have also shown that to exploit the full potential of LPMO-rich cellulase preparations, conditions promoting LPMO activity, in particular the presence of oxygen and reducing equivalents are necessary, as is removal of residual sulfite from the pretreatment step. The use of lignosulfonates as reductants may reduce the costs related to the addition of small molecule reductants in sulfite pretreatment-based biorefineries.

Keywords: Lignocellulose, Sulfite, Pretreatment, Pulping, Cellulase, LPMO, AA9, GH61, Biofuel

Background

Enzymatic degradation of lignocellulose plays a pivotal role in a future biomass-based economy. Woody biomass is a sustainable source of lignocellulosic biomass in many parts of the world, including Scandinavian countries, North and South America, and New Zealand, and its efficient biochemical conversion is of major importance for the transition toward a biomass-based economy [1]. Recalcitrance of lignocellulosic biomass is one of the key factors hampering commercialization of plant biomass-derived fuels and chemicals and woody biomass is particularly challenging due to its physical properties, such as a dense structure and a high lignin content [1, 2]. To overcome biomass recalcitrance, various pretreatment technologies have been introduced. Pretreatments for woody biomass include steam explosion (usually acid-catalyzed), dilute acid, organosolv, and sulfite-pretreatment-to-overcome-recalcitrance-of-lignocellulose

*Correspondence: vincent.eijsink@nmbu.no
[1] Faculty of Chemistry, Biotechnology and Food Science, Norwegian University of Life Sciences, P.O. Box 5003, 1432 Ås, Norway
Full list of author information is available at the end of the article

(SPORL). Generally, pretreatment and subsequent enzymatic digestion of hardwoods is less demanding compared to softwoods, which contain more lignin and have more rigid structure [1, 3]. Relatively efficient pretreatment and enzymatic hydrolysis of softwoods has been demonstrated for organosolv- and SPORL-pretreated biomasses [4, 5]. The BALI™ process is an alternative sulfite-based pretreatment that yields cellulose-rich substrates containing little lignin and hemicellulose relative to, e.g., the SPORL pretreatment, whose enzymatic conversion to glucose is relatively easy [6].

Complete biochemical conversion of cellulose to monomeric sugars requires the synergistic action of several classes of enzymes [7]. By far, the most studied enzymes are cellulose-degrading glycoside hydrolases, known as cellulases. Cellulases include enzymes acting on the ends of the cellulose chain, known as cellulose 1,4-β-cellobiosidases or cellobiohydrolases, cleaving off cellobiose units from reducing and non-reducing chain ends. Endo-β-1,4-glucanases cleave internal glycosidic bonds in the cellulose chains, thus generating novel chain ends, and β-glucosidases convert released soluble cellodextrins and cellobiose to glucose [8]. In natural biomass-converting enzyme systems, the cellulases are accompanied by a wide range of hemicellulose- and lignin-degrading enzymes working in concert to fully decompose lignocellulose.

The discovery of lytic polysaccharide monooxygenases (LPMOs) has challenged the classical model of enzymatic decomposition of polysaccharides. Contrary to cellulases, which rely on hydrolysis to perform catalysis, LPMOs are metalloenzymes breaking down glycosidic bonds by an oxidative mechanism involving molecular oxygen (or, in a recently suggested alternative mechanism, hydrogen peroxide [9]) and an electron donor [10, 11]. LPMO-catalyzed cleavage of the glycosidic bond results in the formation of oxidized chain ends, either at the C1 and/or at the C4 carbon position. So far, LPMOs cleaving glycosidic bonds in chitin [10, 12], cellulose [11, 13–16], cello-oligosaccharides [17], hemicellulose [18, 19], and starch [20, 21] have been identified. LPMOs are classified in the Carbohydrate Active Enzymes (CAZy) database as auxiliary activities families AA9, AA10, AA11, and AA13 [22].

Despite the fact that the boosting effect of LPMOs on cellulase activity was demonstrated early on [13, 15], exploration of the role and performance of LPMOs in applied settings has been limited. Cannella et al. [23] showed that LPMOs present in the commercial cellulase preparation Cellic® CTec2 enhanced hydrolysis of hydrothermally pretreated wheat straw. Müller and colleagues showed the importance of LPMOs in this enzyme preparation for degradation of steam-exploded birch [24]. Using both model cellulosic substrates and industrially relevant lignocellulosic biomass, Hu et al. [25] demonstrated synergy between AA9 LPMOs and cellulases.

In this study, we have investigated the impact of process conditions on the enzymatic degradation of sulfite-pulped softwood biomasses with particular focus on the role of LPMOs, the presence of oxygen and small molecule electron donors, and the effect of residual sulfite. We have utilized commercially available cellulase preparations, Celluclast® 1.5L (supplemented with Novozym 188) and Cellic® CTec3, as well as an in-house produced fungal LPMO from *Thermoascus aurantiacus* (*Ta*LPMO9A) and individual major *Trichoderma reesei* cellulases.

Methods

Cellulosic substrates, pretreatment, and compositional analysis

Norway spruce (*Picea abies*) and loblolly pine (*Pinus taeda*) were subjected to the proprietary pretreatment technology developed by Borregaard AS (Sarpsborg, Norway) [6, 26]. The pretreatment included a sulfite cooking step utilizing calcium or sodium as a counterion, which converts lignin into water-soluble lignosulfonates and removes most of the hemicellulose that is washed out of the remaining cellulose pulp. The composition of the pretreated materials was determined following the National Renewable Energy Laboratory standardized protocol (NREL/TP-510-42618) and is presented in Table 1. The commercially available model cellulosic substrate Avicel PH-101 (Sigma Aldrich, St. Louis, MO, USA) was also used in enzymatic hydrolysis experiments. Dried lignosulfonates were kindly provided by Borregaard AS (Norway, Sarpsborg) and were obtained by the evaporation of the sulfite spent liquor (SSL) that had been stripped of soluble sugars by fermentation [6].

Enzymes

Celluclast® 1.5L, Novozym 188, and Cellic® CTec3 were all kindly provided by Novozymes A/S (Bagsværd,

Table 1 Compositional analysis of pretreated Norway spruce and Loblolly pine

Substrate	Component (% of DM)[a]			
	Glucan[b]	Xylan[b]	Mannan[b]	Acid insoluble lignin[c]
Norway spruce	88.3	4.5	4.8	3.8
Loblolly pine	80.4	3.0	3.0	8.9
Avicel	92.2	2.1	0.1	0.9

Values are presented as a percentage of dry matter

[a] The presence of the potentially inhibitory sugar derivatives furfural and hydroxymethylfurfural was also analyzed; both were not detected

[b] Sugar monomers measured by HPLC and corrected by the hydration factor (*0.9)

[c] Klason lignin (not corrected for ash)

Denmark). The AA9 family LPMO from *Neurospora crassa* (*Nc*LPMO9C), used for generation of C4 oxidized standards, was produced and purified as described previously [24]. *Trichoderma reesei* Cel7A was purified from the culture filtrate of *T. reesei* QM 9414 (VTT Culture Collection, D-74075, Finland) essentially as described in [27]. *Tr*Cel7B and *Tr*Cel6A from *T. reesei* were purified as described by Suurnäkki et al. [28].

The gene encoding *T. aurantiacus* LPMO9A (also known as *Ta*GH61A; [11]) including its native signal sequence was codon optimized for *Pichia pastoris* (GenScript, Piscataway, NJ, USA). A synthetic, *Pichia*-optimized gene encoding *Ta*LPMO9A was excised from the pUC57 vector using *Acc*65I and *Eco*RI restriction enzymes (New England BioLabs [NEB] Inc, MA, USA) and ligated into the same sites of the pPINK-GAP vector [29], yielding the pPINK-GAP_*Ta*LPMO9A plasmid. This plasmid was then linearized with *AflII* and 5 µg of linearized plasmid was transformed into freshly prepared electrocompetent *P. pastoris* PichiaPink™ Strain 4 cells, following the manufacturer's instructions (Thermo Fisher Scientific, Waltham, MA, USA), using a Bio-Rad Gene Pulser II electroporation system (Bio-Rad Laboratories, Hercules, CA, USA) at 1.8 kV, 25 µF, and 200 Ω. Subsequently, the cells were incubated in 1 mL of YPDS medium for 2 h and spread on *Pichia* adenine dropout (PAD) selection plates (Thermo Fisher Scientific, Waltham, MA, USA), followed by incubation for 3–4 days at 30 °C, until white colonies formed. Four colonies were picked and restreaked on fresh PAD plates. Overnight cultures of these four transformants were then screened for protein production in BMGY medium (containing 1% (v/v) glycerol). A glycerol stock was prepared from an overnight culture of the best-producing transformant in BMGY medium by adjusting the cell suspension to 28% (v/v) glycerol.

The best-producing transformant was grown in 20 mL of BMGY medium (containing 1% (v/v) glycerol) in a 100-mL shaken flask at 29 °C and 200 rpm for 16 h. Subsequently, this pre-culture was used to inoculate 4 × 0.5 L of BMGY medium (containing 1% (v/v) glycerol) in 2-L shaken flasks, followed by incubation at 29 °C and 200 rpm for 48 h. After 24 h, the cultures were supplemented with 1% (v/v) glycerol. The cells were harvested by centrifugation at 7.000*g* for 15 min, at 4 °C. The supernatants were collected and dialyzed against 50 mM Bis–Tris buffer, pH 6.5, and concentrated to 150 mL using a VivaFlow 200 tangential crossflow concentrator (MWCO 10 kDa, Sartorius Stedim Biotech Gmbh, Germany). Ammonium sulfate was added to the concentrated supernatant to a final concentration of 1.42 M after which the solution was loaded onto three linearly connected 5-mL HiTrap Phenyl FF columns (GE Healthcare Life Sciences, Uppsala, Sweden) equilibrated with 50 mM Bis–Tris buffer (pH 6.5), containing 1.42 M ammonium sulfate. Proteins bound to the columns were eluted using a 75 mL linear gradient from 1.42 to 0 M ammonium sulfate in 50 mM Bis–Tris buffer (pH 6.5), using a flow rate of 1 mL/min. Collected fractions were analyzed by sodium dodecyl sulfate polyacrylamide gel electrophoresis (SDS-PAGE), and the fractions containing *Ta*LPMO9A were pooled and subsequently concentrated and buffer exchanged to 50 mM Bis–Tris pH 6.5 using Amicon Ultra centrifugal filters (MWCO 10 kDa, Millipore).

Purified *Ta*LPMO9A was saturated with Cu (II) by incubating the enzyme with an excess of $CuSO_4$ (3:1 molar ratio of copper:enzyme) for 90 min at room temperature as described previously [30]. Subsequently, the protein solution was loaded onto a HiLoad 16/60 Superdex 75 size exclusion column (GE Healthcare Life Sciences, Uppsala, Sweden) in 50 mM Bis–Tris buffer (pH 6.5), containing 150 mM NaCl, using a flow rate of 0.75 mL/min. Fractions containing pure protein were identified using SDS-PAGE and subsequently pooled and concentrated using Amicon Ultra centrifugal filters (MWCO 10 kDa, Millipore).

To ensure full copper saturation, the purified LPMO was saturated once more with Cu (II), as described above. This time, after incubating the enzyme with $CuSO_4$, the solution was loaded onto a PD midiTrap G-25 desalting column (GE Healthcare, UK), equilibrated with 20 mM Bis–Tris buffer pH 6.0. Protein-containing fractions, eluted with 1 mL of equilibration buffer, were collected and stored at 4 °C until further use.

All protein concentrations were determined with the Bio-Rad modified Bradford method (Bio-Rad Laboratories, Hercules, CA, USA) utilizing bovine serum albumin (BSA) as a standard [31].

Enzymatic hydrolysis

Saccharification of lignocellulosic biomass and Avicel was conducted in 50-mL rubber sealed glass bottles (Wheaton, Millville, NJ, USA) with 10 mL working volume. The biomass obtained after the sulfite pulping process described above was not washed, milled, or dried prior to the saccharification experiments. Enzymatic hydrolysis was performed with 5% total solids loading in 50 mM sodium acetate buffer pH 5.0 at 50 °C, with 8 mg/g glucan total protein loading of either a 5:1 (w/w) Celluclast®:Novozym 188 mixture or Cellic® CTec3, in the presence or absence of an external electron donor. To facilitate efficient mixing, bottles were rotated at 38 rpm in a Multi RS-60 programmable rotator (Biosan, Riga, Latvia). Anaerobic conditions were reached by vigorously flushing the substrate-buffer suspension with nitrogen (Yara, Trondheim, Norway) for 3 min and addition of 0.025% (w/v, final concentration) of L-cysteine hydrochloride monohydrate (Sigma Aldrich, St. Louis, MO, USA) to ensure complete removal of oxygen. In aerobic conditions, oxygen came from ambient air present in the headspace of the reaction bottles. Reactions were initiated by injection of 800 μL of enzyme preparations, appropriately diluted in 50 mM sodium acetate buffer pH 5.0, through the rubber septum. Reactions were terminated at different time points. To ensure reproducible sampling from the flasks, the entire reaction mixture was diluted threefold with ultrapure water (Merck Millipore, Billerica, MA, USA) and then transferred to new 50-mL Falcon tubes. Reactions were stopped by incubating the Falcon tubes at 100 °C for 15 min in a water bath. Supernatants were collected by centrifugation of the tubes for 15 min at 3803*g* and 4 °C, and these were then transferred to 1.5-mL tubes and stored at −20 °C prior to further analysis.

Product analysis

Glucose and cellobiose released during enzymatic hydrolysis were quantified with High-Performance Liquid Chromatography (HPLC) using a Dionex Ultimate 3000 system (Dionex, Sunnyvale, CA, USA) coupled to a refractive index (RI) detector 101 (Shodex, Tokyo, Japan). Separation of hydrolysis products was achieved utilizing a Rezex ROA-Organic Acid H^+ (8%), 300 × 7.8 mm analytical column equipped with SecureGuard Carbo-H^+ 4 × 3.0 mm guard column (Phenomenex, Torrance, CA, USA), operated at 65 °C, with 5 mm H_2SO_4 as the mobile phase, and a flow rate of 0.6 mL/min. For quantification, the areas of peaks corresponding to glucose and cellobiose were compared to standard curves generated with known concentrations of glucose and cellobiose (in the range of 0.1–10 g/L). Hydrolysis yields were calculated based on detected glucose and cellobiose (typically less than 1% of the total) and expressed as a percentage of the theoretical maximum that would be obtained upon complete conversion of glucan to glucose.

Statistical significance of differences in glucan saccharification yields was determined using two-way ANOVA with Tukey's post hoc test (95% confidence interval) and was carried out using R (R Foundation for Statistical Computing, Vienna, Austria). Statistical significance is shown as follows: $^{*}p < 0.05$, $^{**}p < 0.01$, $^{***}p < 0.001$.

Native cello-oligosaccharides (DP 2 to DP 5) and oxidized sugars were analyzed with High-Performance Anion Exchange Chromatography with Pulsed Amperometric Detection (HPAEC-PAD) using a Dionex ICS 3000 system (Dionex, Sunnyvale, CA, USA) equipped with a CarboPac PA1 2 × 250 mm analytical column with a CarboPac PA1 2 × 50 mm guard column, as described in [15]. Briefly, initial conditions were set to 0.1 M NaOH (100% eluent A), followed by a linear gradient of eluent B (1 M sodium acetate in 0.1 M NaOH), reaching 10% B 10 min after sample injection and 30% B at 35 min after injection. This was followed by a 5-min exponential gradient to 100% eluent B, after which the column was reconditioned by running the initial conditions for 9 min.

C4-oxidized standards were generated with *Nc*LPMO9C by cellopentaose degradation as described in [24]. The data were collected and analyzed using Chromeleon 7.0 software.

Results

Below, several series of hydrolysis experiments are described. Table 2 provides an overview of the hydrolysis yields obtained in these experiments.

Saccharification of a model cellulosic substrate

The model cellulosic substrate Avicel PH-101 was incubated for 48 h with an older generation, LPMO-poor, cellulase mixture (Celluclast® 1.5L supplemented with Novozym 188; 5:1, w/w), as well as a modern, LPMO-containing, cellulase preparation, Cellic® CTec3, under either aerobic or anaerobic conditions, in the presence or absence of an external electron donor (ascorbic acid). These experiments showed that Cellic® CTec3 gives higher hydrolysis yields, but only under aerobic conditions and when adding an external electron donor (Fig. 1a). Indeed, under these conditions only, considerable LPMO activity was detectable (Fig. 1b; the amount of C4-oxidized product amounts to approximately 1.5% of the amount of solubilized glucan). Saccharification of Avicel with the Celluclast® 1.5L:Novozym 188 mixture was hardly affected by the presence of oxygen or the addition of ascorbic acid (Fig. 1c), and under LPMO-promoting conditions, only low product levels were observed (Fig. 1d). Celluclast® 1.5L is a product based on the *T. reesei* secretome, in which three AA9 family LPMOs have been identified, which however, are expressed at low levels [13, 32].

Saccharification of sulfite-pulped softwoods

Sulfite-pulped Norway spruce was incubated with both cellulase preparations for 48 h, under conditions identical to those used for Avicel, with strikingly similar results (Fig. 2a–d). Again, Cellic® CTec3 was the most efficient and, again, only under aerobic conditions and when adding an external source of electrons (Fig. 2a). Saccharification of pretreated Norway spruce with the LPMO-poor cocktail was not improved by aerobic conditions or the presence of electron donor (Fig. 2c). Oxidized products were only observed for reactions with oxygen and externally added electron donor present, and the levels of oxidized product were much higher for Cellic® CTec3 (Fig. 2b), compared

Table 2 Overview of hydrolysis yields for various degradation reactions

Raw material	Enzyme	Conditions	AscA	Sulfite (12.5 mM)	Glucan solubilization after 48 h (% of theoretical)
Avicel PH-101	Cellic® CTec3	Aerobic	+	−	69.0 ± 1.9
		Aerobic	+	+	31.6 ± 0.4
		Anaerobic	+	−	50.0 ± 1.4
		Anaerobic	+	+	44.1 ± 1.2
	Celluclast®:Novozym 188	Aerobic	+	−	55.6 ± 0.9
		Aerobic	+	+	39.5 ± 0.5
		Anaerobic	+	−	57.1 ± 0.7
		Anaerobic	+	+	38.6 ± 4.3
Norway spruce	Cellic® CTec3	Aerobic	+	−	73.7 ± 0.7
		Aerobic	+	+	34.1 ± 0.4
		Anaerobic	+	−	50.6 ± 3.8
		Anaerobic	+	+	55.6 ± 2.3
	Celluclast®:Novozym 188	Aerobic	+	−	64.9 ± 2.1
		Aerobic	+	+	38.7 ± 1.5
		Anaerobic	+	−	60.2 ± 1.4
		Anaerobic	+	+	50.3 ± 2.6
Loblolly pine	Cellic® CTec3	Aerobic	+	−	84.3 ± 1.2
		Anaerobic	+	−	60.3 ± 0.7

to the Celluclast® reaction (Fig. 2d). This result shows that the heavily delignified sulfite-pretreated softwood biomass lacks the potential to activate LPMOs.

Saccharification of sulfite-pulped loblolly pine with Cellic® CTec3 using the same conditions and varying the same parameters showed the same trends as observed for Norway spruce and Avicel. Interestingly, saccharification of pretreated loblolly pine was clearly more efficient than identically pretreated Norway spruce (compare Fig. 2e with a). The higher conversion of loblolly pine was achieved despite a seemingly lower LPMO activity (compare Fig. 2f with b).

The importance of oxygen, which affects LPMO activity, was explored further using extended reaction times (up to 144 h). The results showed that enzymatic hydrolysis of sulfite-pulped Norway spruce under anaerobic conditions never reached the yields obtained under conditions promoting LPMO activity, the final yields being about 65 and 85%, respectively (Fig. 3).

Reduced glutathione and gallic acid can also drive LPMO activity by acting as electron donors [10, 15, 25]. Similarly to the addition of ascorbic acid, addition of each of these reducing agents improved enzymatic degradation of pretreated Norway spruce under aerobic conditions (Additional file 1: Figure S1a, b) albeit less efficiently than ascorbic acid (Fig. 2a).

Repeating the experiments with sulfite-pulped Norway spruce, in the absence of L-cysteine hydrochloride monohydrate (0.025%, w/v), showed that the addition of this compound, used to ensure anaerobic conditions, had no effect on hydrolysis yields (Additional file 2: Figure S2).

Effect of residual sulfite on softwood saccharification

The sulfite-pretreated biomasses used in the experiments described above had been stored for some time and were likely devoid of residual sulfite, which is readily oxidized to sulfate in a reaction that involves oxygen and leads to the generation of various reactive intermediate species [33]. Sulfite pretreatment of lignocelluloses in a factory may result in residual sulfite still being present in the material that is subjected to subsequent enzymatic hydrolysis. To test how residual sulfite would affect enzymatic hydrolysis, we studied the effect of sulfite on hydrolysis of Avicel (Additional file 3: Figure S3) and pretreated Norway spruce (Fig. 4) by Cellic® CTec3 and the Celluclast®-based cocktail. We tested the effect of adding 1000 ppm (12.5 mM) of sulfite ions (in the form of freshly added sodium sulfite), which was considered to be at the higher end of the concentrations that could be expected to be present in freshly pretreated and washed biomass, based on process data from the BALI™ pilot plant at Borregaard AS. Generally, sulfite had a negative effect on enzymatic hydrolysis of all substrates and no LPMO products could be detected (Fig. 4; see also Additional file 1: Figure S1, Additional file 2: Figure S2, Additional file 3: Figure S3). Maximal conversion with Cellic® CTec3 (i.e., aerobic, with AscA, 48 h) was reduced from 69.0 and 73.7% to 31.6 and 34.1% for Avicel and Norway spruce,

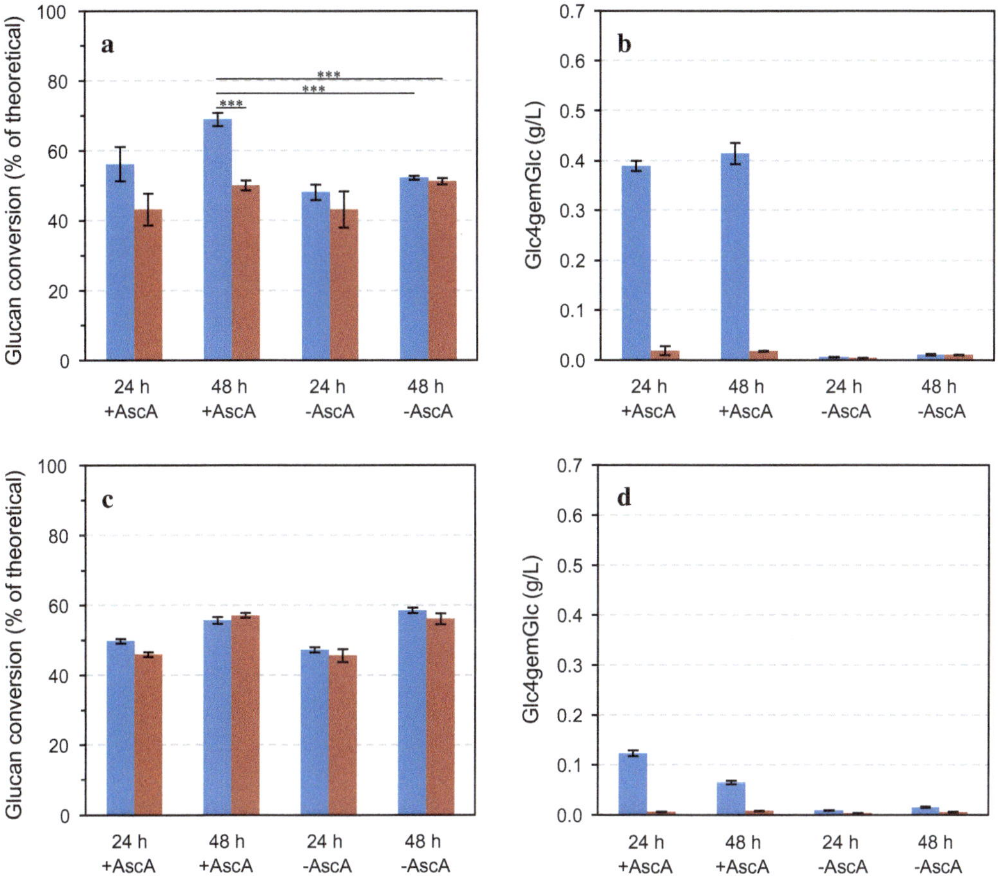

Fig. 1 Saccharification of Avicel PH-101 under aerobic (*blue bars*) or anaerobic (*red bars*) conditions in the presence or absence of 1 mM ascorbic acid (±AscA). The *left panels* show glucan conversion (as a percentage of theoretical glucan conversion) and the *right panels* show the concentration of Glc4gemGlc, at two time points. Enzymatic hydrolysis was carried out with Cellic® CTec3 (**a**, **b**) and Celluclast®:Novozym 188 (mixed at 5:1 ratio, w/w) (**c**, **d**) at 8 mg/g glucan total protein loading, using reaction mixtures containing 5% DM in 50 mM sodium acetate pH 5.0 that were incubated at 50 °C. The data points represent the average value of three independent experiments with one technical replicate per experiment. The *error bars* represent standard deviations of the three independent experiments. The statistical significance of differences in the 48-h saccharification yields was analyzed using two-way ANOVA with Tukey's post hoc test (95% confidence interval) and is indicated as follows: *$p < 0.05$, **$p < 0.01$, ***$p < 0.001$

respectively (Table 2). These decreases in yields are larger than the effects of oxygen and ascorbic acid discussed above and are thus not likely to be only "LPMO effects." Indeed, the efficiency of the LPMO-poor Celluclast® mixture was also affected by sulfite, albeit to a lesser extent (Table 2; Fig. 4; Additional file 3: Figure S3). The effect of sulfite was clearly less pronounced under anaerobic conditions, which could be due to the absence of reactions between oxygen and sulfite that may lead to the formation of various compounds, including the sulfite radical and other radicals [33] that inhibit the enzymes. In line with this, the addition of reducing agent (or "antioxidant") to the aerobic reactions had a clear positive effect on yields, despite the absence of LPMO activity (e.g., Figure 4). In the absence of reducing agent and in the presence of sulfite, hydrolysis yields were generally very low.

Effect of lignosulfonates on degradation of sulfite-pretreated softwoods

Besides a cellulose-rich pulp, sulfite pretreatment also yields lignosulfonates (LS), representing a valuable commodity from the lignin fraction of softwood biomass. It has been shown that LS can improve enzymatic saccharification of pretreated lignocellulosic materials [34, 35]. Using enzymatic hydrolysis of sulfite-pulped Norway spruce as a test case, we assessed whether LS could replace ascorbic acid. It can be observed from Fig. 5 that 0.25% (w/w) LS led to increased hydrolysis yields under aerobic conditions, but to a smaller extent than the addition of ascorbic acid (Fig. 2). The increase in yield obtained by adding LS was accompanied by the production of oxidized products (Fig. 5b), albeit, again, at lower levels compared to reactions with ascorbic acid (Fig. 2b).

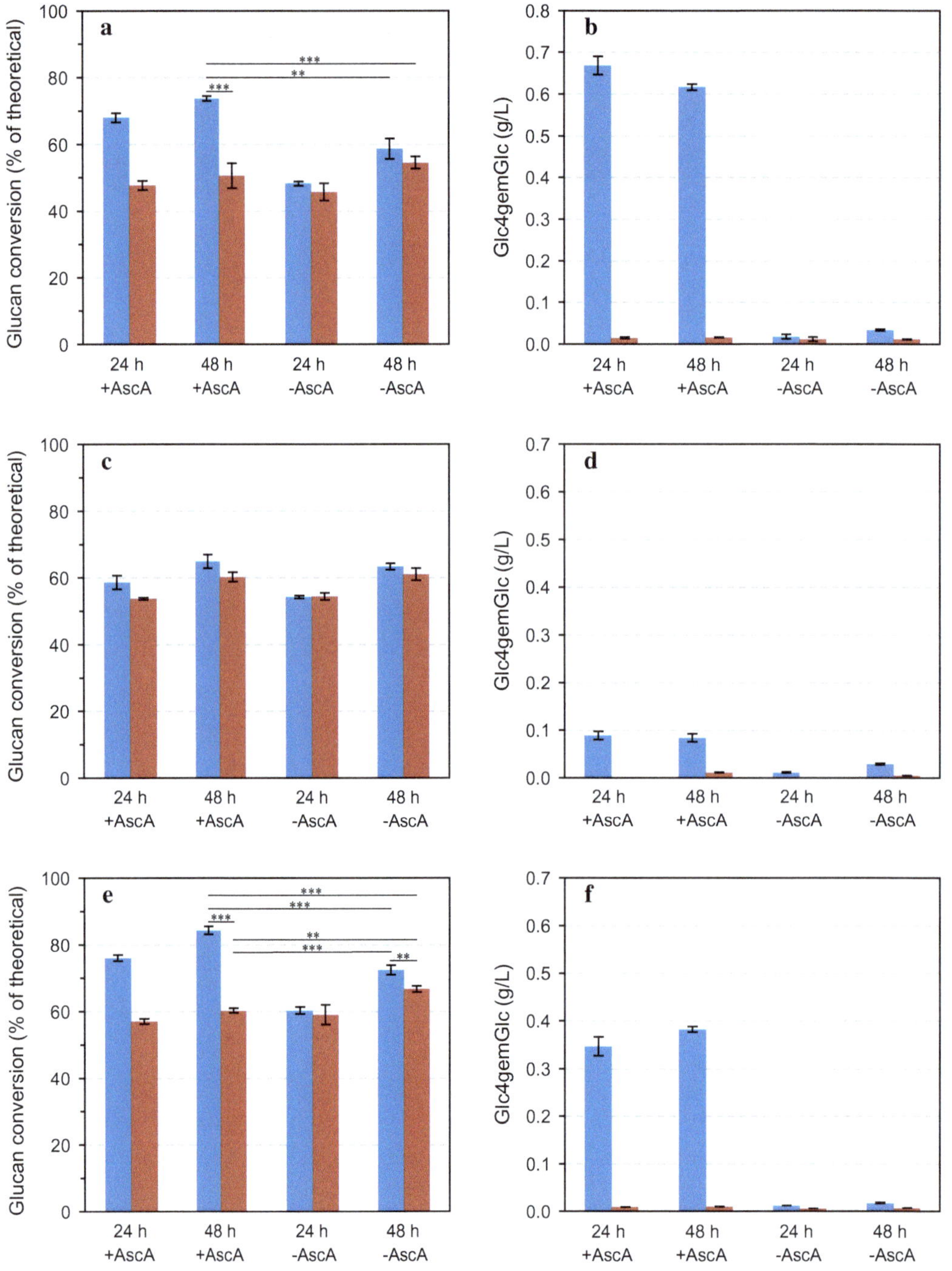

Fig. 2 Saccharification of pretreated Norway spruce (**a**–**d**) and loblolly pine (**e**–**f**) under aerobic (*blue bars*) or anaerobic (*red bars*) conditions in the presence or absence of 1 mM ascorbic acid (±AscA). The *left panels* show glucan conversion (as a percentage of theoretical glucan conversion) and the *right panels* show the concentration of Glc4gemGlc, at two time points. Enzymatic hydrolysis was carried out with Cellic® CTec3 (**a**, **b**, **e**, **f**) and Celluclast®:Novozym 188 (mixed at 5:1 ratio, w/w) (**c**, **d**) at 8 mg/g glucan total protein loading using reaction mixtures containing 5% DM in 50 mM sodium acetate pH 5.0 that were incubated at 50 °C. The data points represent the average value of three independent experiments with one technical replicate per experiment. The *error bars* represent standard deviations of the three independent experiments. The statistical significance of differences in the 48-h saccharification yields was analyzed using two-way ANOVA with Tukey's post hoc test (95% confidence interval) and is indicated as follows: *$p < 0.05$, **$p < 0.01$, ***$p < 0.001$

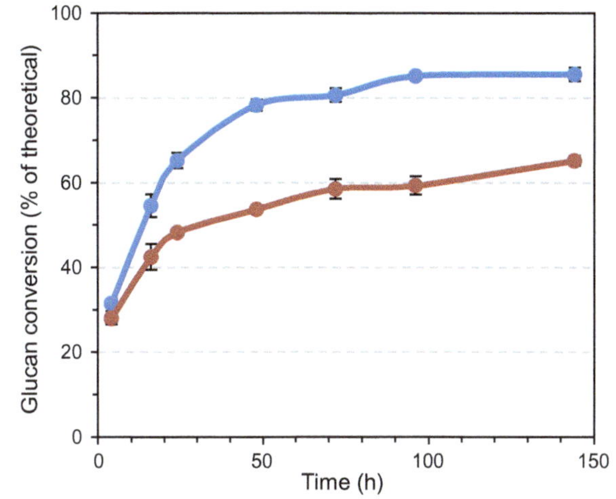

Fig. 3 Saccharification of pretreated Norway spruce under aerobic (*blue line*) or anaerobic (*red line*) conditions in the presence of 1 mM ascorbic acid. Enzymatic hydrolysis was carried out with Cellic® CTec3 at 8 mg/g glucan total protein loading in reaction mixtures containing 5% DM in 50 mM sodium acetate pH 5.0 that were incubated at 50 °C. The data points represent the average value of three independent experiments with one technical replicate per experiment. The *error bars* represent standard deviations of the three independent experiments

Addition of LS helped also to mitigate the negative effect of sulfite ions on hydrolysis efficiency (Fig. 5c), but to a lower extent than ascorbic acid (Fig. 4a) and other tested reducing agents (Additional file 1: Figure S1c, d).

Contribution of individual enzymes to saccharification efficiency

Previous work on the development of minimal enzyme cocktails for conversion of BALI™-pretreated spruce has shown that of the best known individual *T. reesei* cellulases, *Tr*Cel7B is the most important [36]. This study also indicated that a cellulose-active, bacterial LPMO (*Sc*LPMO10C) had little effect on overall saccharification efficiency. Considering these previous results and the clear positive effect of LPMOs contained in Cellic® CTec3 on the saccharification of sulfite-pulped softwoods, we tested the potential of *Ta*LPMO9A and three major *T. reesei* cellulases (*Tr*Cel7A, *Tr*Cel7B, and *Tr*Cel6A) to improve the efficiency of the Celluclast® 1.5L:Novozym 188 mix. We replaced 15% of the enzyme mixture by one of these individual enzymes because previous studies on spiking Celluclast® with *Ta*LPMO9A had indicated that this level of replacement was optimal in terms of increased saccharification yield [24]. Replacement of 15% of total protein by individual cellulases revealed that the addition of *Tr*Cel7B was more beneficial than that of the other individual cellulases and led to a yield comparable with that obtained with 100% of the original cellulase preparation. In line with other results reported in this study, replacement of 15% of the cellulase preparation with *Ta*LPMO9A had a clear beneficial effect on glucan conversion, but only when also ascorbic acid was added (Fig. 6).

Discussion

Since the discovery of the true nature of LPMOs [10], their role in the degradation of cellulose and pretreated lignocellulosic biomasses has been the subject of several studies [13, 23–25, 37]. For example, it has been shown that harnessing LPMO activity improves the saccharification of hydrothermally pretreated wheat straw [23] and steam-exploded birch [24] with an LPMO-rich cellulase preparation, Cellic® CTec2. Both these studies also showed that LPMO activity does not require the presence of externally added small molecule reducing agents, as long as the substrate is rich in lignin. A similar conclusion was drawn by Hu et al. [25] in their study focused on spiking cellulase preparations with an AA9 LPMO in the degradation of several pretreated lignocelluloses. Importantly, the study by Müller et al. (2015) established the importance of the presence of oxygen in harnessing the full potential of LPMOs.

From previous work, it is known that sulfite-pretreated softwoods can be degraded quite efficiently with commercially available cellulase preparations [5, 38]. Sulfite pulping using the BALI™ pretreatment yields cellulose-rich and almost lignin-free materials that are also relatively easy to degrade [6]. So far, the potential role of LPMOs in the degradation of sulfite-pretreated biomasses has not been addressed, nor have the implications of a possible role of LPMOs on process design. The present experiments, using both a model cellulosic substrate and industrially relevant, sulfite-pulped softwoods, shed light on these issues. The data show a clear correlation between the efficiency of saccharification and the detection of LPMO-generated products, underpinning the major contribution of LPMOs to the saccharification process, which amounts to increases in saccharification yield by up to 24% (in the case of sulfite-pulped loblolly pine). The presence of an externally added electron donor and aerobic conditions during enzymatic hydrolysis are both crucial for LPMO activity and for fully exploiting the cellulolytic potential of modern cellulase preparations. The requirement for an external electron donor to harness LPMO activity separates sulfite-pulped lignocelluloses from biomasses subjected to other physicochemical pretreatments and is likely due to extensive lignin removal during sulfite pulping (Table 1). The small amount of remaining lignin bound to the cellulose is most likely sulfonated, which will alter

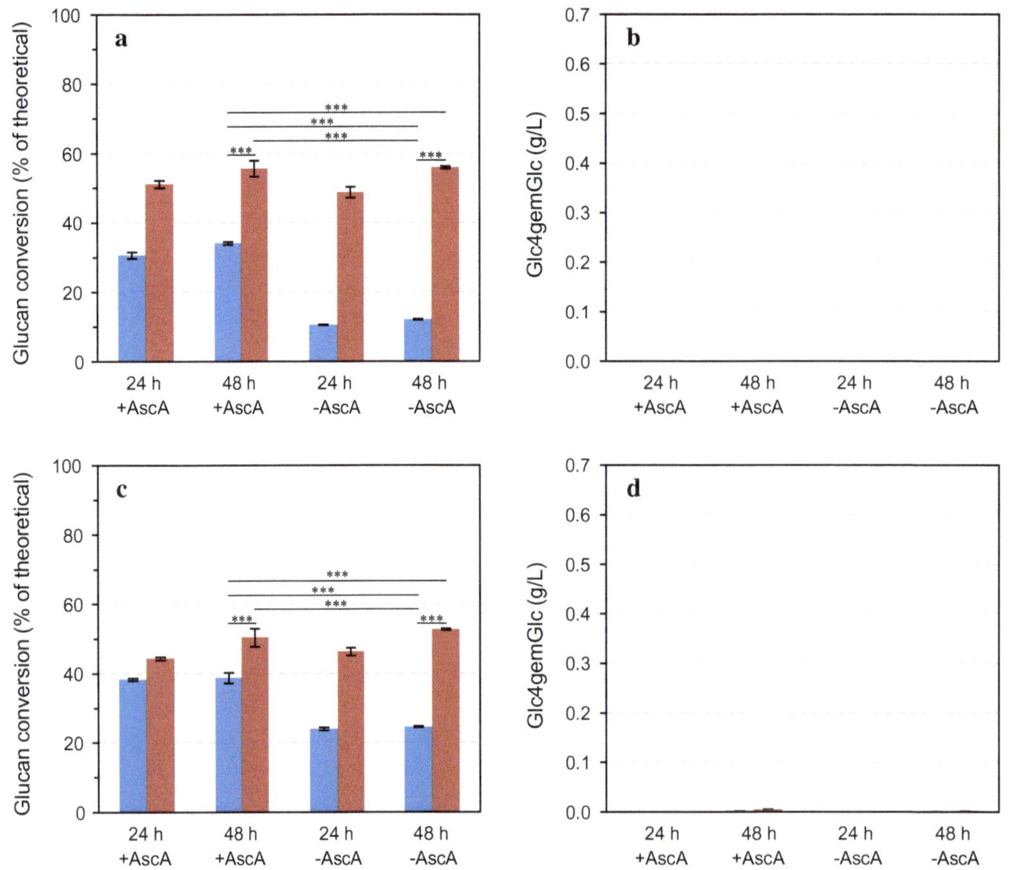

Fig. 4 Saccharification of pretreated Norway spruce in the presence of sulfite under aerobic (*blue bars*) or anaerobic (*red bars*) conditions, in the presence or absence of 1 mM ascorbic acid (±AscA). The *left panels* show glucan conversion (as a percentage of theoretical glucan conversion) and the *right panels* show concentrations of Glc4gemGlc, at two time points. Enzymatic hydrolysis was carried out with Cellic® CTec3 (**a**, **b**) and Celluclast®:Novozym 188 (mixed at 5:1 ratio, w/w) (**c**, **d**) at 8 mg/g glucan total protein loading in reactions containing 5% DM in 50 mM sodium acetate pH 5.0 that were incubated at 50 °C. Data for similar reactions without sulfite are presented in Fig. 2 and data for the effect of sulfite on hydrolysis of Avicel are presented in Additional file 3: Figure S3. The data points represent the average value of three independent experiments with one technical replicate per experiment. The *error bars* represent standard deviations of the three independent experiments. The statistical significance of differences in the 48-h saccharification yields was analyzed using two-way ANOVA with Tukey's post hoc test (95% confidence interval) and is indicated as follows: $*p < 0.05$, $**p < 0.01$, $***p < 0.001$

its properties, e.g., the degree of hydrophilicity, relative to, e.g., the lignin present in steam-exploded materials [3]. Dimarogona et al. [39] and Westereng et al. [40] have demonstrated that lignin is capable of transferring electrons to LPMOs and stimulate their activity in cellulose decomposition, whereas Hu et al. [25] have shown that complete lignin removal from steam-pretreated lodgepole pine nearly completely eliminated the boosting effect of adding an LPMO. Finally, by studying enzymatic saccharification of sugarcane bagasse samples obtained from different pretreatments and with varying degrees of delignification, Rodríguez-Zúñiga et al. [41] showed a positive correlation between lignin content and LPMO activity. All these results confirm early observations by Harris et al. [13], who showed that the enigmatic boosting effect of a GH61 protein (which later turned out to be an LPMO) on cellulase activity depended on the presence of lignin.

In the presence of ascorbate and under aerobic conditions, improvement of the enzymatic degradation of sulfite-pretreated softwoods was correlated with the accumulation of oxidized products. The positive effect of LPMO-promoting conditions was only observed when using LPMO-rich cellulase preparation Cellic® CTec3 and when spiking an older cellulase preparation, LPMO-poor Celluclast® 1.5L, with *Ta*LPMO9A. The dependence on oxygen is evident from both the present study and earlier work on steam-exploded birch by Müller et al. [24]. Interestingly, there are recent indications that LPMO reactions may be driven by H_2O_2 instead of O_2

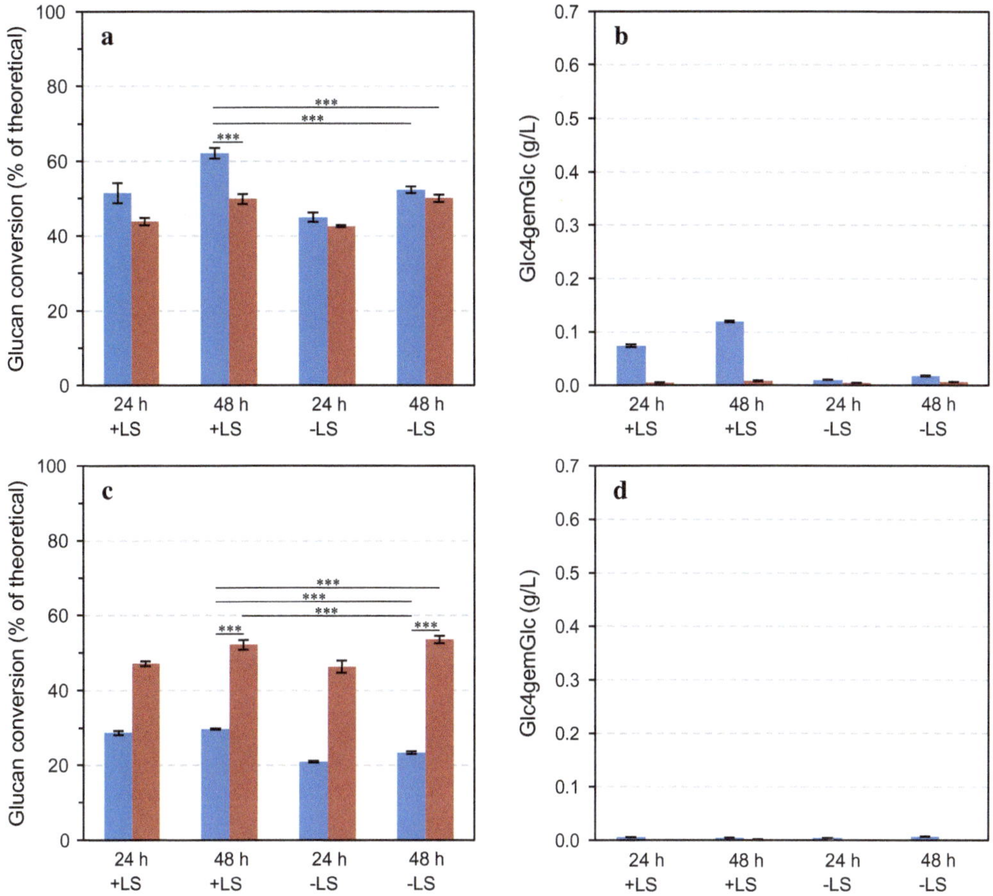

Fig. 5 Saccharification of pretreated Norway spruce in the absence (**a**, **b**) or presence (**c**, **d**) of sulfite under aerobic (*blue bars*) or anaerobic (*red bars*) conditions, in the presence or absence of 0.25% (w/w) lignosulfonates (LS). The *left panels* show glucan conversion (as a percentage of theoretical glucan conversion) and the *right panels* show concentrations of Glc4gemGlc, at two time points. Enzymatic hydrolysis was carried out with Cellic® CTec3 at 8 mg/g glucan total protein loading in reactions containing 5% DM in 50 mM sodium acetate pH 5.0 that were incubated at 50 °C. The data points represent the average value of three independent experiments with one technical replicate per experiment. The *error bars* represent standard deviations of the three independent experiments. The statistical significance of differences in the 48-h saccharification yields was analyzed using two-way ANOVA with Tukey's post hoc test (95% confidence interval) and is indicated as follows: *$p < 0.05$, **$p < 0.01$, ***$p < 0.001$

and that in such a set-up the LPMO requires only sub-stoichiometric amounts of reductant [9]. While further work is needed to study the implications of this novel way to drive LPMO activity, it is worth noting that the use of H_2O_2 may change or even abolish the need for supplying molecular oxygen and reductant in industrial biorefining. Supplying and controlling hydrogen peroxide, which is soluble in water, in high dry matter hydrolysis reactors may be easier, compared to using oxygen.

Careful inspection of Figs. 1 (hydrolysis of Avicel) and 2 (hydrolysis of sulfite-pulped Norway spruce) shows that under conditions that do not lead to LPMO activity, the Celluclast®:Novozym 188 cocktail is more efficient than Cellic® CTec3. Under these conditions, the LPMO fraction of the Cellic® CTec3 enzyme cocktail, perhaps amounting to 15% of the protein, is not active, meaning a reduced effective enzyme load. This observation underpins the importance of harnessing the power of LPMOs in modern cellulase cocktails.

While sulfite pulping generally seems to be a good pretreatment for softwoods, the present data show that residual sulfite that might be carried over from the pretreatment step to the enzymatic hydrolysis step has a negative effect on saccharification efficiency. The oxidation of sulfite to sulfate happens via a multi-step pathway involving the formation of several reactive intermediates [33] that are known to interact with, e.g., proteins [33, 42, 43]. For example, radicals formed during oxidation of sulfite react with methionine and tryptophan in proteins, which could easily lead to enzyme inactivation [33, 43]. Reducing agents, such as ascorbic acid, are commonly used in the food industry to protect from

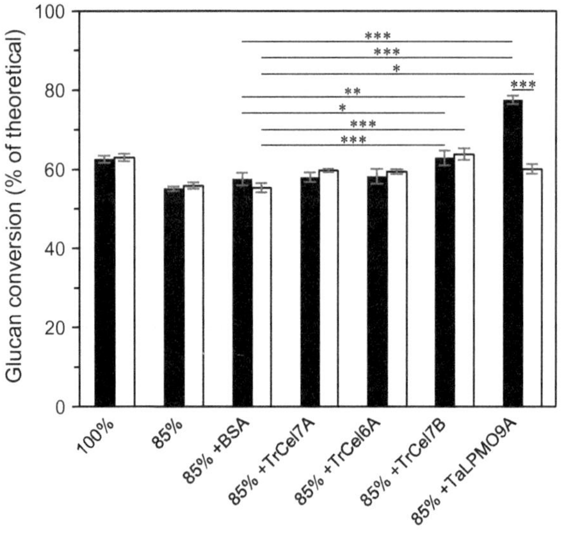

Fig. 6 Saccharification of pretreated Norway spruce in the presence (*black bars*) or absence (*white bars*) of 1 mM ascorbic acid. Enzymatic hydrolysis was carried out with the Celluclast®:Novozym 188 mix (5: 1 w/w) supplemented with major *Trichoderma reesei* cellulases (*Tr*Cel7A, *Tr*Cel6A, and *Tr*Cel7B) or *Ta*LPMO9A. The total protein loading was 8 mg/g glucan, except for the sample labeled "85%," where the total protein loading was at 6.8 mg/g glucan. The indicated individual enzymes comprised 15% of the total protein loading; BSA was used as a control. Experiments were carried out in reactions containing 5% DM in 50 mM sodium acetate pH 5.0 at 50 °C and were incubated at 50 °C for 48 h. The data points represent the average value of three independent experiments with one technical replicate per experiment. The *error bars* represent standard deviations of the three independent experiments. The statistical significance of differences in glucan conversion between the control reactions with BSA and reactions with added individual enzymes (in the presence or absence of 1 mM ascorbic acid) was analyzed using two-way ANOVA with Tukey's post hoc test (95% confidence interval) and is indicated as follows: $^{*}p < 0.05$, $^{**}p < 0.01$, $^{***}p < 0.001$

oxidative damage caused by reactive oxygen species, and indeed our results show that reducing agents prevent the enzyme inactivation happening under aerobic conditions in the presence of sulfite. The lack of LPMO products under these conditions shows that the positive effect of reductants is rather due to the scavenging of free radicals formed during oxidation of sulfite than to the promotion of LPMO activity. Obviously, utilization of large amounts of costly chemicals such as reductants is not economically feasible. Thus, the development of enzymes resistant to high sulfite concentrations or introduction of process steps for removing residual sulfite should be considered. Notably, further research on actual sulfite levels within the biorefinery is needed and so are studies on sulfite stability. The conditions used in this study, where we used high amounts of freshly added sodium sulfite, are not representative of the conditions in a real biorefinery.

It was known from before that lignosulfonates may have a positive effect on the saccharification of sulfite-pretreated lignocellulosic biomasses [34, 35]. These positive effects have been ascribed to the surfactant properties of LS [34] or to the formation of lignosulfonate–cellulase complexes [35], where both effects were thought to lead to reduced non-productive binding of cellulases to the substrate. In the present study, we observed that the positive effect of LS on hydrolysis efficiency was only present under aerobic conditions (Fig. 5a) and was linked to the promotion of LPMO activity (Fig. 5b). Interestingly, we also observed that LS reduced the negative effect of sulfite ions, though to a lesser extent than ascorbate. The exact mechanism of the stimulation of LPMO activity by LS addition is unknown and warrants further research. Notably, this effect could involve the interplay between (soluble) LS and residual lignin bound to the cellulose, which would be analogous to the interplay between solid and soluble lignin described by Westereng et al. [40].

Conclusions

In this study, we have demonstrated efficient enzymatic conversion of sulfite-pretreated softwoods by Cellic® CTec3 and we have demonstrated the importance of LPMOs in this process. We have also shown that, in order to obtain the high saccharification yields that are possible due to LPMO action, it is crucial that process conditions are adapted to the LPMOs, which need oxygen and reducing equivalents. The removal of residual sulfite from pretreatments is of major importance and both assessment of typical sulfite levels and methods to deal with these need further attention. Lignosulfonates have a potential to act as the electron donor for LPMOs, which could reduce the costs related to the addition of small molecule reductants in sulfite pretreatment-based biorefineries.

Additional files

Additional file 1: Figure S1. Saccharification of pretreated Norway spruce in the absence (**a**, **b**) or presence (**c**, **d**) of sulfite under aerobic (*blue bars*) or anaerobic (*red bars*) conditions in the presence or absence of 1 mM gallic acid (+/− GA; *left panels*) or reduced glutathione (+/− RG; *right panels*). Enzymatic hydrolysis was carried out with Cellic® CTec3 at 8 mg/g glucan total protein loading in reaction mixtures containing 5% DM that were incubated in 50 mM sodium acetate pH 5.0, at 50 °C. Reactions with sulfite (*lower panels*) contained 1000 ppm of sulfite ions added as a sodium sulfite (12.5 mM). The data points represent the average value of three independent experiments with one technical replicate per experiment. The *error bars* represent standard deviations of the three independent experiments. The statistical significance of differences in the 48 hour saccharification yields was analyzed using two-way ANOVA with Tukey's post hoc test (95% confidence interval), and is indicated as follows: *, p<0.05; **, p<0.01; ***, p<0.001.

Additional file 2: Figure S2. Saccharification of sulfite-pretreated Norway spruce without addition of L-cysteine hydrochloride monohydrate (0.025 % w/v) in the anaerobic reactions, in the absence (**a**, **b**) or presence (**c**, **d**) of sulfite under aerobic (*blue bars*) or anaerobic (*red bars*) conditions, in the presence or absence of 1 mM ascorbic acid (+/− AscA). The *left panels* show glucan conversion (as a percentage of theoretical glucan conversion) and the *right panels* show the concentration of Glc4gemGlc, at two time points. Enzymatic hydrolysis was carried out with Cellic® CTec3 at 8 mg/g glucan total protein loading in reaction mixtures containing 5% DM in 50 mM sodium acetate pH 5.0 that were incubated at 50 °C. Reactions with sulfite (*lower panels*) contained 1000 ppm of sulfite ions added as a sodium sulfite (12.5 mM). The data points represent the average value of three independent experiments with one technical replicate per experiment. The *error bars* represent standard deviations of the three independent experiments. The statistical significance of differences in the 48 hour saccharification yields was analyzed using two-way ANOVA with Tukey's post hoc test (95% confidence interval), and is indicated as follows: *, $p<0.05$; **, $p<0.01$; ***, $p<0.001$.

Additional file 3: Figure S3. Saccharification of Avicel PH-101 in the presence of sulfite under aerobic (*blue bars*) or anaerobic (*red bars*) conditions, in the presence or absence of 1 mM ascorbic acid (+/− AscA). The *left panels* show glucan conversion (as a percentage of theoretical glucan conversion) and the *right panels* show concentration of Glc4gemGlc, at two time points. Enzymatic hydrolysis was carried out with Cellic® CTec3 (**a**, **b**) and Celluclast:Novozym 188 (mixed at 5:1 ratio, w/w) (**c**, **d**) at 8 mg/g glucan total protein loading in reaction mixtures containing 5 % DM in 50 mM sodium acetate pH 5.0 that were incubated at 50 °C. Data for similar reactions without sulfite are presented in Fig. 1. The data points represent the average value of three independent experiments with one technical replicate per experiment. The *error bars* represent standard deviations of the three independent experiments. The statistical significance of differences in the 48 hour saccharification yields was analyzed using two-way ANOVA with Tukey's post hoc test (95% confidence interval), and is indicated as follows: *, $p<0.05$; **, $p<0.01$; ***, $p<0.001$.

Abbreviations
AA: auxiliary activity; AscA: ascorbic acid; Glc4gemGlc: 4-hydroxy-β-D-xylo-hexopyranosyl-(1→4)-β-D-glucopyranosyl; HPAEC: high-performance anion exchange chromatography; HPLC: high-performance liquid chromatography; LPMO: lytic polysaccharide monooxygenase; LS: lignosulfonates.

Authors' contributions
PC, MD, OB, ML, SJH, and VGHE designed the experiments and interpreted the results; SJH and VGHE supervised the study; PC did most of the experimental work; DP, GM, and MS-A contributed with experiments, materials, and ideas. PC drafted the manuscript, with help from GM, SJH, and VGHE; PC and VGHE finalized the manuscript. All authors read and approved the final manuscript.

Author details
[1] Faculty of Chemistry, Biotechnology and Food Science, Norwegian University of Life Sciences, P.O. Box 5003, 1432 Ås, Norway. [2] Borregaard AS, P.O.Box 162, 1701 Sarpsborg, Norway. [3] VTT Technical Research Centre of Finland, P.O. Box 1000, 02044 VTT Espoo, Finland.

Acknowledgements
The authors would like to acknowledge Kristin Hals (Borregaard AS, Sarpsborg, Norway) for performing compositional analysis of industrial lignocellulosic substrates.

Competing interests
Lersch and Bengtsson are employees of Borregaard AS, Sarpsborg, Norway. Marie Dahlström was at the time of completing this work an employee of Borregaard AS, Sarpsborg, Norway.

Funding
This work was funded by the Research Council of Norway, through Grant 219633.

References
1. Zhu JY, Pan XJ. Woody biomass pretreatment for cellulosic ethanol production: technology and energy consumption evaluation. Bioresour Technol. 2010;101(13):4992–5002. doi:10.1016/j.biortech.2009.11.007.
2. Himmel ME, Ding SY, Johnson DK, Adney WS, Nimlos MR, Brady JW, et al. Biomass recalcitrance: engineering plants and enzymes for biofuels production. Science. 2007;315(5813):804–7. doi:10.1126/science.1137016.
3. Leu SY, Zhu JY. Substrate-related factors affecting enzymatic saccharification of lignocelluloses: our recent understanding. Bioenergy Res. 2013;6(2):405–15. doi:10.1007/s12155-012-9276-1.
4. Pan XJ, Arato C, Gilkes N, Gregg D, Mabee W, Pye K, et al. Biorefining of softwoods using ethanol organosolv pulping: preliminary evaluation of process streams for manufacture of fuel-grade ethanol and co-products. Biotechnol Bioeng. 2005;90(4):473–81. doi:10.1002/bit.20453.
5. Zhu JY, Pan XJ, Wang GS, Gleisner R. Sulfite pretreatment (SPORL) for robust enzymatic saccharification of spruce and red pine. Bioresour Technol. 2009;100(8):2411–8. doi:10.1016/j.biortech.2008.10.057.
6. Rødsrud G, Lersch M, Sjöde A. History and future of world's most advanced biorefinery in operation. Biomass Bioenergy. 2012;46:46–59. doi:10.1016/j.biombioe.2012.03.028.
7. Merino ST, Cherry J. Progress and challenges in enzyme development for biomass utilization. Adv Biochem Eng Biotechnol. 2007;108:95–120. doi:10.1007/10_2007_066.
8. Jørgensen H, Kristensen JB, Felby C. Enzymatic conversion of lignocellulose into fermentable sugars: challenges and opportunities. Biofuel Bioprod Biorefin. 2007;1(2):119–34. doi:10.1002/bbb.4.
9. Bissaro B, Røhr AK, Skaugen M, Forsberg Z, Horn SJ, Vaaje-Kolstad G, et al. Fenton-type chemistry by a copper enzyme: molecular mechanism of polysaccharide oxidative cleavage. cleavage. 2017. doi:10.1101/097022.
10. Vaaje-Kolstad G, Westereng B, Horn SJ, Liu ZL, Zhai H, Sorlie M, et al. An oxidative enzyme boosting the enzymatic conversion of recalcitrant polysaccharides. Science. 2010;330(6001):219–22. doi:10.1126/science.1192231.
11. Quinlan RJ, Sweeney MD, Lo Leggio L, Otten H, Poulsen JC, Johansen KS, et al. Insights into the oxidative degradation of cellulose by a copper metalloenzyme that exploits biomass components. Proc Natl Acad Sci USA. 2011;108(37):15079–84. doi:10.1073/pnas.1105776108.
12. Hemsworth GR, Henrissat B, Davies GJ, Walton PH. Discovery and characterization of a new family of lytic polysaccharide monooxygenases. Nat Chem Biol. 2014;10(2):122–6. doi:10.1038/nchembio.1417.
13. Harris PV, Welner D, McFarland KC, Re E, Navarro Poulsen JC, Brown K, et al. Stimulation of lignocellulosic biomass hydrolysis by proteins of glycoside hydrolase family 61: structure and function of a large, enigmatic family. Biochemistry. 2010;49(15):3305–16. doi:10.1021/bi100009p.
14. Phillips CM, Beeson WT, Cate JH, Marletta MA. Cellobiose dehydrogenase and a copper-dependent polysaccharide monooxygenase potentiate cellulose degradation by *Neurospora crassa*. ACS Chem Biol. 2011;6(12):1399–406. doi:10.1021/cb200351y.
15. Forsberg Z, Vaaje-Kolstad G, Westereng B, Bunæs AC, Stenstrøm Y, MacKenzie A, et al. Cleavage of cellulose by a CBM33 protein. Protein Sci. 2011;20(9):1479–83. doi:10.1002/pro.689.
16. Westereng B, Ishida T, Vaaje-Kolstad G, Wu M, Eijsink VGH, Igarashi K, et al. The putative endoglucanase *Pc*GH61D from *Phanerochaete chrysosporium* is a metal-dependent oxidative enzyme that cleaves cellulose. PLoS ONE. 2011;6(11):e27807. doi:10.1371/journal.pone.0027807.
17. Isaksen T, Westereng B, Aachmann FL, Agger JW, Kracher D, Kittl R, et al. A C4-oxidizing lytic polysaccharide monooxygenase cleaving both cellulose and cello-oligosaccharides. J Biol Chem. 2014;289(5):2632–42. doi:10.1074/jbc.M113.530196.
18. Agger JW, Isaksen T, Várnai A, Vidal-Melgosa S, Willats WGT, Ludwig R, et al. Discovery of LPMO activity on hemicelluloses shows the importance of oxidative processes in plant cell wall degradation. Proc Natl Acad Sci USA. 2014;111(17):6287–92. doi:10.1073/pnas.1323629111.
19. Frommhagen M, Sforza S, Westphal AH, Visser J, Hinz SWA, Koetsier MJ, et al. Discovery of the combined oxidative cleavage of plant xylan and cellulose by a new fungal polysaccharide monooxygenase. Biotechnol Biofuels. 2015;8:101. doi:10.1186/s13068-015-0284-1.
20. Vu VV, Beeson WT, Span EA, Farquhar ER, Marletta MA. A family of starch-active polysaccharide monooxygenases. Proc Natl Acad Sci USA. 2014;111(38):13822–7. doi:10.1073/pnas.1408090111.

21. Lo Leggio L, Simmons TJ, Poulsen JC, Frandsen KE, Hemsworth GR, Stringer MA, et al. Structure and boosting activity of a starch-degrading lytic polysaccharide monooxygenase. Nat Commun. 2015;6:5961. doi:10.1038/ncomms6961.
22. Levasseur A, Drula E, Lombard V, Coutinho PM, Henrissat B. Expansion of the enzymatic repertoire of the CAZy database to integrate auxiliary redox enzymes. Biotechnol Biofuels. 2013;6:41. doi:10.1186/1754-6834-6-41.
23. Cannella D, Hsieh CWC, Felby C, Jørgensen H. Production and effect of aldonic acids during enzymatic hydrolysis of lignocellulose at high dry matter content. Biotechnol Biofuels. 2012;5:26. doi:10.1186/1754-6834-5-26.
24. Müller G, Várnai A, Johansen KS, Eijsink VGH, Horn SJ. Harnessing the potential of LPMO-containing cellulase cocktails poses new demands on processing conditions. Biotechnol Biofuels. 2015;8:187. doi:10.1186/s13068-015-0376-y.
25. Hu JG, Arantes V, Pribowo A, Gourlay K, Saddler JN. Substrate factors that influence the synergistic interaction of AA9 and cellulases during the enzymatic hydrolysis of biomass. Energy Environ Sci. 2014;7(7):2308–15. doi:10.1039/C4EE00891J.
26. Sjöde A, Frölander A, Lersch M, Rødsrud G. Lignocellulosic biomass conversion by sulfite pretreatment. Patent EP2376642 B1. Borregaard AS; 2013.
27. Ståhlberg J, Divne C, Koivula A, Piens K, Claeyssens M, Teeri TT, et al. Activity studies and crystal structures of catalytically deficient mutants of cellobiohydrolase I from *Trichoderma reesei*. J Mol Biol. 1996;264(2):337–49. doi:10.1006/jmbi.1996.0644.
28. Suurnäkki A, Tenkanen M, Siika-aho M, Niku-Paavola ML, Viikari L, Buchert J. *Trichoderma reesei* cellulases and their core domains in the hydrolysis and modification of chemical pulp. Cellulose. 2000;7(2):189–209. doi:10.1023/A:1009280109519.
29. Várnai A, Tang C, Bengtsson O, Atterton A, Mathiesen G, Eijsink VGH. Expression of endoglucanases in *Pichia pastoris* under control of the GAP promoter. Microb Cell Fact. 2014;13:57. doi:10.1186/1475-2859-13-57.
30. Loose JSM, Forsberg Z, Fraaije MW, Eijsink VGH, Vaaje-Kolstad G. A rapid quantitative activity assay shows that the *Vibrio cholerae* colonization factor GbpA is an active lytic polysaccharide monooxygenase. FEBS Lett. 2014;588(18):3435–40. doi:10.1016/j.febslet.2014.07.036.
31. Bradford MM. A rapid and sensitive method for the quantitation of microgram quantities of protein utilizing the principle of protein-dye binding. Anal Biochem. 1976;72:248–54. doi:10.1016/0003-2697(76)90527-3.
32. Adav SS, Chao LT, Sze SK. Quantitative secretomic analysis of *Trichoderma reesei* strains reveals enzymatic composition for lignocellulosic biomass degradation. Mol Cell Proteom. 2012;11(7):M111–012419. doi:10.1074/mcp.M111.012419.
33. Neta P, Huie RE. Free-radical chemistry of sulfite. Environ Health Perspect. 1985;64:209–17. doi:10.2307/3430011.
34. Zhou HF, Lou HM, Yang DJ, Zhu JY, Qiu XQ. Lignosulfonate to enhance enzymatic saccharification of lignocelluloses: role of molecular weight and substrate lignin. Ind Eng Chem Res. 2013;52(25):8464–70. doi:10.1021/ie401085k.
35. Wang ZJ, Zhu JY, Fu YJ, Qin MH, Shao ZY, Jiang JG, et al. Lignosulfonate-mediated cellulase adsorption: enhanced enzymatic saccharification of lignocellulose through weakening nonproductive binding to lignin. Biotechnol Biofuels. 2013;6:156. doi:10.1186/1754-6834-6-156.
36. Chylenski P, Forsberg Z, Ståhlberg J, Várnai A, Lersch M, Bengtsson O, et al. Development of minimal enzyme cocktails for hydrolysis of 1 sulfite-pulped lignocellulosic biomass. J Biotechnol. 2017;246:16–23. doi:10.1016/j.jbiotec.2017.02.009.
37. Hu JG, Chandra R, Arantes V, Gourlay K, van Dyk JS, Saddler JN. The addition of accessory enzymes enhances the hydrolytic performance of cellulase enzymes at high solid loadings. Bioresour Technol. 2015;186:149–53. doi:10.1016/j.biortech.2015.03.055.
38. Wang GS, Pan XJ, Zhu JY, Gleisner R, Rockwood D. Sulfite pretreatment to overcome recalcitrance of lignocellulose (SPORL) for robust enzymatic saccharification of hardwoods. Biotechnol Prog. 2009;25(4):1086–93. doi:10.1002/btpr.206.
39. Dimarogona M, Topakas E, Olsson L, Christakopoulos P. Lignin boosts the cellulase performance of a GH-61 enzyme from *Sporotrichum thermophile*. Bioresour Technol. 2012;110:480–7. doi:10.1016/j.biortech.2012.01.116.
40. Westereng B, Cannella D, Agger JW, Jørgensen H, Andersen ML, Eijsink VGH, et al. Enzymatic cellulose oxidation is linked to lignin by long-range electron transfer. Sci Rep. 2015;5:18561. doi:10.1038/srep18561.
41. Rodríguez-Zúñiga UF, Cannella D, Giordano RD, Giordano RDC, Jørgensen H, Felby C. Lignocellulose pretreatment technologies affect the level of enzymatic cellulose oxidation by LPMO. Green Chem. 2015;17(5):2896–903. doi:10.1039/C4GC02179G.
42. Swan JM. Thiols, disulphides and thiosulphates—some new reactions and possibilities in peptide and protein chemistry. Nature. 1957;180(4587):643–5. doi:10.1038/180643a0.
43. Yang SF. Destruction of tryptophan during aerobic oxidation of sulfite ions. Environ Res. 1973;6(4):395–402. doi:10.1016/0013-9351(73)90055-8.

4

Synthetic CO_2-fixation enzyme cascades immobilized on self-assembled nanostructures that enhance CO_2/O_2 selectivity of RubisCO

Sriram Satagopan[1†], Yuan Sun[2†], Jon R. Parquette[2*] and F. Robert Tabita[1*]

Abstract

Background: With increasing concerns over global warming and depletion of fossil-fuel reserves, it is attractive to develop innovative strategies to assimilate CO_2, a greenhouse gas, into usable organic carbon. Cell-free systems can be designed to operate as catalytic platforms with enzymes that offer exceptional selectivity and efficiency, without the need to support ancillary reactions of metabolic pathways operating in intact cells. Such systems are yet to be exploited for applications involving CO_2 utilization and subsequent conversion to valuable products, including biofuels. The Calvin–Benson–Bassham (CBB) cycle and the enzyme ribulose 1,5-bisphosphate carboxylase/oxygenase (RubisCO) play a pivotal role in global CO_2 fixation.

Results: We hereby demonstrate the co-assembly of two RubisCO-associated multienzyme cascades with self-assembled synthetic amphiphilic peptide nanostructures. The immobilized enzyme cascades sequentially convert either ribose-5-phosphate (R-5-P) or glucose, a simpler substrate, to ribulose 1,5-bisphosphate (RuBP), the acceptor for incoming CO_2 in the carboxylation reaction catalyzed by RubisCO. Protection from proteolytic degradation was observed in nanostructures associated with the small dimeric form of RubisCO and ancillary enzymes. Furthermore, nanostructures associated with a larger variant of RubisCO resulted in a significant enhancement of the enzyme's selectivity towards CO_2, without adversely affecting the catalytic activity.

Conclusions: The ability to assemble a cascade of enzymes for CO_2 capture using self-assembling nanostructure scaffolds with functional enhancements show promise for potentially engineering entire pathways (with RubisCO or other CO_2-fixing enzymes) to redirect carbon from industrial effluents into useful bioproducts.

Keywords: CO_2 fixation, Cell-free systems, Nanostructures, RubisCO, Pathway engineering, Bioconversion

Background

The rapid decline of fossil fuel reserves, emission of greenhouse gases, and potential deleterious effects on the biosphere have been highly publicized. Thus there is a considerable interest in designing strategies to capture atmospheric CO_2 to meet society's fuel and material needs [1–4]. Clearly, usage of fossil reserves is unlikely to be eliminated any time soon, however, recent advances in industrial biotechnology and increasing usage of biochemical paradigms has provided access to a multitude of starting materials that may be potentially utilized in innovative, economical, and sustainable ways to generate both natural and unnatural value-added compounds [5–7]. Enzymes have evolved over billions of years into efficient and highly selective catalysts of chemical reactions, particularly those that are energetically unfavorable under ambient conditions, facilitating their role in the maintenance of life on this planet [8]. Although advances have been made to synthesize products using engineered in vivo biological systems, sustainability, and scale-up capabilities are often limited by the tug-of-war between the cell's objectives and the desired product

*Correspondence: parquette.1@osu.edu; tabita.1@osu.edu
†Sriram Satagopan and Yuan Sun contributed equally to this work
[1] Department of Microbiology, The Ohio State University, 484 West 12th Avenue, Columbus, OH 43210-1292, USA
[2] Department of Chemistry and Biochemistry, The Ohio State University, 100 West 18th Avenue, Columbus, OH 43210-1185, USA

output. Cell-free biomimetic systems have smaller ecological footprints and are convenient alternatives with a potential for rapid design-build-test cycles [5, 8, 9]. Sequential chemical conversions are often achieved in these systems via multienzyme cascades that are spatially organized into three-dimensional structures, which control intermediate flux and enhance overall catalytic efficiency [10–12]. Multienzyme cascades have been constructed on synthetic scaffolds via covalent linkages or within physical compartments to proximally position the enzymes to channel reaction intermediates from one catalytic site to the other [13–19]. However, most of these strategies require a significant synthetic effort to implement and few control three-dimensional structure at the nanoscale. Developments in site-specific protein-nanoparticle conjugation techniques and DNA nanotechnology have enabled the spatial arrangement of proteins into arrays [20–22]. Alternatively, the self-assembly of small molecules offer an expedient strategy to create nanostructured scaffolds to support enzymatic arrays [23–30]. Block copolymers, nanotubes, and DNA nanocages have been used to create nanoscale supports for enzymes [31–34]. However, the potential to co-assemble carbon fixation and associated assimilatory pathway enzymes into nanostructured catalytic platforms remain untested.

RubisCO is the world's most abundant protein and accounts for most of the biological CO_2 fixed on earth. Diverse structural forms of RubisCO were previously characterized, with varying catalytic properties, stabilities, and temperature/pH optima noted [35]. Complex assembly requirements and sensitivity to a variety of inhibitors, effectors, and denaturing conditions have been impediments to the design of stable catalytic platforms using RubisCO. In fact, reconstituting RubisCO activity from inactive and not fully assembled forms requires the action of chaperone and other specialized proteins [36–41]. It has also been a challenge to assemble functional RubisCO into stable, scalable host cells to capture CO_2 for various applications, including improvement of the host cell's primary productivity [42]. Cyanobacteria and some proteobacteria optimize CO_2 fixation by encapsulating RubisCO and carbonic anhydrase (CA) within intracellular compartments called carboxysomes; analogous micro-compartments are employed by eukaryotic algae [43, 44]. In a recent study, functional RubisCO was co-encapsulated with CA in synthetic carboxysome mimics, which could protect the enzyme from proteolytic degradation [45]. With these biological structures in mind, we postulated that RubisCO and other enzymes might be coaxed to self-assemble with nanostructures. Starting with simple di- or tetra-peptide compounds capable of self-assembling into nanotubes or nanofibers, we show here that structurally divergent RubisCO enzymes can form CO_2-fixing nanostructure-enzyme complexes. As a proof of concept for utilizing these nanostructures to assemble entire pathways, we further demonstrated that RubisCO could be coupled with enzymes that catalyze sequential steps of the CBB cycle or the pentose-phosphate pathway within the same nanostructures. Furthermore, the immobilization of one form of RubisCO used in these studies resulted in a significant increase (compared to the unbound enzyme) in the CO_2/O_2 specificity factor, the largest enhancement in specificity of RubisCO thus far achieved.

Results

Nanostructure assembly and optimization of recovered activities using RubisCO

Initial studies were performed by mixing RubisCO (0.1 mg) with a previously described [46] bola amphiphilic naphthalene diimide (NDI-Lys) self-assembling monomer (1 mM) in 1 mL of a buffered solution. These experiments resulted in the precipitation of RubisCO. Part of the incompatibility of RubisCO with these nanotubes arose from the buffered systems necessary to stabilize RubisCO, which screened the charged nanotube head groups, leading to uncontrolled aggregation and precipitation. Thus, two small dipeptide conjugates (compounds **A**/**B**) and a fluorenylmethyloxycarbonyl (Fmoc) tetrapeptide (compound **C**) were chosen based on their ability to self-assemble into stable nanotubes and nanofibers, respectively, with RubisCO in a Bicine buffer (Fig. 1a). Consistently more than 20% of the RubisCO activity present in the unbound enzyme sample could be recovered from these nanostructure-RubisCO complexes. The greater surface charge of the dipeptide monomers, owing to the lysine side-chains, attenuated assembly/aggregation in buffered solvents leading to well-dispersed self-assembled structures. We have previously demonstrated that the dipeptide derivatives (**A**/**B**) self-assemble into nanotubes with diameters ranging from 80 to 120 nm under buffered conditions (Fig. 1a) [26]. Similarly, short peptides, such as **C**, form amyloid-type β-sheet structures that afford stable, one-dimensional nanofibers with 20–30 nm diameters at concentrations as low as 1 mM (Fig. 1a) [47]. Variants of compound **A** and **B** with only one Lys or with the second Lys replaced by either Arg or Glu were also predicted to form similar nanotubes (Fig. 1a). However, when mixed with RubisCO, most of these variants resulted in precipitation and significant loss of RubisCO activity. The mono-lysine variant of compound B could allow for nanostructure formation with RubisCO, but only 7% of the starting level of RubisCO activity was recovered from this complex. Another compound with predominantly acidic amino-acids conjugated to a 7-(diethylamino)-3-coumarin carboxylic acid

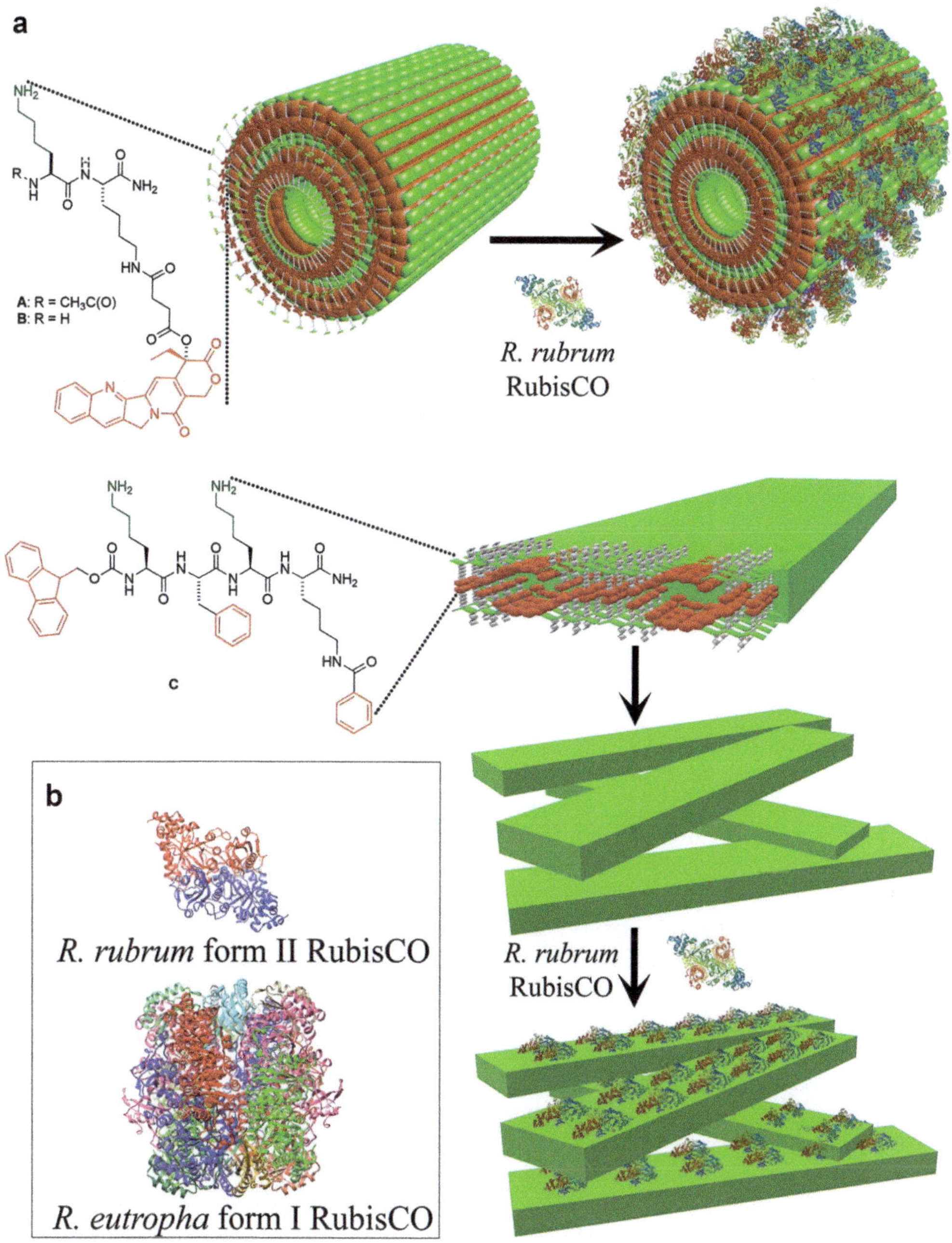

Fig. 1 Illustration of the assembly of nanostructures with RubisCO driven by electrostatic interactions. **a** Structures of compounds that self-assemble into either nanotubes (**A**/**B**) or nanofibers (**C**). **b** Either the dimeric *R. rubrum* form II RubisCO (PDB id—5RUB) or the hexadecameric *R. eutropha* form I RubisCO (PDB id—1BXN) were used in these studies

(Fmoc-EFEK-DAC) also resulted in complete precipitation and loss of RubisCO activity upon mixing. Thus, among various compounds tested for nanostructure-complex formation with RubisCO, compounds **A**, **B**, and **C** seemed most compatible and resulted in complexes with significant RubisCO activity recoveries.

Two structural forms of RubisCO were employed in these studies: the simple form II L_2 dimer RubisCO from *Rhodospirillum rubrum*, which has an elliptical shape with approximate dimensions of 5 × 7 × 10 nm, and the L_8S_8 hexadecameric form I enzyme from *Ralstonia eutropha* that is shaped like a cube with approximately

10 nm sides (Fig. 1b) [48–50]. Accordingly, the dimensions of the nanotubes and nanofibers should readily accommodate both forms of RubisCO. Initial experiments indicated that co-assembly of dipeptides **A** or **B** and the enzyme resulted in low yields of activity in the nanotube-form II RubisCO complexes (equivalent to 0.01–0.05 mg RubisCO per mg nanotube). Further, different preparations of nanotubes co-assembled from different batches of compounds **A** or **B**, and RubisCO resulted in widely different activity recoveries (equivalent to 0.005–0.12 mg of RubisCO per mg nanotube), despite starting with identical concentrations of the constituents.

Although compounds **A** and **B** had been prepared following identical procedures that resulted in highly pure compounds, and stored as lyophilized samples, TEM images of monomer samples from different preparations indicated that some of the older preparations had varying amounts of pre-assembled nanotubes (formed during storage as lyophilized powders) relative to newer preparations (Additional file 1: Figure S1a). Higher RubisCO activity recoveries were typically obtained with these older samples, suggesting that the nanotube precursors (i.e., monomeric compounds **A** or **B**) inhibited the enzyme activity during the co-assembly process. RubisCO activity assays performed by mixing the unbound enzyme with either of the monomeric compounds (fresh preparations), or with fully assembled nanotubes that had been isolated using ultracentrifugation confirmed the inhibitory effect of monomer compounds **A** and **B** (Additional file 1: Figure S1b). Because the pre-assembled nanotubes did not seem to significantly affect the activity of unbound RubisCO, it proved advantageous to assemble enzymes with pre-formed and isolated nanostructures rather than co-assembling them in the presence of monomeric compounds. Indeed, consistently higher recoveries of enzymatic activities (equivalent to 0.1–1 mg RubisCO per mg nanostructure) were obtained by mixing the enzyme with pre-formed nanostructures that had been isolated from unassembled monomers via ultracentrifugation. Re-suspension or washing the nanostructures with the Bicine buffer in the presence or absence of up to 300 mM sodium chloride did not result in any loss of activity, indicating that the enzyme was not weakly bound to the structures (Additional file 1: Table S1).

A range of 1–20 mM of the monomeric compounds **A**–**C** were used in the initial experiments for formation of the respective nanostructures, which were then isolated from unassembled monomers using ultracentrifugation. For each preparation, an appropriate volume of the nanostructure resuspension was mixed with either form I or form II RubisCO in Bicine buffer, resulting in a 2.5- to 5-fold dilution of the nanostructure suspensions in the final sample. Optimal recovery of enzymatic activity for nanotube-RubisCO complexes was obtained with nanotubes prepared from 10 to 20 mM of monomeric compounds (**A** or **B**), for form I and form II enzymes, respectively. The optimal starting concentration of compound **C** for the formation of nanofiber-RubisCO complexes was 2.5 and 5 mM for form I and form II RubisCO, respectively (Additional file 1: Table S2). Lower starting concentrations of nanostructure materials typically resulted in lower yields of the nanostructure-RubisCO complexes, as discernible from the sizes of pellets upon ultracentrifugation. This explains the lower rates of activity recoveries in these cases (Additional file 1: Table S2). Loading experiments were performed by titrating different amounts of RubisCO with fixed amounts of nanomaterials to determine optimal ratios for immobilization. RubisCO concentrations in these initial experiments were in the range 0.1–4 mg/mL in the final preparations. However, for both form I and form II RubisCO enzymes, using more than 2 mg/mL in the final preparation resulted in reduced enzyme activity recovery (Fig. 2). This could be attributed to enzyme precipitation, which resulted in the formation of pellets even with a low-speed centrifugation (8000*g* for 5 min at 4 °C). Notably, no RubisCO remained unassembled at loading concentrations less than or equal to 1 mg/mL, as indicated by the absence of protein or activity in the supernatants recovered after ultracentrifugation. Thus, the lower activity recovery at these concentrations likely reflects the limited surface availability of RubisCO active sites for substrate diffusion in these nanostructures. The addition of a non-specific protein, i.e., bovine serum albumin (BSA), to the samples with lower RubisCO loads resulted in higher activity recoveries typically observed with higher RubisCO loads (Fig. 2). In addition to its well-perceived role in stabilizing proteins, especially in dilute solutions, the smaller size (66.5 kDa, monomer) and the lower pI (4.7) of BSA likely results in displacement of RubisCO molecules from less accessible locations on the nanostructures, making more active sites accessible for activity determination.

Visual characterization of nanostructure-enzyme complexes

Transmission electron microscopy (TEM) was used to visually characterize the nanostructures. Dipeptides **A** and **B** formed uniform cylindrical open-ended nanotubes and compound **C** assembled into a network of β-sheet nanofibers (Fig. 3a). The distribution of enzyme molecules associated with the nanostructures was assessed by TEM using nanogold particles that were specifically bound to the enzymes. This was facilitated by the presence of multiple nickel-nitrilotriacetic acid functionalities

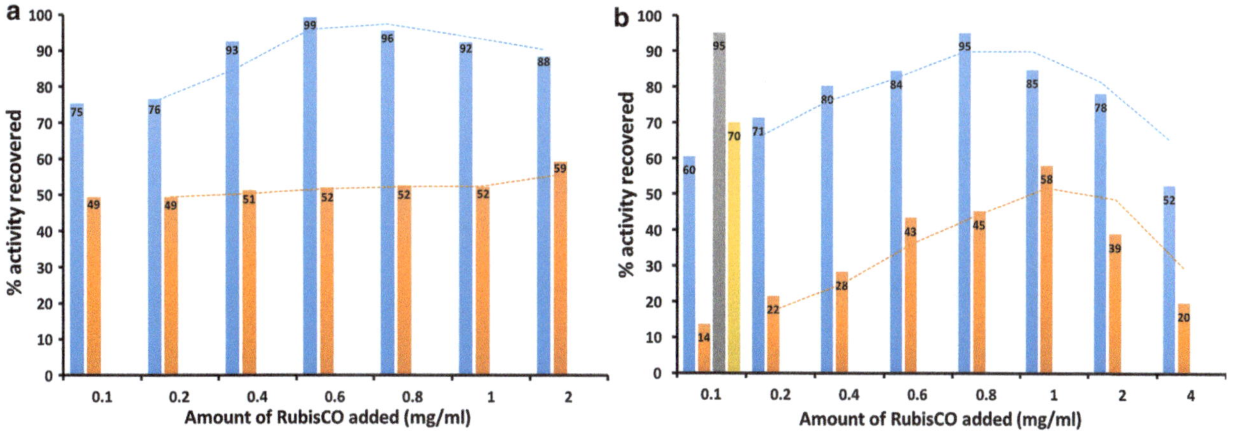

Fig. 2 RubisCO loading with fixed amounts of pre-assembled nanotube **A** (**a** 0.75 mg/mL) or nanofiber **C** (**b** 0.9 mg/mL). Plots show activities measured from nanostructure-RubisCO complexes that had been loaded with varying amounts of either form I (*blue*) or form II (*orange*) RubisCO enzymes. Recovery percentages were calculated relative to the enzyme activities in the corresponding samples with unbound RubisCO. Activities were also measured from identical form I (*gray*) or form II (*yellow*) nanofiber **C** preparations (**b**) that had been supplemented with 1 mg/mL bovine serum albumin (BSA) during RubisCO loading

on the nanogold particles that formed stable, high-resolution complexes with amino-terminal hexa-histidine tags of the recombinant microbial enzymes used in this study. After assembly, these nanostructure-enzyme complexes were visualized by standard TEM without the need for staining or signal enhancement. The density of visible nanogold spots clearly showed that the enzymes were bound to the surface of the nanotubes formed by **A**/**B** and the nanofibers formed by **C** (Fig. 3b).

In-situ production of RuBP within nanostructures using a cascade of enzymes

To establish a nanostructure-supported multienzyme array, we co-assembled enzymes that catalyze additional reactions in either the CBB cycle for CO_2 fixation (pathway 1, Fig. 4a) or the pentose phosphate pathway (pathway 2, Fig. 4b). The partial CBB-pathway reconstruction was accomplished by co-binding the enzymes phosphoriboseisomerase (PRI) and phosphoribulokinase (PRK) to the nanostructures, which would sequentially catalyze the conversion of ribose-5-phosphate (R-5-P) to ribulose-5-phosphate (Ru-5-P), and then to RuBP, with RuBP serving as the substrate for RubisCO in the nanostructures, resulting in the formation of 2 molecules of 3-phosphoglyceric acid (3-PGA) (Fig. 4a). The enzymes PRK and PRI were each added at a concentration of 0.1 mg/mL along with 1.0 mg/mL RubisCO for assembly into nanostructures. Using higher concentrations of PRK or PRI did not improve the reaction flux because the activity of RubisCO was rate-limiting. Pathway 2 (Fig. 4b) was designed to utilize glucose, a simpler and an industrially relevant feedstock, as the starting substrate. This involved the conversion of glucose to Ru-5-P utilizing the enzymes hexokinase (added at 0.05 mg/mL), glucose-6-phosphate dehydrogenase (added at 0.1 mg/mL), and 6-phosphogluconate dehydrogenase (added at 0.7 mg/mL) from the pentose phosphate pathway, and the subsequent conversion of Ru-5-P to 3-PGA using PRK (added at 0.05 mg/mL) and form II RubisCO (added at 0.4 mg/mL) (Fig. 4b). Because bacterial forms of all three CBB enzymes were used in engineered pathway 1, it was possible to express the genes (in *Escherichia coli*) and purify the resultant recombinant proteins containing *N*-terminal hexa-histidine tags, allowing for specific binding with nanogold particles for TEM imaging. The nanostructures associated with either nanogold-PRK or with both RubisCO and PRK each conjugated to different nanoparticles, all resulted in a nanogold-spot distribution pattern that was comparable to what had been observed with just RubisCO in the nanostructures (Additional file 1: Figure S2; Fig. 3). It is thus likely that the overall morphology of the nanostructure-enzyme complexes did not change in response to the type or number of enzymes associated.

Specific formation of stoichiometric amounts of 3-PGA in these reactions was confirmed and quantified using commercial enzymes that couple 3-PGA reduction to glyceraldehyde-3-phosphate formation with the concomitant oxidation of NADH, which could be followed spectrophotometrically. Also, the activities of individual enzymes bound to the nanostructures could be determined in independent assays utilizing the same nanostructure-enzyme complexes (Additional file 1: Table S3). The overall flux through all three enzymatic steps in pathway 1 ranged from 63 to 97% for the

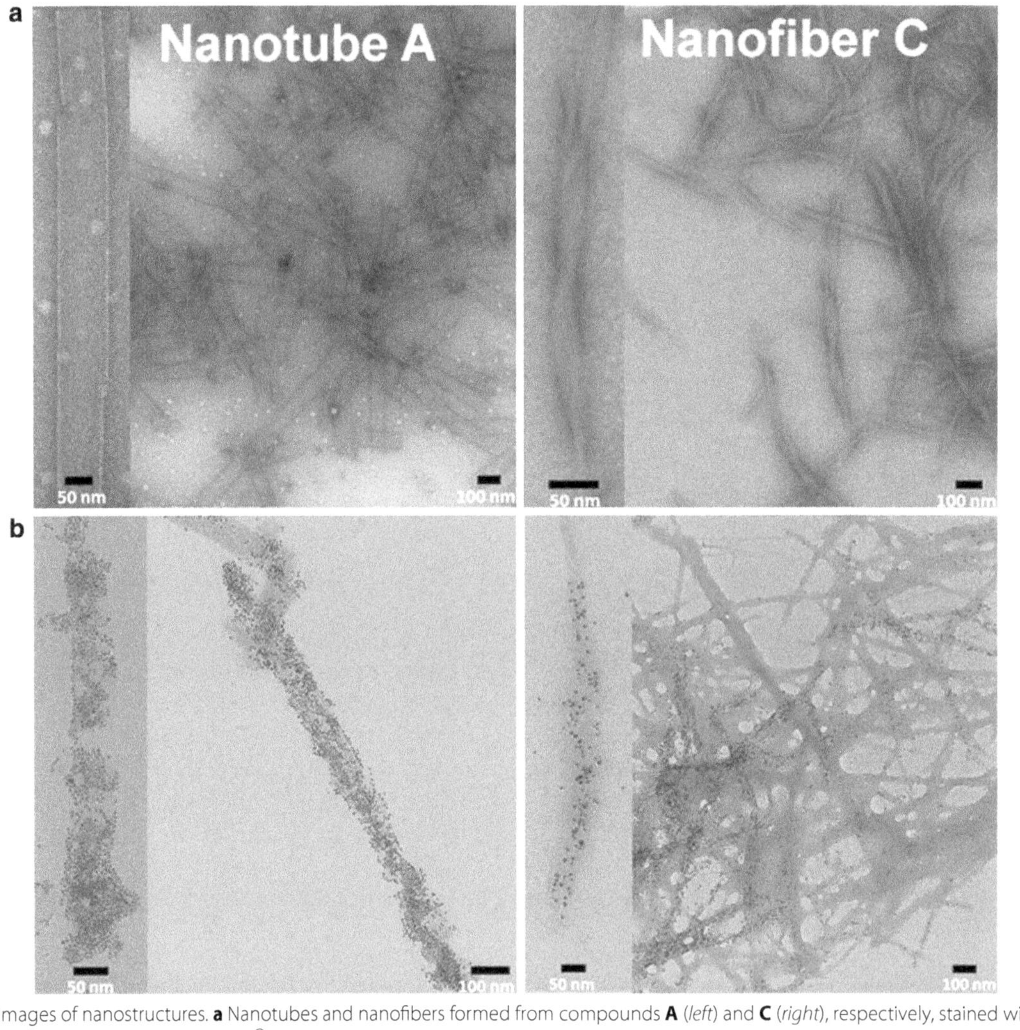

Fig. 3 TEM images of nanostructures. **a** Nanotubes and nanofibers formed from compounds **A** (*left*) and **C** (*right*), respectively, stained with uranyl acetate. **b** TEM images of Ni–NTA Nanogold® particles bound to the hexa-histidine tagged form I RubisCO and associated with either nanotube **A** (*left*) or nanofiber **C** (*right*). Images are representative of multiple samples imaged from independent preparations. For better clarity, a close-up view of a single nanostructure is shown to the *left* of each image

nanostructure-enzyme complexes relative to what was measured using an identical mixture of unbound enzymes (Table 1). Similarly, the pathway fluxes were also determined for the conversion of glucose to 3-PGA (Table 2). Because the steps in pathway 2 resulted in acidification of the reaction mixture, a higher buffer concentration was used. The activity of commercial 6-phosphogluconic dehydrogenase was found to be rate-limiting for this pathway. Although the overall glucose to 3-PGA pathway flux was much lower for the enzyme cascade in both the unbound and the nanostructure-bound samples (Table 2), it can be optimized further by controlling the pH of the reaction mixture and by using enzymes with higher specific activities. Furthermore, two-enzyme nanostructures containing both PRK and RubisCO in the same complex or as separate single-enzyme nanostructure preparations with only RubisCO or PRK, when subsequently mixed, both resulted in combined sequential activity (i.e., the conversion of Ru-5-P to RuBP and then to 3-PGA) (Additional file 1: Table S4). These observations suggest that different sets of enzymes associated with dissimilar nanostructures could be combined as separate interconnected modules to generate specific products.

Protease protection of nanostructure-associated enzymes

The accessibility of the nanostructure-bound enzymes was evaluated by treating the co-assembled structures with a protease (subtilisin). Whereas, none of the nanostructures could protect the larger, more complex form I enzyme from inactivation and presumably proteolytic degradation (Fig. 5a), high levels of activity were retained by the subtilisin-treated nanotube complexes of the

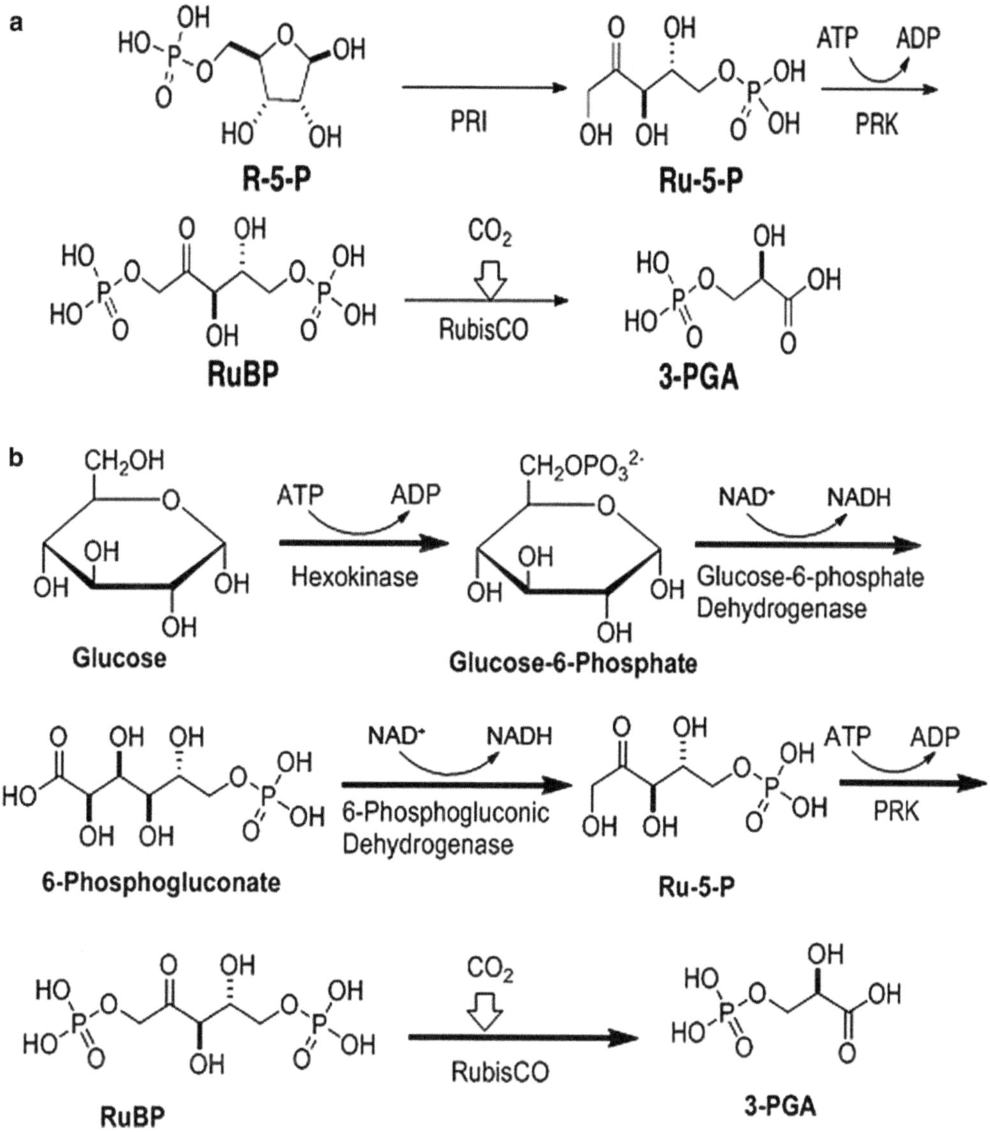

Fig. 4 Schematic of CO_2-fixation pathways assembled in nanostructures. Cascade of enzymatic steps employed to convert either ribose-5-phosphate (R-5-P) (pathway 1 **a**) or glucose (pathway 2 **b**) to 3-PGA

structurally simpler form II RubisCO or PRK enzymes (Fig. 5b, c). The larger form I RubisCO likely assembles with the nanostructures via weaker surface interactions or with a conformation that results in about the same (in the case of nanofiber **C**) or an even greater accessible surface area of the enzyme that is exposed to protease. However, the smaller sizes of the form II RubisCO and PRK enzymes likely ensures that the bulk of the enzyme surface is sequestered via interactions with the nanotube, thereby protecting these smaller proteins from proteolysis. Owing to the dimensions, nanofibers tend to have a greater area exposed relative to the nanotubes, which might increase the susceptibility of associated proteins to proteolytic degradation.

Enzymatic properties of nanostructure-bound RubisCO

The lower levels of activity exhibited by RubisCO typically limits and determines the CBB pathway flux [35]. Several studies have shown that important properties of various enzymes might be enhanced via immobilization or sequestration of some enzymes [33, 34]. Thus, it was of interest to determine if the association with nanostructures could alter the catalytic properties of RubisCO, the rate-limiting enzyme for global biological CO_2 fixation. Indeed, the association with nanostructures caused increases in the substrate K_m values (K_c for CO_2 and K_o for O_2) of the form I enzyme (Table 3). The higher increase in K_o values relative to K_c values resulted in significantly higher CO_2/O_2 specificity

Table 1 Activities measured for the conversion of R-5-P to 3-PGA in nanostructure-multienzyme complexes with *R. eutropha* form I (FI) or *R. rubrum* form II (FII) RubisCO, PRK and PRI

Enzyme cascade	Combined activity[a] (μmoles/min-mg)	% Activity ret.[b]
Unbound FI/PRK/PRI	4.08	100
FI/PRK/PRI-nanotube **A**	3.39	83
FI/PRK/PRI-nanotube **B**	3.49	86
FI/PRK/PRI-nanofiber **C**	3.94	97
Unbound FII, PRK and PRI	3.03	100
FII/PRK/PRI-nanotube **A**	2.45	81
FII/PRK/PRI-nanotube **B**	2.52	83
FII/PRK/PRI-nanofiber **C**	1.90	63

[a] Calculated from stable $^{14}CO_2$ fixed

[b] Activities retained are presented relative to an unbound enzyme mixture comprising all three enzymes. Activity numbers are from one of two independently prepared and assayed nanostructure complexes that yielded similar results

factors (Ω) for the form I enzyme that was associated with the nanostructures (Table 3; Fig. 6 and Additional file 1: Figure S3). Immobilization did not seem to significantly perturb the K_c, K_o, or the Ω values of the nanostructure-associated form II RubisCO (Table 3). Because the enzymes' k_{cat} values were determined based on the amount of enzyme added to the nanostructures, the values obtained for the nanostructure-bound enzymes were generally lower, reflecting the lower rates of enzyme accessibilities and activities recovered (Additional file 1: Table S3c; Table 3).

Discussion

Rising greenhouse gas emissions have prompted chemists, biologists, and engineers alike to explore strategies for the capture of atmospheric CO_2 and its conversion to useful products and biofuels [1, 4, 51–53]. Enzymes offer high fidelities for chemical conversions albeit being limited by structural stabilities and other metabolic constraints when operating in biological cells [5, 8]. Thus, exploring ways to assemble enzymes as part of cell-free biocatalytic systems has become a widespread endeavor, with an aim to engineer directed pathways for producing products of interest. Synthetic short peptide-based amphiphiles are promising precursors that self-assemble into diverse nanostructured materials. Simple and straight-forward procedures for synthesis, biodegradability, and modularity makes them uniquely suited for a multitude of applications. Self-assembling peptides have been used as scaffolds for tissue engineering, drug delivery, gene delivery, biosensing, and biomolecular signaling [30]. Our study demonstrated the utility of these scaffolds for CO_2-capture applications, which is likely to aid industrial strategies for carbon capture.

Enzyme immobilization in these scaffolds seems to occur via weak charge-based associations with the nanostructures. All the scaffolds that could functionally assemble with the enzymes carry a net positive charge at pH 8 used in these experiments. Correspondingly, the pI values of all the enzymes used in this study are less than 8. Using similar scaffolds with no net charge or those with net negative charge precluded functional assembly with enzymes. This generic charge-based self-assembly also allowed easy addition of multiple enzymes at desired loading ratios to the same scaffold. TEM images of gold-nanoparticle-tagged enzymes immobilized into these scaffolds provided visual proof of the enzymes present in these complexes (Fig. 3). The ability to assemble at least two different enzyme cascades into the same set of scaffolds demonstrated the applicability of this system for assembling larger pathways. Because enzymes present in multiple scaffolds could be brought together in assays for obtaining combined sequential activities (Additional file 1: Table S4), it should be possible to spatially separate charge-incompatible enzymes that are part of a pathway by placing them on different scaffolds and bringing them together during assays.

Table 2 Activities measured for the conversion of glucose to 3-PGA in nanostructure-multienzyme complexes with *R. rubrum* form II (FII) RubisCO

Enzyme cascade	Combined activity[a] (μmoles/min-mg)	% Activity ret.[b]
Unbound FII and glucose pathway[c]	0.259	100
FII and glucose pathway[c] in nanotube **A**	0.100	39
FII and glucose pathway[c] in nanotube **B**	0.110	42
FII and glucose pathway[c] in nanofiber **C**	0.115	44

[a] Calculated from stable $^{14}CO_2$ fixed

[b] Activities retained are presented relative to an unbound enzyme mixture comprising all enzymes. Activity numbers are from one of the two assays conducted with independently prepared and assayed nanostructure complexes, which gave similar results

[c] Glucose pathway comprises all enzymes that constitute pathway 2

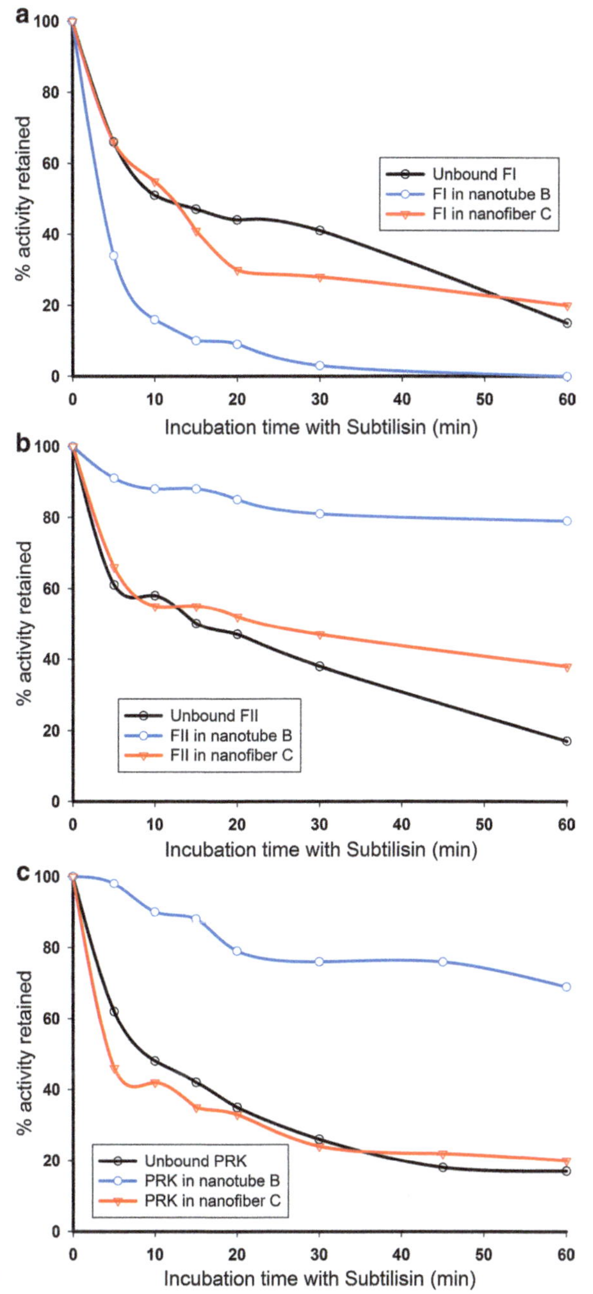

Fig. 5 Proteolytic sensitivities of RubisCO and PRK enzymes in nanostructure complexes. Unbound or nanostructure-associated RubisCO (**a**, **b**) or PRK (**c**) were treated with subtilisin for various times and residual enzymatic activities were measured for each time point. Data shown here is representative of two independent preparations that gave similar results. Nanotubes **A** or **B** provided identical levels of protection to all enzymes and hence the data is shown for only one of them (i.e., nanotube **B**), along with the data for nanofiber **C**

None of the nanostructures could impart improvements to thermal stability for any of the enzymes (data not shown), and smaller enzymes could be protected from proteolysis when associated with nanotubes (Fig. 5). The weak charge-based association that drives both the nanostructure assembly and the formation of nanostructure-enzyme complexes explains the temperature-instability of these complexes. Employing novel composites to capture nanostructures into higher order polymers [52] or the use of crosslinking [54] are potential strategies that could lead to further stabilization of the nanostructure-enzyme complexes. Protection from proteolysis could be attributable to a higher surface packing density and better sequestration of smaller enzymes associated with nanotubes. Presumably, a proportion of the sequestered form II RubisCO or PRK enzyme molecules retain accessibility to substrates and products (i.e., for activity measurements) despite being resistant to proteolytic cleavage. It is intriguing that the nanotube-associated form I enzyme is more susceptible to proteolytic cleavage when compared with the unbound enzyme (Fig. 5a). Whereas the access for subtilisin to unbound form I RubisCO molecules is diffusion controlled, the association with nanotubes and a lower surface packing density likely presents easy surface-access for subtilisin. The differences in surface packing densities may also explain the dissimilar pattern of form I versus form II RubisCO loading onto nanotubes (Fig. 2a). The protective effect is likely diminished for all enzymes associated with nanofiber **C** because the peptide sequence in compound **C** likely results in partial proteolytic degradation of the nanofiber itself (Fig. 5). Although protection from proteolysis seems irrelevant to industrial

Table 3 Catalytic properties of RubisCO enzymes (*R. eutropha* form I (FI) and *R. rubrum* form II (FII) associated with nanostructures

Enzyme[a]	Ω^b (V_cK_o/V_oK_c)	k^b_{cat} (sec^{-1})	K^b_c (μM)	K^b_o (μM)	K_o/K^c_c
Unbound FI	79 ± 1	2.6 ± 0.5	41 ± 4	1093 ± 141	27
FI-nanotube **B**	95 ± 3[d]	1.8 ± 0.1[e]	53 ± 7	1756 ± 341	33
FI-nanofiber **C**	88 ± 1[d]	2.2 ± 0.1[e]	56 ± 8	1693 ± 383	30
Unbound FII	17 ± 2	1.2 ± 0.1	100 ± 7	96 ± 26	1.0
FII-nanotube **B**	17 ± 1	0.6 ± 0.2[e]	120 ± 24	170 ± 41	1.4
FII in nanofiber **C**	17 ± 1	0.6 ± 0.1[e]	96 ± 16	95 ± 32	1.0

[a] Association with either of the nanotubes (i.e., **A** or **B**) imparted similar properties to both form I (FI) and form II (FII) enzymes

[b] Values are the mean ± standard deviation ($n-1$) of at least three independent nanostructure-enzyme complex preparations

[c] Calculated values

[d] An unpaired *t* test with Ω values obtained from three independent experiments gave p values of 0.0006 and 0.0003 for FI-nanotube **B** and FI-nanofiber **C**, respectively, relative to unbound FI

[e] These k_{cat} values were calculated based on the initial amounts of enzyme added to constitute nanostructure-RubisCO complexes. They are not significantly different from the value measured with the unbound enzyme, if the percentage recoveries are factored into the calculations (Additional file 1: Table S3c) for each experiment

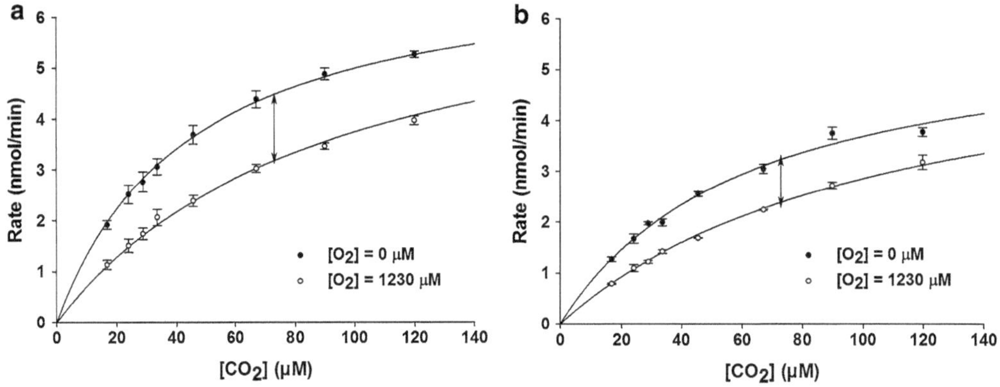

Fig. 6 Reaction velocity measurements of *R. eutropha* form I RubisCO as a function of CO_2 concentration in the presence (*open circles*) or absence (*closed circles*) of saturating levels of oxygen (i.e., 1230 μM). **a** Michaelis–Menten curves for carboxylation activities measured with unbound *R. eutropha* form I RubisCO. **b** Michaelis–Menten curves for carboxylation activities measured with nanotube **B**-form I RubisCO complexes. The extent of oxygen inhibition for each enzyme preparation is indicated (*double-headed arrows*). The K_O (K_i) for O_2 was enhanced about 1.6-fold for the nanotube **B**-form I enzyme complex compared to the unbound enzyme (see Table 3 for kinetic constants). The average value and *error bar* for each data point was calculated with values obtained from two independent assays (with different preparations) performed with identical CO_2 concentrations

scale up applications, our results provide a good measure of enzyme accessibility to other external agents that may be pertinent to specific applications.

A substantial and desirable outcome of immobilizing RubisCO in the nanostructures is the increased K_o of the form I enzyme, which leads to an increased CO_2/O_2 specificity factor (Ω) (Fig. 6; Table 3 and Additional file 1: Figure S6). This is the most significant enhancement of specificity factor thus far reported for any source of RubisCO and greatly exceeds specificity enhancements achieved via mutagenesis and bioselection [55–63]. Moreover, there was little or no compensatory decrease in the k_{cat} for the nanostructure-associated enzyme as often occurs with specificity-enhanced soluble mutant RubisCOs [58, 59]. The apparent increases in the Ω value of nanostructure-associated form I RubisCO is presumably a result of the altered electrochemical microenvironment surrounding the larger RubisCO molecules, which could result in preferentially occluding the paramagnetic O_2 (relative to CO_2) from entering RubisCO's active site. This would also explain the observed increases in the K_o/K_c ratios of the nanostructure-associated form I RubisCO (Table 3). Because the increases in Ω value is only observed with the form I RubisCO and not form II RubisCO, it must be concluded that this is not a result of preferential O_2-occlusion by these nanostructures. Small subunits are known to concentrate RubisCO's catalytic large subunits via ionic and hydrogen-bond interactions involving polar and charged surface residues of both subunits. Based on the structural analysis of several form I RubisCO structures, it is evident that amino acid residues involved in these interactions are directly connected to the active site via secondary structural elements. Thus, the conformational differences occurring during catalysis in the form I enzymes are induced and influenced by small subunits [64]. Other subtle structural differences between the form I and form II RubisCOs, including the presence of a central solvent channel in form I enzymes, have been correlated with differences in their substrate binding and a more "open" structure exhibited by the *R. rubrum* form II RubisCO [64]. Concentration of catalytic large subunits via association with small subunits or through other interactions leading to higher order structures like carboxysomes has been correlated to some form of structural and/or functional advantage [65]. In the absence of small subunits in the *R. rubrum* form II enzyme, surface residues that are usually solvent-exposed could be involved in interactions with the nanostructure, leading to a diluted but well-packed nanostructure-RubisCO matrix. The absence of modulatory structural elements and a more "open" conformation likely explain the unaltered Ω values measured for the form II enzyme present in these nanostructure complexes. Based on these results, it may be fruitful to systematically analyze the effect of immobilization (onto different scaffolds) on structurally divergent RubisCO enzymes. In addition, modulating the chemical structure of nanostructure-forming monomers to preferentially exclude O_2 from these complexes may result in further enhancement of RubisCO's Ω value. Likewise, co-immobilization with carbonic anhydrase might further enhance CO_2 availability around RubisCO molecules as in carboxysomes and analogous micro-compartments [44, 45].

Conclusions

A new application is described for simple self-assembling peptide-based nanostructures. We have demonstrated that the sequential enzymes that are part of a pathway for CO_2 utilization could be co-assembled into catalytically active nanotube- and nanofiber-supported multienzyme complexes. Further, the association with nanostructures appears to improve structure–function properties of the enzymes, with a significant enhancement of form I RubisCO's CO_2/O_2 selectivity attained. Given the fact that RubisCO is a key enzyme that accounts for most CO_2 fixed on earth, these results suggest that stable scaffolds may be prepared to encapsulate structurally complex enzymes along with RubisCO to constitute entire pathways for converting CO_2 into useful products. Further, the use of simple starting materials such as glucose allows for sampling a wide array of compounds, some of which could eventually be employed with alternate nanostructures possessing properties more conducive to efficient product formation. The use of enzymes from thermophilic organisms and cross-linkable nanostructures might also facilitate scale-up for industrial applications.

Methods

Synthesis of peptide conjugates

Peptide conjugates were manually prepared using Fmoc/*t*-Bu solid-phase peptide synthesis on Rink amide resin (loading 0.8 mmol/g). Amide-coupling steps were accomplished with standard techniques for all amino acids: Fmoc-amino acid, 1,3-diisopropylcarbodiimide (DIC), and 1-hydroxybenzotriazole (HOBt) (500 mol% each relative to resin) in 1:1 DMF/DCM for 1.5 h. A solution of 20% piperidine in DMF was used for Fmoc removal and 1% TFA in dichloromethane was used for 4-methyltrityl (Mtt) group de-protection. Peptide conjugates were cleaved from the resin by treatment with TFA/water/triethylsilane (95/1/4) at room temperature for 2 h. Crude peptides were precipitated with cold diethyl ether and purified by reversed-phase HPLC, using a preparative Varian Dynamax C-18 column and eluting with a linear gradient of acetonitrile/water containing 0.1% TFA (10/90 to 100/0 over 30 min). These precipitates were stored as lyophilized powers at 0 °C. Purity was assessed by analytical reverse-phase HPLC and identity confirmed using ESI–TOF mass spectrometry and NMR. All reactions were performed in an atmosphere of argon or nitrogen. ^{1}H NMR was recorded at 400 MHz and ^{13}C NMR spectra at 100 MHz on a Bruker DPX-400 instrument. Dipeptide conjugates **A** and **B** were synthesized as previously reported [26]. The tetrapeptide conjugate **C** was prepared using a similar procedure in which the ε-amino group of the carboxy-terminal lysine was reacted with benzoic anhydride (500 mol%), DIPEA (500 mol%) in DMF for 24 h (Additional file 1: Figure S4). Purity of the resultant compound was verified by HPLC fractionation (Additional file 1: Figure S5).

Characterization of Fmoc-KFKK(Bz)-NH_2 (compound C)

^{1}H NMR (400 MHz, DMSO-d_6) δ 8.46–8.43 (m, 1H), 8.21–8.17 (m, 1H), 7.91–7.81 (m, 5H), 7.74–7.64 (m, 6H), 7.54–7.31 (m, 8H), 7.25–7.12 (m, 4H), 7.05–7.02 (m, 1H), 4.60–4.56 (m, 1H), 4.35–4.16 (m, 4H), 3.93–3.88 (m, 1H), 3.29–3.21 (m, 2H), 3.09–3.04 (m, 1H), 2.83–2.67 (m, 5H), 1.72–1.64 (m, 2H), 1.60–1.38 (m, 9H), 1.38–1.24 (m, 4H); ^{13}C NMR (100 MHz, DMSO-d_6) 173.50, 171.63, 171.04, 170.92, 166.11, 158.27, 157.96, 143.84, 143.68, 140.71, 137.49, 134.62, 130.98, 129.24, 128.19, 127.89, 127.63, 127.09, 125.20, 120.12, 120.09, 65.59, 52.32, 46.66, 38.66, 31.84, 31.29, 28.89, 26.57, 22.81, 22.33, 22.32, 22.12; ESI–MS for $C_{49}H_{63}N_8O_7$ $[M+H]^+$ calculated 875.4820; found 875.4825.

Nanostructure preparation, isolation, and assembly with enzymes

For the preparation of nanostructures, monomeric compounds (**A**, **B**, or **C**) were added to 50 mM Bicine-NaOH, 10 mM $MgCl_2$, pH 8.0 (Bicine buffer) at concentrations of 5–20 mM, sonicated for 15 s using a Model W-385 Sonicator (Heat Systems-Ultrasonics, Inc.) and incubated at room temperature for 64–72 h to promote self-assembly. Nanostructure pellets were obtained by ultracentrifugation at 418,000*g* for 1 h at 4 °C and re-suspended to original volumes in Bicine buffer. These pellets were mixed with enzymes, incubated for 16–20 h at 4 °C, re-isolated using ultracentrifugation, and re-suspended in original volumes of Bicine buffer as before. All enzymes were expressed as recombinant proteins with N-terminal histidine tags using plasmid pET28a (Novagen). *Rhodospirillum rubrum* and *R. eutropha* RubisCOs were purified as described previously [66]. Following a similar procedure, PRK and PRI were purified by Ni–NTA affinity chromatography. The PRK gene was obtained from *Synechococcus* sp. PCC 7942 [67]. The genes encoding RubisCO and PRI enzymes were amplified from *R. eutropha* using appropriate primers from strain H16 (ATCC 17699).

Transmission electron microscopy (TEM) imaging

Images were obtained with a Technai G2 Spirit instrument operating at 80 kV. Samples were freshly diluted in Bicine buffer. Drops (10 μL) of the sample solution in Bicine buffer were applied to carbon coated copper grids (Ted Pella, Inc.) for 2 min before removing the excess solution with filter paper. Images of Ni–NTA Nanogold® particles (Nanoprobes, Inc.) bound to RubisCO or PRK, which were subsequently associated with the nanostructures, were obtained without the need for prior staining.

Nanogold particles were pre-bound to histidine-tagged proteins by following the procedures described elsewhere (Nanoprobes, Inc.). Ni–NTA Nanogold® particles mixed with just the nanostructures in the absence of histidine-tagged proteins provided evidence for the absence of non-specific association. For samples prepared without nanogold particles, uranyl acetate was used to stain the negatively-charged groups. To accomplish this, the grid containing the sample was floated on 10 μL drops of a 2% (w/v) uranyl acetate solution for 1 min before being imaged by TEM. Several replicates were imaged from at least 2 independent preparations to ensure that the imaging pattern was consistent for each sample.

Enzyme assays and proteolysis

All enzyme activities were determined using end-point assays involving radioisotopes as described previously [64]. RubisCO activity measurements were performed via incorporation of $NaH^{14}CO_3$ into acid-stable $[^{14}C]$ 3-PGA, with a 1:1 stoichiometry of substrate to product. The kinetic constants k_{cat}, K_c, and K_o were determined from simultaneous assays performed in vials flushed with either 100% N_2 or 100% O_2, as per the procedure described elsewhere [68] with some modifications. Each reaction received an appropriate amount of enzyme (unbound or in nanostructure complexes) that would fix 2.5–5 nmol of CO_2 per min, based on specific activity calculations. Unbound or nanostructure-bound RubisCO enzymes were pre-activated in identical volumes of Bicine buffer with 10 mM $NaHCO_3$ and 1 mM dithiothreitol (DTT). Each 250-μL reaction mixture comprised of 50 mM Bicine-NaOH, pH 8.0, 10 mM $MgCl_2$, 0.4 mM RuBP, and 8 levels of $NaH^{14}CO_3$, with corresponding CO_2 concentrations ranging from ~10 to 120 μM for form I RubisCO samples and ~25–450 μM for form II RubisCO samples. Reactions were initiated with the addition of activated enzyme and terminated after 1 min with the addition of 200 μL of 3 M formic acid in methanol. The samples were dried in an oven at 80 °C overnight, products re-dissolved in 250 μL 0.25 M HCl, and counted with 5 mL of EcoScint H scintillation cocktail (National Diagnostics). Results were plotted using Sigma Plot 12.0. The k_{cat} and K_c values were derived from Michaelis–Menton plots of data from assays performed anaerobically, i.e., under 100% N_2. The K_o [or K_i (O_2)] values were obtained with data from parallel assays performed in the presence (1.23 mM) and absence (0 mM) of O_2. The k_{cat} values for RubisCOs were calculated by noting the amount of free enzyme added to the nanostructures for forming the complexes because it was not possible to accurately determine the protein concentration within these complexes. The CO_2/O_2 specificity factor measurements involved the incorporation of $[^3H]$-radioactivity from $[1\text{-}^3H]$-RuBP into either a molecule of 3-PGA (carboxylation reaction) or a molecule of 2-phosphoglycolate (2-PG) (oxygenation reaction). These assays were performed with unbound or nanostructure-bound RubisCO (~30–50 μg) under saturating O_2 concentrations (1.23 mM) with 5 mM $NaHCO_3$ in Bicine buffer. The 400-μL reactions were initiated with the addition of $[1\text{-}^3H]$ RuBP (0.65 mM), and incubated at 23 °C for 1 h. The reactions were terminated by injecting 20 μL of 40 mM sodium borohydride. After 15 min, the excess sodium borohydride in the reaction was consumed with the addition of 20 μL of 160 mM glucose. After an additional 15 min, the reaction mixture was diluted with the addition of 1 mL of distilled water. The samples were then de-proteinated using Amicon Ultra Centrifugal filters (50,000 MWCO, Millipore). Reaction products present in 1-mL reaction mixtures were separated using an HPLC system (Shimadzu) on a 1-mL MonoQ column (GE Healthcare), and detected with an in-line scintillation counter (IN/US β-Ram). The ratio of areas in the 3-PGA and 2-PG peaks (corresponding to the V_c/V_o ratio), was used to calculate the specificity values in each case. $[1\text{-}^3H]$ RuBP was synthesized and purified as described elsewhere [69].

For measuring the flux through the pathways, the respective enzymatic activities were coupled to the RubisCO reaction (i.e., by following the incorporation of ^{14}C-label from $NaH^{14}CO_3$ into stable 3-PGA) with the addition of ATP and ribose-5-phosphate (R-5-P) (pathway 1), or with the addition of ATP, NADH, and glucose (pathway 2). Pathway 1 reactions comprised of 3.2 mM R-5-P, 6 mM ATP, 20 mM $NaH^{14}CO_3$, and approximately 3–40 μg of each pathway enzyme (unbound or in nanostructure complexes) in Bicine buffer. Pathway 2 reactions comprised of 100 mM glucose, 2 mM ATP, 2 mM NADH, 20 mM $NaH^{14}CO_3$, 2 mM DTT, and approximately 3–50 μg of each pathway enzyme (unbound or in complex with nanostructures).

For the determination of proteolytic susceptibilities, the nanostructure-RubisCO complexes were incubated with fixed amounts of subtilisin (molar ratios of 1 subtilisin per 2000 RubisCO molecules or 1 per 667 PRK molecules) for various times, followed by the addition of PMSF to arrest proteolysis [70]. The samples were immediately placed on ice and residual RubisCO or PRK activity was measured. PRK activity was measured by following the incorporation of $NaH^{14}CO_3$ into acid-stable $[^{14}C]$ 3-PGA, in coupled reactions with an excess of RubisCO. For each enzyme, the recovered activity percentages were plotted as a function of incubation time with subtilisin.

Additional file

Additional file 1: Figure S1. Inhibition of RubisCO activity with monomeric compounds **A**, **B**, and **C**. **Figure S2.** TEM images of nanogold-labeled histidine-tagged RubisCO and PRK enzymes associated with nanostructures. **Figure S3.** HPLC fractionation profiles of 3H-labeled carboxylation-specific 3-PGA and oxygenation-specific 2-PG measured from samples with unbound and nanostructure-form I RubisCO complex. **Figure S4.** Schematic showing steps involved in the synthesis of compound **C**. **Figure S5.** Purity of compound **C**. **Table S1.** Effect of washing nanostructure-RubisCO complexes with buffer in the presence or absence of salt. **Table S2.** RubisCO activity recoveries obtained with varying concentrations of nanostructures. **Table S3.** Activities of individual enzymes in three-enzyme nanostructure complexes. **Table S4.** Combined activity of PRK and RubisCO enzymes present in the same nanostructure complex, or in mixtures with separate single-enzyme nanostructures.

Abbreviations

RubisCO: ribulose 1,5-bisphosphate carboxylase/oxygenase; RuBP: ribulose 1,5-bisphosphate; R-5-P: ribose-5-phosphate; Ru-5-P: ribulose-5-phosphate; CBB cycle: Calvin–Benson–Bassham cycle; CA: carbonic anhydrase; BSA: bovine serum albumin; TEM: transmission electron microscopy; PRI: phosphoriboseisomerase; PRK: phosphoribulokinase; 3-PGA: 3-phosphoglyceric acid; 2-PG: 2-phosphoglycolate; NADH: reduced nicotinamide adenine dinucleotide; ATP: adenosine triphosphate; K_c: Michaelis constant (K_m) for CO_2; K_o: Michaelis constant for oxygen; K_{RuBP}: Michaelis constant for RuBP; V_c: V_{max} value for carboxylation; V_o: V_{max} value for oxygenation; Ω: RubisCO's CO_2/O_2 specificity factor (V_cK_o/V_oK_c); Fmoc: fluorenylmethyloxycarbonyl; DMF: dimethyl formamide; DCM: dichloromethane; TFA: trifluoroacetic acid; HPLC: high-performance liquid chromatography; ESI–TOF: electrospray ionization time-of-flight; DIPEA: N,N-diisopropylethylamine; Ni–NTA: nickel-nitrilotriacetic acid; PMSF: phenylmethane sulfonyl fluoride.

Authors' contributions

FRT, JRP, and SS conceived and coordinated the study. SS and YS performed all the experiments and analyzed the results with JRP and FRT. The manuscript was written by SS with contributions from all authors. All authors read and approved the final manuscript.

Acknowledgements

We gratefully acknowledge the Center for Applied Plant Sciences (CAPS) at The Ohio State University for partial funding of this work. This work was also financially supported by NIH Grant GM095742 (awarded to FRT) and NSF Grant CHE-1412295 (awarded to JRP).

Competing interests

The authors declare that they have no competing interests.

References

1. Dutcher B, Fan M, Russell AG. Amine-based CO_2 capture technology development from the beginning of 2013—a review. ACS Appl Mater Interfaces. 2015;7:2137–48.
2. McGlade C, Ekins P. The geographical distribution of fossil fuels unused when limiting global warming to 2 °C. Nature. 2015;517:187–90.
3. Ozin GA. Throwing new light on the reduction of CO_2. Adv Mater. 2015;27:1957–63.
4. Schwander T, von Borzyskowski LS, Burgener S, Cortina NS, Erb TJ. A synthetic pathway for the fixation of carbon dioxide in vitro. Science. 2016;354:900–4.
5. Dudley QM, Karim AS, Jewett MC. Cell-free metabolic engineering: biomanufacturing beyond the cell. Biotechnol J. 2015;10:69–82.
6. Fessner WD. Systems biocatalysis: development and engineering of cell-free "artificial metabolisms" for preparative multi-enzymatic synthesis. Nat Biotechnol. 2015;32:658–64.
7. Hodgman CE, Jewett MC. Cell-free synthetic biology: thinking outside the cell. Metab Eng. 2012;14:261–9.
8. Wolfenden R. Massive thermal acceleration of the emergence of primordial chemistry, the incidence of spontaneous mutation, and the evolution of enzymes. J Biol Chem. 2014;289:30198–204.
9. Kay JE, Jewett MC. Lysate of engineered *Escherichia coli* supports high-level conversion of glucose to 2,3-butanediol. Metab Eng. 2015;32:133–42.
10. Agapakis CM, Boyle PM, Silver PA. Natural strategies for the spatial optimization of metabolism in synthetic biology. Nat Chem Biol. 2012;8:527–35.
11. Lopez-Gallego F, Schmidt-Dannert C. Multi-enzymatic synthesis. Curr Opin Chem Biol. 2010;14:174–83.
12. Oroz-Guinea I, Garcia-Junceda E. Enzyme catalyzed tandem reactions. Curr Opin Chem Biol. 2013;17:236–49.
13. Grotzky A, Nauser T, Erdogan H, Schluter AD, Walde P. A fluorescently labeled dendronized polymer–enzyme conjugate carrying multiple copies of two different types of active enzymes. J Am Chem Soc. 2012;134:11392–5.
14. Schoffelen S, van Hest JC. Chemical approaches for the construction of multi-enzyme reaction systems. Curr Opin Struct Biol. 2013;23:613–21.
15. Schoffelen S, Beekwilder J, Debets MF, Bosch D, van Hest JC. Construction of a multifunctional enzyme complex via the strain-promoted azide–alkyne cycloaddition. Bioconjug Chem. 2013;24:987–96.
16. Peters RJ, Marguet M, Marais S, Fraaije MW, van Hest JC, Lecommandoux S. Cascade reactions in multicompartmentalized polymersomes. Angew Chem Int Ed Engl. 2014;53:146–50.
17. Qu R, Shen L, Qu A, Wang R, An Y, Shi L. Artificial peroxidase/oxidase multiple enzyme system based on supramolecular hydrogel and its application as a biocatalyst for cascade reactions. ACS Appl Mater Interfaces. 2015;7:16694–705.
18. Srere PA, Mattiasson B, Mosbach K. An immobilized three-enzyme system: a model for microenvironmental compartmentation in mitochondria. Proc Natl Acad Sci USA. 1973;70:2534–8.
19. Sun J, Ge J, Liu W, Lan M, Zhang H, Wang P, et al. Multi-enzyme co-embedded organic–inorganic hybrid nanoflowers: synthesis and application as a colorimetric sensor. Nanoscale. 2014;6:255–62.
20. Kang W, Liu J, Wang J, Nie Y, Guo Z, Xia J. Cascade biocatalysis by multienzyme-nanoparticle assemblies. Bioconjug Chem. 2014;25:1387–94.
21. Ding S, Cargill AA, Medintz IL, Claussen JC. Increasing the activity of immobilized enzymes with nanoparticle conjugation. Curr Opin Biotechnol. 2015;34:242–50.
22. Shankar S, Soni SK, Daima HK, Selvakannan PR, Khire JM, Bhargava SK, et al. Charge-switchable gold nanoparticles for enhanced enzymatic thermostability. Phys Chem Chem Phys. 2015;17:21517–24.
23. Farokhzad OC, Langer R. Impact of nanotechnology on drug delivery. ACS Nano. 2009;3:16–20.
24. Griffith LG, Naughton G. Tissue engineering—current challenges and expanding opportunities. Science. 2002;295:1009–14.
25. Hartgerink JD, Beniash E, Stupp SI. Self-assembly and mineralization of peptide-amphiphile nanofibers. Science. 2001;294:1684–8.
26. Kim SH, Kaplan JA, Sun Y, Shieh A, Sun HL, Croce CM, et al. The self-assembly of anticancer camptothecin-dipeptide nanotubes: a minimalistic and high drug loading approach to increased efficacy. Chemistry. 2015;21:101–5.
27. Petrov A, Audette GF. Peptide and protein-based nanotubes for nanobiotechnology. Wiley Interdiscip Rev Nanomed Nanobiotechnol. 2012;4:575–85.
28. Sun Y, Kaplan JA, Shieh A, Sun HL, Croce CM, Grinstaff MW, et al. Self-assembly of a 5-fluorouracil-dipeptide hydrogel. Chem Commun. 2016;52:5254–7.
29. Wiradharma N, Tong YW, Yang YY. Self-assembled oligopeptide nanostructures for co-delivery of drug and gene with synergistic therapeutic effect. Biomaterials. 2009;30:3100–9.
30. Seabra AB, Durán N. Biological applications of peptides nanotubes: an overview. Peptides. 2013;39:47–54.
31. Huang A, Qin G, Olsen BD. Highly active biocatalytic coatings from protein-polymer diblock copolymers. ACS Appl Mater Interfaces. 2015;7:14660–9.
32. Kameta N, Masuda M, Shimizu T. Soft nanotube hydrogels functioning as artificial chaperones. ACS Nano. 2012;6:5249–58.

33. Lu Q, Kim Y, Bassim N, Raman N, Collins GE. Catalytic activity and thermal stability of horseradish peroxidase encapsulated in self-assembled organic nanotubes. Analyst. 2016;141:2191–8.
34. Zhao G, Fu J, Dhakal S, Johnson-Buck A, Liu M, Zhang T, et al. Nanocaged enzymes with enhanced catalytic activity and increased stability against protease digestion. Nat Commun. 2016;7:10619.
35. Tabita FR, Satagopan S, Hanson TE, Kreel NE, Scott SS. Distinct form I, II, III, and IV RubisCO proteins from the three kingdoms of life provide clues about RubisCO evolution and structure/function relationships. J Exp Bot. 2008;59:1515–24.
36. Durao P, Aigner H, Nagy P, Mueller-Cajar O, Hartl FU, Hayer-Hartl M. Opposing effects of folding and assembly chaperones on evolvability of RubisCO. Nat Chem Biol. 2015;11:148–55.
37. Goloubinoff P, Christeller JT, Gatenby AA, Lorimer GH. Reconstitution of active dimeric ribulose bisphosphate carboxylase from an unfolded state depends on two chaperonin proteins and Mg-ATP. Nature. 1989;342:884–9.
38. Luo S, Wang ZY, Kobayashi M, Nozawa T. The dimerization of folded monomers of ribulose 1,5-bisphosphate carboxylase/oxygenase. J Biol Chem. 2001;276:7023–6.
39. Studer RA, Christin PA, Williams MA, Orengo CA. Stability-activity tradeoffs constrain the adaptive evolution of RubisCO. Proc Natl Acad Sci USA. 2014;111:2223–8.
40. Wheatley NM, Sundberg CD, Gidaniyan SD, Cascio D, Yeates TO. Structure and identification of a pterin dehydratase-like protein as a ribulose-bisphosphate carboxylase/oxygenase (RubisCO) assembly factor in the α-carboxysome. J Biol Chem. 2014;289:7973–81.
41. Feiz L, Williams-Carrier R, Belcher S, Montano M, Barkan A, Stern DB. A protein with an inactive pterin-4a-carbinolamine dehydratase domain is required for RubisCO biogenesis in plants. Plant J. 2014;80:862–9.
42. Guadalupe-Medina V, Wisselink HW, Luttik MA, de Hulster E, Daran JM, Pronk JT, et al. Carbon dioxide fixation by Calvin-Cycle enzymes improves ethanol yield in yeast. Biotechnol Biofuels. 2013;6:125.
43. Iancu CV, Morris DM, Dou Z, Heinhorst S, Cannon GC, Jensen GJ. Organization, structure, and assembly of α-carboxysomes determined by electron cryotomography of intact cells. J Mol Biol. 2010;396:105–17.
44. Kerfeld CA, Erbilgin O. Bacterial microcompartments and the modular construction of microbial metabolism. Trends Microbiol. 2015;23:22–34.
45. Frey R, Mantri S, Rocca M, Hilvert D. Bottom-up construction of a primordial carboxysome mimic. J Am Chem Soc. 2016;138:10072–5.
46. Shao H, Seifert J, Romano NC, Gao M, Helmus JJ, Jaroniec CP, et al. Amphiphilic self-assembly of an n-type nanotube. Angew Chem Int Ed Engl. 2010;49:7688–91.
47. Kim SH, Parquette JR. A model for the controlled assembly of semiconductor peptides. Nanoscale. 2012;4:6940–7.
48. Schneider G, Lindqvist Y, Lundqvist T. Crystallographic refinement and structure of ribulose-1,5-bisphosphate carboxylase from *Rhodospirillum rubrum* at 1.7 Å resolution. J Mol Biol. 1990;211:989–1008.
49. Branden CI, Schneider G, Lindqvist Y, Andersson I, Knight S, Lorimer GH. X-ray structural studies of RubisCO from *Rhodospirillum rubrum* and spinach. Phil Trans R Soc Lond B. 1986;313:359–65.
50. Hansen S, Vollan VB, Hough E, Andersen K. The crystal structure of RubisCO from Alcaligenes eutrophus reveals a novel central eight-stranded β-barrel formed by β-strands from four subunits. J Mol Biol. 1999;288:609–21.
51. He H, Perman JA, Zhu G, Ma S. Metal-organic frameworks for CO_2 chemical transformations. Small. 2016;12:6309–24.
52. Hery TM, Satagopan S, Northcutt RG, Tabita FR, Sundaresan VB. Polypyrrole membranes as scaffolds for biomolecule immobilization. Smart Mater Struct. 2016;25:125033.
53. Maitland GC. Carbon capture and storage: concluding remarks. Faraday Discuss. 2016;192:581–99.
54. Flood D, Proulx C, Robertson EJ, Battigelli A, Wang S, Schwartzberg AM, et al. Improved chemical and mechanical stability of peptoid nanosheets by photo-crosslinking the hydrophobic core. Chem Commun. 2016;52:4753–6.
55. Bar-Even A, Noor E, Lewis NE, Milo R. Design and analysis of synthetic carbon fixation pathways. Proc Natl Acad Sci USA. 2010;107:8889–94.
56. Mueller-Cajar O, Morell M, Whitney SM. Directed evolution of RubisCO in *Escherichia coli* reveals a specificity-determining hydrogen bond in the form II enzyme. Biochemistry. 2007;46:14067–74.
57. Kreel NE, Tabita FR. Serine 363 of a hydrophobic region of archaeal ribulose 1,5-bisphosphate carboxylase/oxygenase from *Archaeoglobus fulgidus* and *Thermococcus kodakaraensis* affects CO_2/O_2 substrate specificity and oxygen sensitivity. PLoS ONE. 2015;10:e0138351.
58. Madgwick PJ, Parmar S, Parry MA. Effect of mutations of residue 340 in the large subunit polypeptide of RubisCO from *Anacystis nidulans*. Eur J Biochem. 1998;253:476–9.
59. Harpel MR, Hartman FC. Enhanced CO_2/O_2 specificity of a site-directed mutant of ribulose-bisphosphate carboxylase/oxygenase. J Biol Chem. 1992;267:6475–8.
60. Satagopan S, Spreitzer RJ. Plant-like substitutions in the large-subunit carboxy terminus of *Chlamydomonas* RubisCO increase CO_2/O_2 specificity. BMC Plant Biol. 2008;8:85.
61. Smith SA, Tabita FR. Positive and negative selection of mutant forms of prokaryotic (cyanobacterial) ribulose-1,5-bisphosphate carboxylase/oxygenase. J Mol Biol. 2003;331:557–69.
62. Smith SA, Tabita FR. Glycine 176 affects catalytic properties and stability of the *Synechococcus* sp. strain PCC6301 ribulose-1,5-bisphosphate carboxylase/oxygenase. J Biol Chem. 2004;279:25632–7.
63. Spreitzer RJ, Peddi SR, Satagopan S. Phylogenetic engineering at an interface between large and small subunits imparts land-plant kinetic properties to algal RubisCO. Proc Natl Acad Sci USA. 2005;102:17225–30.
64. Gunn LH, Valegard K, Andersson I. A unique structural domain in *Methanococcoides burtonii* RubisCO acts as a small-subunit mimic. J Biol Chem. 2017;292:6838–50.
65. Schneider G, Knight S, Andersson I, Branden C-I, Lindqvist Y, Lundqvist T. Comparison of the crystal structures of L_2 and L_8S_8 RubisCO suggests a functional role for the small subunit. EMBO J. 1990;9:2045–50.
66. Satagopan S, Chan S, Perry LJ, Tabita FR. Structure-function studies with the unique hexameric form II ribulose-1,5-bisphosphate carboxylase/oxygenase (RubisCO) from *Rhodopseudomonas palustris*. J Biol Chem. 2014;289:21433–50.
67. Kobayashi D, Tamoi M, Iwaki T, Shigeoka S, Wadano A. Molecular characterization and redox regulation of phosphoribulokinase from the cyanobacterium *Synechococcus* sp. PCC 7942. Plant Cell Physiol. 2003;44:269.
68. Chen Z, Chastain CJ, Al-Abed SR, Chollet R, Spreitzer RJ. Reduced CO_2/O_2 specificity of ribulose-1,5-bisphosphate carboxylase/oxygenase in a temperature-sensitive chloroplast mutant of *Chlamydomonas reinhardtii*. Proc Natl Acad Sci USA. 1988;85:4696–9.
69. Kuehn GD, Hsu TC. Preparative-scale enzymic synthesis of D-[^{14}C]ribulose 1,5-bisphosphate. Biochem J. 1978;175:909–12.
70. Marin-Navarro J, Moreno J. Modification of the proteolytic fragmentation pattern upon oxidation of cysteines from ribulose 1,5-bisphosphate carboxylase/oxygenase. Biochemistry. 2003;42:14930–8.

Highly efficient methane generation from untreated microalgae biomass

Viktor Klassen[1], Olga Blifernez-Klassen[1], Daniel Wibberg[1], Anika Winkler[1], Jörn Kalinowski[1], Clemens Posten[2] and Olaf Kruse[1*]

Abstract

Background: The fact that microalgae perform very efficiently photosynthetic conversion of sunlight into chemical energy has moved them into the focus of regenerative fuel research. Especially, biogas generation via anaerobic digestion is economically attractive due to the comparably simple apparative process technology and the theoretical possibility of converting the entire algal biomass to biogas/methane. In the last 60 years, intensive research on biogas production from microalgae biomass has revealed the microalgae as a rather challenging substrate for anaerobic digestion due to its high cell wall recalcitrance and unfavorable protein content, which requires additional pretreatment and co-fermentation strategies for sufficient fermentation. However, sustainable fuel generation requires the avoidance of cost/energy intensive biomass pretreatments to achieve positive net-energy process balance.

Results: Cultivation of microalgae in replete and limited nitrogen culture media conditions has led to the formation of protein-rich and low protein biomass, respectively, with the last being especially optimal for continuous fermentation. Anaerobic digestion of nitrogen limited biomass (low-N BM) was characterized by a stable process with low levels of inhibitory substances and resulted in extraordinary high biogas, and subsequently methane productivity [750 ± 15 and 462 ± 9 $mL_N\,g^{-1}$ volatile solids (VS) day^{-1}, respectively], thus corresponding to biomass-to-methane energy conversion efficiency of up to 84%. The microbial community structure within this highly efficient digester revealed a clear predominance of the phyla *Bacteroidetes* and the family *Methanosaetaceae* among the Bacteria and Archaea, respectively. The fermentation of replete nitrogen biomass (replete-N BM), on the contrary, was demonstrated to be less productive (131 ± 33 $mL_N\,CH_4\,g^{-1}VS\,day^{-1}$) and failed completely due to acidosis, caused through high ammonia/ammonium concentrations. The organization of the microbial community of the failed (replete-N) digester differed greatly compared to the stable low-N digester, presenting a clear shift to the phyla *Firmicutes* and *Thermotogae*, and the archaeal population shifted from acetoclastic to hydrogenotrophic methanogenesis.

Conclusions: The present study underlines the importance of cultivation conditions and shows the practicability of microalgae biomass usage as mono-substrate for highly efficient continuous fermentation to methane without any pretreatment with almost maximum practically achievable energy conversion efficiency (biomass to methane).

Keywords: Biofuel, Biogas, Methane, Microalgae mono-substrate, Nitrogen limitation, Continuous anaerobic fermentation/digestion, Maximal energy conversion efficiency, Microbial community, Ammonia/ammonium inhibition

*Correspondence: olaf.kruse@uni-bielefeld.de
[1] Department of Biology/Center for Biotechnology (CeBiTec), Bielefeld University, Universitätsstrasse 27, 33615 Bielefeld, Germany
Full list of author information is available at the end of the article

Background

The steadily increasing global energy demand and limited fossil fuel sources have created tremendous efforts in developing renewable energy sources [1, 2]. Third generation biofuels, specifically derived from microalgae, are considered to be a viable alternative energy resource [3–5] because they can grow on non-arable land using fresh, saline or waste water and produce large amounts of lipids, proteins and carbohydrates over short periods of time, which can be processed into biofuels (e.g., biodiesel, bioethanol, hydrogen, methane) and valuable co-products [5–9]. Microalgae are often described as "lower" plants that never have true stems, roots, and leaves, and grow photoautotrophically by performing oxygenic photosynthesis [10], achieving biomass productivities of up to 91 tons ha^{-1} $year^{-1}$ with relatively simple cultivation systems such as raceway ponds [11, 12]. And although mass production of microalgae is still expensive at the moment, because of their high theoretical and practical areal productivities, microalgae are in focus of research for biofuel production [3, 9, 13]. Nowadays, however, the generation of biofuels such as biodiesel or bioethanol is not economically relevant, due to the currently incurred costs for biomass production and downstream processing [13].

Methane generation via anaerobic fermentation represents an alternative way, generating gaseous fuel from biomass. Anaerobic digestion (AD) is simple in application and highly efficient, since up to 88% conversion efficiency can be reached with the appropriate substrate [14]. AD is widely used for fermentation of the so-called energy crops and organic waste material to gain methane, which is used as fuel or for electricity and heat generation [15–17]. Nevertheless, today microalgae biomass is not regarded as suitable substrate for biogas generation in AD process mainly for two reasons: (1) high recalcitrance towards microbial decomposition mediated by the rigid cell wall, and (2) unfavorable low carbon-to-nitrogen (C/N) ratio of the biomass caused by a high protein content [18, 19]. The resistance of the cell wall can be overcome by application of physical and enzymatical pretreatments [20–24], hereby unfortunately increasing investment costs for biomass processing. Additionally, the continuous fermentation of this pretreated, and thus completely accessible biomass as mono-substrate was shown to be not efficient [19], mainly due to ammonia inhibition of methanogens, released from protein degradation [20, 24–26]. To avoid the C/N imbalance of the substrate, co-fermentation with other carbon-rich substrates represents one possible strategy [27, 28]. Alternatively, some research was performed in the past, for the reduction of the protein content in the biomass by applying limited amounts of nitrogen or phosphate to the culture media [29, 30]. This strategy seem to be favoring not only lower protein content but also the accessibility of algae to microbial communities, which was monitored by methane potential tests and intact cells counting before and after the batch fermentation process [30]. Microalgae belonging to three different genera *Chlamydomonas*, *Chlorella* and *Scenedesmus* revealed with ongoing starvation status higher C/N ratios (24–26, on weight basis) in the biomass and lost subsequently the capability to resist the bacterial degradation, leading consequently to higher methane yields [30] with conversions rates near the theoretical maximum [19].

However, these experiments were performed in batch fermentation mode, allowing conclusions only regarding the accessibility of biomass towards anaerobic degradability and the achievement of maximal possible methane yields. In a regular case (industrial scale), fermentation of biomass is performed in a continuous or semi-continuous mode since this is more efficient regarding volumetric productivity. In this mode, other factors besides biodegradability can play a crucial role, e.g., ammonia or ammonium inhibition (often caused by high protein content), long chain fatty acid inhibition (caused by high lipid content), enrichment of toxic compounds and unbalance of macro/micro nutrients (necessary for growth of microbial community) [13, 15, 31]. Additionally, a variety of process parameters [hydraulic retention time (HRT), organic loading rate (OLR), temperature, pH] has to be considered for optimal performance of the digester, to avoid a complete failure of the process [19, 31].

The present study was aiming to prove the feasibility of microalgae biomass as mono-substrate, derived from nitrogen-limited growth conditions, in a long-term continuous fermentation process.

Results and discussion

Algae cultivation and resulting biomass properties

In previous work, it was elucidated that the composition and the recalcitrance of microalgae biomass strongly depends on the growth conditions, in particular on nutrient availability and harvesting time [30]. To highlight the importance of nutrient availability, microalgae (*Chlamydomonas reinhardtii* CC-1690) biomass for the continuous fermentation was generated using cultivation media with two different nitrogen concentrations (replete-N with 11.77 mM nitrogen and low-N with 3.56 mM nitrogen, supplied as $NaNO_3$). In addition, to avoid changes in biomass characteristics due to storage artifacts, e.g., freezing [32] or drying [33], algae biomass was cultured parallel to the fermentation experiments. The growth of the microalgae biomass in photobioreactors was periodically monitored by measuring organic biomass concentration (Fig. 1). According to the results

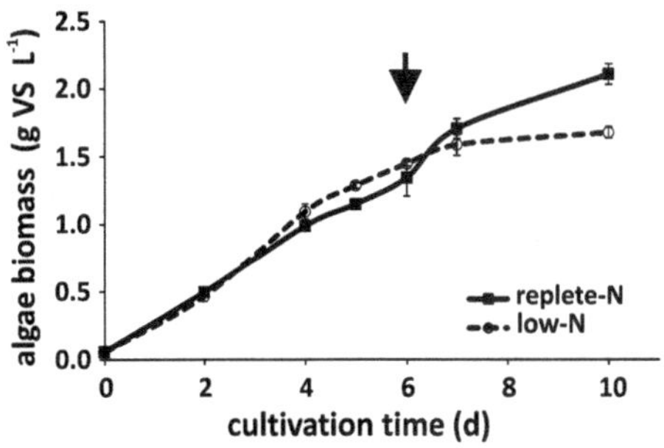

Fig. 1 Photoautotrophic accumulation of algal biomass under replete-N and low-N culture conditions. Harvesting for fermentation experiments was performed at day 6 for both media conditions (indicated by *arrow*). *VS* volatile solids

from previous work [30], biomass harvesting was always performed after 6 days of cultivation for both conditions.

The phototrophic algae, cultivated in culture media with low and replete nitrogen concentrations, showed no significant differences in biomass accumulation rates at the harvesting time (Fig. 1, 6 days). After 7 days of cultivation, an obvious starvation of biomass accumulation could be monitored in low-N media, due to nitrogen depletion. In accordance with the expectation, biomass accumulation was observed in replete-N conditions up to day 10. Conclusively, no obvious disadvantages in biomass productivity (until day 6, harvesting time point) could be observed after the application of nitrogen limiting culturing conditions (Fig. 1).

The biomass composition of *C. reinhardtii* cultivated under replete-N and low-N conditions revealed significant differences regarding the protein and almost no difference in lipid content (Table 1), which is consistent with earlier observations on the total lipid and carbohydrate (mainly starch) content in *C. reinhardtii* CC-1690 under nitrogen deprivation [34, 35]. Consequently, carbohydrates represent the main carbon sink in nitrogen starved *C. reinhardtii* cells.

Based on biomass composition, the theoretical methane potential was calculated using the Buswell equation [36] and empirical formula stated by Heaven et al. [37] and revealed no significant difference with approximately 551 and 549 $mL_N\ g^{-1}$ VS between replete-N and low-N biomass, respectively (Table 1). Furthermore, corresponding to 2.2-fold lower protein content, the concentration of elemental nitrogen in the low-N biomass was decreased to only 2.9 ± 0.2% of dry weight (DW), whereas the nitrogen amount in the replete-N conditions resulted in 7.3 ± 0.7% of DW. This finding has a direct impact on the C/N ratio in the biomass, which is one of the most critical factors for a continuous fermentation process (C/N ratio: replete-N = 6.9 ± 0.7, low-N = 16.3 ± 1.1, Table 1) [38, 39]. In this particular case, the C/N ratio of the biomass, cultured under low-N conditions was within the range of 15–30, which is generally regarded as optimal for fermentation processes [15, 39, 40].

Table 1 Microalgae biomass characteristics

	Replete-N BM	Low-N BM
Proteins (% DW)	61.0 ± 5.1	28.0 ± 3.1
Carbohydrates (% DW)	21.0 ± 3.8	52.9 ± 3.5
Lipids (% DW)	20.1 ± 0.8	21.4 ± 1.2
C (% DW)	50.3 ± 1.6	46.4 ± 1.7
N (% DW)	7.3 ± 0.7	2.9 ± 0.2
Volatile solids (% DW)	95.3 ± 1.0	95.6 ± 0.4
COD (g^{-9} DW)	1.34 ± 0.11	1.31 ± 0.11
C/N ratio	6.9 ± 0.7	16.3 ± 1.1
Theoretical methane potential ($mL_N\ g^{-1}$ VS)	~551	~549

After harvesting for fermentation, important parameters of *C. reinhardtii* biomass were determined and presented as mean values. Error bars represent standard error (SE, $n = 8$)

BM biomass, *DW* dry weight, *N* nitrogen, *C* carbon, *VS* volatile solids, *TMP* theoretical methane potential, *COD* chemical oxygen demand

Anaerobic digestion of microalgae biomass as mono-substrate

The continuous fermentation of algal biomass, generated under replete-N and low-N culture conditions was performed under a constant HRT of 20 days, and the organic loading rate (ORL) was subsequently increased from 1 g VS L^{-1} day^{-1} in the beginning, over 2 g VS L^{-1} day^{-1} and to 4 g VS L^{-1} day^{-1} at the end of the experiment (ORL 1, 2 and 4, respectively, Fig. 2). These loading rates were chosen since ORL 2 and 4 (2 and 4 g VS L^{-1} day^{-1}, respectively) are generally used in biogas plants for continuous wet fermentation processes on industrial scale [15]. Differences in the fermentation performance of these two types of biomass were already obvious in the beginning at OLR 1 (adaptation phase), where the gas productivity was not only lower in the replete-N reactor, but was also coupled to a slower adaptation process (defined by stable biogas production). During the whole OLR 2-period, biogas as well as methane productivities were lower and less constant in the replete-N reactor compared to the low-N reactor. With the start of OLR 4, the gas productivity of the replete-N reactor started to decrease and reached the minimum level of specific biogas productivity of 62 ± 2 mL_N day^{-1} g^{-1} VS, at the end of the experiment. In contrast to replete-N biomass, the biogas as well as methane productivity of the low-N BM reactor remained constantly high (Fig. 2) during the whole experiment (exclusive adaptation period, OLR 1). Despite the significantly lower methane concentration in

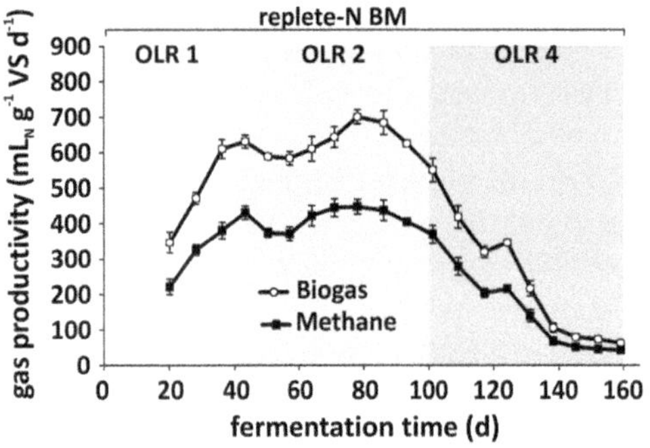

Fig. 2 Biogas and methane productivity via anaerobic fermentation of algal biomass in continuous mode. The biogas productivity was monitored online and methane content was measured weekly (left = replete-N BM, right = low-N BM). Organic loading rate (OLR) is indicated by shades of gray in the background, thereby following biomass concentrations were applied: OLR1 = 1 g VS L^{-1} day^{-1}, OLR2 = 2 g VS L^{-1} day^{-1}, OLR4 = 4 g VS L^{-1} day^{-1}. Error bars represent mean productivity of previous 7 days (SE, $n = 7$). *N* nitrogen, *BM* biomass, *VS* volatile solids

the biogas of low-N digester with 61 ± 0.4% compared to 65 ± 0.9% of replete-N digester (Additional file 1: Figure S1), the overall methane productivity was higher from low-N biomass (Fig. 2) during the complete experimental time course. The overview of the mean biogas and methane productivities, presented in Table 2, underlines that microalgae biomass from replete-N conditions can only efficiently be used at OLR 2 (2 g VS L^{-1} day^{-1}). However, even this organic loading rate of replete-N biomass is already critical since the biogas productivity was not continually stable. The application of a higher loading rate (OLR 4) has a strongly negative effect on the biogas productivity from replete-N biomass (Fig. 2). On the other hand, fermentation of low-N biomass was observed to be stable over both periods OLR 2 and 4, with constantly high methane productivities of 464 ± 9 and 462 ± 9 mL_N g^{-1} VS day^{-1}, respectively (Table 2). The overall achieved methane productivity of low-N algal biomass showed a 36% higher productivity in comparison to maize (Table 2) [41].

Despite of the fact that the theoretical methane potential of replete-N and low-N biomass were quite similar, the specific methane productivity of low-N biomass was significantly higher compared to the biomass derived from replete-N conditions [464 ± 9 mL_N g^{-1} VS day^{-1} vs. 416 ± 11 mL_N g^{-1} VS day^{-1} at OLR 2 and 462 ± 9 mL_N g^{-1} VS day^{-1} vs. 131 ± 33 mL_N g^{-1} VS day^{-1} at OLR 4, respectively (Table 2)]. However, this finding corresponds well to previous observations, where starved biomass showed a higher accessibility and biodegradability compared to biomass from the linear growth phase [30]. To evaluate the possible reasons for the productivity differences between replete-N and low-N biomass, some essential fermentation parameters were analyzed for both reactors (Fig. 3; Additional file 1: Figures S2, S3, S4, Additional file 1: Table S1).

One of the most crucial parameters for the fermentation of protein-rich biomass is nitrogen, which is released during anaerobic decomposition of biomass in form of ammonium into the reactor supernatant [26]. Monitoring of total ammonia nitrogen (TAN) concentration in the reactor revealed indeed a huge difference between the protein-rich (replete-N BM) and low protein (low-N BM) biomass (Fig. 3). The TAN concentrations in low-N reactor were observed to be constantly below 600 mg L^{-1} during the entire experiment. However, the TAN concentration in the replete-N reactor increased at OLR 2 to a value of nearly 1500 mg L^{-1}, which is already close to

Table 2 Overview of mean biogas and methane productivities for the low-N and replete-N reactors

	Specific biogas productivity (mL_N g^{-1} VS day^{-1})			Specific methane productivity (mL_N g^{-1} VS day^{-1})		
	Replete-N BM	Low-N BM	Maize silage	Replete-N BM	Low-N BM	Maize silage
OLR 2 g VS L^{-1} day^{-1}	634 ± 15	761 ± 12	740[a]	416 ± 11	464 ± 9	404[a]
OLR 4 g VS L^{-1} day^{-1}	203 ± 50	750 ± 15	620[a]	131 ± 33	462 ± 9	339[a]

The values were summarized by distinct OLR-phases (OLR 2 = 2 g VS L^{-1} day^{-1}, OLR 4 = 4 g VS L^{-1} day^{-1}). Maize silage productivities were included for comparison as predominantly used renewable substrate for industrial scale fermentation. Error bars represent standard error (SE, $n = 8$)

N nitrogen, *VS* volatile solids

[a] Literature values for maize silage [41]

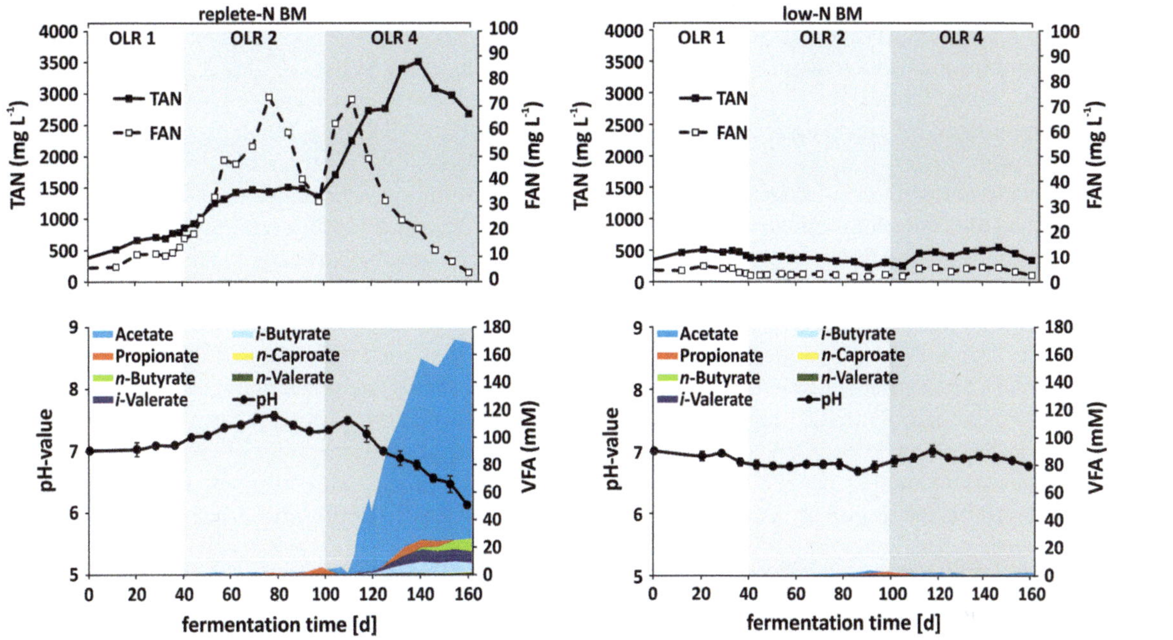

Fig. 3 Analysis of essential fermentation parameters during anaerobic digestion of algal biomass in continuous mode. Left = replete-N BM, Right = low-N BM. Organic loading rate (OLR) is indicated by shades of gray in the background: OLR 1 = 1 g VS L^{-1} day^{-1}, OLR 2 = 2 g VS L^{-1} day^{-1}, OLR 4 = 4 g VS L^{-1} day^{-1}. *Error bars* represent standard deviation (SD, $n = 3$). Detailed VFA concentration values in SI, Additional file 1: Table S1. N = nitrogen, *BM* biomass, *VS* volatile solids, *TAN* total ammonium nitrogen, *FAN* free ammonia nitrogen, *VFA* volatile fatty acids

described inhibitory levels of 1700–1800 mg L^{-1} [26, 42, 43]. These inhibitory levels were exceeded directly after the loading rate of 4 g VS L^{-1} day^{-1} (OLR 4), reaching the maximum of 3507 ± 14 mg L^{-1} at day 140.

Nevertheless, free ammonia is known to be a more efficient inhibitor than ammonium and to have a strong negative effect primarily to the methanogens already at low concentration of 50–100 mg L^{-1} [26]. Indeed, high free ammonia nitrogen (FAN) concentration was observed in the replete-N reactor already at OLR 2 (Fig. 3), which could have had an inhibitory effect on methanogens, indicated by simultaneous decline in methane productivity at days 45–60 (Fig. 2). Yet, despite further increase of FAN to 74 ± 0.06 mg L^{-1} at day 77, the methane productivity remained stable, which may be due to Bacteria or Archaea adaptation to these FAN concentrations, and then the FAN-levels decreased again to 32 ± 0.05 mg L^{-1} (Fig. 3). At the beginning of OLR 4 (day 105), the FAN concentration in replete-N reactor increased again and reached maximal levels at day 112 with 73 ± 0.11 mg L^{-1} comparable to the maximal levels at OLR 2. Additionally, this increase was accompanied by a simultaneous increase of TAN (starting at day 105 as well), followed by a subsequent accumulation of acetate (from day 112, Fig. 3). Nevertheless, the FAN concentration started to decrease after day 112 (Fig. 3, replete-N BM, upper graph), mostly due to a drop of the pH which was caused by the constant increase of the volatile fatty acid (VFA) concentration. Especially, acetate (up to 170 mM) and other intermediate fermentation products (from day 120) such as propionate, *n*-butyrate, *i*-valerate, *i*-butyrate, *n*-caproate, *n*-valerate increased further during the time course of the experiment (Fig. 3, replete-N BM, lower graph; detailed values in SI, Additional file 1: Table S1). It can be assumed that an efficient adaptation of anaerobic microorganisms (especially methanogens) was not possible within the short time period, when the change of crucial factors such as FAN, TAN and VFA occurred. As a consequence, the process inhibition could not be surmounted, resulting in a drastic decrease of methane productivity and finally a complete failure of the fermentation process (Figs. 2, 3, replete-N BM). Similar observations were also made in other continuous fermentation approaches with protein-rich algal biomass as mono-substrate, where high TAN/FAN concentrations, and consequently increasing VFAs have led to decreased methane productivities [20, 24, 25, 29, 44, 45].

On the other hand, the reactor, fed with low-N biomass, did not show any imbalances in fermentation parameters, being constantly low throughout the entire experiment (Fig. 3, low-N BM). Especially, the FAN concentration showed values lower than 5 mg L^{-1} during the complete experimental time, far below inhibitory levels [26]. Furthermore, this observation is also reflected by constantly high methane productivity at different loading rates (Fig. 2, low-N BM, Table 2).

Since the fermentation of microalgae biomass, generated under nitrogen limited conditions was stable and produced constant amounts of methane, it was interesting to evaluate the conversion efficiency level of this process. For this purpose, the theoretical methane potential (TMP) of the biomass was compared to the specific methane productivity reached in the experiments [46]. According to our calculations, the conversion efficiency for low-N biomass to methane reached 84% [calculation specific methane productivity (Table 2) of TMP (Table 1)] for both loading rates (OLR 2 and 4). Having in mind that approximately 12–15% of the organic matter is used for bacterial growth and maintenance requirements during the anaerobic digestion process [14], and therefore being not available for fermentation to methane. The fermentation of low-N biomass within this study reached almost the maximal capacity and represents the most efficient process so far described in the literature for algal biomass as a mono-substrate [19]. For instance, Samson and colleagues observed maximal methane productivity by digestion of *Spirulina maxima* of only 350 mL_N g^{-1} VS day^{-1}, and thus a maximal conversion efficiency of 59%. These results, however, were achieved only under OLR 1 and HRT of 30 days, whereas the productivities decreased significantly when higher loading rates were applied, due to pronounced ammonia inhibition [45]. Even lower maximal productivities of only 267 mL_N CH_4 g^{-1} VS day^{-1} (at OLR 4 and HRT of 20 days) were obtained in another recent study using *Spirulina* biomass [47]. Similar results could be achieved for green algae biomass in other studies, where only 160 mL_N CH_4 g^{-1} VS day^{-1} could be reached for raw *Chlorella vulgaris* biomass, corresponding to 32% conversion efficiency. After thermal pretreatment of the biomass, the yield could be increased by 1.5-fold and still reached only 233 mL_N CH_4 day^{-1} g^{-1} VS corresponding to only 49% of TMP (OLR 0.8, HRT 15) [25]. Very low methane productivities of only 70 mL_N day^{-1} g^{-1} VS were published by Mahdy and co-workers for *C. vulgaris*, corresponding to only 15% conversion efficiency (OLR 1, HRT 15). Nevertheless, parallel digestion of enzymatically pretreated algae biomass was 2.2 times more efficiently digested and resulted in 196 mL_N CH_4 day^{-1} g^{-1} VS corresponding again to only 49% of TMP (OLR 1, HRT 20) [20]. Moreover, in comparison to the fermentation performance with microalgae, the theoretical maximum achieved for macroalgae substrate was in the range of 25–45% [48]. Moreover, the methane productivity from macroalgae fermentation lies often in the range of less than 300 mL CH_4 g^{-1} VS day^{-1} [27, 49, 50], which is significantly lower compared to the productivity of 462 mL_N CH_4 g^{-1} VS day^{-1} achieved in this work with microalgae. Apart from the finding that the methane yield from batch experiments with macroalgae biomass [27] is rather low compared to microalgae, the continuous fermentation under comparable conditions (regarding loading rate) seems also to be less efficient and sensible towards residual salt content in the biomass due to marine origin [49].

Thus, the biomass-to-methane conversion efficiency of 84% demonstrated within this work by the application of low-N algae biomass is not only significantly higher compared to other long-term fermentation trails with untreated biomass but also compared to the results achieved after successful pretreatment of microalgae biomass. Furthermore, this efficiency may represent the maximum practically achievable under the AD conditions [14]. Considering the energy consumption of microbial biomass, the practical efficiency of the fermentation process presented here is at 96–99%, and thus the process may be described as optimal. Based on these "proof of concept" results, this strategy can also be performed under more applied levels. So for instance, the cultivation of microalgae under non-axenic conditions was tested and revealed rather low/negligible contamination levels due to the nature of the photoautotropic culture media (especially low-N conditions) and no negative effect during the fermentation process of this biomass could be observed (unpublished observations). Additionally, other more industrially relevant microalgae species can also be tested in continuous fermentation mode, since our previous batch results for *Parachlorella kessleri* and *Scenedesmus obliquus* were quite promising, exhibiting very similar properties in terms of C/N ratios and methane yields such as *C. reinhardtii* [30]. Moreover, to reduce the cultivations costs of the microalgae and to include more positive environmental aspects to the process, wastewater could be used as nutrition source and flue gas (e.g., biogas after combustion) could be integrated as CO_2 source in the process [27].

Consequence of the fermentation parameter on the microbial community

High-throughput 16S rRNA gene amplicon sequencing was accomplished to investigate how this suboptimal and optimal performance of the replete-N and low-N biomass digesters is reflected on the microbial community. For the comparison of the dynamics of the bacterial community in the different conditions, samples of inoculum (local waste water treatment plant) and the biogas fermenter, fed with replete-N and low-N biomass on the end of OLR 2 (after 100 days) and OLR 4 (after 160 days) were chosen. In all investigated samples, no evidence of eukaryotic plastid 16S rRNA could be found, suggesting that the algal DNA was completely disintegrated during the anaerobic fermentation. Based on the 16S rRNA gene amplicon database (RDP) [51], the biogas producing microbial community was dominated by Bacteria with 99%, and the Archaea was only represented with approximately 1% (Fig. 4). These findings have previously been reported [52–55], and are in agreement with the fact that bacteria are involved in the first three steps of biomass transformation with a high variety of substrate preferences, and Archaea are restricted to a very narrow substrate spectrum in terms of acetate, methyl-group containing compounds as well as CO_2 and H_2.

According to *The prokaryots* [56] and *Bergey's Manual of Systematic Bacteriology* [57, 58], all identified bacterial community members within the investigated samples are typically involved in the anaerobic degradation of the supplied feedstock as they are described to have cellulolytic, saccharolytic, glycolytic, lipolytic, proteolytic and/or acido-/acetogenic capacities. However, many of the bacterial 16S rRNA amplicon reads (27.26 ± 2.75% for inoculum, 28.94 ± 1.37 and 10.39 ± 0.43% for replete-N BM OLR 2 and OLR 4, as well as 48.01 ± 1.77 and 40.58 ± 1.59% for low-N BM OLR 2 and OLR 4, respectively, Fig. 4a) could not be classified at the phylum levels, respectively, confirming that largely bacterial communities in AD reactors remain unknown [59].

The active sludge (inoculum) revealed very high species diversity comprised 603 ± 52 OTUs (Additional file 1: Table S2). Overall, 73% of the identified sequence reads could be assigned to 18 different phyla, with the most abundant among them the members of the phyla *Chloroflexi* (26.78%), *Actinobacteria* (17.96%), *Verrucomicrobia* (7.80%) and *Firmicutes* (7.01%), whereas all other phyla were found only to a minor portion (Fig. 4a). The bacterial diversity dropped significantly during the anaerobic fermentation of algal biomass as mono-substrate, cultivated under replete-N and low-N culture conditions and revealed 178 ± 34 and 111 ± 7 OTUs, as well as 269 ± 20 and 177 ± 2 OTUs for OLR2 and OLR4, respectively (Additional file 1: Table S2). This development indicates that distinct bacteria species begun to dominate due to the selection pressure based on the certain substrate type and amount and other species were extinct. Similar observations were obtained in other studies [60, 61].

Furthermore, in the reactors with no obvious inhibition, the members of the phyla *Bacteroidetes* became dominant in the AD process, followed by *Chlorobi* in the digester with replete-N biomass at OLR 2 or *Spirochaetes* with low-N biomass at OLR 2 and 4 (Fig. 4a). Interestingly, within the phylum *Bacteroidetes*, mainly three different main OTUs were identified (OTU_2, 3 and 26; Additional file 1: Figure S5). OTU_26 is representing the genus *Paludibacter* of the family *Porphyromonadaceae,* which was described to ferment various sugars to acetate and propionate as the major fermentation products [58], and is mostly abundant in the low-N BM digester with high amount of carbohydrates (Table 1). The phyla *Chlorobi* is represented by only one member of the genus *Ignavibacterium* (OTU_36, Additional file 1: Figure S5), which was also described to utilize various carbohydrates [56]. The phyla *Spirochaetes* mainly consists of two OTUs of the order *Spirochaetales* (OTU_8 and 18, Additional file 1: Figure S4), of which OTU_18

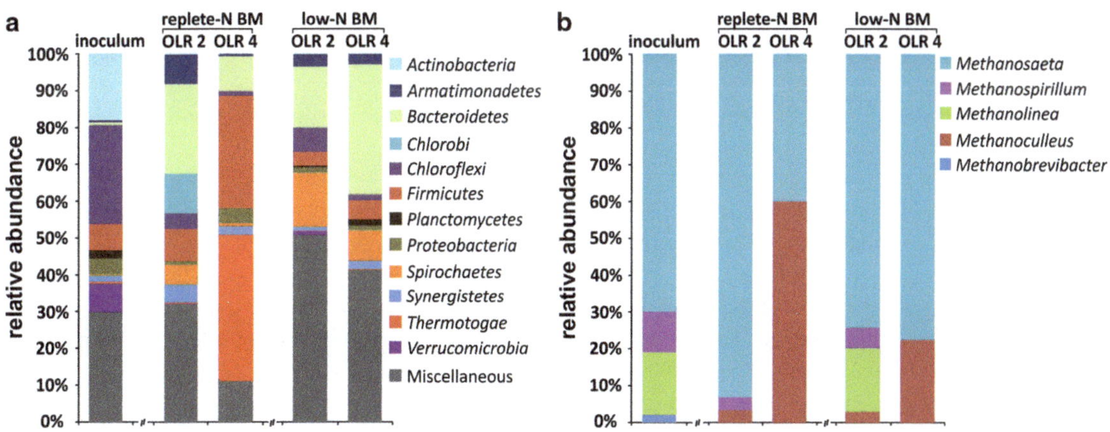

Fig. 4 Bacterial diversity dynamic as assessed by high-throughput 16S rDNA amplicon sequencing. The data is represented at the phyla level for Bacteria (**a**) and family level for Archaea (**b**). The reactors fed with biomass cultivated with replete and low nitrogen content (replete-N BM and low-N BM) were exposed to increasing organic loading rates OLR 2 (2 g VS L^{-1} day^{-1}) and OLR 4 (4 g VS L^{-1} day^{-1}). The inoculum and the sampling periods at the end of each OLR were chosen for microbial community monitoring

could be classified to the genus *Treponema* that utilizes carbohydrates and/or amino acids as carbon and energy source [58]. Interestingly, the digester (replete-N BM, OLR 4), which experienced acidosis because of the high FAN/TAN and VFA concentrations (Fig. 3 replete-N BM), showed a completely different bacterial population, with the members of phyla *Firmicutes* and *Thermotogae* being the most abundant in this samples (Fig. 4a). Thereby, the *Firmicutes* were to 70% represented by the genus *Sporanaerobacter* (OTU_108), and the *Thermotogae* to 99.9% by the species (OTU_125, Additional file 1: Figure S5) similar to *Defluviitoga tunisiensis* [62]. *Sporanaerobacter* was described to be able to utilize some sugars, peptides and various single amino acids into acetate [57, 63]. Moreover, members of *Thermotogae* have been characterized for complex polysaccharide fermentation and hydrogen production [62, 64], what might promote beneficial associations with hydrogenotrophic methanogens [65]. The phyla *Bacteroidetes* is also present in these samples, however, it is in contrast to the well-performing digesters, mainly represented by other members of the family *Porphyromonadaceae* (OTU_78 and 111, Additional file 1: Figure S5). The most members of the family *Porphyromonadaceae* are primarily described to be weakly saccharolytic in contrast to *Paludibacter* observed in well-performing digester, since the bacterial growth was not observed to be significantly affected by carbohydrates, but is enhanced by protein hydrolysates [58], which is also in agreement with the fact that this digester was fed with protein-rich biomass.

In general, archaeal communities were much less diverse than bacterial ones (Fig. 4a, b), with *Methanomicrobiaceae*, *Methanobacteriaceae* and *Methanosaetaceae* being the dominant families. The members of *Euryarchaeota* in the inoculum (active sludge of the local waste water treatment plant) are present to 1.18% ± 0.13 and are consistent on the genus level of *Methanobrevibacter*, *Methanolinea* and *Methanospirillum* and *Methanosaeta*, with the last being the most abundant of the methanogenic community. This finding is also in agreement with the general consideration of the acetoclastic activity being the dominant methanogenic pathway [66, 67]. Distribution, similar to the inoculum, could be observed in the well-performing (replete-N BM OLR 2 and low-N BM OLR 2, 4) digesters, with *Methanosaeta* sp. representing the most abundant Archaea in the methanogenic community, followed by *Methanoculleus* sp. and *Methanospirillum* sp. and *Methanolinea* sp (Fig. 4b). On the other hand, the archaeal community in replete-N BM digester OLR 4 is dominated by *Methanoculleus* sp. and to lesser extent by *Methanosaeta* sp., suggesting an apparent redirection from the acetoclastic towards hydrogenotrophic methanogenesis. The increased abundance of *Methanoculleus* sp. could possibly be attributable to the sensitivity of acetoclastic Archaea towards volatile fatty acid intoxication (acidosis) and/or higher availability of hydrogen provided by certain bacterial species [68] like the members of the phyla *Thermotogae*. Similar behavior could be also observed in other studies, whereby the authors suggested that the replacement of the dominant *Methanosaeta* sp. by *Methanoculleus* sp. might be a potential warning indicator of acidosis within the fermenter [60, 61, 69].

Conclusions

Biogas generation from microalgae biomass has been researched for approximately 60 years with the major outcome that microalgae represent a rather challenging substrate for anaerobic digestion due to high cell wall recalcitrance and unfavorable C/N ratio, owing to its high protein content [18, 19]. The present study investigated the anaerobic digestion from microalgae biomass generated in replete-N as well as naturally occurring (nitrogen limitation, low-N) conditions. The use of algal biomass from replete nitrogen conditions, especially at OLR 4 have led to an inhibition of the digester, caused by high TAN/FAN and VFA concentrations, and thus to fermentation failure with very low methane productivity. In the failed reactor (replete-N biomass, OLR 4), a clear shift could be observed in the bacterial community to the phyla *Firmicutes* and *Thermotogae* and archaeal population changed from acetoclastic to hydrogenotrophic methanogenesis.

In contrast to fermentation of replete-N biomass, the application of nitrogen limitation during the microalgae cultivation resulted in generation of biomass with significant changes in the composition (highly accessible biomass with two times lower protein content), and thus in an optimal mono-substrate for efficient AD process in continuous manner. The fermentation process was characterized by stable process parameters with very low levels of main inhibitory compounds. The investigation of the microbial communities revealed *Bacteroidetes* phyla as subsequently dominating the efficiently preforming digester, indicating that these members adapted most efficiently to the microalgae mono-substrate. Furthermore, among the methanogens, the family of *Methanosaeta* sp. was predominant, suggesting the acetoclastic methanogenesis to be the main pathway during the successful anaerobic degradation of microalgae. The productivity of methane was constantly on a high level (464 ± 9 and 462 ± 9 $mL_N\ g^{-1}$ VS day^{-1} at OLR of 2 and 4, respectively), thus corresponding to an energy conversion efficiency (biomass to methane) of 84%. Taken into account the amount of organic matter used to form new microbial cells and energy for cell metabolism was 12–15% [14], algae substrate conversion efficiency reached in this study almost the practically achievable maximum of 96–99%. According to these considerations, algae biomass can be used highly efficiently for AD without any energy or cost intensive pretreatments. Thus, the presented results of the efficient continuous fermentation of low-N biomass are moving the industrial application of biofuel generation from algal biomass in a more economically feasible direction, especially because the generation of algae biomass under these conditions saves significantly expensive fertilizers (e.g., nitrogen).

Methods

Strains and growth conditions

Chlamydomonas reinhardtii strain CC-1690 from the Chlamydomonas Center (Duke University, Durham NC, USA) was used for all experiments. Liquid algal cultures were grown photoautotrophically under continuous white light (300 μmol photons m^{-2} s^{-1}; Osram L 36 W/865, Osram Germany). Cultivations were conducted in glass bottles (DURAN®, max. capacity 3.5 L, outer diameter 110 mm and 450 mm high, Schott Germany) with 3 L of algae culture, under continuous agitation on a magnetic stirrer. Carbon supply was achieved by bubbling with moisture pre-saturated, carbon dioxide-enriched air (3% v/v) with a flow rate of 5 L h^{-1}. Nutrients were provided by a modified Provasoli based minimal medium [70]. For replete nitrogen culture conditions, the following components and concentrations were applied: K_2HPO_4 0.57 mM; H_3BO_3 0.16 mM; $MgSO_4$ 4.87 mM; KCl 21.46 mM; $NaNO_3$ 11.77 mM; $CaCl_2$ * $2H_2O$ 2.72 mM; $FeCl_3$ * 6 H_2O 12.2 μM; Na_2-EDTA 12.5 μM; EDTA 103 μM; $ZnCl_2$ 2.2 μM; $MnCl_2$ * $4H_20$ 16.7 μM; $CoCl_2$ * 6 H_2O 50.4 nM; $CuCl_2$ * $2H_2O$ 17.6 nM; Na_2MoO_4-* $2H_2O$ 24.8 nM. Low-nitrogen cultivation conditions were realized according to previous work [30] by applying a limited amount of nitrogen (3.56 mM $NaNO_3$ equals to 50 mg of nitrogen per liter culture).

Determination of algal biomass concentration

The biomass concentration was determined by centrifugation of 15 mL of cell culture (3000×*g* for 5 min, at least three technical replicates per sample) and drying of the cell pellet in a pre-weighted glass tube at 105 °C for 24 h. To determine the organic biomass fraction, the sample tubes were subsequently incubated at 550 °C for 5 h and the residual ash determined by weighing. The amount of organic biomass (dry weight minus the ash content) was calculated and expressed as volatile solids (VS, g L^{-1}).

Measurement of elemental N and C content in the biomass (C/N ratio)

Total carbon (C) and nitrogen (N) content of the algal biomass was determined via an element analyzer (VARIO EL III, Elementar Analysesysteme, Hanau, Germany) as described before [71].

Anaerobic fermentation and quantitative biogas measurement

The continuous fermentation of algae biomass was performed according to the VDI 4630 guideline [46]. Fermentation was performed in B Braun glass fermenters, maximal capacity of 2 L. Fermentation temperature of 38 °C was provided by external tempered water bath via water circulation thought a build-in water jacket in the fermenter. Reactor content was stirred at 100 rpm via slices stirring system (in a 15 min. ON- and 15 min OFF-mode). The digester was operated with 1 L working volume (inoculated with microbial community from anaerobic digester of a local waste water treatment plant Bielefeld–Heepen, Germany) and constant hydraulic retention time (HRT) of 20 days. Feeding/withdrawing was performed manually with a syringe (first 50 mL rector content out, thereafter 50 mL algae substrate in) daily (despite of semi-continuous feeding mode, the fermentation is designated here, in agreement to VDI guideline [46] as a continuous process). Biogas (water free after condensations column) evolution was measured continuously by a MilliGascounter® (MGC-1 V 3.0, 3, 2 mL, Ritter, Germany) and evaluated by applying RIGAMO Software (Ritter, Germany), followed by normalization of the gas volume to standard temperature of 0 °C. Organic loading rate (OLR) was increased subsequently and simultaneously in both digesters [fed with replete-N and low-N biomass (BM)], from day 0 to 40 (OLR 1 = 1 g VS L^{-1} day^{-1}), from day 40 to 100 (OLR 2 = 2 g VS L^{-1} day^{-1}) and from day 100 to 160 (OLR 4 = 4 g VS L^{-1} day^{-1}). The fresh algal substrate was obtained by centrifugation of the cultures at 3000×*g* for 5 min and removal of the supernatant. To avoid freezing or drying artifacts, biomass was diluted by addition of H_2O to required concentration and stored by 2 °C prior feed (max. 2 weeks long).

Methane content measurement via gas chromatography (GC)

The determination of the methane content within the biogas was performed by GC analysis weekly in nine technical replicates. Biogas from the fermenter was sampled with a gas tight syringe (5 mL) through a rubber seal and injected into a gas chromatograph GCM MicroBox III (Elster GmbH, Germany) equipped with an Micropacked HayeSep A-Column (Length: 65 cm, inner diameter: 0.3 mm) and a thermal conductivity detector (TCD). Column temperature in the first 50 s was at 50 °C with following linear increase 4 °C s^{-1} to 165 °C witch was hold constant till the end by 120 s. Argon was used as the carrier gas and the calibration of the GC was performed with defined gas (Linde, Germany) containing O_2(0.103%), H_2S(0.208%), H_2(0.498%), CH_4(59.4%), CO_2(34.4%) and N_2(5.391%), mixed according to DIN EN ISO 6141.3.

Determination of biomass composition and theoretical methane potential

Determination of lipid fraction was performed in two technical and four biological replicates from 50 mg of lyophilized biomass each. After homogenization (3 × 30 s at 6500 rpm using a Precellys 24, Peqlab, Erlangen, Germany), the total lipid fraction was extracted according to a modified Folch protocol [72] using a total of 4 mL of methanol and 8 mL of chloroform. Contaminants were removed by washing the extract with 3 mL of deionized water. After evaporation of solvents under nitrogen atmosphere, lipid fraction was determined via gravimetrical measurement. The total cellular protein amount was determined using Bio-Rad DC Protein assay (Bio-Rad, CA, USA). The amount of total carbohydrates was determined using the protocol according to Dubios et al. [73]. The theoretical methane potential was calculated in accordance with Buswell equation and empirical formula stated by Heaven et al. with TMP`s of 446, 415 and 1014 $mL_N\ g^{-1}$ VS for proteins, carbohydrates and lipids, respectively. Within the formula *P* stays for protein, *C* for carbohydrate and *L* for lipid content on VS basis.

$$\mathrm{TMP}\left(\mathrm{mL_N g^{-1} VS}\right) = \frac{P\%}{\sum(P\% + C\% + L\%)} \times 446 + \frac{C\%}{\sum(P\% + C\% + L\%)} \times 415 + \frac{L\%}{\sum(P\% + C\% + L\%)} \times 1014 .$$

Determination of the fermentation parameters

Total ammonium nitrogen (TAN) was determined using colorimetric verification via cuvette tests LCK302 (Hach Lange GmbH, Germany). Free ammonia nitrogen (FAN) was calculated from TAN value in respect to temperature and pH according to the formula given by Astals and colleagues [74]. Total organic- and inorganic-carbon (TOC and TIC, respectively) were measured via LCK381, total nitrogen was determined via LCK 338, (Hach Lange GmbH, Germany). The determination of volatile fatty acid (VFA) concentrations was performed via GC-FID analysis. Sample preparation was done according to the 5560D procedure [75] and analyzed using a Shimadzu GC-2010 plus Gas Chromatograph equipped with a Macherey–Nagel OPTIMA® FFAPplus (Length: 30 m, inner diameter: 0.25 mm) column (Macherey–Nagel, Germany) and coupled to an FID detector (supplied with H_2 and synthetic air). Analysis was performed under constant pressure of 231.9 kPa with He as carrier gas and N_2 as makeup gas with constant flow rate of 60 cm s^{-1}. Column temperature in the first 2 min was at 100 °C with following linear increase to 175 °C within 15 min. VFA-Mix standard (46975-U, Supelco Analytical, Sigma-Aldrich, USA) at concentrations of 0.1, 1 and 10 mM was used for calibration.

Microbial monitoring by high-throughput 16S rRNA amplicon sequencing

Genomic DNA was extracted as previously described by Zhou et al. [76]. For the determination of the taxonomic profile of the biogas community, high-throughput sequencing of the hypervariable V3–V4 regions of the 16S rRNA gene was performed on the Illumina MiSeq system by applying the paired-end protocol, according to the manufacturer's instructions and using of the Illumina recommended gene specific primer sequences [77]. For the data processing and analysis, an amplicon analysis pipeline was used as recently described [59, 78]. Briefly, raw sequences were merged by FLASH [79] and further processed and analyzed using the UPARSE pipeline [80] based on Usearch 8.0 [81] with default settings. Processed operational taxonomic units (OTU) were taxonomically classified using the RDP classifier 2.7 [51].

Additional file

Additional file 1: Figure S1. Methane concentration in the biogas, produced during the fermentation of replete-N and low-N algae biomass (replete-N BM and low-N BM, respectlively). Statistics: two-sample t-test with 95% confidence interval. **Figure S2.** Concentration of total carbon and nitrogen during the experimental time course in replete-N BM digester (A) and low-N BM digester (B). Concentration of total organic and inorganic carbon (TOC and TIC) is shown for replete-N BM digester (C) and low-N BM digester (D). Measurements were performed in three replicates; error bars represent standard deviation (SD). **Figure S3.** Concentration of volatile and total solids (VS and TS, respectively) during the experimental time course in replete-N BM digester (A) and low-N BM digester (B). Measurements were performed in at least three replicates; error bars represent standard deviation (SD). **Figure S4.** Concentration of chemical oxygen demand (COD) during the experimental time course in replete-N BM digester (A) and low-N BM digester (B). Measurements were performed in three technical replicates; error bars represent standard deviation (SD). **Figure S5.** Bacterial diversity dynamics as assessed by high-throughput 16S rRNA amplicon sequencing and represented at the OTU level. The reactors, fed with biomass cultivated in media with replete and low nitrogen content (replete-N BM and low-N BM) were exposed to increasing organic loading rates of 2 and 4 g VS L^{-1} d^{-1} (OLR 2 and OLR 4, respectively). The inoculum and the sampling periods at the end of each OLR were chosen for microbial community monitoring. **Table S1.** Analysis of the volatile fatty acid (VFA) content during the time course of the experiment. The identification and quantification of the intermediate fermentation products (mM) was determined via GC-FID. The indicated error (±) represents standard deviation (SD, n = 2). **Table S2.** Filtered sequences during amplicon processing. OTU=operational taxonomic unit, N=nitrogen, sd=standard deviation, OLR=organic loading rate, rep=replicate.

Abbreviations

AD: anaerobic digestion; C/N: carbon-to-nitrogen ratio; DW: dry weight; BM: biomass; SE: standard error; SD: standard deviation; FAN: free ammonia nitrogen; TAN: total ammonium nitrogen; OLR: organic loading rate; HRT: hydraulic retention time; VFA: volatile fatty acids; C: carbon; N: nitrogen; TMP: theoretical methane potential; VS: volatile solids.

Authors' contributions

VK, OBK and OK designed the experiments; VK performed algae biomass cultivation and fermentation with all concomitant measurements; OBK and AW preformed 16S rDNA amplicon sequencing, DW performed the bioinformatic analysis; interpretation and visualization of all data was done by VK and OBK; VK, OBK, CP, JK and OK wrote and edited the manuscript. All authors read and approved the final manuscript.

Author details

[1] Department of Biology/Center for Biotechnology (CeBiTec), Bielefeld University, Universitätsstrasse 27, 33615 Bielefeld, Germany. [2] Institute of Life Science Engineering (KIT), Bioprocess Engineering, University of Karlsruhe, Fritz-Haber-Weg 2, 76131 Karlsruhe, Germany.

Acknowledgements

OK and VK gratefully acknowledge the Deutsche Forschungsgemeinschaft [DFG Grant No. KR 1586/7-1] and DW the "Bielefeld-Gießen Resource Center for Microbial Bioinformatics"—BiGi [BMBF Grant No. 031A533] for support. The authors would like to thank Swapnil Chaudhari for providing technical assistance for lipid extraction.

Competing interests

VK, OBK and OK are inventors of a EP-Patent application (No. 17152658.5—1501, applicant Universität Bielefeld). The authors declare that they have no further/other competing interests.

References

1. Martinot E, Dienst C, Weiliang L, Qimin C. Renewable energy futures: targets, scenarios, and pathways. Annu Rev Environ Resour. 2007;32:205–39.
2. REN21. Renewables 2015 global status Report. REN21 renewables. Paris; 2015.
3. Stephens E, Ross IL, King Z, Mussgnug JH, Kruse O, Posten C, et al. An economic and technical evaluation of microalgal biofuels. Nat Biotechnol. 2010;28:126–8.
4. Borowitzka MA, Moheimani NR. Algae for biofuels and energy. Berlin: Springer; 2013.
5. Formighieri C. Solar-to-fuel conversion in algae and cyanobacteria. 1st ed. Cham: Springer International Publishing; 2015. http://dx.doi.org/10.1007/978-3-319-16730-5_2.
6. Brennan L, Owende P. Biofuels from microalgae-A review of technologies for production, processing, and extractions of biofuels and co-products. Renew Sustain Energy Rev. 2010;14(2):557–77.
7. Kruse O, Hankamer B. Microalgal hydrogen production. Curr Opin Biotechnol. 2010;21(3):238–43.
8. Grewe S, Ballottari M, Alcocer M, D'Andrea C, Blifernez-Klassen O, Hankamer B, et al. Light-harvesting complex protein LHCBM9 is critical for photosystem II activity and hydrogen production in *Chlamydomonas reinhardtii*. Plant Cell. 2014; 26: 1598–611. http://www.ncbi.nlm.nih.gov/pubmed/24706511.
9. Georgianna DR, Mayfield SP. Exploiting diversity and synthetic biology for the production of algal biofuels. Nature. 2012; 488: 329–35. http://www.nature.com/doifinder/10.1038/nature11479.
10. Hallmann A, Hallman A. Algal transgenics and biotechnology. Transgenic Plant J. 2007; 1: 81–98. http://www.uni-bielefeld.de/biologie/Zellbiologie/publik/paper/2007tpj.pdf.
11. Chisti Y. Raceways-based production of algal crude oil. Green. 2013;3:195–216.
12. Wolf J, Stephens E, Steinbusch S, Yarnold J, Ross IL, Steinweg C, et al. Multifactorial comparison of photobioreactor geometries in parallel microalgae cultivations. Algal Res. 2016;15:187–201.
13. Bux F, Chisti Y, editors. Algae biotechnology: products and processes. Cham: Springer International Publishing; 2016.
14. Raposo F, Fernández-Cegrí V, De la Rubia MA, Borja R, Béline F, Cavinato C, et al. Biochemical methane potential (BMP) of solid organic substrates: evaluation of anaerobic biodegradability using data from an international interlaboratory study. J Chem Technol Biotechnol. 2011;86:1088–98.
15. Weiland P. Biogas production: current state and perspectives. Appl Microbiol Biotechnol. 2010;85(4):849–60.
16. Thi NBD, Sen B, Chen CC, Kumar G, Lin CY. Food waste to bioenergy via anaerobic processes. Energy Procedia. 2014;61:307–12.
17. Thi NBD, Lin C-Y, Kumar G. Waste-to-wealth for valorization of food waste to hydrogen and methane towards creating a sustainable ideal source of bioenergy. J Clean Prod Elsevier. 2016;122:29–41.
18. Golueke CG, Oswald WJ, Gotaas HB. Anaerobic digestion of Algae. Appl Microbiol. 1957;5:47–55.
19. Klassen V, Blifernez-Klassen O, Wobbe L, Schlüter A, Kruse O, Mussgnug JH. Efficiency and biotechnological aspects of biogas production from microalgal substrates. J Biotechnol. 2016;234:7–26.
20. Mahdy A, Mendez L, Ballesteros M, González-Fernández C. Protease pretreated *Chlorella vulgaris* biomass bioconversion to methane via semi-continuous anaerobic digestion. Fuel. 2015;158:35–41.
21. Mahdy A, Mendez L, Ballesteros M, González-Fernández C. Autohydrolysis and alkaline pretreatment effect on *Chlorella vulgaris* and *Scenedesmus* sp. methane production. Energy. 2014;78:48–52.
22. Marsolek MD, Kendall E, Thompson PL, Shuman TR. Thermal pretreatment of algae for anaerobic digestion. Bioresour Technol. 2014;151:373–7.
23. Mendez L, Mahdy A, Demuez M, Ballesteros M, González-Fernández C. Effect of high pressure thermal pretreatment on Chlorella vulgaris biomass: organic matter solubilisation and biochemical methane potential. Fuel. 2014;117:674–9.
24. Schwede S, Rehman ZU, Gerber M, Theiss C, Span R. Effects of thermal pretreatment on anaerobic digestion of *Nannochloropsis salina* biomass. Bioresour Technol. 2013;143:505–11.
25. Mendez L, Mahdy A, Ballesteros M, González-Fernández C. Biomethane production using fresh and thermally pretreated *Chlorella vulgaris* biomass: a comparison of batch and semi-continuous feeding mode. Ecol Eng. 2015;84:273–7.
26. Yenigün O, Demirel B. Ammonia inhibition in anaerobic digestion: a review. Process Biochem. 2013;48(5):901–11.
27. Dębowski M, Zieliński M, Grala A, Dudek M. Algae biomass as an alternative substrate in biogas production technologies—Review. Renew Sustain Energy Rev. 2013;27:596–604.
28. Zhen G, Lu X, Kobayashi T, Kumar G, Xu K. Anaerobic co-digestion on improving methane production from mixed microalgae (*Scenedesmus* sp., *Chlorella* sp.) and food waste: kinetic modeling and synergistic impact evaluation. Chem Eng J. 2016;299:332–41.
29. Markou G, Angelidaki I, Georgakakis D. Carbohydrate-enriched cyanobacterial biomass as feedstock for bio-methane production through anaerobic digestion. Fuel. 2013;111:872–9.
30. Klassen V, Blifernez-Klassen O, Hoekzema Y, Mussgnug JH, Kruse O. A novel one-stage cultivation/fermentation strategy for improved biogas production with microalgal biomass. J Biotechnol. 2015;215:44–51.
31. Speece RE. Anaerobic biotechnology for industrial wastewater treatment. Environ Sci Technol. 1983;17:416A–27A.
32. Gruber-Brunhumer MR, Jerney J, Zohar E, Nussbaumer M, Hieger C, Bochmann G, et al. Acutodesmus obliquus as a benchmark strain for evaluating methane production from microalgae: influence of different storage and pretreatment methods on biogas yield. Algal Res. 2015;12:230–8.
33. Mussgnug JH, Klassen V, Schlüter A, Kruse O. Microalgae as substrates for fermentative biogas production in a combined biorefinery concept. J Biotechnol. 2010;150:51–6.
34. Bogen C, Al-Dilaimi A, Albersmeier A, Wichmann J, Grundmann M, Rupp O, et al. Reconstruction of the lipid metabolism for the microalga Monoraphidium neglectum from its genome sequence reveals characteristics suitable for biofuel production. BMC Genom. 2013;14:926.
35. Li Y, Han D, Hu G, Sommerfeld M, Hu Q. Inhibition of starch synthesis results in overproduction of lipids in *Chlamydomonas reinhardtii*. Biotechnol Bioeng. 2010;107:258–68.
36. Symons GE, Buswell AM. The methane fermentation of carbohydrates. J Am Chem Soc. 1933;55:2028–36.
37. Heaven S, Milledge J, Zhang Y. Comments on "Anaerobic digestion of microalgae as a necessary step to make microalgal biodiesel sustainable". Biotechnol Adv. 2011;29:164–7.
38. Bohutskyi P, Bouwer E. Biogas production from algae and cyanobacteria through anaerobic digestion: a review, analysis, and research needs.

In: Lee JW, editor. Adv biofuels Bioprod. New York: Springer; 2013. p. 873–975.
39. Zubr J. Methanogenic fermentation of fresh and ensiled plant materials. Biomass. 1986;11:159–71.
40. Braun R. Biogas-Methangärung organischer Abfallstoffe. Wien: Springer; 1982.
41. Mähnert P, Linke B. Kinetic study of biogas production from energy crops and animal waste slurry: effect of organic loading rate and reactor size. Environ Technol. 2009;30:93–9.
42. Orris EA. Ammonia nitrogen and the anaerobic environment on JSTOR. Water Pollut Control Fed. 1961;33:978–95.
43. Melbinger NR, Donnellon J, Zablatzky HR. Toxic effects of ammonia nitrogen in high-rate digestion. Water Pollut Control Fed. 1971;43:1658–70.
44. Jegede Abiodun. Anaerobic digestion of cyanobacteria and chlorella to produce methane for biofuel. Int J Agric Biol Eng. 2012;5:68–74.
45. Samson R, LeDuy A. Detailed study of anaerobic digestion of *Spirulina maxima* algal biomass. Biotechnol Bioeng. 1986;28:1014–23.
46. VDI 4630. Fermentation of Organic Compounds. Düsseldorf: Verein Deutscher Ingenieure, VDI Handbook; 2006. p. 1–93.
47. Nolla-Ardèvol V, Strous M, Tegetmeyer HE. Anaerobic digestion of the microalga Spirulina at extreme alkaline conditions: biogas production, metagenome, and metatranscriptome. Front Microbiol. 2015;6:597.
48. Wise DL, Augenstein DC, Ryther JH. Methane fermentation of aquatic biomass. Resour Recover Conserv. 1979;4:217–37.
49. Sun MT, Fan XL, Zhao XX, Fu SF, He S, Manasa MRK, et al. Effects of organic loading rate on biogas production from macroalgae: Performance and microbial community structure. Bioresour Technol. 2017; 235: 292–300. http://linkinghub.elsevier.com/retrieve/pii/S0960852417303474.
50. Vergara-Fernández A, Vargas G, Alarcón N, Velasco A. Evaluation of marine algae as a source of biogas in a two-stage anaerobic reactor system. Biomass Bioenerg. 2008;32:338–44.
51. Wang Q, Garrity GM, Tiedje JM, Cole JR. Naive Bayesian classifier for rapid assignment of rRNA sequences into the new bacterial taxonomy. Appl Environ Microbiol. 2007;73:5261–7.
52. Carballa M, Smits M, Etchebehere C, Boon N, Verstraete W. Correlations between molecular and operational parameters in continuous lab-scale anaerobic reactors. Appl Microbiol Biotechnol. 2011;89:303–14.
53. Liu FH, Wang SB, Zhang JS, Zhang J, Yan X, Zhou HK, et al. The structure of the bacterial and archaeal community in a biogas digester as revealed by denaturing gradient gel electrophoresis and 16S rDNA sequencing analysis. J Appl Microbiol. 2009;106:952–66.
54. Regueiro L, Veiga P, Figueroa M, Alonso-Gutierrez J, Stams AJM, Lema JM, et al. Relationship between microbial activity and microbial community structure in six full-scale anaerobic digesters. Microbiol Res. 2012;167:581–9.
55. Sundberg C, Al-Soud WA, Larsson M, Alm E, Yekta SS, Svensson BH, et al. 454 pyrosequencing analyses of bacterial and archaeal richness in 21 full-scale biogas digesters. FEMS Microbiol Ecol. 2013;85:612–26.
56. Rosenberg E, De Long EF, Lory S, Stackebrandt E, Thompson F, editors. The prokaryotes. 4th ed. Berlin Heidelberg: Springer; 2014.
57. Vos P, Garrity G, Jones D, Krieg N, Ludwig W, Rainey FA, et al., editors. Bergey's manual of systematic bacteriology, vol. 3. 2nd ed. New York: Springer; 2009.
58. Krieg NR, Ludwig W, Whitman WB, Hedlund BP, Paster BJ, Staley JT, et al., editors. Bergey's manual of systematic bacteriology, vol. 4. 2nd ed. New York: Springer; 2010.
59. Theuerl S, Kohrs F, Benndorf D, Maus I, Wibberg D, Schlueter A, et al. Community shifts in a well-operating agricultural biogas plant: how process variations are handled by the microbiome. Appl Microbiol Biotechnol. 2015;99:7791–803.
60. Blume F, Bergmann I, Nettmann E, Schelle H, Rehde G, Mundt K, et al. Methanogenic population dynamics during semi-continuous biogas fermentation and acidification by overloading. J Appl Microbiol. 2010;109:441–50.
61. Carballa M, Regueiro L, Lema JM. Microbial management of anaerobic digestion: Exploiting the microbiome-functionality nexus. Curr. Opin. Biotechnol. 2015;33:103–11.
62. Maus I, Cibis KG, Wibberg D, Winkler A, Stolze Y, König H, et al. Complete genome sequence of the strain Defluviitoga tunisiensis L3, isolated from a thermophilic, production-scale biogas plant. J Biotechnol. 2015;203:17–8.
63. Hernandez-Eugenio G, Fardeau ML, Cayol JL, Patel BKC, Thomas P, Macarie H, et al. Sporanaerobacter acetigenes gen. nov., sp. nov., a novel acetogenic, facultatively sulfur-reducing bacterium. Int J Syst Evol Microbiol. 2002;52:1217–23.
64. Conners SB, Mongodin EF, Johnson MR, Montero CI, Nelson KE, Kelly RM. Microbial biochemistry, physiology, and biotechnology of hyperthermophilic *Thermotoga* species. FEMS Microbiol Rev. 2006;30:872–905.
65. Muralidharan V, Rinker KD, Hirsh IS, Bouwer EJ, Kelly RM. Hydrogen transfer between methanogens and fermentative heterotrophs in hyperthermophilic cocultures. Biotechnol Bioeng. 1997;56:268–78.
66. Gujer W, Zehnder AJ. Conversion processes in anaerobic digestion. Water Sci Technol. 1983;15:127–67.
67. Pavlostathis SG, Giraldo-Gomez E. Kinetics of anaerobic treatment—a critical review. CRC Crit Rev Environ Control. 1991;21:411–90.
68. Karakashev D, Batstone DJ, Trably E, Angelidaki I. Acetate oxidation is the dominant methanogenic pathway from acetate in the absence of Methanosaetaceae. Appl Environ Microbiol. 2006;72:5138–41.
69. Goux X, Calusinska M, Lemaigre S, Marynowska M, Klocke M, Udelhoven T, et al. Microbial community dynamics in replicate anaerobic digesters exposed sequentially to increasing organic loading rate, acidosis, and process recovery. Biotechnol Biofuels. 2015;8:122.
70. Provasoli L, McLaughlin JJ, Droop MR. The development of artificial media for marine algae. Arch Mikrobiol. 1957;25:392–428.
71. Platner C, Piñol J, Sanders D, Espadaler X. Trophic diversity in a Mediterranean food web-Stable isotope analysis of an ant community of an organic citrus grove. Basic Appl Ecol. 2012;13:587–96.
72. Folch J, Lees M, Stanley GHS. A simple method for the isolation and purification of total lipids from animal tissues. J Biol Chem. 1957;226:497–509.
73. DuBois M, Giles KA, Hamilton JK, Rebers PA, Smith F. Colorimetric method for determination of sugars and related substances. Anal Chem. 1956;28:350–6.
74. Astals S, Nolla-Ardèvol V, Mata-Alvarez J. Anaerobic co-digestion of pig manure and crude glycerol at mesophilic conditions: biogas and digestate. Bioresour Technol. 2012;110:63–70.
75. Clescerl LS, Greenberg AE, Eaton AD. Organic and volatile acids. In: Clescerl LS, Greenberg AE, Eaton AD, editors. Stand. Methods Exam. Water Wastewater. 20th ed. Baltimore: American Public Health Association, American Water Works Association, Water Environment Federation; 1999. p. 1325.
76. Zhou J, Bruns MA, Tiedje JM. DNA recovery from soils of diverse composition. Appl Environ Microbiol. 1996;62:316–22.
77. Klindworth A, Pruesse E, Schweer T, Peplies J, Quast C, Horn M, et al. Evaluation of general 16S ribosomal RNA gene PCR primers for classical and next-generation sequencing-based diversity studies. Nucleic Acids Res. 2013;41:e1.
78. Liebe S, Wibberg D, Winkler A, Puehler A, Schlueter A, Varrelmann M. Taxonomic analysis of the microbial community in stored sugar beets using high-throughput sequencing of different marker genes. FEMS Microbiol Ecol. 2016;92:1–12.
79. Magoč T, Salzberg SL. FLASH: fast length adjustment of short reads to improve genome assemblies. Bioinformatics. 2011;27:2957–63.
80. Edgar RC. UPARSE: highly accurate OTU sequences from microbial amplicon reads. Nat Methods. 2013;10:996–8.
81. Edgar RC. Search and clustering orders of magnitude faster than BLAST. Bioinformatics. 2010;26:2460–1.

6

High-throughput analysis of chemical components and theoretical ethanol yield of dedicated bioenergy sorghum using dual-optimized partial least squares calibration models

Meng Li[1,2], Jun Wang[1,2], Fu Du[1,2], Boubacar Diallo[1,2] and Guang Hui Xie[1,2*]

Abstract

Background: Due to its chemical composition and abundance, lignocellulosic biomass is an attractive feedstock source for global bioenergy production. However, chemical composition variations interfere with the success of any single methodology for efficient bioenergy extraction from diverse lignocellulosic biomass sources. Although chemical component distributions could guide process design, they are difficult to obtain and vary widely among lignocellulosic biomass types. Therefore, expensive and laborious "one-size-fits-all" processes are still widely used. Here, a non-destructive and rapid analytical technology, near-infrared spectroscopy (NIRS) coupled with multivariate calibration, shows promise for addressing these challenges. Recent advances in molecular spectroscopy analysis have led to methodologies for dual-optimized NIRS using sample subset partitioning and variable selection, which could significantly enhance the robustness and accuracy of partial least squares (PLS) calibration models. Using this methodology, chemical components and theoretical ethanol yield (TEY) values were determined for 70 sweet and 77 biomass sorghum samples from six sweet and six biomass sorghum varieties grown in 2013 and 2014 at two study sites in northern China.

Results: Chemical components and TEY of the 147 bioenergy sorghum samples were initially analyzed and compared using wet chemistry methods. Based on linear discriminant analysis, a correct classification assignment rate (either sweet or biomass type) of 99.3% was obtained using 20 principal components. Next, detailed statistical analysis demonstrated that partial optimization using sample set partitioning based on joint *X*–*Y* distances (SPXY) for sample subset partitioning enhanced the robustness and accuracy of PLS calibration models. Finally, comparisons between five dual-optimized strategies indicated that competitive adaptive reweighted sampling coupled with the SPXY (CARS-SPXY) was the most efficient and effective method for improving predictive performance of PLS multivariate calibrations.

Conclusions: As a dual-optimized methodology, sample subset partitioning combined with variable selection is an efficient and straightforward strategy to enhance the accuracy and robustness of NIRS models. This knowledge should facilitate generation of improved lignocellulosic biomass feedstocks for bioethanol production. Moreover, methods described here should have wider applicability for use with feedstocks incorporating multispecies biomass resource streams.

*Correspondence: xiegh@cau.edu.cn
[1] College of Agronomy and Biotechnology, China Agricultural University, Beijing 100193, China
Full list of author information is available at the end of the article

Keywords: Bioenergy sorghum, Near infrared spectroscopy, Linear discriminant analysis model, Chemical components, Theoretical ethanol yield, Optimal sample subset partitioning, Optimal variable selection, Partial least squares model

Background

Interest in biomass as an alternative energy resource has been increasing dramatically in step with increases in fossil energy costs worldwide [1]. As a promising advanced biofuel, bioethanol has become one of the most practical solutions for carbon emission reduction and food security in recent decades [2, 3]. Moreover, production of second-generation bioethanol derived from lignocellulosic agricultural wastes offers farmers and rural communities more economic opportunities than did first-generation bioethanol production [4]. Consequently, optimization of lignocellulosic feedstocks for second-generation bioethanol production is currently under intensive development globally [5]. Among candidate biofuel feedstocks, sorghum (*Sorghum bicolor* L.) contains exceptional levels of soluble sugars (in the stem) and starch (in the grain), both of which are directly fermentable into bioethanol [1, 6]. Meanwhile, a large amount of degradable lignocellulose from stem is available for second-generation bioethanol production after the harvest season. As a typical C4 crop, sorghum offers several advantages: rapid growth rate, high tolerance to drought and saline/alkaline soil conditions, and worldwide adaptability. Nowadays, sorghum is the fifth most widely produced cereal crop in the world and is grown not only for grain production but also for fiber, forage, and sugar production [7, 8]. For energy purposes, sorghum may be divided into two specific categories: sweet sorghum and biomass sorghum. Sweet sorghum accumulates high levels of sugar in the stem of the plant, while biomass sorghum contains abundant structural carbohydrates that are produced in sufficient quantities to meet future energy demands [8]. To date, the application of sorghum to bioethanol production has rarely been reported. Therefore, the first essential step would be to compare energy potentials between these two sorghum categories.

Bioenergy sorghum cell walls consist mainly of cellulose, hemicellulose, lignin, and ash. Cellulose, which consists of β-1,4 linked linear glucans, is the most abundant biopolymer and carbon sink on earth [9]. Hemicellulose is a class of heterogeneous polysaccharides that constitutes between 15 and 35% of the total biomass of hardwoods and herbaceous plants [10]. Lignin is a hydrophobic polymer consisting of three major phenolic components: hydroxyphenyl, guaiacyl, and syringyl units [11]. Ash-forming elements are often embedded in biomass and include potassium, calcium, magnesium, silicon, sodium, phosphorus, sulfur, and chlorine. Knowledge of both ash level and composition are prerequisites for effective industrial biomass process design [12, 13], but these parameters are difficult to determine. As a consequence, the diversity of chemical components and their varied proportions among lignocellulosic feedstocks frequently leads to the so-called "biomass recalcitrance," requiring expensive and time-consuming bioenergy extraction processing steps [14]. Therefore, high-throughput determinations of chemical composition and theoretical ethanol yield (TEY) are increasingly necessary in order to screen large numbers of lignocellulosic feedstocks. Successful screening will guide development of plant breeding and genetic modification programs toward the ultimate goal of achieving high biofuel yields at low cost [4].

Near-infrared spectroscopy (NIRS) technology is based on the absorption of electromagnetic radiation by chemical covalent bonds (primarily C–H, N–H, O–H, and S–H) combined with multivariate calibration [15]. As a non-destructive and rapid analytical method, NIRS has been widely used for qualitative analysis of various organic feedstocks and for physical and chemical quantitative analyses across the food, pharmaceutical, and agricultural industries [16, 17]. With regard to qualitative analysis, numerous promising techniques for necessary preprocessing of NIRS data have been developed during the last few decades including principal component analysis, K-nearest neighbor analysis, genetic algorithms, partial least squares discriminant analysis, and others [5, 18]. As part of these analyses, linear discriminant analysis is a well-known and frequently used computational tool [15, 19]. In general, bioenergy extraction processes are sensitive to defined distributions of physical and chemical characteristics, e.g., "Biomass Variability" or "Feedstock Quality," that vary greatly among plant species impact biofuel conversion performance [20, 21]. As an ideal high-throughput biomass classification technique, NIRS qualitative analysis has been viewed as increasingly valuable to the bioenergy industry [5, 13, 22, 23]. However, very little research has yet been done to evaluate NIRS for lignocellulosic biomass feedstock classification in the preparation for bioenergy production from sorghum [1, 4, 6, 23–26].

For quantitative analysis in the bioenergy sector, earlier research mainly focused on the development of spectral preprocessing steps and multivariate calibration algorithms [22, 23]. As two crucial quantitative modeling steps, sample subset partitioning and spectra variable selection have offered great advantages for improving the accuracy and robustness of prediction models [27–30]. On one hand, the predictive performance of the NIRS model heavily depends on the internal connection between spectral features and various analytical properties of the sample subset [28, 30, 31]. In this context, several studies have addressed the problem of sample subset partitioning. Galvão et al. [32] reported a stepwise procedure for selecting samples according to their differences in both x (instrumental responses) and y (predicted parameter) spaces. Subsequently, the use of a method incorporating sample set partitioning based on joint X–Y distances (SPXY) has gained wide acceptance as an advantageous alternative to existing sample subset partitioning strategies [28, 32]. On the other hand, the selection of spectral variables allows for the selection of the optimal variables subset that would greatly improve prediction performance by improving calibration reliability prior to inverse calibration (for primarily PLS). Because PLS does not completely solve the over-fitting problem observed in multivariate calibration and pattern recognition without variable selection [29], the following efficient variable selection methods have been devised: competitive adaptive reweighted sampling, selectivity ratio, variable importance for projection, Monte Carlo-uninformative variable elimination, and uninformative variable elimination [33–37]. Neither sample subset partitioning nor spectra variable selection had exhibited enough promise in previous studies to encourage their use for the determination of lignocellulosic properties [5, 13, 22, 23, 38]. Therefore, from a dual-optimization standpoint, it appears essential to identify the enhancement of both sample subset partitioning and variable selection that affects the performance of NIRS predictive models.

In the present study, we made comparisons between 70 sweet sorghum and 77 biomass sorghum samples by focusing on the main chemical components and bioethanol potentials of stem. Next, a reliable and accurate qualitative method was presented for lignocellulosic feedstock classification into sweet or biomass category using linear discriminant analysis. SPXY was then employed to partition calibration and validation subsets for 147 bioenergy sorghum samples. Next, six partial-optimized PLS models were developed for the prediction of chemical components and TEY. Ultimately, the main objective of this study was to compare five dual-optimized strategies for improving the predictive performance of PLS multivariate calibrations. In summary, this work provides powerful qualitative and quantitative tools to guide future feedstock selection, bioenergy crop breeding, and genetic modification programs to achieve more efficient biofuel production from diverse lignocellulosic feedstocks.

Results and discussion

Diversity of soluble sugar, cell wall components, and TEY

Sample diversity was reflected by the levels of soluble sugar, cellulose, hemicellulose, lignin, ash, and TEY shown in Fig. 1. In general, sweet sorghum samples exhibited much higher soluble sugar levels (10.8–47.6%) than did biomass sorghum samples (5.4–35.4%) (Fig. 1a; Additional file 1: Table A1). Conversely, biomass sorghum samples were relatively richer in cellulose, hemicellulose, and lignin. However, both sorghum categories displayed almost the same ash level, averaging 3.8%, which was relatively lower than that observed in other bioenergy crops [13]. To confirm the energy biofuel potential of bioenergy sorghum stem, we further carried out TEY originating from both hexose (C6) and pentose (C5) in this work (Fig. 1b; Additional file 1: Table A2). Due to the high levels of soluble sugar, sweet sorghum exhibited relatively higher TEY from C6 than did biomass sorghum, with the average value of 255.3 g/kg. By contrast, biomass sorghum exhibited more TEY from C5, with the average value of 112.1 g/kg. As a consequence, the total TEY levels of both sorghum categories were nearly the same. In particular, the total TEY of all 147 bioenergy sorghum samples ranged from 287.1 to 429.3 g/kg, with an average value of 340.9 g/kg. These results were similar to that of sorghum grain (312.0 g/kg) and show that bioenergy sorghum stem is a viable biofuel feedstock [1].

Because representative bioenergy sorghum samples displayed large population variability and broad distributions of sample characteristics, we further performed correlation analyses between chemical components and TEY values (Fig. 1c, d). Notably, a significantly positive correlation was observed between soluble sugar and TEY in both sweet sorghum and biomass sorghum samples ($p < 0.01$), indicating that soluble sugar greatly contributes to TEY in bioenergy sorghum. A high level of cellulose also exhibited a significantly positive correlation ($p < 0.01$) with TEY. Conversely, the hemicellulose level negatively correlated with TEY in the present work. Meanwhile, it was found that lignin and ash, known to contribute to biomass recalcitrance, strongly and negatively were correlated with TEY. Generally speaking, a higher hemicellulose level could generate more bioethanol. However, it exhibited significantly positive correlations with bioconversion barriers (e.g., ash and lignin) and negative correlations with soluble and insoluble carbohydrates in bioenergy sorghum. Therefore, reduction in lignin and/or ash contents could greatly enhance bioethanol potential. Moreover, it indicated to select ideal bioenergy sorghum varieties with relatively lower hemicellulose level, which could lead to higher cellulose and soluble sugar levels for producing a higher TEY level.

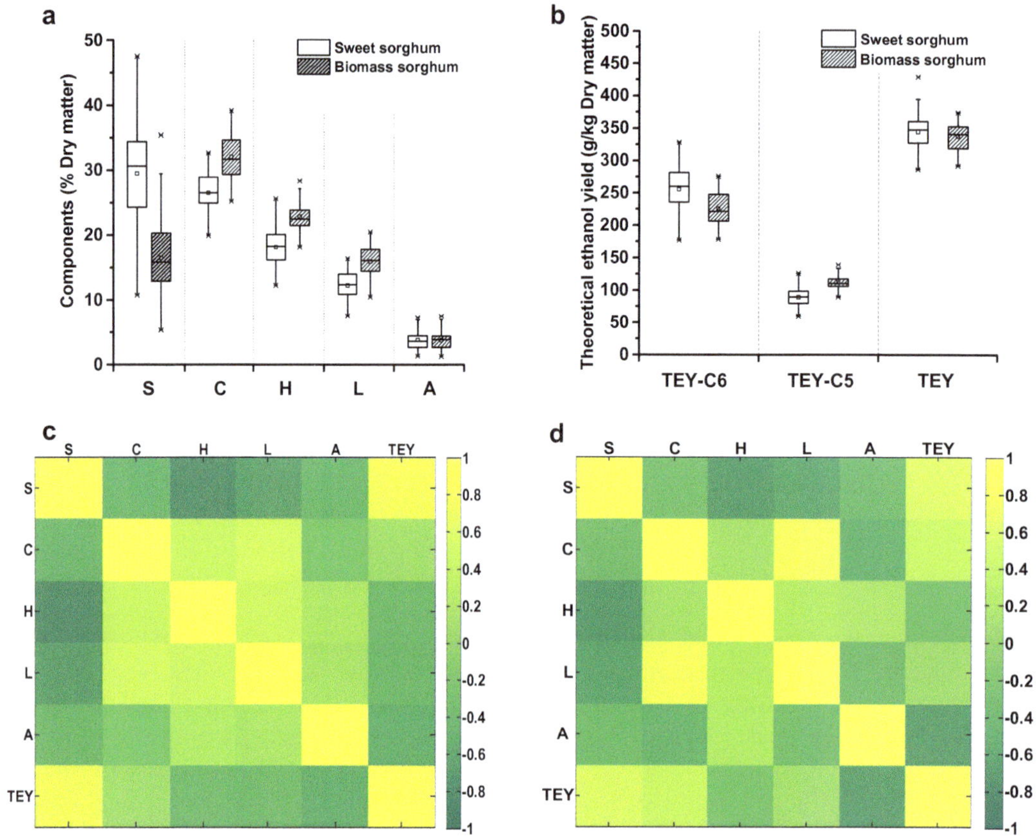

Fig. 1 Diversity of chemical components and theoretical ethanol yield, and their correlations in 147 bioenergy sorghums. *S* soluble sugar, *C* cellulose, *H* hemicellulose, *L* lignin, *A* ash, *TEY* theoretical ethanol yield, *TEY-C6* TEY from hexose, and *TEY-C5* TEY from pentose. **a** Chemical components of 147 bioenergy sorghum samples, **b** TEY of 147 bioenergy sorghum samples, **c** correlations among chemical components and TEY of 70 sweet sorghum samples, **d** correlations among chemical components and TEY of 77 biomass sorghum samples

Diversity of NIRS and qualitative classification of bioenergy sorghum samples

A briefly step-by-step flowchart for understanding the high-throughput qualitative and quantitative methodology is available in Fig. 2. Firstly, NIRS of 147 bioenergy sorghum samples was performed using wave numbers from 4000 to 10,000 with a resolution of 8 cm^{-1} (Fig. 3a). The main absorption band peaks occurred in the range from 4000 to 7400 cm^{-1} (Table 1). In previous studies, the strong peak at approximately 5150–5195 cm^{-1} was primarily attributed to O–H asymmetric stretching and O–H deformation bands of water [4, 6, 24]. For soluble sugar, the most important spectral regions were 5150–5195, 5776–5796, and 6775–6822 cm^{-1} [4]. The band deformation (O–H, C–H, and C–H_2), band stretching vibration (O–H, C–H, C–H_2, C–O, and C–C), and first overtone stretching (O–H) of cellulose greatly contributed to the absorption band peaks around 4015–4022, 4392–4412, 4760–4780, 5776–5796, 6329–6336, 6775–6822, and 7305–7328 cm^{-1} [4, 38]. The C–H and C–H_2 deformation and stretching vibration, and the first overtone O–H stretching band of hemicellulose were indicated by spectral changes within wavenumber ranges of 4285–4296, 4392–4412, and 6775–6822 cm^{-1} [4, 38]. Lignin could be identified at 4015–4022, 4392–4412, and 5776–5796 cm^{-1} by its stretching vibration (O–H, C–H, C–O, and C–C) and the overtone stretching band of O–H [24, 38]. Ash, an inorganic component, could not be detected directly using NIRS, but could be determined from its association with the organic component of bioenergy sorghum [13, 39]. Finally, as previously reported, TEY could be calibrated and predicted using absorption band peaks at around 7060, 5230, 4440, and 4330 cm^{-1} [4].

Based on representative reconstructed spectra variables, principal component analysis was primarily developed to identify outliers, as well as for sample comparison and classification [5]. As shown in Fig. 3b, we initially constructed a principal component analysis model whereby the 3D score plot of NIRS from 147 bioenergy sorghum samples displayed a relatively diverse and symmetrical distribution. The score points of sweet

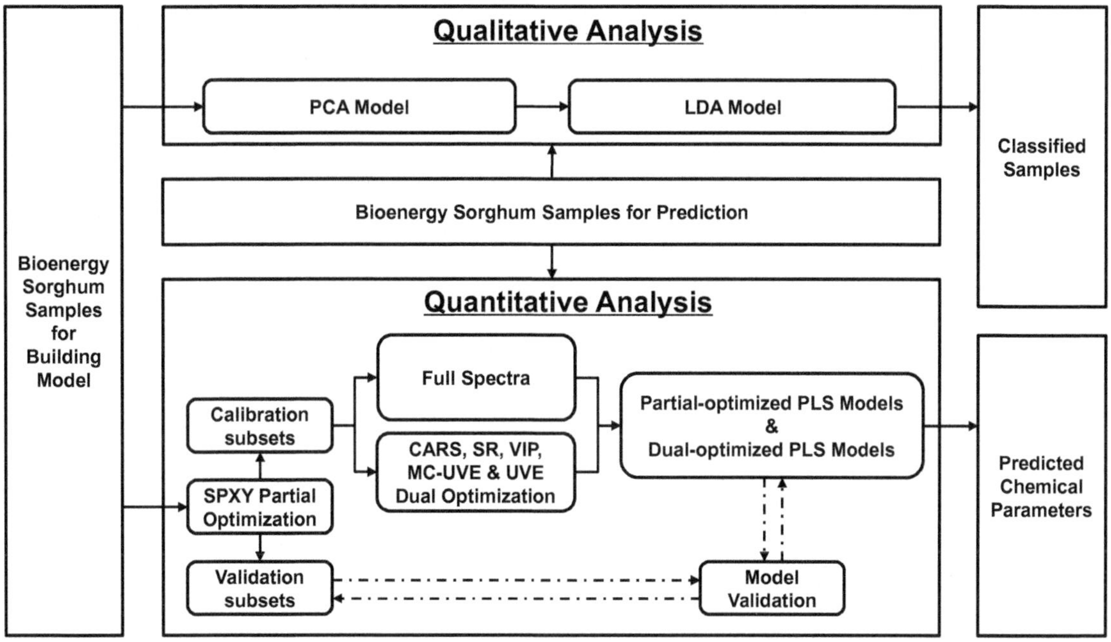

Fig. 2 Flowchart of NIRS qualitative and quantitative analyses. *PCA* principal component analysis, *LDA* linear discriminant analysis, *SPXY* sample set partitioning based on joint *X–Y* distances, *CARS* competitive adaptive reweighted sampling, *SR* selectivity ratio, *VIP* variable importance for projection, *UVE* uninformative variable elimination, *MC-UVE* UVE couple with the principle of Monte Carlo, *PLS* partial least squares

sorghum and biomass sorghum samples were not distinctly separated using three principal components (accounting for 95.9% of spectra variance), consistent with a previous bamboo compositional prediction study [24]. Using further qualitative analysis, linear discriminant analysis, which is based on principal component analysis and Mahalanobis distances, was employed to differentiate between the two categories of bioenergy sorghum samples in this study. Figure 3c shows the correct classification rate and variance of the total sample set after considering a maximum number of 40 principal components. Generally, both correct classification rate and variance increased sharply in step with increases observed for principal components (from 0 to 20), before reaching peaks at 99.3 and 100%, respectively. Meanwhile, Fig. 3d demonstrates the Mahalanobis distance from each sample to the models for sweet sorghum and biomass sorghum. It was obvious that only one sample was mismatched while 20 principal components were taken into account. Therefore, the results demonstrate that linear discriminant analysis could be a powerful and efficient tool for biomass feedstock classification for optimization of bioenergy processing [15]. In addition, the broad range of high absorbance and symmetrical distribution of score points indicated that the spectra of bioenergy sorghum samples possessed good sample representation and functioned well during calibration and prediction.

Optimization of calibration and validation subset partitioning by SPXY

It is well known that the accuracy and robustness of NIRS quantitative analysis models rely heavily on calibration and validation subsets [30]. In this case, a representative calibration subset must be selected on the basis of NIRS and analytical properties that were extracted from a pool of real samples [32]. Meanwhile, representative external validation samples should also be selected to assess the quality of the quantitative analysis model [13]. Nonetheless, studies of sample subset partitioning have been largely overlooked in the bioenergy sector [5, 22, 23, 38]. In recent years, many strategies have been developed to address this problem from an optimization perspective, such as random sampling and Kennard–Stone. Random sampling is a popular technique because of its simplicity. But it does not guarantee the representativity of the sample subset, nor does it prevent extrapolation problems. To solve these problems, Kennard–Stone was developed to cover the multidimensional space in a uniform manner by maximizing the Euclidean distances between the instrumental response vectors (x) of the selected samples. However, the predictive performance of the NIRS model heavily depends on the internal connection between instrumental response vectors (x) and various analytical properties (y) of the sample subset. Hence, a uniquely advantageous method, designated SPXY, was initially employed to sort 147 bioenergy sorghum samples into

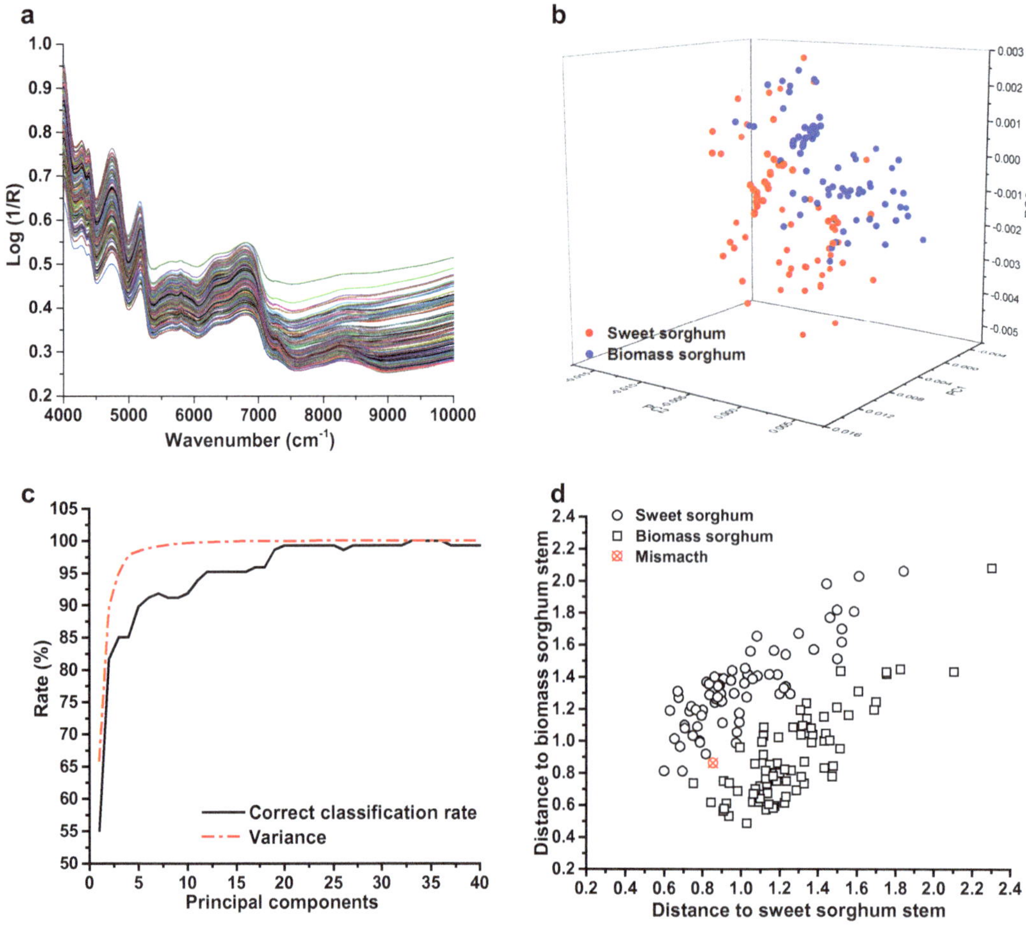

Fig. 3 Description and qualitative classification of 147 bioenergy sorghum samples. **a** Original spectra of 147 bioenergy sorghum samples, **b** 3D plot of the principal component analysis scores of 147 bioenergy sorghum samples, **c** the correct classification rate and variance obtained by linear discriminant analysis, **d** Mahalanobis distance of 147 bioenergy sorghum samples (principal components = 20)

Table 1 The main absorption band peak location of chemical components in bioenergy sorghum

Wavenumber (cm^{-1})	Component	Bond vibration
4015–4022	C, L	C–H$_{str}$, C–C$_{str}$
4285–4296	H	C–H$_{str}$, C–H$_{def}$
4392–4412	C, H, L	C–H$_{str}$, C–H$_{2str}$, O–H$_{str}$, C–O$_{str}$, C–H$_{def}$, C–H$_{2def}$
4760–4780	C	O–H$_{def}$, C–H$_{def}$, O–H$_{str}$
5150–5195	W, S	O–H$_{asym}$, O–H$_{str}$, O–H$_{def}$
5776–5796	S, C, L	1st OT C–H$_{str}$
6329–6336	C	1st OT C–H$_{str}$
6775–6822	S, C, H	1st OT C–H$_{str}$
7305–7328	C	1st OT C–H$_{str}$, C–H$_{def}$

C cellulose, *H* hemicellulose, *L* lignin, *S* soluble sugar, *W* water

calibration and validation subsets for multivariate calibration in the present paper. Notably, this method extends the Kennard–Stone algorithm by encompassing the variability in both the x and y dimensions for the calculation of inter-sample distances [31]. As a consequence, based on the value of predicted objects, one of every four samples was included in the validation subset, while the remaining samples were used for the calibration subset.

As shown in Fig. 4, histograms of calibration and validation subsets for soluble sugar, cellulose, hemicellulose, lignin, ash, and TEY are presented. The dashed lines overlaid upon each histogram represent normal distributions and were used to embody the discrepancy between each histogram and normality. In general, all subsets optimized using SPXY displayed a relatively wide and symmetrical distribution. Despite the fact that the 147 bioenergy sorghum samples originated from a multispecies feedstock population that included various genotypes and phenotypes, almost all the calibration and validation subsets for each property exhibited relatively wide and symmetrical distributions in the present

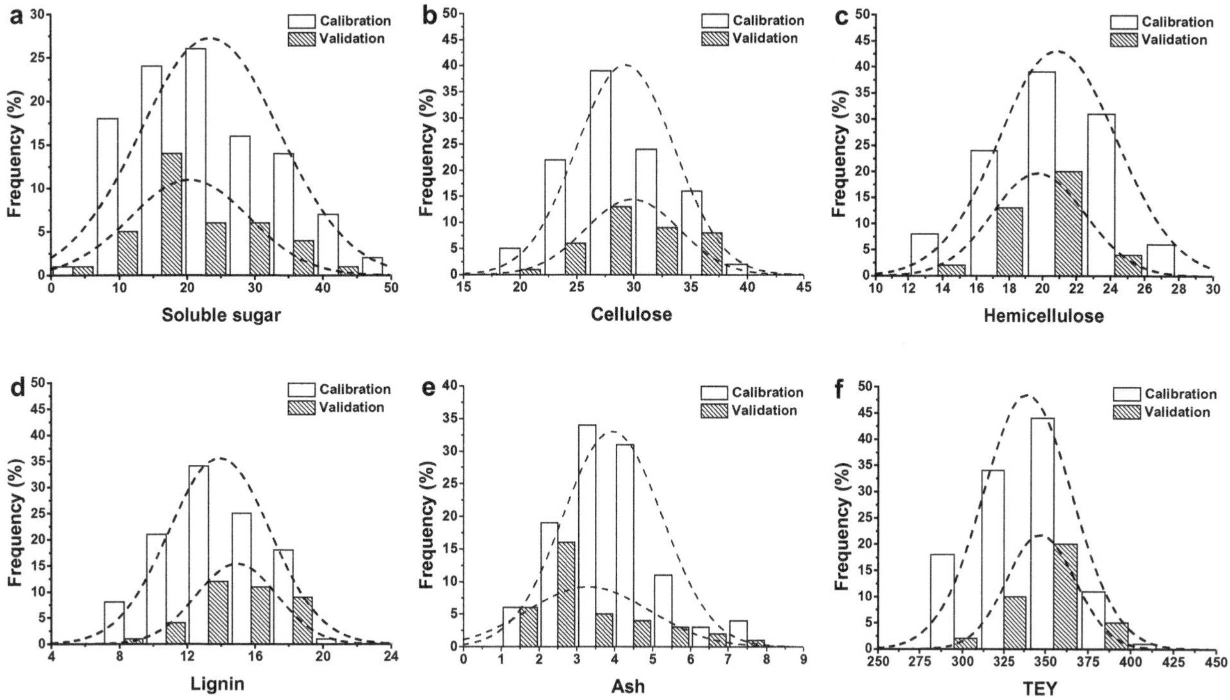

Fig. 4 Histograms of chemical components and theoretical ethanol yield. Soluble sugar (**a**), cellulose (**b**), hemicellulose (**c**), lignin (**d**), ash (**e**), and theoretical ethanol yield (**f**) for calibration and external validation subsets that were partitioned by sample set partitioning based on joint *X*–*Y* distances

study (Figs. 3a, b, 4). Additionally, each calibration subset represented nearly the same distributions as that of the corresponding validation subset; no distinct bimodal, skewed, or uniform distributions were observed (Fig. 4). In order to further demonstrate the effect of optimization of SPXY on calibration and validation subset partitioning, six principal component analysis models were established for soluble sugar, cellulose, hemicellulose, lignin, ash, and TEY (Fig. 5). Similarly, the 3D score plots of NIRS from both calibration and validation subsets were well mixed and displayed relatively symmetrical distributions. Hence, these results indicate that the SPXY method could efficiently optimize NIRS results as well as the analytical properties of both the calibration subset and related validation subset. These optimizations thus led to an accurate and robust NIRS quantitative analysis model.

Partial-optimized PLS models

Based on the SPXY-optimized calibration and validation subsets, six partial-optimized PLS models were developed using full spectra for the prediction of soluble sugar, cellulose, hemicellulose, lignin, ash, and TEY. To reduce any multiplicative and additive effects resulting from instrument settings or variations stemming from sample and environmental conditions, the spectra averaged from 64 scans were subjected stepwise to multiplicative scatter correction, then Norris derivative filtering, then first derivative analysis. In addition, all six models were fully cross-validated using the "leave-one-out" method and the optimal number of principal components (from 6 to 9) for each model was determined using the root mean square error of calibration (RMSEC) and the root mean square error of cross-validation (RMSECV) [13]. The numbers of samples for each of the previously described calibration and validation subsets used for chemical components and TEY determinations were further reduced by the removal of sample outliers, as described by Xu et al. [26].

Summary statistics of the partial-optimized PLS model are shown in Table 2. Based on the data pretreatment discussed above, a fair prediction of the soluble sugar, cellulose, hemicellulose, lignin, ash, and TEY was obtained using the strongest model, which exhibited low RMSEC (0.65–12.95) and RMSECV (0.74–14.18) relative to the other models. The coefficients of determination for calibration (R^2_C) and cross validation (R^2_{CV}) were each within the high ranges of 0.72–0.95 and 0.61–0.94, respectively. These results demonstrate the effect of SPXY enhancement on calibration of relevant bioenergy sorghum sample properties. In addition,

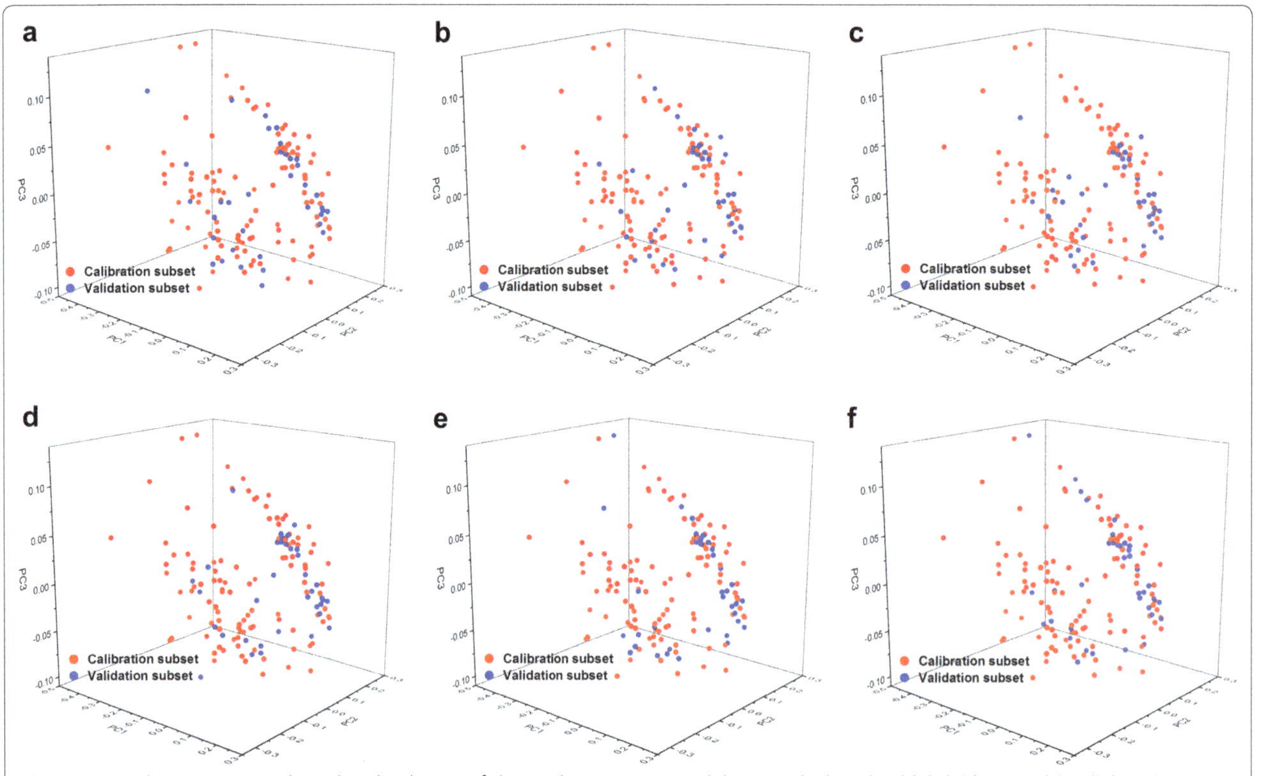

Fig. 5 Principal component analysis plots distribution of chemical components and theoretical ethanol yield. Soluble sugar (**a**), cellulose (**b**), hemicellulose (**c**), lignin (**d**), ash (**e**), and theoretical ethanol yield (**f**) for calibration and external validation subsets that partitioned by sample set partitioning based on joint *X*–*Y* distances

Table 2 Summary statistics of partial-optimized partial least squares models for chemical components and theoretical ethanol yield (TEY) of bioenergy sorghum

Parameter	Calibration				Cross validation		External validation				
	Number	PCs	RMSEC	R_C^2	RMSECV	R_{CV}^2	Number	RMSEP	R_V^2	RPD	RER
Soluble sugar	108	7	2.27	0.95	2.66	0.93	37	2.57	0.91	3.39	16.42
Cellulose	108	9	0.93	0.95	1.26	0.91	37	1.18	0.91	3.46	16.30
Hemicellulose	108	7	0.75	0.95	0.85	0.94	37	0.64	0.95	4.33	25.15
Lignin	107	8	0.65	0.95	0.74	0.94	37	0.65	0.92	3.66	19.72
Ash	108	7	0.68	0.72	0.82	0.61	37	0.49	0.90	3.25	12.58
TEY	108	6	12.95	0.76	14.18	0.72	37	8.08	0.84	3.42	17.61

PCs principal components

a summary of published literature using the ratio of performance to deviation (RPD) suggested that excellent calibration models must exhibit a RPD value greater than 3, while RPD values between 2.4 and 1.5 are considered acceptable [4, 24]. Moreover, the American Association of Cereal Chemists Method 39-00 demonstrated that any model that has a range error ratio (RER) ≥10 is acceptable for quality control; if the RER of a model is greater than 15, the model is considered very good for research quantification [24]. In the present study, almost all RPD and RER values were greater than 3 and 15, respectively (except for the RER of ash) (Table 2). As a consequence, Fig. 6 illustrates that good coefficients were obtained using the square of the correlations between predicted and reference values for soluble sugar, cellulose, hemicellulose, lignin, ash, and TEY after external validation (R_V^2 > 0.839). All validation subsets were well predicted using the model with a relatively low root mean square error of prediction (RMSEP), which further

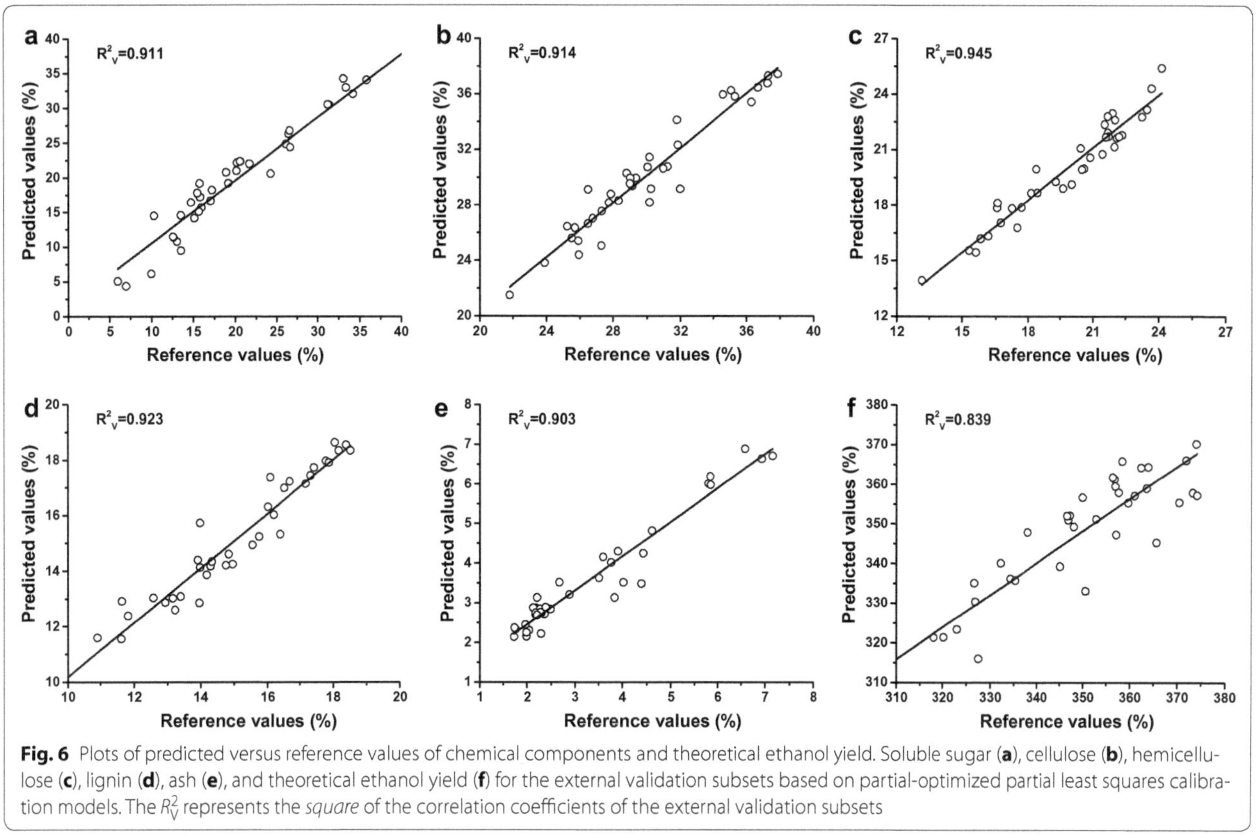

Fig. 6 Plots of predicted versus reference values of chemical components and theoretical ethanol yield. Soluble sugar (**a**), cellulose (**b**), hemicellulose (**c**), lignin (**d**), ash (**e**), and theoretical ethanol yield (**f**) for the external validation subsets based on partial-optimized partial least squares calibration models. The R^2_V represents the *square* of the correlation coefficients of the external validation subsets

demonstrates the utility of sample subsets optimization for prediction of multispecies feedstock properties. For the predictions for sorghum samples, partial-optimized PLS models were better than previous models for soluble sugar predication [6, 26, 40] and were superior to a series of studies that focused on cell wall components [4, 25, 26, 41] and TEY prediction [4]. In addition, a review of NIRS research published in recent years suggested that in most cases, partial-optimized PLS models are also effective for studying cellulose (glucan), hemicellulose (mainly xylan), lignin, and ash from multispecies feedstock when comparing values of R^2_V, RMSEP, RPD, and RER [5, 13, 16, 23, 24, 39, 42]. Therefore, past results coupled with the results of this study collectively indicate that sample subsets optimization can enhance the accuracy and robustness of NIRS quantitative analysis models for study of multispecies biomass resource streams.

Optimization of the variable selection by CARS, SR, VIP, MC-UVE, and UVE

Due to its potential for extracting chemical information from overdetermined systems, multivariate calibration has been extensively used for applications in the analysis of multi-component spectroscopic data, especially in the NIRS field [33]. Additionally, recent approaches have shown that proper spectral variable selection has proved to be a critical step for multivariate spectroscopic calibration; many strategies have been developed to address this problem from an optimization perspective in recent years [29, 37, 43]. Competitive adaptive reweighted sampling (CARS) was proposed in 2009 by Li et al. [33]. This specialized strategy for variable selection allows for the selection of an optimal variable subset existing in the full spectra coupled with PLS regression using the simple but effective principle "survival of the fittest," upon which Darwin's Evolution Theory is based [33]. The selectivity ratio (SR) method, proposed first for biomarker discovery, is obtained by calculating the ratio of explained to residual variance of the *X* variables upon the *Y* target-projected component [34]. The variable importance for projection (VIP) method selected those *X* variables that contribute most to the underlying variation in the *X* variables [35]. Currently, one of the most general variable selection methods is the uninformative variable elimination (UVE) method, which evaluates the reliability of each variable in the model using stability criteria and eliminates uninformative variables. Moreover, Cai et al. [44] in 2008 reported a modified method combining UVE with Monte Carlo principles (MC-UVE) to achieve satisfactory prediction results in comparison to many other methods of wavelengths selection (MC-UVE). Although the methods listed above have been applied widely in analytical chemistry, none has yet been used to optimize the multivariate calibration model of lignocellulosic components analysis, especially for bioenergy sorghum samples. Therefore, it would be worthwhile to apply variable selection to biomass multivariate calibration modeling and conduct a comparison of five standard methods for NIRS model optimization.

As shown in Fig. 7, the optimal variable subsets of soluble sugar, cellulose, hemicellulose, lignin, ash, and TEY were selected using algorithms mentioned above. First, it was possible to demonstrate that the variable numbers selected by each method could be ranked in the order CARS < SR < VIP < MU-UVE < UVE for soluble sugar, cellulose, hemicellulose, and lignin (Fig. 7a–d). Meanwhile, the order of variable numbers for both ash and TEY was SR < CARS < VIP < MU-UVE < UVE (Fig. 7e, f). As a result, CARS and SR exhibited higher efficiency, as measured using "informative variables" selection, than did the other algorithms. Additionally, it could be observed that the spectral range of about 4000–7400 cm^{-1} was most effective for variable selection, consistent with our previous finding showing that to be a strong absorbance region (Figs. 3a, 7).

In order to perform a thorough analysis, we designated the characteristic variables that were selected using 5, 4, 3, 2, 1, or 0 methods as follows: "most important," "very important," "important," "normal," "less important," and "uninformative" variables, respectively. Summary statistics for the characteristic variables of soluble sugar, cellulose, hemicellulose, lignin, ash, and TEY are provided in Table 3. Due to the basic principles that generally apply to characteristic variables, "most important" variables and "very important" variables could be considered to be the most important informative variables within the full spectra. In the present study, organic properties (soluble sugar, cellulose, hemicellulose, lignin, and TEY) achieved large numbers of "most important" variable and "very important" variable, ranging from 4 to 7 and from 37 to 64, respectively. By contrast, the number of them from ash were much lower than those of the other five properties. This might be due to the fact that ash, an inorganic mixture, is not easily or directly measurable by NIRS, as are organic properties [13, 39]. As previous studies have confirmed, the stability and accuracy of the multivariate calibration model relied heavily on strong informative variables [28]. Although the "most important" variable and "very important" variable make up only a small proportion of the full spectra, numbers

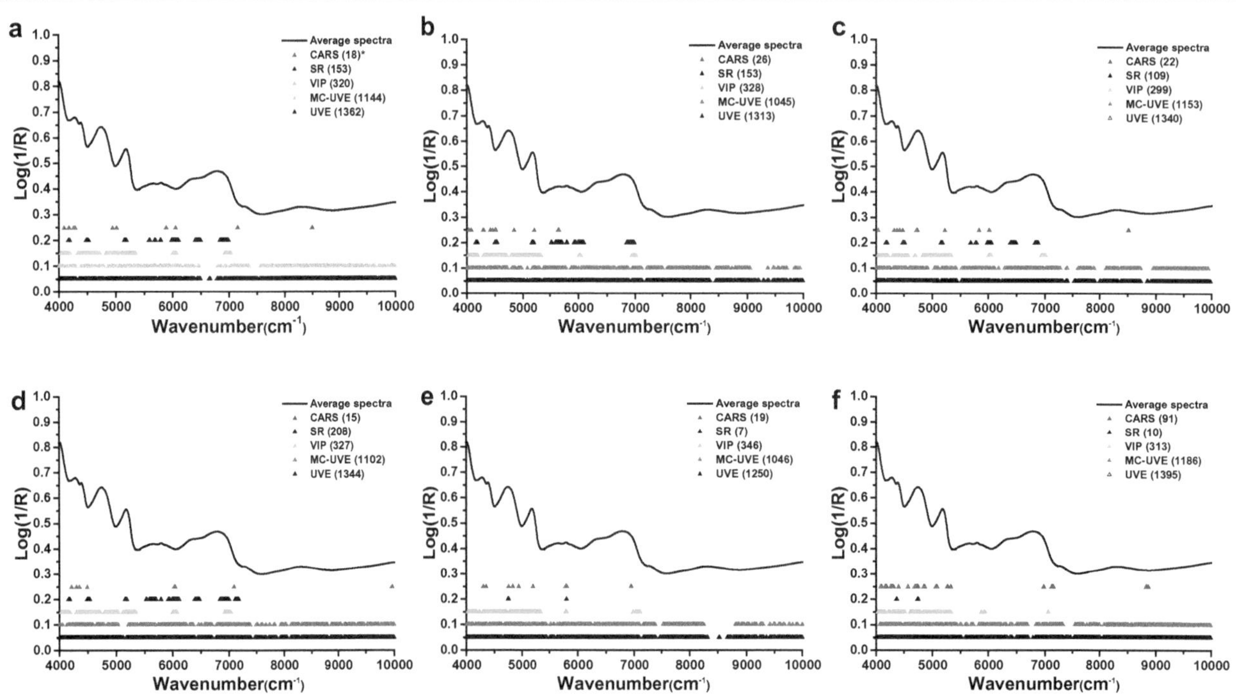

Fig. 7 The optimal variable selection in calibration subset by CARS, SR, VIP, MC-UVE, and UVE. *CARS* competitive adaptive reweighted sampling, *SR* selectivity ratio, *VIP* variable importance for projection, *UVE* uninformative variable elimination, and *MC-UVE* UVE couple with the principle of Monte Carlo. Soluble sugar (**a**), cellulose (**b**), hemicellulose (**c**), lignin (**d**), ash (**e**), and theoretical ethanol yield (**f**). *Asterisk* refers to the number of variables selected by each method

Table 3 The statistics of characteristic variables for chemical components and theoretical ethanol yield (TEY)

Parameter	MIV		VIV		IV		NV		LIV		UIV	
	N	*P* (%)	*N*	*P* (%)	*N*	*P* (%)	*N*	*P* (%)	*N*	*P* (%)	*N*	*P* (%)
Soluble sugar	4	0.3	64	4.1	230	14.8	908	58.3	215	13.8	136	8.7
Cellulose	5	0.3	56	3.6	240	15.4	824	52.9	248	15.9	184	11.8
Hemicellulose	4	0.3	37	2.4	251	16.1	918	59.0	166	10.7	181	11.6
Lignin	7	0.5	62	4.0	276	17.7	826	53.1	233	15.0	153	9.8
Ash	1	0.1	14	1.0	269	17.3	796	51.1	208	13.4	269	17.3
TEY	4	0.3	55	3.5	234	15.0	921	59.2	211	13.6	132	8.5

MIV most important variable, *VIV* very important variable, *IV* important variable, *NV* normal variable, *LIV* less important variable, *UIV* uninformative variable, *N* number of selected variables, *P* the proportion of selected variables in total variables

of these variables unquestionably could affect the predictive performance of each property for multivariate calibration. On the other hand, judicious elimination of uninformative variables can also bolster the effectiveness of the calibration model and improve its predictive performance [45]. In this study, the "less important" variable and "uninformative" variable numbers for each property ranged from 166 to 248 and from 132 to 269, accounting for 10.7–15.9% and 8.5–17.3% of the total 1557 variables observed, respectively (Table 3). Therefore, variable selection, an effective optimization strategy, could be an indispensable step for NIRS modeling of lignocellulosic feedstock. Furthermore, the five optimal variable subsets selected here provide very useful information to facilitate further biomass components prediction (Fig. 7).

Dual-optimized PLS models

Coupled with SPXY-optimized sample subsets, five methods for variable selection were employed to develop 30 dual-optimized PLS models for quantitative analysis of soluble sugar, cellulose, hemicellulose, lignin, ash, and TEY. Meanwhile, the six partial-optimized PLS models described above served as controls. Summary statistics for a total of 36 defined models are presented in Table 2 and Additional file 1: Tables A3–A7. Compared with other dual-optimized models, CARS-SPXY exhibited the greatest positive impact upon the stability of PLS models when comparing values of principal components (5–8), R^2_C (0.73–0.96), RMSEC (0.65–11.50), R^2_{CV} (0.69–0.95), and RMSECV (0.68–13.20) (Fig. 8 and Additional file 1: Table A3). Furthermore, lower RMSEP (0.48–7.41) and higher RPD (3.34–4.73) and RER (12.94–27.45) values were obtained using the robust CARS-SPXY dual-optimized PLS models, which were significantly better than those of four other dual-optimized models for the most relevant properties. Notably, the use of one in four samples for external validation led to strong correlations between predicted and reference values (Additional file 2: Fig. A1). These results demonstrate that CARS-SPXY was the best method to clearly improve upon the accuracies of the prediction performance of the PLS model among these five dual-optimized methods. These results are consistent with our previously stated results that CARS displayed relatively higher efficiency over other methods in the selection of informative variables as well as for judicious elimination of uninformative variables (Fig. 7).

Meanwhile, the sample subsets and spectral variable subsets were optimized using VIP-SPXY, MC-UVE-SPXY, and UVE-SPXY models using the following parameters: a low number of principal components; high R^2_C, R^2_{CV}, R^2_V; low RMSEC, RMSECV, RMSEP; and high RPD, RER (Additional file 1: Tables A5–A7). Additional file 2: Figures A3–A5 demonstrate that all relevant properties were successfully predicted by the three models mentioned above, which generally displayed relatively higher correlations between predicted and reference values than controls (Fig. 6). These results indicated that VIP-SPXY, MC-UVE-SPXY, and UVE-SPXY models could achieve higher performance for predicting soluble sugar, cellulose, hemicellulose, lignin, ash, and TEY (Fig. 8). Conversely, it was found that SR-SPXY significantly decreased the stability of the calibration models because the RMSEC and RMSECV for each property sharply increased. Meanwhile, significant decreases in R^2_C and R^2_{CV} were observed in the present study which were generally lower than controls (Table 2; Additional file 1: Table A4). As a consequence, RPD and RER decreased significantly and relatively lower R^2_V values were obtained for the validation set of each relevant property. These results probably stem from the inability of the SR method to achieve informative variable selection and uninformative variables elimination (Figs. 7, 8; Additional file 1: Table A4 and Additional file 2: Figure A2).

In summary, dual-optimized PLS models generally display significantly higher robustness and accuracy than partial-optimized PLS models for the prediction of relevant properties by utilizing dual optimization of sample subset partitioning and variable selection. Additionally, CARS-SPXY was the most effective dual-optimization method for improving the predictive performance of the PLS model in this study (Fig. 8).

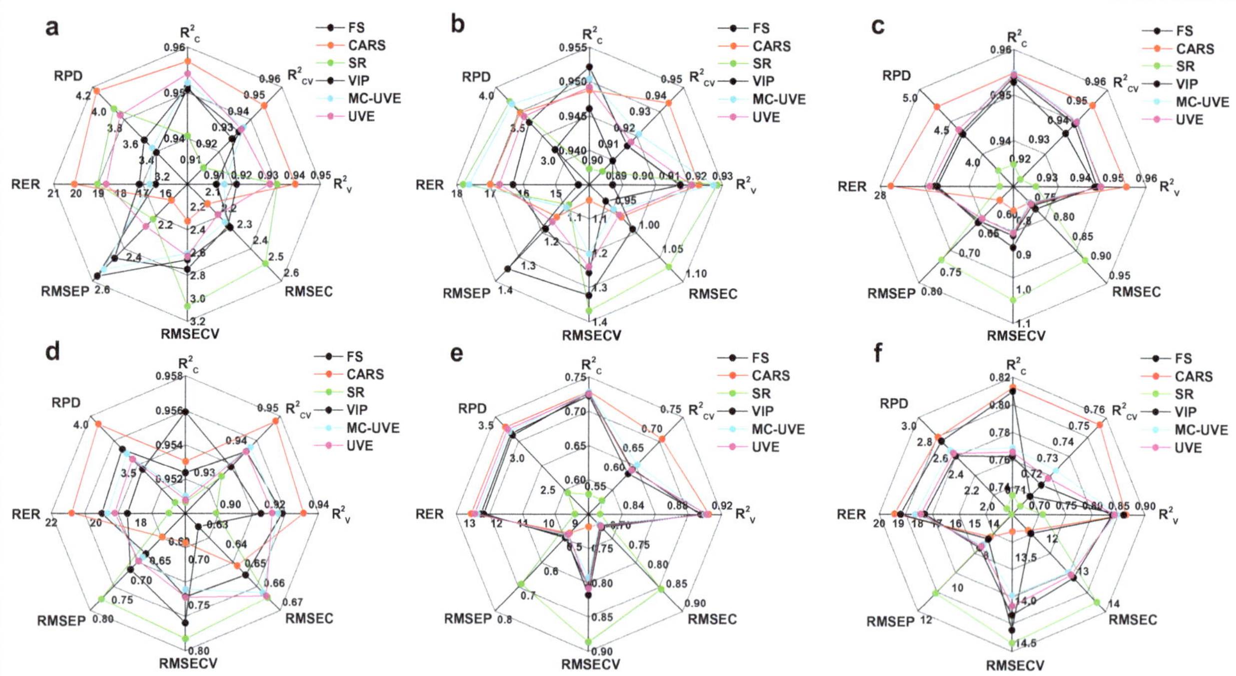

Fig. 8 The performance of 30 dual-optimized PLS models and 6 partial-optimized PLS models. *FS* full spectra, *CARS* competitive adaptive reweighted sampling, *SR* selectivity ratio, *VIP* variable importance for projection, *UVE* uninformative variable elimination, *MC-UVE* UVE couple with the principle of Monte Carlo. Soluble sugar (**a**), cellulose (**b**), hemicellulose (**c**), lignin (**d**), ash (**e**), theoretical ethanol yield (**f**)

Conclusions

A large number of bioenergy sorghum samples, including 70 sweet sorghum and 77 biomass sorghum samples, exhibited diverse chemical components and high potential for bioethanol production. Using the NIRS of total 147 samples, a qualitative analysis between sweet sorghum and biomass sorghum was conducted via the linear discriminant analysis model and 99.3% of samples were correctly classified using a set of 20 principal components. Meanwhile, sample subset partitioning and variable selection could substantially enhance the predictive performance of a PLS model for determination of chemical components and TEY. These dual-optimized models generally achieved high R^2_C, R^2_{CV}, R^2_V, RPD, and RER values and low RESEC, RESECV, and RMSEP values relative to partial-optimized models, demonstrating relatively higher enhancement of both robustness and accuracy of prediction models. Finally, comparative analyses of five dual-optimized methods indicate that CARS-SPXY was the most efficient method for improving the predictive performance of PLS multivariate calibration. Therefore, high-throughput dual-optimized NIRS models could facilitate feedstock selection, bioenergy crop breeding, and genetic modification for efficient biofuel production.

Methods

Sample selection and preparation

To select bioenergy sorghum samples with diverse genetic, environment, and cultivation background is of importance for accuracy and robustness model building. Since 2006, National Energy R&D Center for Non-food Biomass (NECB) has collected nationwide sorghum germplasm resources to make hybrid breeding for a bioethanol purpose. Till 2013, the most representative six sweet type and six biomass type hybrids have been developed by NECB for commercial scale production. To evaluate the yield performance and energy potential, we planted these twelve bioenergy sorghum varieties under different cultivation treatments including plant spacing and fertilizer rate, in Zhuozhou, Hebei and Uxin, Inner Mongolia in 2013 and 2014. The two study sites were some 800 km apart in northern China. A total of 147 representative samples were taken at the two sites for the high-throughput NIRS models building in this study.

All sorghum samples were harvested on their dates of physiological maturity after a growth period of between 110 and 150 days. Preparation of samples was performed according to the process described by Li et al. [3] with minor modifications. In this study, stem samples were ground using a crusher mill into particles 1–2 cm in size. The particles were first dried at 45 °C for 48 h after inactivation at 105 °C for 20 min. Next, the dried particles were ground into powders and passed through a combined −40/+80 mesh screen. A total of 147 mesh-screened samples (as dry matter) were stored in a dry container until use.

Biomass compositional analysis and TEY calculation

The soluble sugar was extracted with distilled water and determined by the anthrone/H_2SO_4 method using a UV–VIS spectrometer (TU-1901, Beijing Purkinje Instruments Co. Ltd., Beijing, China) according to Li et al. [3] The standard curve was plotted using D-glucose as the standard (purchased from Xilong Scientific Co., Ltd., China).

The structural carbohydrates (i.e., glucose, xylose, and arabinose) and lignin (acid-soluble lignin and acid-insoluble lignin) were extracted using a two-step sulfuric acid hydrolysis process. Sample quantity and composition were measured using an HPLC system (1260 series, Agilent Technologies, Santa Clara, CA, USA) equipped with an Aminex HPX-87H chromatography column (300 × 7.8 mm, particle size 9 µm, Bio-Rad Laboratories, Hercules, CA, USA), a UV–VIS spectrometer (TU-1901, Beijing Purkinje Instruments Co. Ltd.) and a muffle furnace (VULCAN 3-550, Densply International Inc., York, PA, USA) according to the Laboratory Analytical Procedures from National Renewable Energy Laboratory [46]. Dry matter (2 g per sample) was added to ceramic crucibles (30 mL volume) to determine ash content after incineration in the muffle furnace. Cellulose content was calculated from the glucose content. Hemicellulose was calculated from the sum of xylose and arabinose contents. Lignin was calculated from the sum of acid-soluble lignin and acid-insoluble lignin contents.

Theoretical ethanol yield (TEY), as reported by Zhao et al. [47], was calculated from the C6 sugar (soluble sugar and cellulose), C5 sugar (hemicellulose), and total sugar (C6 sugar and C5 sugar) in the dry matter, with minor modifications. In order to obtain the TEY results in g/kg, the equation below was modified and the term dry biomass (t/ha) was removed.

$$\text{TEY (g/kg)} = \text{TEY-C6 (g/kg)} + \text{TEY-C5 (g/kg)}$$

The results of TEY values could be found in Additional file 1: Table A2.

NIRS measurement and pretreatments

All dry matter was scanned and recorded from three separate samplings using a Thermo Antaris II FT-NIR (Thermo Scientific Inc., Madison, WI, USA) equipped with a diffuse reflectance accessory. Each spectrum was averaged from 64 scans at a resolution of 8 cm^{-1} in the wavenumber range of 4000–10,000 cm^{-1}, including 1557 spectral variables. The spectrometer was controlled and data were acquired using TQ Analyst software (ver. 9.3.107, Thermo Scientific Inc.). Spectra were first adjusted using multiplicative scatter correction to correct spectra for scatter. Next, a Norris derivative filter was used to reduce random noise. The first derivative was used to resolve spectra peak overlap and eliminate linear baseline drift [26]. The purpose of the corrections above was to remove multiplicative and additive effects stemming from instrument settings or variations caused by sample and environmental conditions [25].

Qualitative analysis

A principal component analysis model was calculated using TQ Analyst 9.3 to evaluate the spectral distribution of bioenergy sorghum samples based on 3D plots (Fig. 3b). The principal component analysis models included 147 samples and incorporated a maximum number of 40 principal components. As a well-known and frequently used technique, linear discriminant analysis was employed for data classification in this study. The Mahalanobis distance was determined between the samples and each sorghum class center. Meanwhile, the correct classification rate was calculated to obtain a classification result according to He et al. [15].

$$\begin{aligned}\text{TEY-C6 (g/kg)} = {} & [\text{soluble sugar (\%)} + \text{cellulose (\%)} \times 1.11\ (\text{conversion factor of sugar from cellulose})] \\ & \times 0.51\,(\text{conversion factor of ethanol from sugar}) \\ & \times 0.85\,(\text{process efficiency of ethanol from sugar}) \\ & \times 1000/0.79\left(\text{specific gravity of ethanol, g mL}^{-1}\right)\end{aligned}$$

$$\begin{aligned}\text{TEY-C5 (g/kg)} = {} & \text{hemicellulose(\%)} \times 1.11\ (\text{conversion factor of sugar from hemicellulose}) \\ & \times 0.51\,(\text{conversion factor of ethanol from sugar}) \\ & \times 0.85\,(\text{process efficiency of ethanol from sugar}) \\ & \times 1000/0.79\left(\text{specific gravity of ethanol, g mL}^{-1}\right)\end{aligned}$$

Quantitative analysis

In the present study, a stepwise procedure reported by Galvão et al. [32] was employed to select samples according to their differences in both measured properties and NIRS. Based on this method, one of every four samples was sorted into external validation subsets using ChemDataSolution, ver. 2.0 (Dalian ChemData Solution Technology Co. Ltd., Dalian, China) to obtain a fair multivariate prediction [48] and the remaining samples were used for the calibration subsets.

The previously described NIRS variables were selected using the software ChemDataSolution ver. 2.0. Before determination of quantitative chemical components and TEY, CARS, SR, VIP, MC-UVE, and UVE models were employed for NIRS optimization in this study.

A total of 36 PLS multivariate calibrations were developed using TQ Analyst software (ver. 9.3.107) [49], which predicted one property at a time based on partial optimization or dual optimization in the present study. The "leave-one-out" method was recommended for cross-validation when developing PLS models, in order to select the optimal number of factors and to avoid over-fitting [13, 25]. The performance of multivariate calibrations was evaluated using RMSEC, RMSECV, RMSEP, and R^2_C, R^2_{CV} , and R^2_V. Furthermore, the RPD and the RER were also calculated to ascertain the potential of the NIRS models for application to breeding or industry screening [1, 24, 25, 50].

Statistical analysis

All chemical assays were conducted in triplicate and the average values are presented as percentage of dry weight. Correlation coefficients were calculated by performing Pearson's rank correlation analysis using IBM SPSS Statistics V.22. This analysis used mean values and coefficient coefficients of variation were calculated from all original determinations and defined as the ratio of the standard deviation to the mean value.

Additional files

Additional file 1: Table A1. Descriptive statistics of chemical components (%) for the 147 bioenergy sorghum samples. **Table A2.** Descriptive statistics of theoretical ethanol yield (TEY, g/kg) for the 147 bioenergy sorghum samples. **Table A3.** Summary statistics of CARS-SPXY dual optimized PLS model for the determination of soluble sugar, cellulose, hemicellulose, lignin and theoretical ethanol yield (TEY). **Table A4.** Summary statistics of SR-SPXY dual optimized PLS model for the determination of soluble sugar, cellulose, hemicellulose, lignin and theoretical ethanol yield (TEY). **Table A5.** Summary statistics of VIP-SPXY dual optimized PLS model for the determination of soluble sugar, cellulose, hemicellulose, lignin and theoretical ethanol yield (TEY). **Table A6.** Summary statistics of MC-UVE-SPXY dual optimized PLS model for the determination of soluble sugar, cellulose, hemicellulose, lignin and theoretical ethanol yield (TEY). **Table A7.** Summary statistics of UVE-SPXY dual optimized PLS model for the determination of soluble sugar, cellulose, hemicellulose, lignin and theoretical ethanol yield (TEY).

Additional file 2: Figure A1. Plots of predicted versus measured values of parameters. Soluble sugar (a), cellulose (b), hemicellulose (c), lignin (d), ash (e), and theoretical ethanol yield (f) for the external validation subsets based on CARS-SPXY dual-optimized PLS models. The R^2_V represents the square of the correlation coefficients of the external validation subsets. **Figure A2.** Plots of predicted versus measured value of parameters. Soluble sugar (a), cellulose (b), hemicellulose (c), lignin (d), ash (e), and theoretical ethanol yield (f) for the external validation subsets based on SR-SPXY dual-optimized PLS models. The R^2_V represents the square of the correlation coefficients of the external validation subsets. **Figure A3.** Plots of predicted versus measured value of parameters. Soluble sugar (a), cellulose (b), hemicellulose (c), lignin (d), ash (e), and theoretical ethanol yield (f) for the external validation subsets based on VIP-SPXY dual-optimized PLS models. The R^2_V represents the square of the correlation coefficients of the external validation subsets. **Figure A4.** Plots of predicted versus measured value of parameters. Soluble sugar (a), cellulose (b), hemicellulose (c), lignin (d), ash (e), and theoretical ethanol yield (f) for the external validation subsets based on MC-UVE-SPXY dual-optimized PLS models. The R^2_V represents the square of the correlation coefficients of the external validation subsets. **Figure A5.** Plots of predicted versus measured value of parameters. Soluble sugar (a), cellulose (b), hemicellulose (c), lignin (d), ash (e), and theoretical ethanol yield (f) for the external validation subsets based on UVE-SPXY dual-optimized PLS models. The R^2_V represents the square of the correlation coefficients of the external validation subsets.

Abbreviations

TEY: theoretical ethanol yield; NIRS: near infrared spectroscopy; SPXY: sample set partitioning based on joint *X–Y* distances method; PLS: partial least squares; CARS: competitive adaptive reweighted sampling; SR: selectivity ratio; VIP: variable importance for projection; UVE: uninformative variable elimination; MC-UVE: UVE couple with the principle of Monte Carlo.

Authors' contributions

GHX organized this research as the laboratory chief. ML selected bioenergy sorghum samples for biomass compositional analysis and theoretical ethanol yield calculation inclusion in this study and drafted the manuscript. ML, JW, FD, and BD measured the chemical components of total 147 samples. ML and JW built 36 models with TQ Analyst 9.3 and ChemDataSolution 2.0. ML performed the statistical analysis with IBM SPSS Statistics 22 and drafted figures and tables. All authors read and approved the final manuscript.

Author details

[1] College of Agronomy and Biotechnology, China Agricultural University, Beijing 100193, China. [2] National Energy R&D Center for Non-food Biomass, China Agricultural University, Beijing 100193, China.

Acknowledgements

The authors would like to acknowledge Wei Zhang, Chuandong Sun, Xu Chen, and Baosen Jin for bioenergy sorghum sampling, and Mingxing Xue, Lei Chen, and Yan Li for chemical composition analysis of the samples. This work was supported by Department of Energy Conservation and Technology Equipment of China's National Energy Administration (Science and Technology Department, No. [2012] 32) and funded by China Datang New Energy Co. Ltd and Henan Tianguan Group Co., Ltd.

Competing interests

The authors declare that they have no competing interests.

References

1. Li J, Danao M, Chen S, Li S, Singh V, Brown P. Prediction of starch content and ethanol yields of sorghum grain using near infrared spectroscopy. J Near Infrared Spectrosc. 2015;23:85.
2. Li M, Feng S, Wu L, Li Y, Fan C, Zhang R, Zou W, Tu Y, Jing HC, Li S, Peng L. Sugar-rich sweet sorghum is distinctively affected by wall polymer features for biomass digestibility and ethanol fermentation in bagasse. Bioresour Technol. 2014;167:14–23.
3. Li M, Wang J, Yang Y, Xie G. Alkali-based pretreatments distinctively extract lignin and pectin for enhancing biomass saccharification by alter ing cellulose features in sugar-rich Jerusalem artichoke stem. Bioresour Technol. 2016;208:31–41.
4. Guimarães CC, Simeone MLF, Parrella RAC, Sena MM. Use of NIRS to predict composition and bioethanol yield from cell wall structural components of sweet sorghum biomass. Microchem J. 2014;117:194–201.
5. Xiao L, Wei H, Himmel ME, Jameel H, Kelley SS. NIR and Py-mbms coupled with multivariate data analysis as a high-throughput biomass characterization technique: a review. Front Plant Sci. 2014;5:1–10.
6. Chen S, Danao MC, Singh V, Brown PJ. Determining sucrose and glucose levels in dual-purpose sorghum stalks by Fourier transform near infrared (FT-NIR) spectroscopy. J Sci Food Agric. 2014;94:2569–76.
7. Zegada-Lizarazu W, Monti A. Are we ready to cultivate sweet sorghum as a bioenergy feedstock? A review on field management practices. Biomass Bioenerg. 2012;40:1–12.
8. Rooney WL, Blumenthal J, Bean B, Mullet JE. Designing sorghum as a dedicated bioenergy feedstock. Biofuels Bioprod Biorefin. 2007;1:147–57.
9. Arioli T, Peng L, Betzner AS, Burn J, Wittke W, Herth W, Camilleri C, Höfte H, Plazinski J, Birch R, et al. Molecular analysis of cellulose biosynthesis in Arabidopsis. Science. 1998;279:717–20.
10. Scheller HV, Ulvskov P. Hemicelluloses. Annu Rev Plant Biol. 2010;61:263–89.
11. Chen F, Dixon RA. Lignin modification improves fermentable sugar yields for biofuel production. Nat Biotechnol. 2007;25:759–61.
12. Wei M, Zhu W, Xie G, Lestander TA, Wang J, Xiong S. Ash composition in Cassava stems originating from different locations, varieties, and harvest times. Energ Fuel. 2014;28:5086–94.
13. Payne CE, Wolfrum EJ. Rapid analysis of composition and reactivity in cellulosic biomass feedstocks with near-infrared spectroscopy. Biotechnol Biofuels. 2015;8:43.
14. Martin AP, Palmer WM, Byrt CS, Furbank RT, Grof CP. A holistic high-throughput screening framework for biofuel feedstock assessment that characterises variations in soluble sugars and cell wall composition in Sorghum bicolor. Biotechnol Biofuels. 2013;6:186.
15. He C, Chen L, Yang Z, Huang G, Liao N, Han L. A rapid and accurate method for on-line measurement of straw–coal blends using near infrared spectroscopy. Bioresour Technol. 2012;110:314–20.
16. Fagan CC, Everard CD, McDonnell K. Prediction of moisture, calorific value, ash and carbon content of two dedicated bioenergy crops using near-infrared spectroscopy. Bioresour Technol. 2011;102:5200–6.
17. Zhao N, Wu Z, Zhang Q, Shi X, Ma Q, Qiao Y. Optimization of parameter selection for partial least squares model development. Sci Rep-UK. 2015;5:11647.
18. Ghasemi-Varnamkhasti M, Forina M. NIR spectroscopy coupled with multivariate computational tools for qualitative characterization of the aging of beer. Comput Electron Agric. 2014;100:34–40.
19. Brown SDTR. Comprehensive chemometrics: chemical and biochemical data analysis. Amsterdam: Elsevier; 2009.
20. Williams CL, Westover TL, Emerson RM, Tumuluru JS, Li C. Sources of biomass feedstock variability and the potential impact on biofuels pro duction. Bioenerg Res. 2016;9:1–14.
21. Li C, Aston JE, Lacey JA, Thompson VS, Thompson DN. Impact of feedstock quality and variation on biochemical and thermochemical conver sion. Renew Sustain Energy Rev. 2016;65:525–36.
22. Chadwick DT, McDonnell KP, Brennan LP, Fagan CC, Everard CD. Evaluation of infrared techniques for the assessment of biomass and biofuel quality parameters and conversion technology processes: a review. Renew Sustain Energy Rev. 2014;30:672–81.
23. Xu F, Yu J, Tesso T, Dowell F, Wang D. Qualitative and quantitative analysis of lignocellulosic biomass using infrared techniques: a mini-review. Appl Energ. 2013;104:801–9.
24. Yang Z, Li K, Zhang M, Xin D, Zhang J. Rapid determination of chemical composition and classification of bamboo fractions using visible–near infrared spectroscopy coupled with multivariate data analysis. Biotechnol Biofuels. 2016;9:35.
25. Wu L, Li M, Huang J, Zhang H, Zou W, Hu S, Li Y, Fan C, Zhang R, Jing H, et al. A near infrared spectroscopic assay for stalk soluble sugars, bagasse enzymatic saccharification and wall polymers in sweet sorghum. Bioresour Technol. 2015;177:118–24.
26. Xu F, Zhou L, Zhang K, Yu J, Wang D. Rapid determination of both structural polysaccharides and soluble sugars in sorghum biomass using near-infrared spectroscopy. Bioenerg Res. 2015;8:130–6.
27. Zhu X, Shan Y, Li G, Huang A, Zhang Z. Prediction of wood property in Chinese Fir based on visible/near-infrared spectroscopy and least square-support vector machine. Spectrochim Acta Part A Mol Biomol Spectrosc. 2009;74:344–8.
28. Tan C, Qin X, Li M. An ensemble method based on a self-organizing map for near-infrared spectral calibration of complex beverage samples. Anal Bioanal Chem. 2008;392:515–21.
29. Yun YH, Liang YZ, Xie GX, Li HD, Cao DS, Xu QS. A perspective demonstration on the importance of variable selection in inverse calibration for complex analytical systems. Analyst. 2013;138:6412–21.
30. Lucà F, Conforti M, Castrignanò A, Matteucci G, Buttafuoco G. Effect of calibration set size on prediction at local scale of soil carbon by Vis-NIR spectroscopy. Geoderma. 2017;288:175–83.
31. Siano GG, Goicoechea HC. Representative subset selection and standardization techniques. A comparative study using NIR and a simulated fermentative process UV data. Chemometr Intell Lab. 2007;88:204–12.
32. Galvao R, Araujo M, Jose G, Pontes M, Silva E, Saldanha T. A method for calibration and validation subset partitioning. Talanta. 2005;67:736–40.
33. Li H, Liang Y, Xu Q, Cao D. Key wavelengths screening using competitive adaptive reweighted sampling method for multivariate calibration. Anal Chim Acta. 2009;648:77–84.
34. Rajalahti T, Arneberg R, Berven FS, Myhr K, Ulvik RJ, Kvalheim OM. Biomarker discovery in mass spectral profiles by means of selectivity ratio plot. Chemometr Intell Lab. 2009;95:35–48.
35. Farrés M, Platikanov S, Tsakovski S, Tauler R. Comparison of the variable importance in projection (VIP) and of the selectivity ratio (SR) methods for variable selection and interpretation. J Chemometr. 2015;29:528–36.
36. Yan-kun L. Determination of diesel cetane number by consensus modeling based on uninformative variable elimination. Anal Methods. 2012;4:254–8.
37. Li J, Huang W, Chen L, Fan S, Zhang B, Guo Z, Zhao C. Variable selection in visible and near-infrared spectral analysis for noninvasive determination of soluble solids content of 'Ya' Pear. Food Anal Method. 2014;7:1891–902.
38. Schwanninger M, Rodrigues J, Fackler K. A review of band assignments in near infrared spectra of wood and wood components. J Near Infrared Spectrosc. 2011;19:287.
39. Sanderson MA, Agblevor F, Collins M, Johnson DK. Compositional analysis of biomass feedstocks by near infrared reflectance spectroscopy. Biomass Bioenerg. 1996;11:365–70.
40. Guimarães CC, Assis C, Simeone MLF, Sena MM. Use of near-infrared spectroscopy, partial least-squares, and ordered predictors selection to predict four quality parameters of sweet sorghum juice used to produce bioethanol. Energ Fuel. 2016;30:4137–44.
41. Wolfrum EPCST. Multivariate calibration models for sorghum composition using near-infrared spectroscopy. Technical Report NREL/TP-510056838. Golden: National Renewable Energy Laboratory (NREL); 2013.
42. Hames BR, Thomas SR, Sluiter AD, Roth CJ, Templeton DW. Rapid biomass analysis. In Biotechnology for fuels and chemicals: The Twenty-Fourth Symposium. Totowa, NJ: Humana Press; 2003. p. 5–16.
43. Yun Y, Cao D, Tan M, Yan J, Ren D, Xu Q, Yu L, Liang Y. A simple idea on applying large regression coefficient to improve the genetic

algorithm-PLS for variable selection in multivariate calibration. Chemometr Intell Lab. 2014;130:76–83.
44. Cai W, Li Y, Shao X. A variable selection method based on uninformative variable elimination for multivariate calibration of near-infrared spectra. Chemometr Intell Lab. 2008;90:188–94.
45. Balabin RM, Smirnov SV. Variable selection in near-infrared spectroscopy: benchmarking of feature selection methods on biodiesel data. Anal Chim Acta. 2011;692:63–72.
46. Sluiter A, Hames B, Ruiz R, Scarlata C, Sluiter J, Templeton D, Crocker D. Determination of structural carbohydrates and lignin in biomass laboratory analytical procedure. Golden: Natl Renew Energy Lab; 2008.
47. Zhao YL, Dolat A, Steinberger Y, Wang X, Osman A, Xie GH. Biomass yield and changes in chemical composition of sweet sorghum cultivars grown for biofuel. Field Crop Res. 2009;111:55–64.
48. Huang J, Xia T, Li A, Yu B, Li Q, Tu Y, Zhang W, Yi Z, Peng L. A rapid and consistent near infrared spectroscopic assay for biomass enzymatic digestibility upon various physical and chemical pretreatments in Miscanthus. Bioresour Technol. 2012;121:274–81.
49. Wold S, Sj Str MM, Eriksson L. PLS-regression: a basic tool of chemometrics. Chemometr Intell Lab. 2001;58:109–30.
50. Fox GP, O Donnell NH, Stewart PN, Gleadow RM. Estimating hydrogen cyanide in forage Sorghum (*Sorghum bicolor*) by near-infrared spectroscopy. J Agric Food Chem. 2012;60:6183–7.

Screening and evolution of a novel protist xylose isomerase from the termite *Reticulitermes speratus* for efficient xylose fermentation in *Saccharomyces cerevisiae*

Satoshi Katahira[1], Nobuhiko Muramoto[1], Shigeharu Moriya[2], Risa Nagura[1], Nobuki Tada[3], Noriko Yasutani[3], Moriya Ohkuma[4,5], Toru Onishi[3] and Kenro Tokuhiro[1*]

Abstract

Background: The yeast *Saccharomyces cerevisiae*, a promising host for lignocellulosic bioethanol production, is unable to metabolize xylose. In attempts to confer xylose utilization ability in *S. cerevisiae*, a number of xylose isomerase (XI) genes have been expressed heterologously in this yeast. Although several of these XI encoding genes were functionally expressed in *S. cerevisiae*, the need still exists for a *S. cerevisiae* strain with improved xylose utilization ability for use in the commercial production of bioethanol. Although currently much effort has been devoted to achieve the objective, one of the solutions is to search for a new XI gene that would confer superior xylose utilization in *S. cerevisiae*. Here, we searched for novel XI genes from the protists residing in the hindgut of the termite *Reticulitermes speratus*.

Results: Eight novel XI genes were obtained from a cDNA library, prepared from the protists of the *R. speratus* hindgut, by PCR amplification using degenerated primers based on highly conserved regions of amino acid sequences of different XIs. Phylogenetic analysis classified these cloned XIs into two groups, one showed relatively high similarities to *Bacteroidetes* and the other was comparatively similar to *Firmicutes*. The growth rate and the xylose consumption rate of the *S. cerevisiae* strain expressing the novel XI, which exhibited highest XI activity among the eight XIs, were superior to those exhibited by the strain expressing the XI gene from *Piromyces* sp. E2. Substitution of the asparagine residue at position 337 of the novel XI with a cysteine further improved the xylose utilization ability of the yeast strain. Interestingly, introducing point mutations in the corresponding asparagine residues in XIs originated from other organisms, such as *Piromyces* sp. E2 or *Clostridium phytofermentans*, similarly improved xylose utilization in *S. cerevisiae*.

Conclusions: A novel XI gene conferring superior xylose utilization in *S. cerevisiae* was successfully isolated from the protists in the termite hindgut. Isolation of this XI gene and identification of the point mutation described in this study might contribute to improving the productivity of industrial bioethanol.

Keywords: Xylose isomerase, Termite, Cloning and functional expression of XI gene, *Saccharomyces cerevisiae*, Xylose fermentation, Ethanol production, Mutagenesis

*Correspondence: tokuhiro@mosk.tytlabs.co.jp
[1] Bioinspired Systems Research-Domain, Toyota Central R&D Labs., Inc., 41-1, Yokomichi, Nagakute, Aichi 480-1192, Japan
Full list of author information is available at the end of the article

Background

Bioethanol produced from regenerative biomass resources is a promising sustainable alternative to fossil fuels in order to reduce greenhouse gas emitted from the automobiles. Because of increasing global public concern about food security and environment, it is preferable to use lignocellulosic biomass, instead of edible sugar, as the substrate for bioethanol fermentation. Lignocellulosic biomass, such as corn stover, wheat straw, and rice straw among others, is an attractive substrate because it is cheap and abundantly found as agricultural and forest residues.

Yeast *Saccharomyces cerevisiae* is widely used as a candidate host organism for bioethanol production due to its robustness, inhibitor tolerance, and high ethanol productivity [1]. However, to obtain the highest ethanol yield from each biomass substrate, co-fermentation of all of sugars contained in the lignocellulosic biomass hydrolysate is necessary. Since wild-type *S. cerevisiae* strains cannot utilize D-xylose, the second major sugar present in lignocellulosic hydrolysate (up to 40%), numerous attempts have been made to recombinantly create yeast strains that can efficiently ferment xylose [2–6]. Two different pathways have been proposed for the anaerobic fermentation of xylose by *S. cerevisiae*. The first pathway consists of sequential redox reaction of xylose, catalyzed by xylose reductase (XR) and xylitol dehydrogenase (XDH), as a result of which D-xylulose is produced. The resulting xylulose is then phosphorylated by xylulokinase (XK), and subsequently utilized by the endogenous pentose phosphate pathway followed by glycolysis to produce ethanol [7–11]. However, because of the differences in the cofactor preferences of NADPH-dependent XR and NAD^+-dependent XDH, substantial amount of xylitol is accumulated, and thus, only suboptimal yield of ethanol is obtained when a recombinant yeast harboring the XR-XDH pathway is used for this purpose.

The second pathway consists of one step conversion of xylose to fermentable xylulose by xylose isomerase (XI) without the involvement of redox cofactors. A large number of XIs have been heterologously expressed in *S. cerevisiae*, and several of them were found to be functional [2, 12–27]. A XI gene that expressed highly functional XI activity in *S. cerevisiae* was first cloned from the anaerobic fungus *Piromyces* sp. strain E2, isolated from the feces of an Indian elephant [13, 28]. The *Piromyces* XI gene was overexpressed in a *S. cerevisiae* strain [29], and after long-term adaptation to xylose and following extensive genetic engineering of the strain [30], the recombinant strain yielded high amount of ethanol from xylose without any accumulation of xylitol. Thereafter, two XI genes, one from the cattle rumen fungus *Orpinomyces* sp. strain *ukk1* and another from the anaerobic bacterium *Clostridium phytofermentans*, were functionally expressed in *S. cerevisiae*. The *Orpinomyces* XI was highly homologous to the *Piromyces* XI (94% amino acid sequence identity), and both exhibited similar specific enzyme activities [14]. The *Clostridium* XI, on the other hand, exhibited lower sequence identity (<60%) to the *Piromyces* XI. The *Clostridium* XI was reported to be less inhibited by xylitol compared to the *Piromyces* XI [12].

Recently, two other XI genes, isolated from a soil metagenomic library, were shown to be functionally active in *S. cerevisiae* [18]. Although the enzyme activities of these XIs, expressed in *S. cerevisiae*, were comparable to that of the *Piromyces* XI, the aerobic growth rate of the recombinant yeast strain, expressing either one of these XI genes, on xylose was much lower than the growth rate of the yeast strain expressing the *Piromyces* XI. More recently, a new XI gene isolated from the bovine rumen metagenomics library was functionally expressed in *S. cerevisiae* and the expressed XI activity was improved through mutagenesis [27].

In this paper, we first report the cloning of a novel XI gene (cDNA) from the protists residing in the termite hindgut and then show that the xylose fermentation capacity of the recombinant *S. cerevisiae* strain expressing this clone is superior to previously known XI clones. Analysis of the kinetic properties of the protist XI reveals that the substrate specificity of this XI for xylose is higher than that of the other known XIs, suggesting that the expressed XI may rapidly consume low concentration of xylose from the fermentation medium. In addition, we discovered that a point mutation in the XI gene leads to improved cell growth on xylose and also leads to an increased xylose consumption rate for the yeast expressing this mutant XI. Furthermore, we show that introduction of the same mutation in XIs from various other organisms also improves their respective performances.

Results

Isolation of novel XI genes and their sequence analysis

Novel XI genes were isolated from a cDNA library, which was constructed previously from the protists residing in the *Reticulitermes speratus* hindgut [31], using a PCR-based cloning method. Briefly, degenerate primers, designed based on the alignment of DNA sequences of known XI genes, were used to amplify fragments with XI gene-like partial sequences. Next, the 5′- and 3′-regions of these XI gene-like fragments were amplified by PCR from the same cDNA library using primers that were designed based on the internal sequences of these XI gene-like fragments and the vector sequences flanking the 5′- and 3′-regions of the cDNA inserts. Finally, the entire open reading frame containing DNA fragment of each clone was amplified from the cDNA library using primers that were designed based on the DNA sequences of 5′- and 3′-regions of the XI gene-like fragments. DNA sequencing analysis of these clones led to the identification of eight novel XI genes and they were named as *RsXI-A* (GenBank Accession Number: LC272084), *-B* (GenBank Accession Number: LC272085), *-C1* (GenBank Accession Number: HV438106.1), *-C2* (GenBank Accession Number: HV438107.1), *-D1* (GenBank Accession Number: HV438108.1), *-D2* (GenBank Accession Number: LC272086), *-E* (GenBank Accession Number: LC272087), and *-F* (GenBank Accession Number: HV438109.1).

The amino acid sequences of eight novel XI genes were derived from their respective DNA sequences, and sequence homologies among these proteins were compared. Based on their sequence identities, these genes were divided into two groups. Whereas RsXI-A and RsXI-B exhibited 68.9% amino acid identity between them, RsXI-C1, RsXI-C2, RsXI-D1, RsXI-D2, RsXI-E, and RsXI-F revealed 81–92% amino acid identity among themselves. However, the degree of identity between these two groups was low (52% or less).

Next, we performed phylogenetic analysis of these newly cloned XIs and XIs cloned from other sources, including those that are known to be functional in *S. cerevisiae*, such as XI genes of *Piromyces* sp. E2 and *C. phytofermentans*. As shown in Fig. 1, RsXI-A and RsXI-B showed relatively high similarities to the XIs from the *Bacteroidetes* family. The XIs from the *Fungi* family, including those from *Piromyces* sp. E2, were close to this phylum. However, RsXI-C1, RsXI-C2, RsXI-D1, RsXI-D2, RsXI-E, and RsXI-F were comparatively similar to the XIs from the *Firmicutes* family. These six novel XIs exhibited 63.4, 60.9, 62.6, 62.9, 60.1, and 62.2% identity, respectively, to the XI from *C. phytofermentans* (Additional file 1: Table S1). The results of phylogenetic analysis suggested that these XIs from the protists may belong to a phylogenetic affiliation that is different from the previously reported XIs.

XI activities of novel XIs in recombinant yeasts

To determine whether these novel XI genes could be functionally expressed in the yeast *S. cerevisiae*, we assayed for xylose isomerase activity in cell extracts of *S. cerevisiae* harboring plasmids expressing these XI genes. The yeast strains used in this experiment are listed in Additional file 2: Table S2. In this assay, xylose is isomerized to xylulose by the xylose isomerase present in the crude cell extracts, and the amount of xylulose produced is determined by the cysteine-carbazole-sulfuric acid method [32].

As shown in Fig. 2, no XI activity was observed when the cell extract was prepared from the control strain carrying the empty vector. Similarly, XI activity was also not observed when the cell extract was prepared from the strain harboring the plasmid expressing *RsXI-A*, *RsXI-B*, *RsXI-D2*, or *RsXI-E*. XI activity was, however, detected in the cell extract prepared from the MP110 strain that harbored the XI gene from *Piromyces* sp. E2 (*PiXI*). More importantly, significant level of XI activity was also detected in the cell extracts prepared from the recombinant strain harboring the plasmid expressing *RsXI-C1*, *RsXI-C2*, *RsXI-D1*, or *RsXI-F*. Among these, the XI activity observed in the cell extract of MR310 strain, which expressed *RsXI-C1*, was equivalent to that of the MP110. Because the XI activity was highest in the yeast expressing *RsXI-C1*, we chose this clone for further characterization.

Determination of the origin of *RsXI-C1*

Next, we determined the origin of *RsXI-C1*. Although it is likely that the origin of this gene is the symbiotic eukaryotic protists of *R. speratus* because the cDNA library, from which this gene was isolated, was constructed by the poly-T selection-based method, it is however possible to get contaminated by genes from the symbiotic bacteria. Therefore, we used RT-PCR and in situ hybridization techniques to determine the organism from which this gene was originated.

For this purpose, we first fractionated symbiotic protists and bacteria present in the termite hindgut by low-speed centrifugation and then performed RT-PCR on each fraction to determine which fraction would contain *RsXI-C1*. Accordingly, *RsXI-C1* was detected only in the protistan fraction (Additional file 3: Figure S1).

To further confirm that *RsXI-C1* is only expressed in protists, we performed in situ hybridization analysis. The hybridization signal was observed only with the antisense probe (Fig. 3a, b), but not with the sense probe (Fig. 3c, d), in small number of protists, morphologically which were thought to be of the genus *Pyrsonympha* from their appearances (spiral cell and asymmetric shape along anterior–posterior axis [33]). This result suggests that either the *RsXI-C1* expressing protists are quite rare or they express *RsXI-C1* only under specific condition.

Kinetic properties of RsXI-C1

For the kinetic studies, we used two recombinant *S. cerevisiae* strains, WR311 and WP111, expressing codon-optimized RsXI-C1 and PiXI, respectively. Codon optimization was carried out to improve the expression levels of these two genes in *S. cerevisiae*. Kinetic properties of RsXI-C1 and PiXI were determined using the crude cell extracts of these strains and various xylose concentrations. The kinetic parameters were calculated from the Lineweaver–Burk plot (xylose isomerase activity

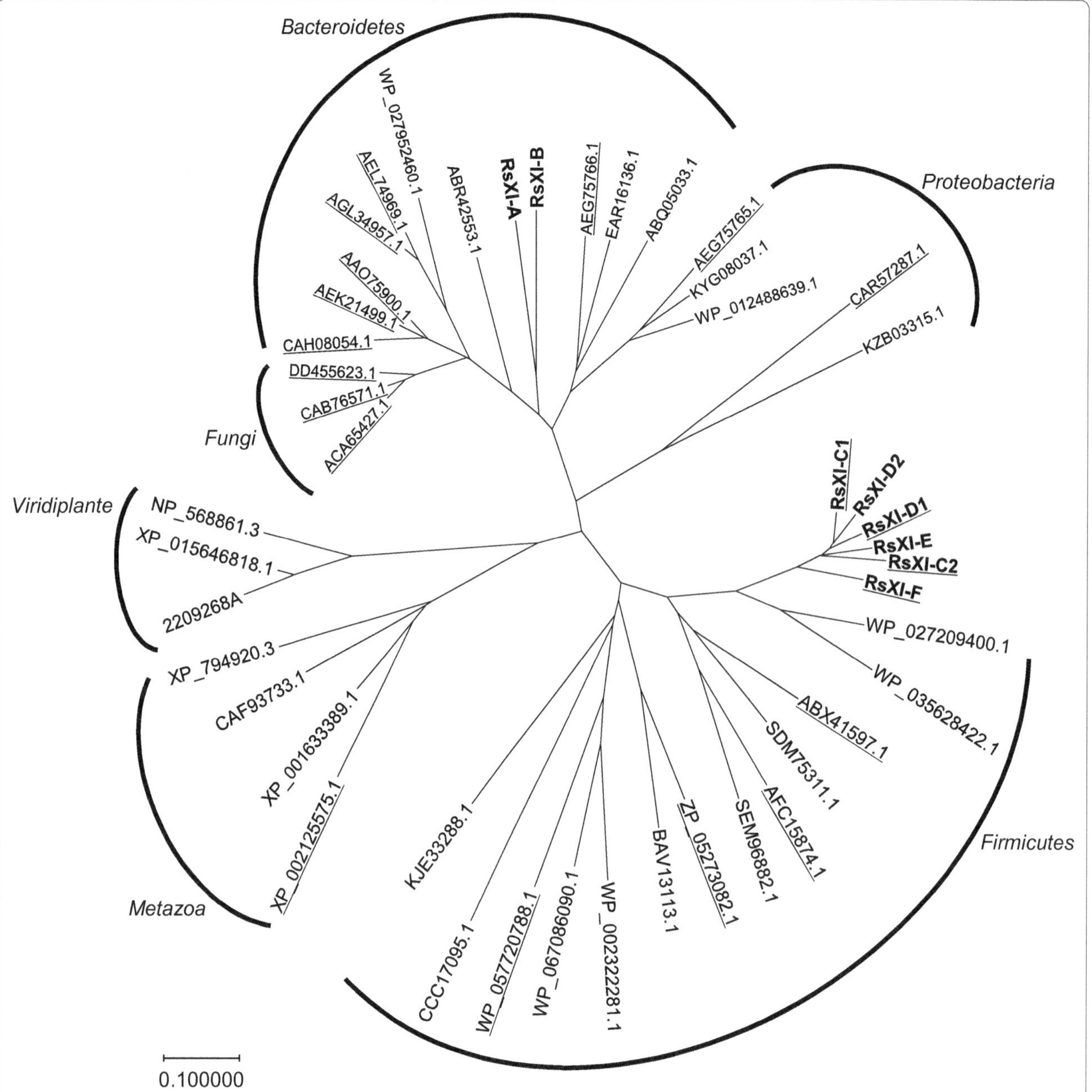

Fig. 1 Phylogenetic tree. Phylogenetic analysis based on the amino acid sequences of the clones isolated in this study (*bold*) and other known XIs. Organism sources were selected based on their functional expression in *S. cerevisiae* (*underlined*) and sequence similarity. Protein names or protein accession numbers are given

vs. xylose concentration). The V_{max} of PiXI (0.0103 μmol/min mg/protein) was about 1.4 times higher than that of the RsXI-C1 (0.0074 μmol/min mg/protein), but the K_m value of RsXI-C1 for xylose was 30% lower than that of the PiXI, thus suggesting that the affinity of the RsXI-C1 for xylose (10.52 ± 1.10 mM) was higher than that of PiXI (14.90 ± 0.57 mM).

Growth of the yeast strain expressing *RsXI-C1O* on xylose

To evaluate the growth performance, growths of strains WR311 (RsXI-C1), WP111 (PiXI), and WC111 (expressing codon-optimized XI gene from *C. phytofermentans*, CpXI) were monitored in a medium containing xylose as the sole carbon source (SX medium). The results of the growth assay are shown in Fig. 4. As can be seen, OD_{660}

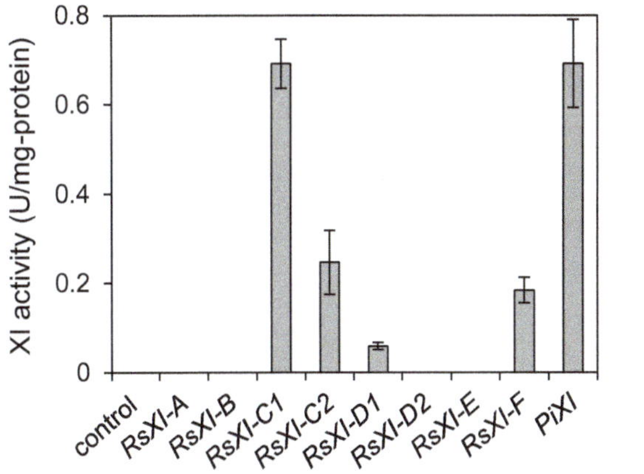

Fig. 2 Xylose isomerase activities of the novel XI genes and *Piromyces* sp. E2 XI gene. Each gene was expressed in *S. cerevisiae* PP600 strain and xylose isomerase activity in the cell extract of each recombinant yeast strain was determined as described in "Methods". *Error bars* represent standard deviations of 4 independent experiments

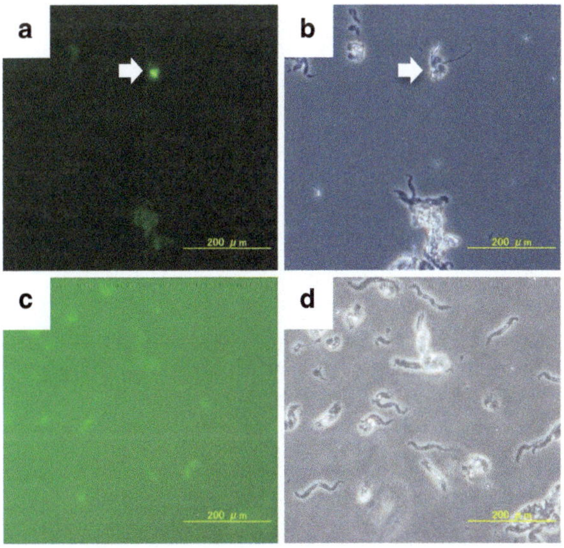

Fig. 3 In situ hybridization. **a** Fluorescent image of in situ hybridization performed using the anti-sense probe. The picture was taken in 14-s exposure. Hybridization signal was indicated by the *arrowhead*. **b** Optical image of the same specimen shown in *panel* **a**. *Arrowhead* shows the individual that exhibited positive signal in *panel* **a**. **c** Fluorescent image of in situ hybridization performed using the sense probe. The picture was taken in 60-s exposure. No hybridization signal was observed. **d** Optical image of the same specimen shown in *panel* **c**. *Bar* 200 μm

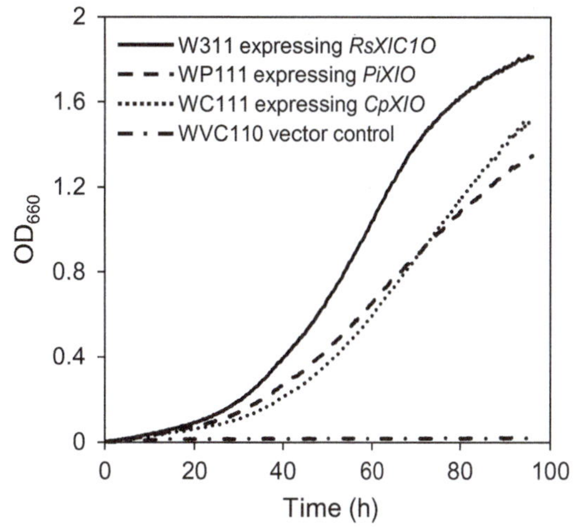

Fig. 4 Aerobic growth of recombinant *S. cerevisiae* strains expressing different XIs on xylose. Yeast strains were cultivated aerobically at 30 °C in L-shaped test tubes in the xylose minimal medium containing 50 g/l xylose as the sole carbon source (see "Methods"). The data points shown are averages of three independent experiments. Deviations were always below 10% of the average

of the control cell culture (WVC110 strain harboring control vector) did not increase, but the OD_{660} of the cultures of all the strains harboring the XI genes increased with culture time. The yeast strain WR311 (RsXI-C1) grew faster than the strains WP111 (PiXI) and WC111 (CpXI). The maximal specific growth rate of the WR311 strain (0.071 h^{-1}) was about 1.2 times higher than the specific growth rates of the WP111 and WC111 strains (0.058 and 0.057 h^{-1}, respectively).

Fermentation ability of the yeast strain harboring the *RsXI-C1O* expression plasmid

Xylose fermentation abilities of the recombinant yeast strains were estimated from the amount of xylose consumed and ethanol produced in a medium that either contained 50 g/l xylose as the sole carbon source or contained a mixture of 30 g/l glucose and 20 g/l xylose.

In the xylose fermentation assay, the initial xylose consumption rate of WR311 was twice as high as the xylose consumption rates of WP111 and WC111 (Fig. 5). After 72-h cultivation, xylose in the culture medium of WR311 was depleted, whereas 15 g/l or more xylose remained in the culture media of WP111 and WC111. The xylose consumption rates of WR311, WP111, and WC111 were 0.71, 0.51, and 0.51 g xylose/l h, respectively. In all cases, accumulation of only a little amount of xylitol (0.64 g/l or less) and some glycerol (2.9 g/l or less) was observed after 72-h fermentation. As a result, the ethanol yields of WR311, WP111, and WC111 were 0.39, 0.38, and 0.38 g of ethanol produced per g of consumed xylose, respectively.

The time course of glucose/xylose fermentation by the recombinant yeast strains is shown in Fig. 6. As shown, glucose was preferably consumed at the initial stage of fermentation by all transformants, and xylose concentration only began to decrease as the glucose was depleted

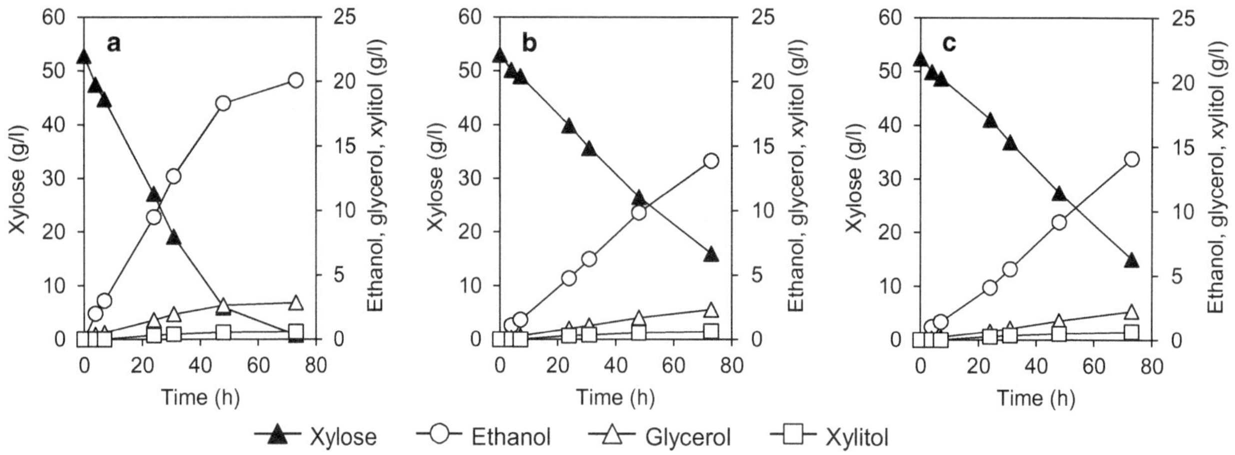

Fig. 5 Anaerobic batch fermentation on xylose. Fermentation of xylose by WR311 (**a**), WP111 (**b**), and WC111 (**c**) strains under anaerobic condition was carried out in SX medium containing 50 g/l xylose as the sole carbon source (see "Methods"). Amounts of fermentation products (ethanol, glycerol, and xylitol) are indicated on the right-hand *Y-axis*. The data points shown are averages of three independent experiments. Deviations were always below 10% of the average

from the culture medium. In all cases, we found almost no accumulation of the by-product xylitol. WP111 and WC111 consumed less xylose; thus, 10 g/l or more xylose remained in the culture media of WP111 (Fig. 6b) and WC111 (Fig. 6c) after 72-h fermentation. On the other hand, the xylose consumption rate of the WR311 strain was about twice that of the other strains, and almost all the xylose was consumed after 72-h fermentation (Fig. 6a).

To ensure that the higher xylose fermentation ability of WR311 was indeed due to the expression of the XI gene and was not influenced by the host strain, we used another host strain, PP600, and created recombinant strains expressing codon-optimized *RsXI-C1*, *PiXI*, and *CpXI*. We then repeated the fermentation experiment using these recombinant PP600 strains under microaerobic condition under xylose minimal medium and the results are shown in Additional file 4: Figure S2. As was observed with the recombinant W600W strains, the recombinant PP600 strain expressing *RsXI-C1* exhibited higher xylose consumption rate than the other recombinant strains, thus suggesting that the observed higher xylose isomerase activity of RsXI-C1 was not host strain dependent.

It is also believed that the background status of host strain, such as the expression levels of genes related to PPP or sugar transport, might influence the xylose fermenting ability. To test this possibility, we analyzed the expression levels of these genes in each recombinant W600W strain (i.e., WR311, WP111, and WC111) as well as in control strain by quantitative PCR. Results shown in Additional file 5: Figure S3 revealed that although there were some differences in the expression levels of these genes, the observed differences were not significant, indicating that the background strains are at the same status. Collectively, these results suggest that the RsXI-C1 expressed in yeast can utilize xylose more efficiently than that either by PiXI or by CpXI.

Screening and identification of xylose isomerase mutants with improved growth on xylose

We performed random mutagenesis of *RsXI-C1O* followed by growth-based screening on xylose to select for a mutant XI gene that would confer enhanced ability to ferment xylose in the recombinant *S. cerevisiae* strain. Mutated DNA fragments and the native *RsXI-C1O* were transformed separately into the W600W strain using a low-copy-number centromeric vector pRS316GAP. For growth-based screening, yeast transformants, harboring a plasmid expressing either the wild-type or a mutant *RsXI-C1O*, were cultured in the SX medium. After long-term cultivation, cells were harvested and spread on a SX agar plate. Next, 20 colonies, growing faster than the control strain and harboring mutant *RsXI-C1O*, were selected from the SX agar plate for the growth experiment in SX medium (Additional file 6: Figure S4). Based on the specific growth rates of these 20 colonies, top 10 colonies with better growth rates were selected and used for plasmid preparation.

Results of the sequencing analysis revealed 5 different mutant sequences. Notably, each mutant XI gene contained a single amino acid substitution at one of the following amino acid residues of RsXI-C1: T76I, E125G, I286F, N337T, and K384E. Genes corresponding to these mutants were designated as *RsXIC1O-T76I*, *RsXIC1O-E125G*, *RsXIC1O-I286F*, *RsXIC1O-N337T*, and *RsXIC1O-K384E*, respectively. Plasmids carrying each one

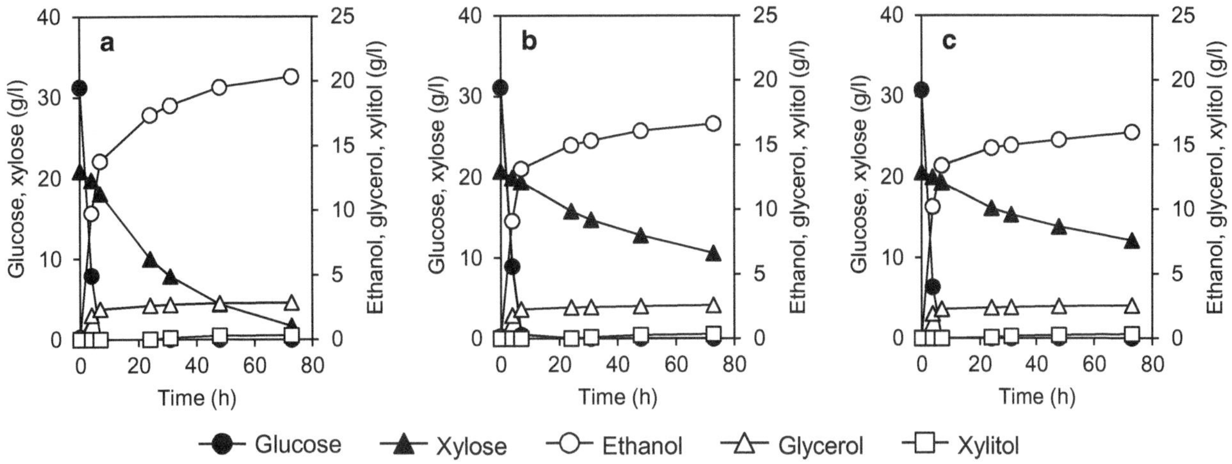

Fig. 6 Anaerobic batch fermentation on glucose/xylose mixture. Fermentation of glucose/xylose by yeast strains WR311 (**a**), WP111 (**b**), and WC111 (**c**) was carried out in medium containing 30 g/l glucose and 20 g/l xylose under anaerobic condition. Amounts of ethanol, glycerol, and xylitol produced as a result of fermentation are indicated on the right-hand *Y-axis*. The data points shown are averages of three independent experiments. Deviations were always below 10% of the average

of these mutated XI genes were transformed into the W600W strain, and the resulting strains are summarized in Additional file 2: Table S2. A growth assay was performed to evaluate the growths of these recombinant yeasts on xylose. The specific growth rates of these mutants are compared in Fig. 7. Among these, only the specific growth rate of the WR324T (RsXIC1O-N337T) strain was 1.6-fold higher than that of the control strain.

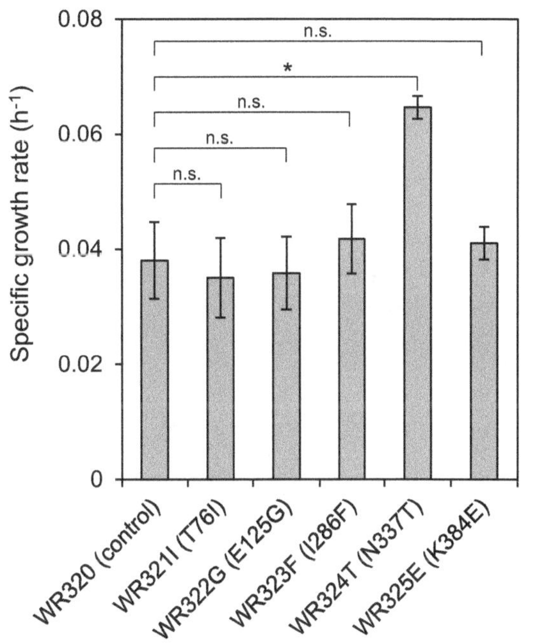

Fig. 7 Specific growth rates of the yeast strains expressing mutated RsXI-C1s. Growth of cells, harboring either the wild-type RsXI-C1 (control) or point mutants of RsXI-C1 (T76I, E125G, I286F, N337T, or K384E), was carried out in SX medium under aerobic condition. Growth rates were determined as described in "Methods". *Error bars* indicate standard deviations of biological quintuplicates. Statistical significance was determined using Student's *t* test. *$P < 0.05$, significant difference. *n.s.* no significant difference

Fermentation performance of the N337T XI mutant expressing strain

As mentioned above, the yeast strain expressing the N337T mutant of RsXI-C1 (WR324T) exhibited enhanced growth rate on xylose. Therefore, to evaluate the xylose fermentation ability of the WR324T (N337T) strain, xylose fermentation experiments were carried out under anaerobic condition in 50 ml of SX medium containing 50 g/l xylose (initial $OD_{600} = 10$). Figure 8 shows time-dependent changes in xylose and ethanol concentrations in the fermentation medium of the strains WR324T and WR320 (control strain). As shown, the amount of xylose consumed by WR320 in 72 h was 12.9 g/l and that by WR324T during the same period was 32.0 g/l. Thus, the xylose consumption and the ethanol production rates for the WR324T strain were 2.5 times higher than those for the WR320 strain.

Construction and evaluation of saturated mutagenesis library

Next, we examined whether replacement of the N337 residue of RsXI-C1 with other amino acids would improve the xylose utilization ability. For this purpose, a site-saturation mutagenesis library was constructed by site-directed mutagenesis using pRS316GAP-RsXIC1O as the template. The resulting plasmids, each carrying a

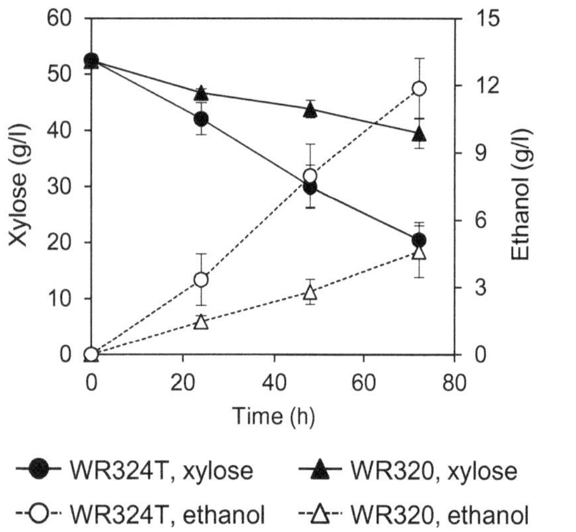

Fig. 8 Time course of anaerobic batch fermentation on xylose. *S. cerevisiae* strains WR324T (expressing *RsXIC1O-N337T* mutant) and WR320 (expressing wild-type *RsXI-C1O*) were cultivated in the fermentation medium that contained 50 g/l xylose as the sole carbon source under anaerobic condition. Concentrations of xylose remaining in the medium (*closed symbols*) and ethanol produced during the fermentation (*open symbols*) at the indicated times were determined by HPLC. *Error bars* represent standard deviations of four independent experiments

mutant *RsXI-C1O* in which the N337 residues was substituted with one of the remaining amino acid residues, were separately introduced into the W600W strain to generate eighteen mutant strains (Additional file 2: Table S2). These mutant strains were then tested for their abilities to ferment xylose using the microaerobic fermentation assay. Figure 9 shows xylose consumption by these mutant and control yeast strains after 72-h incubation. Xylose consumed by the WR320 strain (control) at 72 h was 2.9 g/l, and that by the WR324T (N337T) strain at 72 h was 12.0 g/l. Out of 18 different single amino acid substitution mutants, WR324C (N337C), WR324V (N337V), and WR324A (N337A) consumed more xylose (13.9, 7.8, and 7.3 g/l, respectively) than WR320. Among these mutants, the WR324C strain consumed the highest amount of xylose (4.8-fold more than WR320). From these results, it became clear that the xylose utilization ability of yeast could be improved by substituting the asparagine at position 337 of RsXI with threonine, cysteine, valine, or alanine.

Effect of mutation on the xylose utilization abilities of other XIs

We next examined whether introduction of the same point mutation would similarly improve the xylose utilization abilities of other XIs, such as PiXI, CpXI, and the XI from *Lactococcus lactis* (LlXI) [15], all of which were functionally expressed in *S. cerevisiae.*

Alignment of the amino acid sequences of RsXI-C1, PiXI, CpXI, and LlXI (Fig. 10a) revealed that the amino acid residue corresponding to the asparagine 337 of RsXI-C1 was also an asparagine residue in all of them; interestingly, amino acid sequences adjacent to this asparagine were found to be highly conserved. Therefore, we substituted the corresponding asparagine residue in

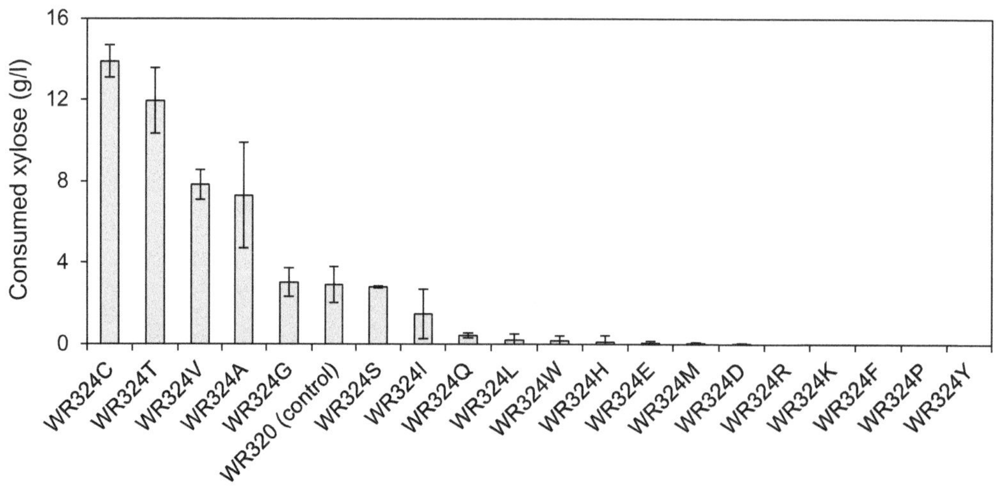

Fig. 9 Xylose fermentation by recombinant yeast strains expressing single amino acid substitution mutants of RsXI-C1. Xylose consumed by *S. cerevisiae* cells expressing the indicated RsXI-C1 point mutants, obtained by site-saturation mutagenesis of Asn-337, after 72-h fermentation under microaerobic condition. Xylose fermentation was carried out as described in "Methods". In all cases, initial cell density (absorbance at 600 nm, OD_{600}) was 1.0. *Error bars* indicate standard deviations of biological duplicates or triplicates

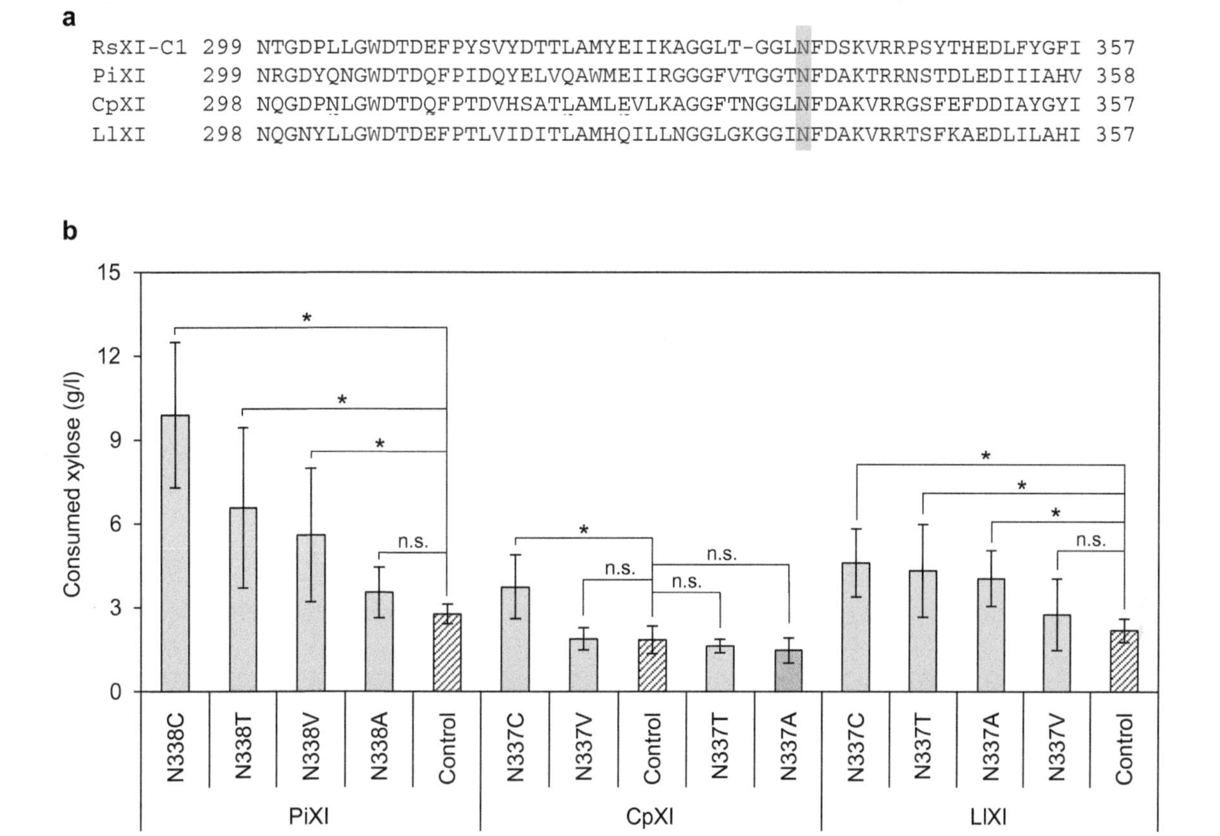

Fig. 10 **a** Alignment of amino acid sequences of different XIs. RsXI-C1, novel XI cloned from the cDNA library of symbiotic protists of *R. speratus*; PiXI, XI of *Piromyces* sp. E2; CpXI, XI of *C. phytofermentans*; LlXI, XI of *L. lactis*. Amino acid sequence alignment was carried out using the genetic information analysis software GENETIX version 10. The *shaded column* indicates the point mutation site. **b** Xylose fermentation by recombinant yeast strains expressing mutated PiXI, CpXI, and LlXI. Xylose consumed after 72-h fermentation under microaerobic condition was shown. Xylose fermentation was carried out as described in "Methods". In all cases, initial cell density (absorbance at 600 nm, OD_{600}) was 10.0. *Error bars* represent standard deviations of biological tetraplicates. Statistical significance was determined using Student's *t* test. *$P < 0.05$, significant difference. *n.s.* no significant difference

these three XIs with an alanine, a cysteine, a threonine, or a valine residue. The generated mutant genes and each native wild-type XI gene were transformed into the W600W strain using the low-copy vector pRS316GAP. The resulting plasmids are listed in Additional file 7: Table S3, and the resulting recombinant yeasts are listed in Additional file 2: Table S2.

Xylose utilization abilities of yeast strains harboring these mutant plasmids were assessed as before. Figure 10b shows the amount of xylose consumed by each one of these recombinant yeast strains after 72-h incubation. Xylose consumed by WP120 (expressing wild-type PiXI) after 72-h fermentation was 2.8 g/l; in contrast, xylose consumed by the PiXI mutant strains WP124C (N338C), WP124T (N338T), and WP124V (N338V) were 9.9, 6.6, and 5.6 g/l, respectively, indicating their improved xylose utilization ability. Similarly, strains harboring cysteine substitution mutant of CpXI (N337C), and cysteine, threonine, or alanine substitution mutant of LlXI (N337C, N337T, or N337A) also exhibited improved xylose utilization ability.

Taken together, our results suggest that the N337 of RsXI-C1 as well as the corresponding asparagine residues of PiXI, CpXI, and LlXI might play an important role in recognizing xylose, because mutagenesis of this residue conferred improved xylose utilization in yeast.

Discussion

Yeast *S. cerevisiae* has been used for industrial bioethanol production because of its superior capacity, such as robust fermentation, inhibitor tolerance, and high ethanol productivity [1, 34]. The XI pathway, which can circumvent the redox cofactor imbalance, has been proposed as the most efficient pathway in *S. cerevisiae* to produce ethanol from xylose [29]. Several XI genes have been reported to be functional in *S. cerevisiae*. Thus, the recombinant yeast strains expressing the XI genes isolated from *Piromyces* sp. E2, *C. phytofermentans*, *Orpinomyces*, and *Bacteroides* fermented xylose efficiently after long-term adaptation to xylose and extensive genetic engineering [2, 14, 26, 35–38]. Here, for the first time, we report isolation of novel XI clones from the protists residing in the termite hindgut; these newly cloned XI genes exhibited low sequence identity with other known XI genes, several of which were functionally expressed in *S. cerevisiae*. Although we performed multiple homology searches using the amino acid sequences of the newly found XIs, we failed to identify any distinctive feature that would explain why some of the XIs are active while the others are inactive when expressed in *S. cerevisiae*. The amino acid sequence derived from *RsXI-C1*, one of the newly isolated XI genes that exhibited the highest XI activity when expressed in yeast (Fig. 2), shared about 47–63% sequence identity with the amino acid sequences of other functional XIs (Additional file 1: Table S1). Although these novel XIs from protists were comparatively similar to the XIs from *Firmicutes*, they formed a slightly different cluster in the phylogenetic tree analysis (Fig. 1). This may be the result of evolutionary changes in the XI genes under a unique environment, such as the termite gut.

Our RT-PCR assay and in situ hybridization results showed that the *RsXI-C1* expressing protists might be of genus *Pyrsonympha*. However, in situ hybridization signal was not observed in most *Pyrsonympha*—signals were observed only from a few small size individuals. It is known that *Pyrsonympha* have several morphologically uncharacterized taxonomy units [39]; *Pyrsonympha* that expressed *RsXI-C1* might be one of such morphologically uncharacterized species.

The kinetic analyses revealed that the V_{max} value of RsXI-C1 was 30% lower than that of PiXI, and the K_m value of RsXI-C1 for xylose (10.52 ± 1.10 mM) was 30% lower than that of PiXI (14.90 ± 0.57 mM). The lower K_m value of RsXI-C1 suggests that the affinity for xylose was higher for RsXI-C1 than that for PiXI. As *S. cerevisiae* is known not to have any specific xylose transporter, this yeast is believed to take up xylose into the cell via the non-specific glucose transporters. Considering that the glucose transporters exhibit a significantly lower affinity for xylose (K_m = 49–300 mM) than for glucose (K_m = 1–28 mM) [7], intracellular xylose concentration would be low, especially when the xylose concentration becomes lower as the xylose gets consumed. Because of the high affinity of XI for xylose, the yeast could convert xylose to xylulose more efficiently at such low xylose concentration. Thus, rapid consumption of xylose would be expected in yeast expressing the higher affinity XI. Indeed, the recombinant yeast strain expressing RsXI-C1 consumed xylose with higher rate under anaerobic batch condition in xylose minimal medium (Fig. 5) as well as in mixed glucose and xylose medium (Fig. 6), and also under microaerobic fermentation condition (Additional file 8: Figure S5) than the strain expressing either PiXI or CpXI. Since only small amount of xylitol was produced, the yield of ethanol was quite high (0.39 g/g in xylose medium and 0.40 g/g in mixed glucose and xylose medium). Clearly, enhanced ability to ferment xylose led to higher ethanol titer and shorter fermentation time that can greatly reduce the production cost of bioethanol. It is noteworthy that expression of *RsXI-C1O* in the yeast strain W600W, in which XK and non-oxidative pentose phosphate pathway genes have been overexpressed and endogenous aldose reductase gene *GRE3* has been deleted, led to marked increase in cell growth and fermentation efficiency in xylose minimal medium without the evolutionary adaptation process.

So far, many attempts have been made to improve the activity of recombinantly expressed xylose isomerase in *S. cerevisiae* [15, 19, 27, 40–42]. To obtain a RsXI-C1 that would confer enhanced xylose fermentation in recombinant yeast, we carried out random mutagenesis of *RsXI-C1O* coupled with growth-based screening on xylose. Consequently, we isolated the N337T mutant of RsXI-C1, which when expressed in yeast exhibited 1.6-fold higher specific growth rate on xylose (Fig. 7) and consumed 2.5 times more xylose (Fig. 8) than the yeast strain expressing the wild-type RsXI-C1. Moreover, analysis of substitution mutants of the asparagine 337 residue of RsXI-C1 revealed that substitution of the asparagine 337 to cysteine further improved the xylose utilization ability of the recombinant yeast (Fig. 9). In addition, substitution of the corresponding asparagine residues of PiXI, CpXI, and LlXI with alanine, cysteine, threonine, or valine improved their abilities to utilize xylose (Fig. 10).

Previous reports showed that introduction of mutations in the XIs from *Piromyces* sp. E2 and bovine rumen metagenomic library improved their enzyme activities [27, 40]. It is thought that these mutations might increase the activity of XI by increasing the monomer-binding contacts or substrate enzyme interactions. Additional file 9: Figure S6 shows the predicted three-dimensional models of the active sites of RsXI-C1 and N337C mutant of RsXI-C1. Although the residue Asn337, which was subjected to mutagenesis, is distantly located from the active site residues (Phe101, His102, Asp103, and Lys235), our modeling result suggested that this residue is sitting close to Asp339 at the metal binding site. The side chain of Asn337 is also close to the metal biding site. It seems that the change occurred in the metal binding site, as a result of replacement of the asparagine residue by the cysteine residue, may have contributed to improved XI performance.

These findings demonstrate the importance of the N337 residue in improving the xylose consumption capacity of RsXI-C1, as XIs from several other organisms, all of which were functionally expressed in *S. cerevisiae*, also shared the same characteristics.

In this study, we reported cloning of novel XI genes from the protists residing in the termite hindgut and successfully expressed some of these clones in *S. cerevisiae*. One of these XI genes, named here as *RsXI-C1*, expressed high XI activity in yeast, and the expressed RsXI-C1 exhibited higher substrate affinity for xylose than those reported earlier for other XI genes expressed in yeast. The recombinant strain expressing RsXI-C1 fermented xylose efficiently. To further improve xylose fermentation, it might be necessary to carry out strain engineering including strong and stable expression of *RsXI-C1* from a multicopy genomic integration vector targeted to the rDNA cluster, introduction of an efficient sugar transporter, tuning of the genes related to xylose metabolism, use of industrial diploid yeast strains, and perform long-term adaptive evolution on xylose [5, 6, 37, 38], individually or in combination.

Conclusions

In this study, novel XI coding genes were successfully isolated from a cDNA library of protists residing in the *R. speratus* hindgut. The recombinant *S. cerevisiae* strain expressing *RsXI-C1* exhibited better growth rate and xylose fermentation ability than those strains expressing other known XI genes, including the XI genes from *Piromyces* and *Clostridium*. Furthermore, a single amino acid substitution mutant of RsXI-C1, obtained by random mutagenesis of *RsXI-C1* combined with growth-based screening on xylose, improved xylose fermentation by the yeast strain expressing this mutant. In addition, mutants carrying similar single amino acid substitutions at the corresponding asparagine residues of several other XIs, previously cloned from different organisms, also conferred improved ability to ferment xylose in *S. cerevisiae*.

Methods

Strain and media

The *Escherichia coli* DH5α (Nippon gene, Japan) was used for all recombinant DNA manipulations. All the yeast strains used and constructed in this study are summarized in Additional file 2: Table S2. *S. cerevisiae* strains MT8-1 [43] and W303-1B [44] were used as the host strains for the gene expression and fermentation, respectively. *E. coli* was grown in Luria–Bertani medium (10 g/l tryptone, 5 g/l yeast extract, 5 g/l sodium chloride) containing 100 mg/l ampicillin or 50 mg/l kanamycin. *S. cerevisiae* strains were aerobically cultivated at 30 °C in synthetic medium [SD medium; 20 g/l glucose and 6.7 g/l yeast nitrogen base without amino acids (Difco/BD Diagnostic Systems, MD, USA) with appropriate supplements.

Adaptation to xylose is known to greatly improve the xylose fermentation abilities of the recombinant yeast strains [29, 35]. Thus, to exclude the effect of adaptation, we used glucose as the carbon source to prepare the precultures used for the enzyme and growth assays, and also in fermentation experiments.

PCR amplification and cloning procedure

PCR reactions were carried out either with PrimeSTAR HS DNA polymerase or with Ex Taq HS DNA polymerase (Takara Bio Inc., Japan). The PCR primers used in this study are listed in Additional file 10: Table S4. PCR products were cloned into pCR2.1-TOPO using the TOPO TA cloning kit or pCR-BluntII-TOPO using the ZERO Blunt TOPO PCR cloning kit (Thermo Fisher Scientific Inc., MA, USA). For generating plasmids, the DNA fragments were cloned into appropriate vectors using the In-Fusion Advantage PCR Cloning kit (Takara Bio Inc.).

Isolation of novel XI genes from cDNA library

A cDNA library, previously prepared from the protists residing in the *R. speratus* hindgut [31], was used in this study for the cloning of novel XI genes. In order to obtain a xylose isomerase gene from this cDNA library, degenerate primers mXI-F1 and mXI-R1, which were designed based on the alignment of DNA sequences of known XI genes obtained from GenBank, were synthesized. Using these degenerate primers, an approximately 400-bp DNA fragment was amplified from this cDNA library. The purified fragment was cloned and sequenced. Consequently, various plasmids containing inserts exhibiting partial sequence similarities to known XI genes were obtained. The 5′- and 3′-regions of these sequences were amplified from the cDNA library by PCR using appropriate combinations of the following primer pairs: RsA-F, RsB-R, RsC1-R, RsC2-F, RsD-R, RsE-F, and RsF-R, which were designed from the revealed partial sequences of inserts, and Lib-F and Lib-R, which were designed from the vector sequence of the 5′- and 3′-flanking regions of the cDNA insert. Amplified DNA fragments were cloned and sequenced. Next, full-length DNA fragments were obtained by PCR using the cDNA library as the template and appropriate primer pairs (RsA-termR, RsB-termF, RsC1-termF, RsC2-termR, RsD-termF, RsE-termR, and RsF-termF, all of which were designed from the 5′- or 3′-regions of the gene sequence, and Lib-F and Lib-R). Amplified DNA fragments were cloned and sequenced. Finally, the entire cDNA sequence of each clone was amplified by PCR from the cDNA library using one of the following appropriate primer pairs: RsXI-A-IF-F/RsXI-A-IF-R, RsXI-B-IF-F/RsXI-B-IF-R, RsXI-C1-IF-F/

RsXI-C1-IF-R, RsXI-C2-IF-F/RsXI-C2-IF-R, RsXI-D-IF-F/RsXI-D-IF-R, RsXI-E-IF-F/RsXI-E-IF-R, or RsXI-F-IF-F/RsXI-F-IF-R. Each amplified fragment was then cloned into the *Sac*II/*Xho*I site of pRS436GAP (DDBJ Accession Number: AB3048629), and complete nucleotide sequence of each amplified DNA fragment was obtained. A general map of the resulting plasmid, created as above by inserting a RsXI gene into pRS436GAP, is shown in Additional file 11: Figure S7a. All the generated plasmids are listed in Additional file 7: Table S3.

Phylogenetic analysis

The amino acid sequences derived from the full-length cDNA clones and DNA sequences of previously reported XIs were aligned and the phylogenetic tree was plotted using the genetic information analysis software GENETYX version 10 (GENETYX, Japan). The phylogenetic tree was constructed with 1000 boot strap replicates using the neighbor-joining method [45].

RT-PCR procedure

To determine the origin of *RsXI-C1*, we fractionated symbiotic protists and bacteria, and performed RT-PCR on cDNA prepared from each fraction.

Excised hindguts from 30 individuals of *R. speratus* were placed into a 1.5-ml microfuge tube along with 100 μl of solution U [46] and disrupted by plastic pestle on ice. The resulting suspension was then filtered through a 100-μm pore size nylon mesh to remove the gut debris. The filtrate was centrifuged at 500 rpm for 5 min using a swing bucket rotor. The supernatant was used as the bacterial fraction and the pellet was used as the protistan fraction. To clean up the bacterial fraction, the supernatant was collected after centrifugation at 500 rpm for 5 min, and this cycle was repeated 4 times. The resulting supernatant was subsequently centrifuged at 15,000 rpm for 1 min and the resulting bacterial pellet was saved and the supernatant was discarded. To clean up the protistan fraction, the protistan pellet (see above) was resuspended in 100 μl of solution U, centrifuged at 500 rpm for 5 min, and the supernatant was discarded after centrifugation. This cycle was repeated 4 times. All centrifugation steps were performed at 4 °C.

Total RNA was purified from each fraction by RNeasy mini kit (QIAGEN, Hilder, Germany). RNA was eluted from the spin column with 30 μl of RNase-free water included in the kit. Eluted RNA was used as a template for RT-PCR.

The purified total RNA was reverse transcribed with the degenerated complementary primer RsXI_C1_Cend_comp, which was designed based on the C-terminus amino acid sequence of RsXI-C1. The transcribed mixture was then used in a PCR amplification reaction using the degenerate primer RsXI_C1_Cend_comp and the degenerated primer RsXI_C1_Nend, which was designed based on the N-terminus amino acid sequence of RsXI-C1.

Reverse transcription reaction was performed at 50 °C for 10 min using 200 units of Superscript IV (Thermo Fisher Scientific Inc.) in a total volume of 20 μl that contained 2 pmol of each primer, 1× Superscript IV buffer, 0.5 mM dNTP, and 5 mM DTT, following which the reverse transcriptase was inactivated by incubating the reaction mixture at 80 °C for 10 min. After the reaction, the remaining RNA was removed by treatment with RNase H (2 units per reaction) at 37 °C for 30 min.

The reaction product, obtained as above, was used for the PCR amplification. PCR reaction was carried out in 50 μl total volume and each reaction mixture contained 1× ExTaq buffer (Takara Bio Inc.), 50 pmol of each primer, 0.2 mM dNTP, 5 units ExTaq (Takara Bio Inc.), and 5 μl of transcribe product. The PCR reaction cycle used for amplification was as follows: 94 °C for 30 s, 50 °C for 45 s, and 72 °C for 2 min (total 30 cycles). The obtained amplicon was analyzed by 1% agarose electrophoresis.

In situ hybridization

In situ hybridization was performed as described previously [47] with some modification. Excised hindguts isolated from 10 termites were placed into a 1.5-ml microfuge tube along with 100 μl of solution U and disrupted by plastic pestle on ice. The resulting suspension was then filtered through a 100-μm pore size nylon mesh to remove the gut debris.

Symbiotic protists were collected from the resulting suspension by centrifugation at 500 rpm for 3 min at 4 °C and the pellet was washed 3-times with 100 μl of solution U. Washed protists were fixed by incubating with solution U containing 4% formaldehyde at room temperature for 15 min. After fixation, suspended protists were dehydrated by sequentially washing with 50% ethanol for 5 min, 80% ethanol for 5 min, and finally with 100% ethanol for 5 min. After dehydration, suspended protists were fixed on glass slides by spotting and drying.

Spotted specimen was treated with pre-hybridization solution (2 pmol/μl RsXI_C1_Nend, 300 mM NaCl, 30 mM sodium citrate, and 0.01% SDS) at 37 °C for 30 min. After the removal of pre-hybridization solution, hybridization was carried out in hybridization solution that contained 2 pmol/μl fluorescein-labeled C_nested_comp (anti-sense) probe or fluorescent-labeled C_nested (sense) probe, 300 mM NaCl, 30 mM sodium citrate, and 0.01% SDS, and slides were incubated with this solution at 37 °C for 1 h. 5′ end fluorescein-labeled oligonucleotide probes were synthesized by Eurofins genomics K.K.

(Tokyo, Japan). Probe sequences are listed in Additional file 10: Table S4.

After hybridization, slides were washed twice with PBS-T (0.4% NaCl, 0.01% KCl, 0.145% $Na_2HPO_4 \cdot 12H_2O$, 0.01% KH_2PO_4, 0.2% Tween 20) at 37 °C for 15 min to remove excess probes. Slides were then sealed with Vector Shield (Vector Laboratories, CA, USA) and observed under a fluorescent microscope (Model IX70, Olympus, Japan).

Construction of plasmids

Expression plasmids were constructed to overexpress the following genes of the non-oxidative pentose phosphate pathway: *XKS1* encoding xylulokinase, *RKI1* encoding ribulose 5-phosphate isomerase, *RPE1* encoding ribulose 5-phosphate epimerase, *TKL1* encoding transketolase, and *TAL1* encoding transaldolase. The integration vector pXhisHph-HOR7p-ScXK (Additional file 11: Figure S7b), which is targeted to the *HIS3* loci in chromosome XV, was used for the construction of *XKS1* overexpression plasmid. The integration vector pXAd3H-HOR7p-ScTAL1-ScTKL1 (Additional file 11: Figure S7c), which is targeted to the upstream region of *ADH3* in chromosome XIII, was used for the construction of *TAL1* and *TKL1* overexpression plasmids. The integration vector pXGr3L-HOR7p-ScRPE1-ScRKI1 (Additional file 11: Figure S7d), which is targeted to the *GRE3* loci in chromosome VIII, was used for the construction of *RPE1* and *RKI1* overexpression plasmids. All the genes were expressed under the control of the *HOR7* promoter and the *CYC1* terminator.

A codon-optimized *RsXI-C1* (*RsXI-C1O*) was synthesized based on the amino acid sequence of RsXI-C1. The XI gene from the *Piromyces* sp. E2 (*PiXI*) was synthesized based on the DNA sequence acquired from the GenBank (GenBank Accession Number: AJ249909.1). The codon-optimized version of *PiXI* (*PiXIO*) and codon-optimized XI gene from *C. phytofermentans* DSM18823 (*CpXIO*) were synthesized based on the amino acid sequences acquired from the GenBank (GenBank Accession Numbers: CAB76571.1 and ABX41597.1, respectively). These synthesized DNA fragments, which were custom synthesized from GenScript (NJ, USA), were individually cloned into the *Sac*II/*Xho*I site of pRS436GAP, and the resulting plasmids were designated as pRS436GAP-PiXI, pRS436GAP-PiXIO, pRS436GAP-RsXIC1O, and pRS436GAP-CpXIO, respectively (Additional file 7: Table S3).

Yeast transformation

Yeasts were transformed with a Frozen-EZ Yeast transformation II kit (Zymo Research, CA, USA). About 1 μg of episomal plasmid or about 5 μg of linearized integration vector was used for the transformation, followed by selection on selective SD agar plates. Resulting yeast strains are summarized in Additional file 2: Table S2.

Strain construction

To create a strain that would overexpress PPP-related genes, plasmids pXhisHph-HOR7p-ScXK, pXAd3H-HOR7p-ScTAL1-ScTKL1, and pXGr3L-HOR7p-ScRPE1-ScRKI1, each digested with restriction enzyme Sse8387I, were introduced into the strain MT8-1 and the resulting strain was named PP600.

W303-1B-based PPP overexpression strain was constructed as follows. First, the *ADE2* (GenBank Accession Number: M59824) was amplified by PCR from the genomic DNA purified from *S. cerevisiae* S288C and using the primer pair ADE2 + 1F and ADE2 + 1716R. The resulting amplification product was used to transform the W303-1B strain that complemented the *ade2* mutation. The resulting strain, which was named W303-1BA, was transformed with linearized DNA fragments obtained from the Sse8387I digested plasmids pXhisHph-HOR7p-ScXK, pXAd3H-HOR7p-ScTAL1-ScTKL1, and pXGr3L-HOR7p-ScRPE1-ScRKI1. The resulting strain was named W600.

Finally, the *TRP1* (GenBank Accession Number: 851570) and its neighboring region were amplified by PCR from the genomic DNA purified from *S. cerevisiae* S288C and using the primer pair TRP1M-F and TRP1M-R. The resulting amplification product was used to transform W600 strain to complement the *trp1* mutation, and the resulting strain was named W600W. Strains PP600 and W600W were transformed with the XI gene expression plasmids. Descriptions of the strains created in this study are listed in Additional file 2: Table S2.

Enzyme assays

Yeast transformants were cultivated at 30 °C for 24 h in SD medium, following which cells were harvested by centrifugation, washed twice with sterile distilled water, and then washed twice with 100 mM phosphate buffer (pH 7.0). Washed cell pellets were disrupted with glass beads (diameter, 0.3 mm, Yasui Kikai, Japan) using Micro mixer E-36 (Taitec Corporation, Japan) or Multi-beads shocker (Yasui Kikai, Japan). Cell extracts were centrifuged at 12,000 rpm at 4 °C for 5 min, and the resulting supernatants were collected as crude cell extracts. The total protein concentration in the crude cell extract was determined using the Quick Start protein assay kit (Bio-Rad, CA, USA).

XI activities in the crude cell extracts of recombinant yeast strains were determined by two different assay methods. One method, which has been modified from a previously described method [32], was used for measuring the XI activity of novel XI gene products. In brief, the assay was carried out in a reaction mixture that contained 50 mM maleic acid (pH 6.85), 10 mM $MgSO_4$, 1 mM $CoCl_2$, 1 mM $MnCl_2$, and 10 mM xylose. The reaction was initiated by the addition of crude cell extract, and then the reaction mixture was incubated at 30 °C for 30 min. After the incubation period was over, cysteine-carbazole-sulfuric acid regent, containing 2.7 mM cysteine, 0.22 mM carbazole, and 66% H_2SO_4, was added to the reaction mixture. The mixture was incubated for 20 min at room temperature, following which its absorbance at 540 nm was measured to determine the amount of D-xylulose produced using a D-xylulose standard curve. One unit (U) of enzyme activity was defined as the activity that produced one μmol D-xylulose per min at 30 °C.

The other method was used to determine the kinetic properties of XIs as described previously [12]. In this method, the XI activity was determined spectrophotometrically by measuring the oxidation of NADH at 340 nm. The assay mixture contained 100 mM Tris–HCl (pH 7.5), 10 mM $MgSO_4$, 0.15 mM NADH, 2 U sorbitol dehydrogenase (Wako, Japan), and crude cell extract. The reaction was performed at 30 °C, and it was initiated by the addition of D-xylose to a final concentration of 5–250 mM.

Growth assays

Yeast strains were grown on SD medium at 30 °C for 24 h. After cultivation, cells were collected, washed with sterile distilled water, and then inoculated into SX medium that contained 6.7 g/l of yeast nitrogen base without amino acids and 20 g/l of xylose as the sole carbon source. The initial cell density was adjusted to an optical density at 600 nm (OD_{600}) of 0.05. For the growth assay, cells were grown in L-shaped test tubes at 30 °C with shaking (70 rpm) in a Bio-photorecorder (model TVS062CA, ADVANTEC, Japan).

Anaerobic batch fermentation

Anaerobic batch fermentations were carried out in 100-ml medium bottles sealed with caps equipped with gas check valves. Yeast strains were aerobically pre-cultivated in SD medium for 3 days at 30 °C. Each pre-culture was separately inoculated into SD medium and aerobically cultivated for 24 h at 30 °C. Cells were collected and washed with distilled water. For xylose fermentation assay, these strains were inoculated into 50 ml fermentation medium, which contained 6.7 g/l of yeast nitrogen base without amino acids and 50 g/l of xylose. For glucose/xylose fermentation assay, these strains were inoculated into 50 ml fermentation medium, which contained 6.7 g/l of yeast nitrogen base without amino acids, 30 g/l of glucose, and 20 g/l of xylose. The initial cell density in the fermentation medium was adjusted to OD_{600} of 10. All fermentation experiments were performed at 30 °C with agitation (100 rpm).

Concentrations of glucose, xylose, glycerol, xylitol, and ethanol in the fermentation medium were analyzed by high-performance liquid chromatography (HPLC) (Prominence, Shimadzu, Japan) on a HPX-87H column (Bio-Rad), used together with a refractive index detector (model RID-10A, Shimadzu). The HPLC system was operated at 60 °C using 0.05% H_2SO_4 as the mobile phase (flow rate, 0.6 ml/min).

Microaerobic fermentation

Microaerobic fermentations were carried out in 96-well Storage Blocks (Corning Incorporated, NY, USA). Yeast strains were aerobically pre-cultured in 96-well Storage Blocks (1 ml of SD medium per well), sealed with sterile breathable Nunc Sealing Tape (Thermo Fisher Scientific Inc.), with shaking at 1500 rpm in an M-BR-022UP constant temperature incubator shaker (Taitec Corporation) for 24 h at 30 °C. Next, 200 μl of the pre-culture was used to inoculate 1 ml of SD medium and cultured under the same condition for 24 h. Cells were harvested, washed twice with 1 ml of sterile water, and then suspended in 0.2 ml of sterile water to prepare yeast suspensions. Fermentation test was performed under the following conditions. One milliliter of SX medium, containing 20 g/l of xylose as the sole carbon source, was placed into each well of a 96-well storage block, to which yeast suspension was added to a final OD_{600} of 1.0 (for comparing xylose fermentation by the mutants generated by saturated mutagenesis) or to a final OD_{600} of 10.0 (for comparing xylose fermentation by other mutated XIs). For establishing a microaerobic condition, each well was hermetically sealed with Titer Stick HC film (Kajixx Co., Ltd., Japan), and fermentation was carried out in the M-BR-022UP incubator at 30 °C with shaking at 1500 rpm. An aliquot of the fermented liquid was taken out at different times as indicated and analyzed by HPLC.

Introduction of mutation in RsXI-C1 and growth-based screening of mutants

To introduce random mutations in the XI genes, an error-prone PCR was carried out with GeneMorph II kit (Agilent technologies, CA, USA) using pRS436GAP-RsXIC1O as the template and primer pair pRSSacII-AAA-ATG-F4 and pRSXhoI-TAA-R3; sequences of these primers are listed in Additional file 10: Table S4. Using this method, we were able to introduce on average 3 mutations per 1000 bases in the DNA fragment (error rate 0.3%) by error-prone PCR.

A yeast centromeric vector pRS316GAP was constructed for introducing the mutated XI genes into yeast. The P_{TDH3}-T_{CYC1} region was amplified from the plasmid pRS436GAP by PCR using the primers TDH3p-CYC1t-IF-F and TDH3p-CYC1t-IF-R (Additional file 10: Table S4). The amplified DNA fragment was inserted into the PvuII-digested pRS316 (NBRP Accession Number: BYP562) to generate pRS316GAP (Additional file 11: Figure S7e). In this plasmid, the cloned genes are under the control of the *TDH3* promoter and *CYC1* terminator.

The mutated *RsXI-C1O* library, produced by error-prone PCR (see above), was mixed with *Sac*II and *Xho*I digested pRS316GAP, and directly introduced into the W600W strain using the gap repair cloning protocol [48]. The resulting yeast transformants were cultured in SD medium for 2 days. The non-mutated DNA fragment (control), amplified by PCR using pRS436GAP-RsXIC1O as the template, was introduced into the W600W strain in the same way as described above.

One hundred microliter of the mutant library culture was added to 5 ml of SX medium, and incubated at 30 °C with shaking at 70 rpm in the Bio-photorecorder. After 7 days of cultivation, cells were harvested from the culture medium, and resuspended in 5 ml of fresh SX medium to a final OD_{600} of 0.1. After culturing for 7 days in the Bio-photorecorder, cells were recovered from the culture medium and then spread on an SX agar plate and incubated at 30 °C. Colonies growing faster than the control were selected and cultured on an SD agar plate. Twenty clones were selected for the growth experiment in SX medium, from which top 10 strains exhibiting better growth rates than the control strain were chosen for plasmid extraction. Plasmids were extracted using the Zymoprep Yeast Plasmid Minipreparation kit (Zymo research). DNA inserts of these extracted plasmids were sequenced completely to identify the introduced mutation. These plasmids were subsequently used to retransform W600W.

Construction of the saturation mutagenesis library

An amino acid point mutation library, targeting the asparagine 337 (N337) residue of RsXI, was constructed by site-directed mutagenesis using the plasmid pRS316GAP-RsXIC1O as the template, primers listed in Additional file 10: Table S4, and a Quick Change Lightning MultiSite-Directed Mutagenesis kit (Agilent Technologies) following the protocol provided with the kit. Consequently, 18 plasmids, each expressing a different single amino acid substitution mutant (except asparagine and threonine), were obtained (Additional file 7: Table S3). Eighteen different strains were generated by transforming the W600W strain separately with each mutant plasmid. Resulting strains are listed in Additional file 2: Table S2.

Introduction of mutation into other XIs

Preparation of DNA templates for introducing a point mutation was carried out as follows. The coding regions of *PiXIO* and *CpXIO* were amplified from the plasmids pRS436GAP-PiXIO and pRS436GAP-CpXIO, respectively, by PCR. The resulting DNA fragments were inserted into the *Sac*II/*Xho*I digested pRS316GAP plasmid to construct pRS316GAP-PiXIO and pRS316GAP-CpXIO, respectively. A synthetic *LlXI* (*LlXIO*), codons optimized for expression in yeast, was custom synthesized (GenScript Corporation); the amino acid sequence of LlXI was acquired from the GenBank (RefSeq Accession Number: WP_057720788.1). The resulting DNA fragment was cloned into the *Sac*II/*Xho*I digested pRS316GAP to construct pRS316GAP-LlXIO.

Plasmids pRS316GAP-PiXIO, pRS316GAP-CpXIO, and pRS316GAP-LlXIO were used as templates for the PCR-based site-directed mutagenesis. For each XI gene, four different single amino acid substitution mutants were created; thus, the respective asparagine residue in each XI was substituted with cysteine, threonine, valine, and alanine. Corresponding plasmids harboring these substitution mutants are listed in Additional file 7: Table S3. The resulting plasmids were individually introduced into the W600W strain to create the XI mutant expressing strains (Additional file 2: Table S2).

Protein structure prediction

The three-dimensional predicted model of RsXI-C1 and N337C mutant of RsXI-C1 were generated based on the crystal structure of *Bacteroides thetaiotaomicron* xylose isomerase (Protein Data Bank code 4XKM) [49] using Discovery Studio (Dassault Systèmes, France).

Additional files

Additional file 1: Table S1. Identities between novel XIs and other functional XIs.

Additional file 2: Table S2. List of strains.

Additional file 3: Figure S1. RT-PCR analysis of fractionated symbiotic bacteria and protists. Symbiotic bacteria and protists in termite hindgut were fractionated by slow centrifugation and each fraction was used for RT-PCR analysis as described in "Methods". Amplified products were analyzed by electrophoresis on 1% agarose gel. Left lane, standard DNA markers: 23,130 bp, 9416 bp, 6557 bp, 4361 bp, 2322 bp, 2027 bp, 1353 bp, 1078 bp, 872 bp, and 603 bp. Arrowhead, amplified fragment from the *RsXI-C1*.

Additional file 4: Figure S2. Microaerobic xylose fermentation by recombinant PP600 strains. *S. cerevisiae* PP600 expressing RsXI-C1 (MR311), PiXI (MP111), and CpXI (MC111) were cultivated under microaerobic fermentation condition in SX medium as described in "Methods". The initial cell density was same ($OD_{600} = 5$) in all cases. Amount of xylose consumed by each strain was determined by HPLC. Xylose consumed after 120 h fermentation was shown. Error bars represent standard deviations of biological tetraplicates. Statistical significance was determined using Student's *t* test. *$P < 0.05$, significant difference.

Additional file 5: Figure S3. Relative expression of xylose utilization related genes in recombinant yeast strains. Recombinant strains (WR311, WP111, WC111, and WVC110) were aerobically pre-cultivated in SD medium at 30 °C for 24 h. Each pre-culture was separately inoculated into SD medium. The initial cell density was adjusted to OD_{600} of 0.05, and aerobically cultivated at 30 °C. After 24 h cultivation, cells were lysed and total RNA was extracted using High Pure RNA Isolation Kit (Roche, Switzerland) according to the manufacturer's instructions. Reverse transcription of extracted RNA was carried out using high capacity RNA-to-cDNA Kit (Thermo Fisher Scientific Inc.) according to the manufacturer's instructions. Quantitative PCR was carried out using a qPCR detection system (ABI PRISM 7000 sequence detection system, Thermo Fisher Scientific Inc.) and power SYBR Green Master Mix (Thermo Fisher Scientific Inc.). Primer sequences used in this experiment were listed in Additional file 10: Table S4. Relative gene expression values were calculated by the &&CT method and normalized by housekeeping gene *TUB6*. Error bars represent standard deviations of biological triplicates. *XKS1*, xylulokinase; *TKL1*, transketolase 1; *TKL2*, transketolase 2; *TAL1*, transaldolase; *RKI1*, ribose 5-phosphate isomerase; *RPE1*, ribulose 5-phosphate epimerase; *GND2*, 6-phosphogluconate dehydrogenase; *SOL4*, 6-phosphogluconolactonase; *ZWF1*, glucose-6-phosphate dehydrogenase; *HXT1*, hexose transporter 1; *HXT5*, hexose transporter 5; *HXT7*, hexose transporter 7; *GAL2*, : galactose permease.

Additional file 6: Figure S4. Specific growth rates of mutant strains isolated from growth-based screening on xylose. Growth assays in liquid culture of 20 mutant strains were carried out in SX medium under aerobic condition. Growth rates were determined as described in "Methods".

Additional file 7: Table S3. List of plasmids.

Additional file 8: Figure S5. Xylose fermentation by recombinant W600W strains. *S. cerevisiae* W600W expressing RsXI-C1 (WR320), PiXI (WP120), LIXI (WL120), and CpXI (WC120) were cultivated under microaerobic fermentation condition in SX medium as described in "Methods". The initial cell density was same ($OD_{600} = 10$) in all cases. Amount of xylose consumed by each strain was determined by HPLC. Xylose consumed after 72 h fermentation was shown. Error bars represent standard deviations of biological duplicates.

Additional file 9: Figure S6. Predicted structures of the active sites of RsXI-C1 and N337C mutant of RsXI-C1. Predicted three-dimensional structures of the active sites of (**a**) wild-type RsXI-C1 and (**b**) N337C mutant of RsXI-C1 (see text for the details regarding model building). Positions of the Asn337 (wild-type) and Cys337 (N337C mutant) residues are indicated. Also shown are the active site residues (Phe101, His102, Asp103, and Lys235), metal ion binding residues (Glu233, Glu269, Asp297, and Asp339), and manganese ion (blue sphere).

Additional file 10: Table S4. Primers used in this study.

Additional file 11: Figure S7. Maps of plasmid vectors used in this study. (**a**) The multicopy plasmid for the expression of XI genes. (**b**) The low-copy centromeric plasmid for the expression of XI genes. (**c**) The integration plasmid targeted to the HIS3 loci in chromosome XV for the expression of *XKS1*. (**d**) The integration plasmid targeted to the upstream region of *ADH3* in chromosome XIII for the expression of *TAL1* and *TKL1*. (**e**) The integration plasmid targeted to the *GRE3* loci in chromosome VIII for the expression of *RPE1* and *RKI1*.

Abbreviations

XI: xylose isomerase; XR: xylose reductase; XDH: xylitol dehydrogenase; XK: xylulokinase; PPP: pentose phosphate pathway; PiXI: XI from *Piromyces* sp. E2; RsXI: XI from the protists in the hindgut of *Reticulitermes speratus*; CpXI: XI from *Clostridium phytofermentans*; LIXI: XI from *Lactococcus lactis*; HPLC: high-performance liquid chromatography; OD_{600}: optical density at 600 nm; OD_{660}: optical density at 660 nm.

Authors' contributions

SK, NM, SM, TO, and KT designed the experiments. SK, NM, RN, SM, and KT carried out the experiments. NT and NY provided helpful advice and support for the experimental design and data analysis. SM and MO provided academic advice. SK, SM, and KT wrote the manuscript. All authors read and approved the final manuscript.

Author details

[1] Bioinspired Systems Research-Domain, Toyota Central R&D Labs., Inc., 41-1, Yokomichi, Nagakute, Aichi 480-1192, Japan. [2] RIKEN Center for Sustainable Resource Science, 1-7-22 Suehiro-cho, Tsurumi-ku, Yokohama, Kanagawa 230-0045, Japan. [3] Biotechnology and Afforestation Laboratory, New Business Planning Div, Toyota Motor Corporation, 1099 Marune, Kurozasa-cho, Miyoshi, Aichi 470-0201, Japan. [4] RIKEN Center for Sustainable Resource Science, 3-1-1 Koyadai, Tsukuba, Ibaraki 305-0074, Japan. [5] Japan Collection of Microorganisms, RIKEN BioResource Center, 3-1-1 Koyadai, Tsukuba, Ibaraki 305-0074, Japan.

Acknowledgements

We thank Chie Imamura, Akinori Ikeuchi (TOYOTA CENTRAL R&D LABS., INC.), and Nemuri Todaka (RIKEN) for helpful discussions and technical assistance, and also thank Nobuhiro Ishida, Norihiro Mitsukawa, and Haruo Takahashi (TOYOTA CENTRAL R&D LABS., INC.) for valuable comments.

Competing interests

The authors declare competing interests. This work was mainly supported by TOYOTA CENTRAL R&D LABS., INC. (Japan) and Toyota Motor Corporation (Japan). SK, RN, and KT are employees of TOYOTA CENTRAL R&D LABS., INC. NT, NY, and TO are employees of Toyota Motor Corporation. TOYOTA CENTRAL R&D LABS., INC. has submitted two patent applications on xylose isomerase from the protists in the termite hindgut (WO2011078262 and WO2014156194). RIKEN is co-applicant of WO2011078262 and Toyota Motor Corporation is co-applicant of WO2014156194. All authors are named as inventors.

Funding

Not applicable.

References

1. Hahn-Hägerdal B, Karhumaa K, Fonseca C, Spencer-Martins I, Gorwa-Grauslund MF. Towards industrial pentose-fermenting yeast strains. Appl Microbiol Biotechnol. 2007;74:937–53.
2. van Maris AJ, Winkler AA, Kuyper M, de Laat WT, van Dijken JP, Pronk JT. Development of efficient xylose fermentation in *Saccharomyces cerevisiae*: xylose isomerase as a key component. Adv Biochem Eng Biotechnol. 2007;108:179–204.
3. Van Vleet JH, Jeffries TW. Yeast metabolic engineering for hemicellulosic ethanol production. Curr Opin Biotechnol. 2009;20:300–6.
4. Cai Z, Zhang B, Li Y. Engineering *Saccharomyces cerevisiae* for efficient anaerobic xylose fermentation: reflections and perspectives. Biotechnol J. 2011;7:34–46.

5. Moysés DN, Reis VC, de Almeida JR, de Moraes LM, Torres FA. Xylose fermentation by *Saccharomyces cerevisiae*: challenges and prospects. Int J Mol Sci. 2016;17:207.
6. Kwak S, Jin YS. Production of fuels and chemicals from xylose by engineered *Saccharomyces cerevisiae*: a review and perspective. Microb Cell Fact. 2017;16:82.
7. Jeffries TW, Jin YS. Metabolic engineering for improved fermentation of pentoses by yeasts. Appl Microbiol Biotechnol. 2004;63:495–509.
8. Ho NW, Chen Z, Brainard AP, Sedlak M. Successful design and development of genetically engineered *Saccharomyces* yeasts for effective cofermentation of glucose and xylose from cellulosic biomass to fuel ethanol. Adv Bichem Eng Biotechnol. 1999;65:163–92.
9. Hahn-Hägerdal B, Karhumaa K, Jeppsson M, Gorwa-Grauslund MF. Metabolic engineering for pentose utilization in *Saccharomyces cerevisiae*. Adv Biochem Eng Biotechnol. 2007;108:147–77.
10. Bera AK, Ho NW, Khan A, Sedlak M. A genetic overhaul of *Saccharomyces cerevisiae* 424A(LNH-ST) to improve xylose fermentation. J Ind Microbiol Biotechnol. 2011;38:617–26.
11. Konishi J, Fukuda A, Mutaguchi K, Uemura T. Xylose fermentation by *Saccharomyces cerevisiae* using endogenous xylose-assimilating genes. Biotechnol Lett. 2015;37:1623–30.
12. Brat D, Boles E, Wiedemann B. Functional expression of a bacterial xylose isomerase in *Saccharomyces cerevisiae*. Appl Environ Microbiol. 2009;75:2304–11.
13. Kuyper M, Harhangi HR, Stave AK, Winkler AA, Jetten MS, de Laat WT, den Ridder JJ, den Camp HJ, van Dijken JP, Pronk JT. High-level functional expression of a fungal xylose isomerase: the key to efficient ethanolic fermentation of xylose by *Saccharomyces cerevisiae*? FEMS Yeast Res. 2003;4:69–78.
14. Madhavan A, Tamalampudi S, Ushida K, Kanai D, Katahira S, Srivastava A, Fukuda H, Bisaria VS, Kondo A. Xylose isomerase from polycentric fungus *Orpinomyces*: gene sequencing, cloning, and expression in *Saccharomyces cerevisiae* for bioconversion of xylose to ethanol. Appl Microbiol Biotechnol. 2009;82:1067–78.
15. Rønnow B, Andersen TH, Sibbesen O. Microorganism expressing xylose isomerase. WO2010070549; 2010.
16. Teunissen AWRH, De Bont JAM. Xylose isomerase genes and their use in fermentation of pentose sugars. WO2010074577; 2010.
17. Ha SJ, Kim SR, Choi JH, Park MS, Jin YS. Xylitol does not inhibit xylose fermentation by engineered *Saccharomyces cerevisiae* expressing *xylA* as severely as it inhibits xylose isomerase reaction in vitro. Appl Microbiol Biotechnol. 2011;92:77–84.
18. Parachin NS, Gorwa-Grauslund MF. Isolation of xylose isomerases by sequence- and function-based screening from a soil metagenomic library. Biotechnol Biofuels. 2011;4:9.
19. Aeling KA, Salmon KA, Laplaza JM, Li L, Headman JR, Hutagalung AH, Picataggio S. Co-fermentation of xylose and cellobiose by an engineered *Saccharomyces cerevisiae*. J Ind Microbiol Biotechnol. 2012;39:1597–604.
20. de Figueiredo Vilela L, de Mello VM, Reis VC, Bon EP, Gonçalves Torres FA, Neves BC, Eleutherio EC. Functional expression of *Burkholderia cenocepacia* xylose isomerase in yeast increases ethanol production from a glucose-xylose blend. Bioresour Technol. 2013;128:792–6.
21. Hector RE, Dien BS, Cotta MA, Mertens JA. Growth and fermentation of D-xylose by *Saccharomyces cerevisiae* expressing a novel D-xylose isomerase originating from the bacterium *Prevotella ruminicola* TC2-24. Biotechnol Biofuels. 2013;6:84.
22. Klaassen P, de Jong RM, van Suylekom GP. A pentose sugar fermenting cell. WO2013017644; 2013.
23. Dragovic Z, Gamauf C, Reisinger C, Kettling U. Pentose fermenting microorganisms. WO2013117631; 2013.
24. Chen Z, Kelly KJ, Ye RW. Cow rumen xylose isomerases active in yeast cells. WO2014164392; 2014.
25. Chen Z, Kelly KJ, Ye RW. Bacterial xylose isomerases active in yeast cells. WO2014164410; 2014.
26. Peng B, Huang S, Liu T, Geng A. Bacterial xylose isomerases from the mammal gut Bacteroidetes cluster function in *Saccharomyces cerevisiae* for effective xylose fermentation. Microb Cell Fact. 2015;14:70.
27. Hou J, Shen Y, Jiao C, Ge R, Zhang X, Bao X. Characterization and evolution of xylose isomerase screened from the bovine rumen metagenome in *Saccharomyces cerevisiae*. J Biosci Bioeng. 2016;121:160–5.
28. Harhangi HR, Akhmanova AS, Emmens R, van der Drift C, de Laat WT, van Dijken JP, Jetten MS, Pronk JT, den Camp HJ. Xylose metabolism in the anaerobic fungus *Piromyces* sp. strain E2 follows the bacterial pathway. Arch Microbiol. 2003;180:134–41.
29. Kuyper M, Winkler AA, van Dijken JP, Pronk JT. Minimal metabolic engineering of *Saccharomyces cerevisiae* for efficient anaerobic xylose fermentation: a proof of principle. FEMS Yeast Res. 2004;4:655–64.
30. Kuyper M, Hartog MMP, Toirkens MJ, Almering MJH, Winkler AA, van Dijken JP, Pronk JT. Metabolic engineering of a xylose-isomerase-expressing *Saccharomyces cerevisiae* strain for rapid anaerobic xylose fermentation. FEMS Yeast Res. 2005;5:399.
31. Todaka N, Inoue T, Saita K, Ohkuma M, Nalepa CA, Lenz M, Kudo T, Moriya S. Phylogenetic analysis of cellulolytic enzyme genes from representative lineages of termites and a related cockroach. PLoS ONE. 2010;5:e8636.
32. Belfaquih N, Penninckx MJ. A bifunctional beta-xylosidase-xylose isomerase from *Streptomyces* sp. EC 10. Enzyme Microb Technol. 2000;27:114–21.
33. Koidzumi M. Studies on the intestinal protozoa found in the termites of Japan. Parasitology. 1921;13:235–309.
34. Hahn-Hägerdal B, Wahlbom CF, Gárdonyi M, van Zyl WH, Cordero Otero RR, Jönsson LJ. Metabolic engineering of *Saccharomyces cerevisiae* for xylose utilization. Adv Biochem Eng Biotechnol. 2001;73:53–84.
35. Kuyper M, Toirkens MJ, Diderich JA, Winkler AA, van Dijken JP, Pronk JT. Evolutionary engineering of mixed-sugar utilization by a xylose-fermenting *Saccharomyces cerevisiae* strain. FEMS Yeast Res. 2005;5:925–34.
36. Zhou H, Cheng JS, Wang B, Fink GR, Stephanopoulos G. Xylose isomerase overexpression along with engineering of the pentose phosphate pathway and evolutionary engineering enable rapid xylose utilization and ethanol production by *Saccharomyces cerevisiae*. Metab Eng. 2012;14:611–22.
37. Demeke MM, Dietz H, Li Y, Foulquié-Moreno MR, Mutturi S, Deprez S, Den Abt T, Bonini BM, Liden G, Dumortier F, et al. Development of a D-xylose fermenting and inhibitor tolerant industrial *Saccharomyces cerevisiae* strain with high performance in lignocellulose hydrolysates using metabolic and evolutionary engineering. Biotechnol Biofuels. 2013;6:89.
38. Demeke MM, Dumortier F, Li Y, Broeckx T, Foulquié-Moreno MR, Thevelein JM. Combining inhibitor tolerance and D-xylose fermentation in industrial *Saccharomyces cerevisiae* for efficient lignocellulose-based bioethanol production. Biotechnol Biofuels. 2013;6:120.
39. Moriya S, Dacks JB, Takagi A, Noda S, Ohkuma M, Doolittle WF, Kudo T. Molecular phylogeny of three oxymonad genera: *Pyrsonympha*, *Dinenympha* and *Oxymonas*. J Eukaryot Microbiol. 2003;50:190–7.
40. Lee SM, Jellison T, Alper HS. Directed evolution of xylose isomerase for improved xylose catabolism and fermentation in the yeast *Saccharomyces cerevisiae*. Appl Environ Microbiol. 2012;78:5708–16.
41. Subbian E, Zhang X, Mijts B, Cho CM. Pentose fermentation by a recombinant microorganism. WO2011150313; 2011.
42. Froehlich A, Henningsen B, Lengyel Z. Chimeric polypeptides having xylose isomerase activity. WO2016024218; 2016.
43. Tajima M, Nogi Y, Fukasawa T. Primary structure of the *Saccharomyces cerevisiae GAL7* gene. Yeast. 1985;1:67–77.
44. Thomas BJ, Rothstein R. Elevated recombination rates in transcriptionally active DNA. Cell. 1989;56:619–30.
45. Saitou N, Nei M. The neighbor-joining method: a new method for reconstructing phylogenetic trees. Mol Biol Evol. 1987;4:406–25.
46. Trager W. The cultivation of a cellulose-digesting flagellate, *Trichomonas termopsidis*, and of certain other termite protozoa. Biol Bull. 1934;66:182–90.
47. Moriya S, Noda S, Ohkuma M, Kudo T. *In situ* hybridization of termite microbes. Methods Mol Biol. 2007;353:105–12.
48. Lundblad V, Zhou H. Manipulation of plasmids from yeast cells. Curr Protoc Mol Biol. 2001. doi:10.1002/0471142727.
49. Han BG, Bong SM, Cho JW, Kim MD, Kim SJ, Lee BI. Crystal structure of a class 2 D-xylose isomerase from the human intestinal tract microbe *Bacteroides thetaiotaomicron*. Biodesign. 2015;3:41–7.

Genome sequencing and transcriptome analysis of *Trichoderma reesei* QM9978 strain reveals a distal chromosome translocation to be responsible for loss of *vib1* expression and loss of cellulase induction

Christa Ivanova[1,6*], Jonas Ramoni[2], Thiziri Aouam[1], Alexa Frischmann[2], Bernhard Seiboth[2], Scott E. Baker[3], Stéphane Le Crom[4], Sophie Lemoine[5], Antoine Margeot[1] and Frédérique Bidard[1*]

Abstract

Background: The hydrolysis of biomass to simple sugars used for the production of biofuels in biorefineries requires the action of cellulolytic enzyme mixtures. During the last 50 years, the ascomycete *Trichoderma reesei*, the main source of industrial cellulase and hemicellulase cocktails, has been subjected to several rounds of classical mutagenesis with the aim to obtain higher production levels. During these random genetic events, strains unable to produce cellulases were generated. Here, whole genome sequencing and transcriptomic analyses of the cellulase-negative strain QM9978 were used for the identification of mutations underlying this cellulase-negative phenotype.

Results: Sequence comparison of the cellulase-negative strain QM9978 to the reference strain QM6a identified a total of 43 mutations, of which 33 were located either close to or in coding regions. From those, we identified 23 single-nucleotide variants, nine InDels, and one translocation. The translocation occurred between chromosomes V and VII, is located upstream of the putative transcription factor *vib1*, and abolishes its expression in QM9978 as detected during the transcriptomic analyses. Ectopic expression of *vib1* under the control of its native promoter as well as overexpression of *vib1* under the control of a strong constitutive promoter restored cellulase expression in QM9978, thus confirming that the translocation event is the reason for the cellulase-negative phenotype. Gene deletion of *vib1* in the moderate producer strain QM9414 and in the high producer strain Rut-C30 reduced cellulase expression in both cases. Overexpression of *vib1* in QM9414 and Rut-C30 had no effect on cellulase production, most likely because *vib1* is already expressed at an optimal level under normal conditions.

Conclusion: We were able to establish a link between a chromosomal translocation in QM9978 and the cellulase-negative phenotype of the strain. We identified the transcription factor *vib1* as a key regulator of cellulases in *T. reesei* whose expression is absent in QM9978. We propose that in *T. reesei*, as in *Neurospora crassa*, *vib1* is involved in cellulase induction, although the exact mechanism remains to be elucidated. The data presented here show an example of a combined genome sequencing and transcriptomic approach to explain a specific trait, in this case the QM9978 cellulase-negative phenotype, and how it helps to better understand the mechanisms during cellulase gene regulation. When focusing on mutations on the single base-pair level, changes on the chromosome level can be easily

*Correspondence: christa.ivanova@pasteur.fr; frederique.bidard-michelot@ifpen.fr
[1] IFP Energies Nouvelles, 1-4 Avenue de Bois-Préau, 92852 Rueil-Malmaison, France
Full list of author information is available at the end of the article

overlooked and through this work we provide an example that stresses the importance of the big picture of the genomic landscape during analysis of sequencing data.

Keywords: *Trichoderma reesei*, Genome analysis, Transcriptome, *vib1*, Cellulase production, Promoter, Translocation

Background

In order to bypass the current dependence on fossil resources and to reduce carbon dioxide emissions, cellulosic and hemicellulosic polymers are considered as environmentally clean and renewable energy sources and are used for the production of bioethanol and platform chemicals. Lignocellulosic material as found in agricultural crop residues, grasses, and wood must undergo physical and chemical pre-treatment before fermentable sugars can be released by the action of cellulosic enzymes. The production costs of those enzyme cocktails are still very high and present a limiting factor in the production of second-generation biofuels [1].

Trichoderma reesei, a filamentous fungus naturally found growing on decaying wood, is well known for its high capacity to secrete large amounts of cellulolytic enzymes and some strains are commonly used as industrial protein production hosts [2, 3]. All *T. reesei* mutants which are currently used for industrial enzyme production or academic research originate from a single wild-type strain (QM6a) isolated on the Solomon Islands during the second world war due to its capability to degrade the cotton fabric of tents of the US army troops stationed there [4]. The strain was part of the US Army Quartermaster (QM) culture collection, and its cellulase system has been characterized over several decades [5]. During the first oil crisis in the 1970s, improved *T. reesei* strains were chosen for industrial enzyme production [5–7].

Since then, different mutagenesis programs have produced several strain lineages of *T. reesei,* some with higher enzyme production capabilities and some with lower or absent cellulase production [8] (Fig. 1). Most notably, the strain lineage developed at the Rutgers University in the 1970s produced strain Rut-C30, with the highest cellulase production levels available for academic research [7]. Strains QM9414 and QM9978 are both a product of mutagenesis programs at the US Army Natick laboratories. Whereas QM9414 serves as a common reference strain due to its moderately improved cellulase production, QM9978, QM9979, and QM9136 have cellulase-negative phenotypes providing an opportunity for understanding the mechanisms underlying cellulase regulation [9]. Indeed, the genetic causes leading to cellulase overproduction remain relatively uncharacterized. Only a few examples exist where a specific genetic change could be linked to an improved or decreased cellulase production [8, 10–16]. Whole genome sequence analysis of several strains has contributed to the understanding of the involved mechanisms [13, 14, 16–18] (Fig. 1).

In this work, we chose the strain QM9978 as a model to understand cellulase production as it is a cellulase-negative strain and the cause for it was yet unknown.

Earlier studies on strain QM9978 suggested that there is a defect in the cellulase induction process but not in the basal expression of cellulase genes [9]. The uptake of cellobiose and sophorose is not affected suggesting that its cellulase-negative phenotype is most likely a result of a defective component specifically involved in the regulation of cellulase expression rather than during induction [19]. Another study using gene expression analysis of QM9414 and QM9978 on sophorose performed by the rapid subtraction hybridization (RaSH) method led to the identification of 18 differentially expressed genes, among which was the homologue of the *N. crassa* clock modulator *vivid*, the translation elongation factor 1α, and the transcriptional activator Hap1 [20]. Phenotypically, QM9978 forms less conidia than QM6a when grown on potato dextrose agar (PDA), fails to produce sufficient cellulases on any tested medium, and shows no growth when cultivated in liquid medium with cellulose as the sole carbon source [9, 20, 21] (Fig. 2).

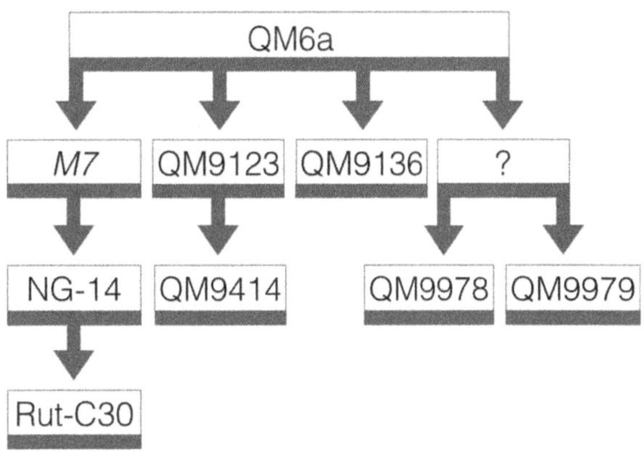

Fig. 1 Pedigree of the *T. reesei* strain QM9978 and its relationship to the wild-type isolate QM6a. The progeny of the original isolate QM6a was derived by classical mutagenesis using UV light, irradiation by linear particle accelerators, and/or *N*-methyl-*N*′-nitro-*N*-nitrosoguanidine (NTG). Strain M7 is indicated in *italics*, as it is no longer available. Strains QM9136, QM9978, and QM9979 are deficient in cellulase production. Sequence analysis identified QM9978 to originate from a strain that is unknown. It is therefore pictured with a question mark

In an effort to identify new components of the cellulase gene regulation network of *T. reesei*, we used a whole genome sequencing and transcriptomic analysis approach comparing the cellulase-negative strain QM9978 to the wild-type strain QM6a. We show that—among a few other mutations that were unlikely to have a significant impact on cellulase formation—this strain contains a chromosomal translocation upstream of the promoter region of the *N. crassa* homologue *vib-1* (vegetative incompatibility blocked) and that *vib1* expression in QM9978 is absent. The cellulase production in this strain can be restored by introduction of an additional *vib1* copy and by *vib1* overexpression. We further investigated the role of *vib1* in the improved cellulase producer strains QM9414 and Rut-C30 and show that cellulase gene regulation in the three strains follows different regulatory pathways.

Results

Genome analysis of QM9978

To identify the genetic differences between QM6a and QM9978 leading to the cellulase-negative phenotype, we performed whole-genome high-throughput Illumina sequencing of QM9978 (for details, see "Methods" section). The comparative sequence analysis identified a total of 43 mutations in QM9978, of which 33 were found in the coding region of genes or in the assumed promoter (800 bp upstream) or terminator (200 bp downstream) region. Of those, 23 were characterized as SNVs (single-nucleotide variants), eight as InDels (small insertions and deletions), and two as deletions with 7 and 47 bp missing, respectively (Table 1). They were located on 22 different scaffolds. The number of mutations in QM9978 is about three times higher than in QM9136, another QM6a derivative cellulase-negative strain used for the study of the cellulolytic system of *T. reesei* [18] (Fig. 1). Whole genome sequence analysis of QM9136 identified a truncation in the major cellulase regulator *xyr1* to be responsible for the cellulase-negative phenotype of this strain [18]. The gene encoding *xyr1* in QM9978 is intact, and therefore another cause must exist for its disability to degrade cellulose.

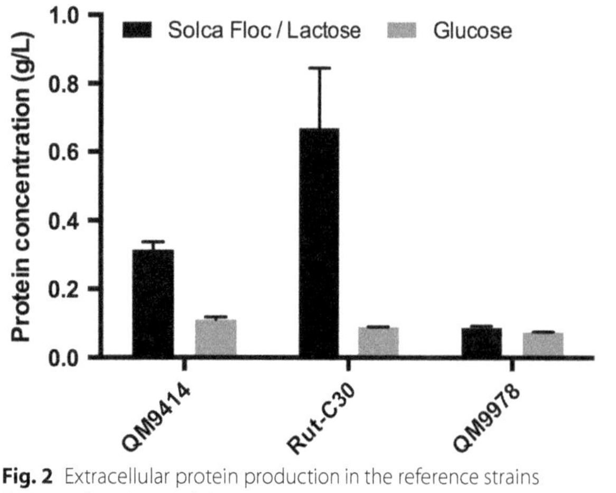

Fig. 2 Extracellular protein production in the reference strains QM9414, Rut-C30, and QM9978. Protein concentration in the supernatant was determined from cultures on liquid minimal medium supplemented with 10 g/L Solka-floc cellulose/lactose or glucose. Supernatant was harvested after 3 days and analyzed by the Lowry method using BSA as a standard. Data of two biological and three technical replicates are shown

Trichoderma reesei QM9123 was, until now, commonly believed to be the parental strain of QM9978. Since the genome sequence of QM9123 was published in 2010, we performed a comparison to QM9978 and showed that none of the 20 mutations identified in QM9123 is shared by QM9978 [16]. We therefore exclude the possibility that QM9123 is the parental strain of QM9978 and present an up-dated version of the *T. reesei* strain lineage in Fig. 1. No detailed record on the generation of the strain QM9978 and the closely related strain QM9979, which is also cellulase negative, exists, but we assume that the high number of mutations is a result of two or more subsequent mutagenesis events. Sequence comparison of QM9978 to QM9979 showed common mutations between those two strains indicating that they originate from the same parental strain (F.B., personal communication). Compared to high cellulase producer strains like *T. reesei* NG14 (one InDel, 78 SNVs, and nine large structural variations) and Rut-C30 (one deletion, two InDels, and 106 SNVs additional to NG14), the number of mutations is comparably low, probably because of alternative mutagenesis methods, and number and severity of treatments [14, 16, 17, 22].

From the 19 genes where a mutation occurred in the coding region, only the role of the blue-light receptor BLR1 in cellulase regulation has been studied in detail. Although BLR1 and BLR2 are the key regulators of the blue-light response in fungi and conidiation, it was shown that *blr1* is not essential for cellulase expression [23] (Table 1).

Our research focused on the identified deletions and those mutations were subsequently verified by PCR analysis using primers flanking the deleted fragment. For two predicted deletions (on scaffold 1 and scaffold 16), we could not confirm the results obtained by Illumina sequencing as no PCR product was obtained. The recent genome assembly of *T. reesei* using chromosomal contact data in a computational approach identified scaffolds 1 and 16 to be located on chromosomes number five and seven, respectively [24–26]. Upon close examination of our sequencing data using the Integrative genome viewer program (http://software.broadinstitute.org/software/igv/) combined with the information on

Table 1 Mutations found in *T. reesei* QM9978 compared to WT QM6a

SNP_id	Mutation	Event	Position	Transcript	Element	Annotation
1_2509832	SNV	A > G		Trire2:119839	Intron	Putative cyanamide hydratase
1_753158	SNV	T > G		Trire2:21170	Intron	Putative 40S ribosomal protein S12
1_753159	SNV	C > T		Trire2:119626	Prom	Putative 40S ribosomal protein S22
1_96633	Transloc	t(1;16)(96,633;631,551)		Trire2:54675	Upstream	Putative transcription factor VIB1
10_21554	InDel	−1:G	+1671	Trire2:121962	Exon	Blue-light regulator 1 BRL1
11_325040	SNV	G > C	+583	Trire2:4231	Exon	Putative siderophore regulation protein
12_146369	InDel	−1:T		Trire2:108419	Prom	Putative protein of unknown function
12_146369	InDel	−1:T		Trire2:122376	Prom	Putative protein of unknown function
12_778518	SNV	T > C		Trire2:108592	Intron	Putative leucine aminopeptidase
12_842068	SNV	G > A	+776	Trire2:79153	Exon	Putative 5-oxoprolinase OXP1
13_429746	SNV	G > A		Trire2:79300	Intron	Putative sphingolipid long chain base-responsive protein
13_839848	SNV	T > C		No		
14_189528	InDel	1:A		Trire2:79485	Term	Putative peptide *N*-glycanase degradation
15_427120	SNV	C > G		Trire2:4767	Prom	Putative biotin synthase
16_631551	Transloc	t(1;16)(96,633;631,551)		Trire2:80028	Upstream	Putative ABC transporter
17_646809	SNV	G > T		Trire2:65965	Prom	Putative S-adenosyl-L-methionine-dependent methyltransferase
20_145726	SNV	T > C	+293	Trire2:55671	Exon	Putative NAD(P)-binding Rossmann-fold domain protein
2_618372	SNV	C > T	+2098	Trire2:66913	Exon	Putative phosphatidylinositol 3-kinase
24_496758	InDel	−1:G		Trire2:123658	Intron	Putative guanylate-binding domain protein
26_169591	SNV	CG > AC	+2901	Trire2:51893	Exon	Putative lipid body protein ppoC
26_331506	SNV	G > A		Trire2:68803	Term	Putative RTA-1 domain protein
28_280867	InDel	−1:T		No		
29_224118	SNV	T > C		No		
3_1061204	InDel	−7:ATATCAT		No		
3_1353739	SNV	G > A	+635	Trire2:104458	Exon	Putative protein of unknown function
31_146257	SNV	A > G	+427	Trire2:69881	Exon	Putative pyoverdine/dityrosine biosynthesis protein
33_52840	InDel	−1:T		No		
37_85324	SNV	G > A		Trire2:124181	Intron	Putative DML1 protein
4_1236402	SNV	A > T		No		
4_1680892	SNV	G > A	+127	Trire2:41761	Exon	Putative vacuolar iron transporter
4_890445	SNV	A > T		Trire2:75921	Prom	Putative tRNA-dihydrouridine synthase 1
4_890446	SNV	A > T		Trire2:75921	Prom	Putative tRNA-dihydrouridine synthase 1
43_42477	SNV	G > A	+1144	Trire2:6057	Exon	Putative ubiquinol cytochrome-c reductase assembly protein
6_289426	SNV	G > T		No		
6_289431	SNV	C > G		No		
6_622255	SNV	G > C		Trire2:105968	Prom	Putative PcbC family protein
8_1174800	InDel	−7:GGAGGTC		Trire2:39911	Intron	Putative S-adenosylmethionine-homocysteine methyltransferase
8_1290905	InDel	−1:A	+452	Trire2:47897	Exon	Putative ABC transporter
9_1123734	SNV	G > T		No		
9_223955	InDel	−1:A		Trire2:77836	Prom	Putative protein of unknown function
9_475049	InDel	−1:G		No		

The SNP_id number gives the position on the scaffold as obtained by Illumina sequencing. The presented transcript ID corresponds to the protein ID in the JGI database (http://genome.jgi.doe.gov)

the Joint Genome Institute homepage (http://genome.jgi.doe.gov), it appeared that what we originally believed to be a deletion event could in reality be a translocation resulting from a chromosome break 1095 bp upstream of the ATG start codon of Trire2:54675 located on chromosome VII (Fig. 3a). To verify this hypothesis, we designed oligonucleotides flanking the assumed DNA break. Only PCRs with the primer combination 80028_MutaC_R

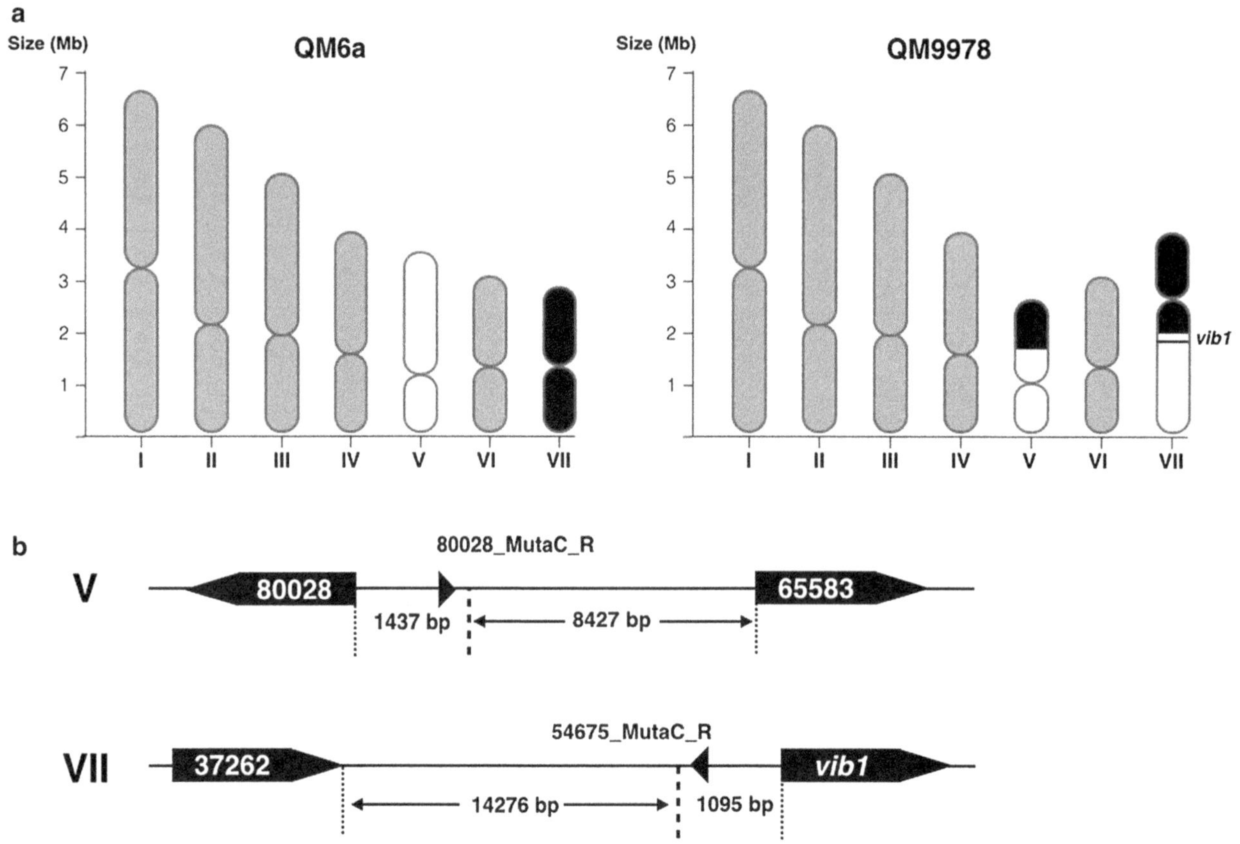

Fig. 3 Analysis of the chromosomal translocation in *T. reesei* QM9978. **a** Schematic depiction of the seven chromosomes in *T. reesei* QM6a and QM9978 showing the translocation. **b** Verification of the translocation by PCR was performed with oligonucleotides 800_MutaC_R and 54675_MutaC_R binding outside of the DNA break. The resulting DNA-fragment was analyzed by Sanger sequencing in order to identify the exact translocation point

and 54675_MutaC_R binding on the two different chromosomes resulted in an amplicon (Fig. 3b). Sequencing of this amplicon identified the region where the DNA breaks occurred. As a result from the translocation, the gene Trire2:37262 (a putative ubiquitin-protein ligase) upstream from Trire2:54675 on chromosome VII is replaced by Trire2:80028, a putative ABC transporter on chromosome V (Fig. 3b).

Gene expression profiling in *T. reesei* QM9978

Changes in gene expression between QM9978 and QM6a were analyzed in order to define the impact of mutations at the transcriptional level. Strains were grown on D-glucose for approximately 24 h until the carbon source was depleted from the medium (less than 3 g/L glucose remaining). Then, a lactose pulse of 10 g/L was added to the medium as lactose is known to be a strong inducer of cellulases [27]. Total RNA of two biological replicates was extracted at three different time points: 0 h (after carbon source depletion and before lactose addition), 24 and 48 h after the lactose pulse. Statistical treatments and differential analyses were performed using DESeq 1.8.3. A threshold of two for the log2 ratios with an adjusted p value <0.001 was used to identify genes differentially expressed (DE). Gene functions were identified by the aid of a manually annotated *T. reesei* genome database proprietary to IFPEN.

Comparison of the gene expression in QM9978 to QM6a identified a total of 785 genes differentially expressed (DE) in at least one time point. The number of down-regulated genes slightly increases during the time course, whereas the number of up-regulated genes remains stable. Interestingly, DE genes are mainly down-regulated, meaning their expression is lower in QM9978 then in QM6a. We compared the gene expression data with our sequencing results and identified five DE genes, which were affected by mutations. The genes were characterized by homology search using the NCBI BLASTp tool (https://blast.ncbi.nlm.nih.gov/Blast.cgi) and the Joint Genome Institute genomics resource database (http://genome.jgi.doe.gov). The annotation was additionally verified manually using the FUNGIpath database (http://embg.igmors.u-psud.fr/fungipath/). This way we identified ID51893—an orthologue of the lipid body protein ppoC (*A. nidulans*) involved in coordinating sexual and asexual sporulation [28], ID54675—a homologue of the transcription factor *vib-1* involved in vegetative incompatibility and cellulose utilization in *N. crassa* [29], ID68803—a predicted RTA1-domain protein, and ID108418 and ID122376—two hypothetical proteins of unknown function.

From the total of 785 genes, only 153 genes (23%) were DE at all time points. In this gene set, we find 14 glycoside hydrolases (including *cbh2*, *xyn2*, and *agl1*), eight putative transporters (including *crt1*), and four transcription factors including *vib1* (ID54675) and *azf1/sah-1* (ID61055) as identified by homology search as described above [30]. The *N. crassa* homologues of these two transcription factors have been shown to be involved in glucose signaling and to be up-regulated during growth on the silvergrass *Miscanthus* [31]. Of these, 153 genes two are impacted by mutation: *vib1* and the putative RTA1-domain protein (ID68803).

Among the four genes surrounding the translocation on chromosomes V and VII identified during our sequence analysis, only *vib1* and Trire2:80028, a putative ABC transporter whose expression is diminished in QM9978, show altered expression levels (Fig. 3).

Blastp analysis showed that *T. reesei vib1* shows 49% similarity to *N. crassa vib-1* [29, 32]. Since *N. crassa vib-1* has been shown to be essential for cellulase production in *N. crassa*, we investigated whether a link could be established between the translocation, *vib1* expression, and the loss of cellulase production in the QM9978 strain.

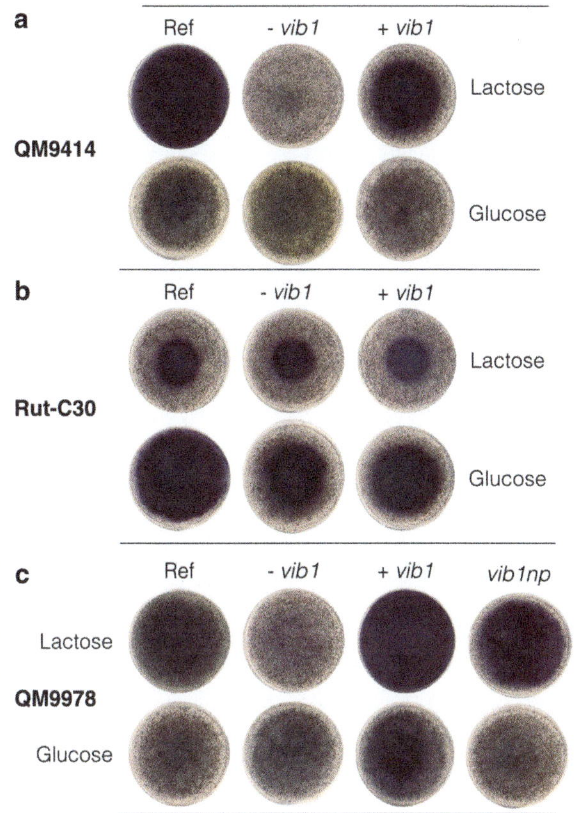

Fig. 4 Detection of cellulase production by *vib1* mutants on AZCL-HE-Cellulose. Comparison of *vib1* deletion (−*vib1*) and overexpression (+*vib1*) strains of **a** QM9414, **b** Rut-C30, and **c** QM9978 on minimal medium supplemented with glucose or lactose (1% w/v) and AZCL-HE-Cellulose (0,1% w/v). *Blue* halos on the agar plates result from degradation of the AZCL-HE-Cellulose substrate releasing a *blue dye* into the agar. Pictures were taken from the *bottom* side of the plates. The high enzyme production of Rut-C30 on glucose is due to the bigger colony size of this strain on glucose. For a comparison of colony size to enzyme production see Additional file 1: Figure S1, depicting the same plates from above. Pictures were taken after 4 days of incubation. For each strain, two biological replicates (individual clones) were assayed and one is shown representatively

Analysis of the role of *vib1* in cellulase induction

Constitutive expression of vib1 restores cellulose utilization in QM9978

To determine whether *vib1* is involved in cellulose deconstruction by *T. reesei*, we constructed a *vib1* overexpression QM9978 strain (QM9978 +*vib1*) by ectopic integration of the gene under the control of the strong constitutive promoter of *T. reesei* glyceraldehyde-3-phosphate dehydrogenase (*gpd*) [33]. For two independent clones, we monitored enzyme production during 4 days using a fast screening assay based on azurine-crosslinked cellulose (AZCL-HE-cellulose) that releases a soluble blue dye into the agar medium after enzymatic hydrolysis by endoglucanases, in which the size of the blue halo on an agar plate corresponds to the amount of secreted cellulases. Plates supplemented with 10 g/L glucose or lactose (for cellulase induction) and 1 g/L AZCL-HE-cellulose were used and the results of one clone per strain are representatively shown in Fig. 4 and Additional file 1: Figure S1. Strain QM9978 failed to secrete a sufficient amount of cellulases on lactose to decompose the AZCL-HE polymer, whereas the agar of the QM9978 +*vib1* strain was stained blue due to enzymatic hydrolysis. To verify that no residual VIB1 activity remains in QM9978, we constructed the *vib1* deletion strain QM9978 −*vib1*. Like QM9978, this strain was unable to decompose AZCL-HE-cellulose, indicating that the mutation in QM9978 equals a null phenotype (Fig. 3).

Since the data from the plate assay suggested a strong impact of *vib1* on cellulase production, we produced liquid cultures grown on 10 g/L lactose/Solka-floc (cellulase induction) and on 10 g/L glucose (repression) to determine the protein content in the supernatant and the volumetric enzyme activity (see "Methods" for details on the protocol). Both the extracellular protein concentration and the filter paper activity were found to be significantly increased in strain QM9978 +*vib1* compared to QM9978 ($p < 0.0001$ and $p < 0.01$, respectively) (Fig. 5).

The translocation in QM9978 occurred 1095 bp upstream of the *vib1* ATG start codon. Therefore, the indicated promoter of approximately 1 kb remains unaffected [34, 35]. To test whether the abolished *vib1* expression in QM9978 is due to the missing 5′ upstream region, we constructed a QM9978 strain expressing *vib1* under the control of 850 bp of its native promoter, ectopically integrated. This allowed us to investigate the role of the 5′ region absent in QM9978, for example, if activator-binding sites were affected. The resulting strain QM9978_*vib1np* had an enzyme secretion capacity comparable to QM9978 +*vib1* on plate test using AZCL-HE-cellulose (Fig. 4). These results were confirmed by real-time qPCR analysis, where we detected a significant accumulation of *cbh1* (*cel7a*) transcript, both for the overexpression strain QM9978 +*vib1* and for QM9978_*vib1np* (Fig. 6a). Expression levels in both strains remained lower compared to strain QM9414, probably due to the fact that strain QM9414 has undergone a different mutagenesis treatment during strain engineering leading to higher expression of *chb1* than a cellulase-restored QM9978 strain.

Analysis of *vib1* transcription in QM9978 +*vib1* and QM9978_*vib1np* confirmed that *vib1* transcripts were present at levels equivalent to QM9414, a strain with unimpaired *vib1* expression (Fig. 6b).

vib1 deletion, but not overexpression, has an effect on cellulase induction in higher producer strains

Strain QM9414 has been extensively used as a reference strain in the literature when comparing different mutations due to its capacity to produce a measurable amount of cellulases (QM6a producing cellulase levels below the detection limit of most assays). In addition to QM9414, we chose strain Rut-C30 as this is the highest enzyme producer strain (reported to produce 30 g/L enzymes, depending on culture conditions) available for public research and is also used in industrial applications [36]. We performed our study in Rut-C30 to show if an industrial strain originating from the Rut lineage can be further improved by *vib1* overexpression.

To gain a better understanding of the role that *vib1* plays in cellulase induction we made use of the different genetic backgrounds of strain QM9414 and Rut-C30 and constructed the strains QM9414 ±*vib1* and Rut-C30 ±*vib1* by deleting *vib1* or ectopically introducing an additional gene copy under the control of the strong constitutive *gpd1* promoter, equivalent to the approach used in QM9978.

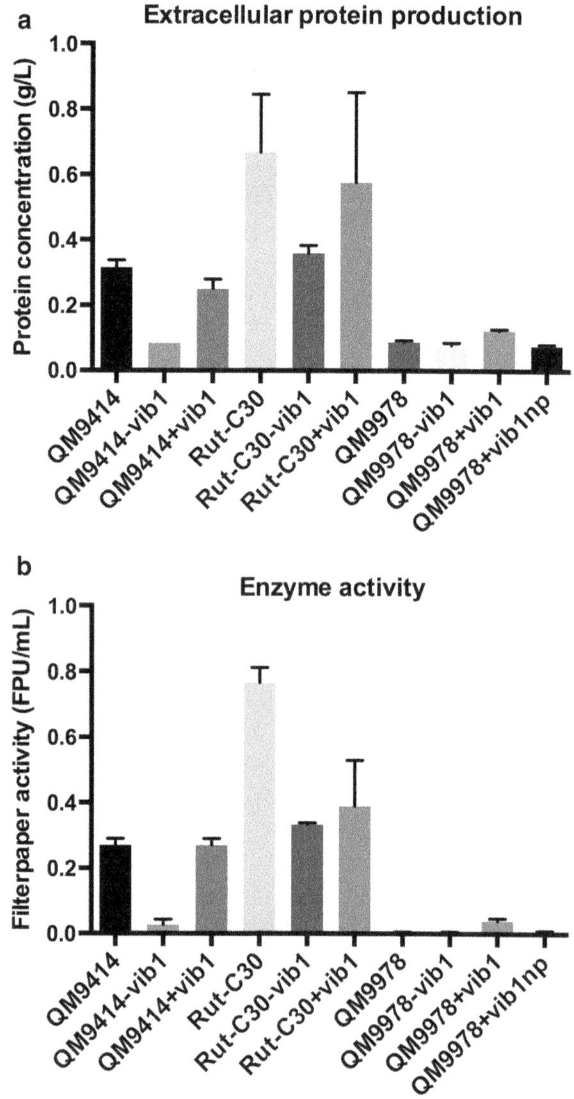

Fig. 5 Comparison of extracellular protein production and cellulase activity in *vib1* deletion and overexpression strains. **a** Protein concentration was determined by the Lowry method and is shown for −/+*vib1* strains and their respective reference strains QM9414, Rut-C30, and QM9978 using a BSA standard. **b** Filter paper activity was determined for the same samples, allowing a direct comparison between protein content and enzyme activity in the supernatant. The samples were prepared from liquid minimal medium supplemented with 10 g/L Solka-floc cellulose/lactose as carbon source. Supernatant was harvested after 3 days. Data of two biological and three technical replicates are shown

Screening on AZCL-HE-Cellulose revealed that *vib1* deletion had a negative effect on cellulase production in QM9414 and Rut-C30, whereas no effects could be observed for the overexpression (Fig. 4). This finding was further confirmed by the measurement of extracellular protein concentration and enzyme activity in liquid

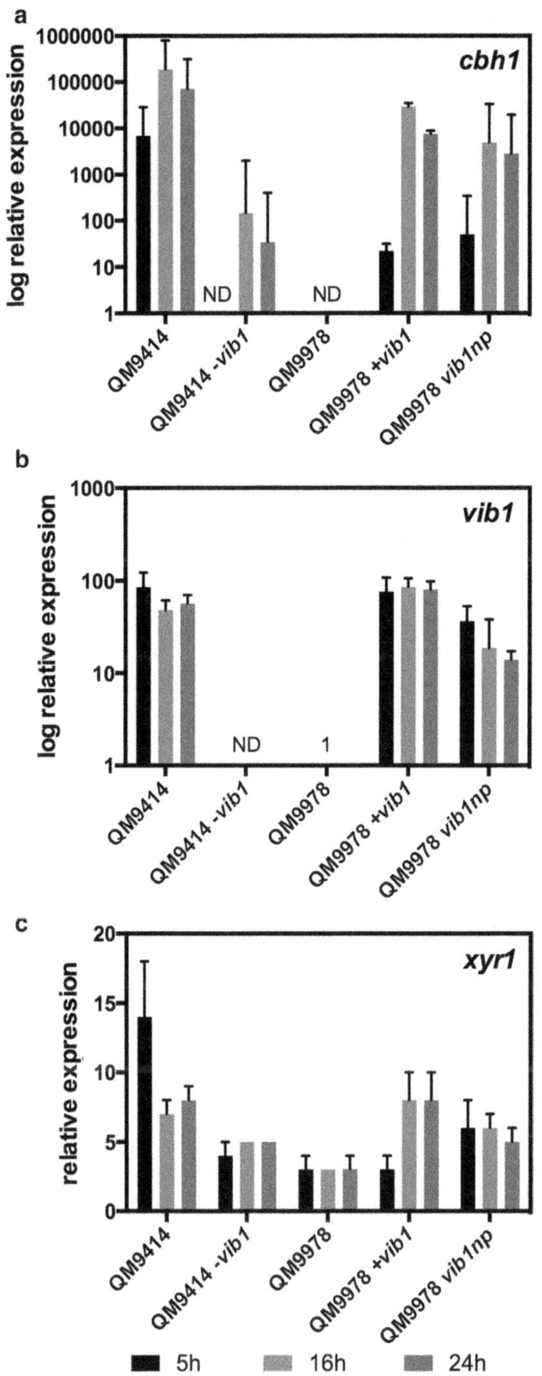

Fig. 6 VIB1 is required for cellulase production and signaling. Expression levels for the cellulase gene *cbh1* **a**, the transcription factor *vib1* **b**, and the main positive regulator *xyr1* **c** were assessed in −/+*vib1* strains versus WT after a shift from glycerol to 1% (v/w) lactose. Gene expression levels were measured by relative quantitative PCR using *tef1* and *sar1* as a control and normalized to expression level at time point zero (24 h growth on glycerol) for *cbh1* and *xyr1* or to expression level in QM9978 for *vib1*. Samples were taken 5 h (*black*), 16 h (*light gray*), and 24 h (*dark gray*) after replacement. Mean values ± standard error of triplicate data from two biological replicates are shown. *ND* not detected

culture (minimal media supplemented with Solka-floc and lactose) and reflected in the reduced expression of *cbh1* (*cel7a*) in the *Δvib1* strains (Figs. 5, 6). We therefore conclude that overexpression of *vib1* does not additionally improve cellulase production in those two strains.

Since previous studies have identified XYR1 as a major regulator of cellulase gene expression and *xyr1* deficiency to be linked to a cellulase-negative phenotype [37], we determined *xyr1* transcript levels in the QM9978 and QM9414 strains. Deletion of *vib1* in QM9414 significantly reduced *xyr1* expression, whereas over-/expression of *vib1* increased *xyr1* levels in QM9978 ($p < 0.001$) (Fig. 6c). This finding indicates that XYR1 could function downstream of VIB1 and the reduced *xyr1* expression could contribute to the cellulase-negative phenotype of −*vib1 T. reesei* strains.

Discussion

The manipulation of *T. reesei* strains by random mutagenesis and subsequent selection procedures has led to the isolation of strains with cellulase production capacities surpassing 100 g/L in industrial culture conditions [38]. The exact identification of the genetic changes that are responsible for the cellulase overproduction relied for a long time on laborious complementation studies, which were not very successful. Advances in high-throughput sequencing and genomic profiling have significantly contributed to a lower cost of these technologies and thus offer a feasible alternative for the identification of mutations [14, 16–18]. Strains obtained by random mutagenesis often contain a multitude of mutations, and genome analysis of QM9978 resulted in the identification of 33 mutations found in the coding region of genes or their promoters, plus four genes found in the region of a chromosomal rearrangement. In order to try and reduce the number of candidate genes responsible for the cellulase-negative phenotype, we combined the genome sequence analysis with a transcriptomic study in cellulase inducing conditions. This tandem approach and clues from existing literature about the role of *N. crassa vib-1* in cellulase regulation led us to select *vib1* as the main candidate, despite the fact that the coding region and the close-by intergenic regions of the gene were not affected by the mutagenesis [39]. A sequencing approach alone probably would not have resulted in *vib1* being selected as prime candidate as the identified mutation is far away from the start codon. Considering distal genomic events and combining genome sequencing with a transcriptome analysis seems a robust approach for microbial strain "reverse-engineering."

Indeed, analysis of the genomic landscape led to the identification of a translocation between chromosomes V and VII. It has been shown that DNA damage such as double-strand breaks can lead to incompletely cleared sites of repair causing gene silencing through epigenetic events [40, 41]. Histone modifications such as methylation, acetylation, or displacement are a common mechanism of gene regulation in filamentous fungi [42]. We speculated that the chromosome break led to histone modifications no longer favoring active transcription, since the translocation took place only 1 kb upstream of the *vib1* start codon. *Vib1* and Trire2:80028, a putative ABC transporter whose expression is also diminished in QM9978 when compared to QM6a, show altered expression levels. The other two genes involved (Trire2:65583 and Trire2:37262) are located at longer distances and therefore may be too far away to see their expression impacted (8427 and 14,276 bp, respectively, as compared to 1095 and 1437 bp). Another explanation would involve the rearrangement of regulator binding sequences (e.g., inhibitors of transcription) in the close proximity of the affected genes, thus having an unfavorable effect on gene expression.

Vib1 was initially identified in *N. crassa*, where it was shown to be required for vegetative incompatibility [29]. Since then, several studies have shown that *vib1* and its homologs in other organisms are involved in the regulation of a diverse spectrum of metabolic processes including programmed cell death, antifungal compound production, and the response to carbon starvation, among others [43–46]. During a recent screening of a transcription factor deletion library in *N. crassa*, mutants unable to grow on cellulosic biomass were identified. Among them, *vib-1* was shown to be essential for cellulose degradation. This finding led us to the assumption that *vib1* may be involved in cellulase regulation in *T. reesei*, and the observed down-regulation in the QM9978 strain could be due to the relatively distal chromosomal translocation identified upstream. To rule out the possibility that the discovered balanced translocation deleted activator-binding sites positioned further upstream and which are required for active transcription of *vib1*, we re-introduced *vib1* +850 bp as promoter sequence, which are not affected by the translocation. Since this approach could restore cellulase production, we conclude that despite a functional promoter and gene, the new genomic environment arising from the rearrangement is responsible for the loss of *vib1* expression, either by epigenetic modifications or through other mechanisms involving, for example, the binding of distal transcriptional repressors.

Overexpression of *vib1* in QM9978 restored enzyme secretion in this strain, and gene deletion in QM9414 abolished cellulase production, confirming that *vib1* is required for cellulose degradation. Gene deletion in the industrial strain Rut-C30 resulted in reduced although not completely abolished cellulase production, indicating that additional factors contribute to cellulase induction, especially since overexpression of *vib1* had no effect on cellulase production for QM9414 and Rut-C30. We therefore conclude that *vib1* is not suitable for the amelioration of hyper-producing strains, as they probably already accumulated other mutations masking the effect of *vib1* overexpression.

The exact role of the different transcription factors in *T. reesei* during cellulase induction is not fully understood yet. Although most of the cellulase genes are regulated in a coordinated way, their relative expression levels differ in higher producers [47]. Overexpression or constitutive activation of *xyr1* lead to a significant elevation/deregulation of xylanase transcription levels and had an impact on cellulase induction in other studies [48, 49].

Our transcriptional analysis showed differential regulation of the main positive regulator *xyr1* depending on *vib1* expression. We observed no significant changes in *xyr1* expression during cultivation on lactose, which is consistent with recent findings where *xyr1* induction does not require metabolism of D-galactose [50, 51]. The significantly reduced expression of *xyr1* in −*vib1* strains could contribute to the cellulase-negative phenotype of QM9978, as *xyr1* deletion has been shown to abolish cellulase induction on cellulose and sophorose and to impair the induction of hemicellulase genes involved in xylan and arabinan degradation in *T. reesei*, *N. crassa*, and *A. niger* [52–54]. Since *xyr1* expression in QM9978 and QM9414 −*vib1* is only reduced and not completely abolished, other factors besides VIB1 must be involved in its regulation and in cellulase gene expression.

A recent model for cellulase regulation in *N. crassa* proposes the involvement of several conserved transcription factors and shows that VIB-1 functions upstream of the inducing signal, repressing CRE1 and COL26 and thus inducing CLR-1 and CLR-2 [39, 55, 56] to promote cellulase production. The *T. reesei* homologue of CLR-2 was found to be significantly up-regulated during growth on lactose but a detailed analysis of its function is still missing [50, 57]. Overexpression of *clr2* in QM9414 has shown that *clr2* (gene 26163 in the respective study) had only a minor effect on cellulase production in *T. reesei*, and a homologue of *clr-1*, which is essential for cellulase production on *N. crassa*, is missing in *T. reesei* [13]. BglR, the COL26 homologue of *T. reesei*, was not regulated during growth on lactose in our study, neither in the QM6a wild type nor in QM9978. A recent study identified *col-26* to be involved in starch signaling in *N. crassa* but no data are available yet for *T. reesei* [58]. Our transcriptomic analysis showed no difference in *bglR*-expression in strains expressing/not expressing *vib1*. We therefore cannot conclude that *bglR* plays a major role in *vib1* downstream signaling in *T. reesei.* Whether a synergy between *cre1*, *bglR*, and *vib1* exists in *T. reesei*, like it does in *N. crassa*, remains for the moment unclear. Transcriptomic analysis performed previously by our lab on strains NG14 and RUT-C30 shows an enhanced expression of *vib1* in RUT-C30 during growth on lactose [59]. Since the major difference between NG14 and RUT-C30 is a partial deletion of *cre1* in RUT-C30, it may be speculated that in *T. reesei*, contrary to *N. crassa*, *vib1* acts downstream of the inducing signal, which would also explain why *bglR* is not differentially expressed in a *Δvib1* background.

In a study performed by Schmoll et al., 18 genes specifically expressed in QM9978 compared to QM9414 were identified [20]. From those genes, we find six which are differentially expressed in our study. The differences in experimental conditions and reference strains may explain why a different gene set was identified by Schmoll et al. compared to our analysis and why *vib1* was not detected as differentially expressed.

Conclusions

Our analysis used genomic data in combination with gene expression analysis and revealed a chromosomal recombination event to be responsible for the cellulase-negative phenotype of QM9978. We identified VIB1 as a transcription factor required for the degradation of lignocellulosic biomass whose expression is absent in QM9978. Deletion of *vib1* leads to a defect in cellulase production in different improved cellulase producer strains, whereas overexpression had no influence on cellulase expression. These results confirm that *vib1* is an important component of the cellulase regulon in *N. crassa* and *T. reesei*, but the precise function of *vib1* in the regulatory network of *T. reesei* needs to be further examined ([39], this study). However, since straightforward overexpression did not increase the production, a more detailed understanding of the regulatory mechanisms would be required to make this transcription factor interesting for industrial applications by gene engineering strain improvement. Our study demonstrates the importance of studying cellulase-negative strains complementary to high-producing strains for the understanding of the cellulase gene regulation process. It is, to our knowledge, the first report of a distal chromosomal rearrangement in a biotechnology-relevant strain affecting an otherwise intact promoter-gene system, thus provoking a dramatic phenotype. These phenomena should not be overlooked during genome-wide analysis of strains, which often focus on point mutations and InDels.

Methods

Fungal strains, culture conditions, and measurement of cellulase formation

Trichoderma reesei wild-type strain QM6a (ATCC 13631), the early cellulase producing mutant *T. reesei* QM9414 (ATCC 26921), the cellulase-negative mutant QM9978 (obtained originally from Dr. K.O' Donnell, U.S. Department of Agriculture, Peoria, IL), and the hypercellulosic strain Rut-C30 (ATCC 56765) were used throughout this study and kept on potato dextrose agar (Sigma, St. Louis, MO) at 30 °C (Additional file 2: Table S1). *T. reesei* was grown in liquid culture in 250-mL Erlenmeyer flasks on a rotary shaker (250 rpm) at 30 °C in 50 mL minimal medium as described [60] using frozen spores as inoculum. Carbon sources used are specified in the legends to the respective figures, and were used as 1% (w/v) in batch cultures. To assess the cellulase formation on different carbon sources using a colorimetric assay based on AZCL-HE-cellulose, strains were grown on minimal medium plates supplemented with 5 g/L of the respective carbon source in addition to AZCL-HE-cellulose (Megazyme International, Ireland) to a concentration of 1 g/L. Determination of protein content in culture supernatants was assessed by the modified Lowry protein assay as described using three biological and at least two technical replicates [61]. Global cellulase activity was measured using the IUPAC standard Filter Paper Assay [62], after a 15-fold miniaturization similar to [63] which allowed working in 2-mL Eppendorf tubes. In order to surround the desired 4% conversion yield, four enzyme dilutions were tested for each sample. This miniaturized protocol was validated by comparison with the standard protocol. Statistical analysis of the results was performed with the Prism GraphPad software (https://www.graphpad.com/scientific-software/prism/).

For transcriptomic studies 50 mL of minimal medium (5 g/L $(NH_4)_2SO_4$, 5 g/L KH_2PO_4, 11.7 g/L trisodium citrate, 0.6 g/L $MgSO_4$, 0.6 g $CaCl_2$, 20 g/L glucose) were prepared in a 250-mL Erlenmeyer flask. The medium was supplemented with the following trace elements: 0.5 mL/L $FeSO4{\cdot}7H_2O$, 2 mL/L $CoCl_2$, 0.16 mL/L $MnSO_4$, and 0.14 mL/L $ZnSO_4$. Prior to sterilization, the pH of the medium was adjusted to 5.4. Strains QM6a and QM9978 were inoculated in duplicate with approx 10^6 conidia/mL. Cultivation was carried out at 30 °C at 150 rpm. When the glucose was close to depletion (<3 g/L), 5 mL of a lactose solution was added (25 g/L lactose, 6 g/L $(NH_4)_2SO_4$, 5 g/L trisodium citrate). Glucose concentration was assessed by enzymatic reaction using an analox glucose analyzer GM10 (Imlab). Extracellular protein concentration was measured against a BSA standard (0–1.5 g/L range with second-order regression) using the Quick Start Bradford Protein Assay kit (Bio-Rad).

To monitor gene expression levels via real-time qPCR, the replacement technique described by Sternberg and Mandels [64] was used: mycelia, grown for 24 h on 1% (w/v) glycerol as the carbon source under otherwise similar conditions as described above, were cautiously filtered through Miracloth (Calbiochem, EMD Biosciences, La Jolla, CA), washed with tap water, and resuspended in minimal medium supplemented with the respective carbon source at 1% (w/v) and the incubation continued for 5, 16, and 24 h. To guarantee equal treatment of the cultures, they were done in parallel and results of three technical replicates were obtained.

Escherichia coli NEB 10-beta competent cells (New England Biolabs, France) were used for the propagation of vector molecules and DNA manipulations.

Vector construction for generation of gene deletion and overexpression mutants

Deletion cassettes consisting of 1–1.5 kb fragments of the gene specific flanking regions interrupted by the hygromycin resistance cassette were assembled by yeast recombinational cloning [50, 65]. Oligonucleotides (10 mM) 5F + 3R and 3F + 3R were used for amplification of the individual flanking regions from genomic *T. reesei* DNA using Phusion polymerase (Thermo Fisher Scientific). Approximately 19 bp that overlap with the pRS426 (URA +) yeast shuttle vector or the *hph* gene were introduced by PCR at each flanking end to allow homologous recombination. The *hph* marker cassette was amplified from pLHhph1 with oligonucleotides hphF and hphR [65]. A PCR touchdown program ranging from 62 to 58 °C for annealing was used for amplification. Oligonucleotide sequences are shown in Additional file 3: Table S2. Yeast transformation was performed as described [66] using the lithium chloride/polyethylene glycol procedure. Yeast strain WW-YH10 (ATCC 208405) was transformed with both flanking regions, the *hph* marker cassette and an *EcoR*I/*Xba*I-digested plasmid pRS426 [67]. Transformants were selected on synthetic complete dropout medium (SC-URA with uracil dropped out). Following total DNA isolation from liquid SC-URA medium, plasmids were introduced into chemically competent NEB 10-beta *E. coli* cells (New England Biolabs) [68]. The outside primers 5F + 3R were used to synthesize the complete deletion cassette from plasmids isolated from *S. cerevisiae*.

For construction of overexpressing strains, the coding region of *vib1* was amplified using oligonucleotides VIB1 for and VIB1 rev and placed under the control of the strong constitutive *gpd* promoter (oligonucleotides promGPD for and PromGPD rev) followed by the *gpd* terminator (termGPD for and termGPD rev) and the *hph* resistance cassette (hph for and hph rev, see Additional file 3: Table S2) all inserted into the *Xba*I/*EcoR*I-linearized plasmid pUC19 yielding plasmid pUC_vib1_OE. Oligonucleotides K7 VIB1 for and K7 VIB1 rev were used to amplify the complete overexpression cassette from pUC_vib1_OE.

For reintroduction of *vib1* under the control of the native promoter into the strain QM9978, the coding region plus approximately 850 bp of the promoter region (after the translocation breakpoint) together with 500 bp of the terminator region were amplified using oligonucleotides VIB1_F and VIB1_R (Additional file 3: Table S2) and introduced in the *Sac*I/*BamH*I-linearized plasmid pLHhph1 using the Gibson Assembly Cloning Kit (New England Biolabs) [65]. Five micrograms of the circular plasmid were used for protoplast transformation of *T. reesei*.

Transformation of *T. reesei* and analysis of transformants

Trichoderma reesei QM9414, QM9978, and Rut-C30 were used for the generation of *vib1* deletion and overexpression strains. Protoplast preparation and transformation were performed as previously described [69]. The deletion cassettes were purified after PCR (QIAquick PCR Purification kit, QIAGEN) and concentrations were determined (Nanodrop Spectrophotometer, Peqlab). After transformation protoplasts were stabilized and regenerated on selection medium supplemented with 100 mg/mL hygromycin B. For sporulation, transformants were transferred to 12-well plates and purified by plating conidiospores onto plates with 0.1% Triton X-100 as colony restrictor. Single colonies were transferred to selective medium and screened for correct integration of the deletion or overexpression cassettes by PCR analysis (primer abbreviated with "ch" in Additional file 3: Table S2). Insertion of the *gpd-vib1* overexpression cassette was verified with primer pairs TR46-verif 5′vib1 for + TR47-verif 5′vib1 rev, TR48-verif 3′ vib1 for + TR49-verif 3′vib1 rev, and TR50-verif hph vib1 for + TR51-verif hph vib1 rev amplifying the 5′or 3′ region and the *hph* marker gene, respectively. Three or four independent clones were isolated for each strain construction and analyzed as biological replicates.

Gene expression analysis by real-time qPCR

DNase treated (DNase I, RNase free; Fermentas) RNA (5 mg) was reverse transcribed with the RevertAidTM First Strand cDNA Kit (Fermentas, Germany) according to the manufacturer's protocol using a combination (1:1) of the provided oligo-dT and random hexamer primers. All assays were carried out in 96-well plates, which were covered with optical tape, as previously described [70].

Primers, amplification efficiency, and R-square values are given in Additional file 4: Table S3. The amplification efficiency was determined using triplicate reactions from a dilution series of cDNA, and then calculated from the given slopes in the IQ5 Optical system Software v2.0. Expression ratios were calculated using the REST 2009 Software (Quiagen, Germany). All samples were analyzed in two independent experiments with three replicates in each run.

RNA sample preparation

At each time point, 5 mL culture medium containing mycelia was filtrated on 1.2-μL GF/C glass microfiber membranes and flash frozen in liquid nitrogen. Frozen mycelia were disrupted and homogenized in Lysing Matrix C (Fast RNA Pro Red Kit, MP Bio) in the presence of 700 μL Lysis buffer RLT (RNeasy Mini Kit, Qiagen) complemented with 7 μL of β-mercaptoethanol on a rotor–stator homogenizer (FastPrep, MP Bio, Santa Ana, CA, USA). After centrifugation, the lysate was transferred on a Qiashredder spin column (Qiagen) and centrifuged for 2 min to eliminate all cell debris. The flow-through was then mixed with 0.5 volumes of pure ethanol and transferred to an RNeasy spin column. All further steps, including an on-column DNAse digestion, were realized following the manufacturer's instructions. RNA quantification was performed with Qubit 2.0 (Life Technologies, Carlsbad, CA, USA). The quantity and quality of the total RNA was determined using a NanoDrop ND-1000 spectrophotometer (Nanodrop Technologies, Wilmington, USA) and by electrophoresis on 1% agarose gel. An additional quality control was performed on the Bioanalyzer 2100 system (Agilent, Santa Clara, USA).

RNA-seq library preparation and analysis

Library preparation and Illumina sequencing were performed at the Ecole normale superieure genomic platform (Paris, France). Messenger (polyA+) RNAs were purified from 1 μg of total RNA using oligo(dT) primer. Libraries were prepared using the strand non-specific RNA-Seq library preparation TruSeq RNA Sample Prep Kits v1 (Illumina). Libraries were multiplexed by four on one single flowcell lane and subjected to 50-bp paired-end read sequencing on a HiSeq 1500 device. A mean of 52 ± 10 million passing Illumina quality filter reads was obtained for each of all the samples. We performed RNA-seq analysis using the Eoulsan pipeline [71]. Only read 1 was considered for the further analyses. Before mapping, poly N read tails were trimmed, reads ≤11 bases were removed, and reads with quality mean ≤12 were discarded. Reads were then aligned against the *T. reesei genome* (version 2 from the JGI website) using the Bowtie mapper (version 0.12.7) [72]. Alignments from reads matching more than once on the reference genome were removed using Java version of samtools [73]. To compute gene expression, the *T. reesei* genome annotation from JGI (version 2) was used. All overlapping regions between alignments and referenced exons were counted. The sample counts were normalized using DESeq 1.8.3 [74]. Statistical treatments and differential analyses were also performed using DESeq 1.8.3. A threshold of two for the log2 ratios with an adjusted p value <0.001 was used to identify genes differentially expressed (DE). Gene functions were identified by the aid of a manually annotated *T. reesei* genome database proprietary to IFPEN.

The RNA-Seq gene expression data and raw fastq files are available at the GEO repository (http://www.ncbi.nlm.nih.gov.gate1.inist.fr/geo/) under Accession Number GSE89199.

Illumina genome sequencing of *T. reesei* QM9978

Chromosomal DNA from *T. reesei* QM9978 was prepared as previously described [14]. Fragment libraries were prepared according to the TruSeq® DNA Sample Preparation Guide from Illumina (http://www.illumina.com). These libraries were then loaded onto the cluster generation station for single-molecule bridge amplification using the Standard Cluster Generation kit from Illumina. The slide with amplified clusters was then subjected to sequencing on the Illumina Genome Analyser I (GAI) for single reads using the 36 cycle Sequencing Kit version 1 from Illumina.

The whole genome sequence of *T. reesei* QM9978 will be deposited at the Sequence Read Archive (SRA; http://www.ncbi.nlm.nih.gov/sra).

Sequence alignment and analyses

Illumina short reads from QM9978 were mapped onto the *T. reesei* genome (http://genome.jgi-psf.org/Trire2/), using the Maq 0.6.6 software solution [75]. Mapping was done with two maximum mismatches. InDels and SNVs were identified using Maq 0.6.6. Homozygous mutations were selected and filtered on genomic context (complexity = 1, uniqueness >15.8, GC percentage between 0.31 and 0.74). Each mutation was manually checked using the Integrative Genome Viewer and compared to the sequence of the Rut lineage to remove the false mutations coming from initial QM6a error sequencing. Large structural variations were searched using the BreakDancer software and filtered on the reads number covering the genomic variations [76].

We evaluated the location of SNVs and deletions according to gene annotations using the "filtered models" from the JGI website. From this annotation, we calculated the position of intron, promoter (using an 800 bp upstream region), and terminator (using a 200-bp downstream region).

Functions of genes were predicted by similarity to orthologous genes when found in other fungal taxa using the FUNGIpath database (http://embg.igmors.u-psud.fr/fungipath/). If no orthologous gene function could be identified but a gene with a homology e-value superior to e-10 was identified, the product of this gene was annotated as "similar to." Conserved domains were identified either by interproscan (http://www.ebi.ac.uk/interpro/search/sequence-search) or by NCBI BLASTp search (https://blast.ncbi.nlm.nih.gov/Blast.cgi).

To validate the translocation between chromosomes V and VII, we amplified the region around genes encoded by the protein IDs 80028 and 54675 by PCR using oligonucleotides up and downstream of the coding region (Primer 54675_MutaC_F + 54675_MutaC_R and 80028_MutaC_F + 80028_MutaC_R Fig. 3b; Additional file 3: Table S2). PCR amplicons were sequenced with adequate primers using Sanger cycle sequencing/capillary electrophoresis (Microsynth AG, Austria).

Additional files

Additional file 1: Figure S1. Growth analysis of *vib1* mutant strains on AZCL-HE-Cellulose. The images correspond to Fig. 4 showing the enzyme production by the strains on AZCL-HE-Cellulose supplemented with 10 g/L lactose or glucose. Comparison of colony size to cellulose degradation shows that strains QM9978 and QM9414 -*vib1* impaired in cellulase production are still able to grow on lactose. Pictures were taken after 4 days of incubation. For each strain two biological replicates (individual clones) were assayed and one is shown representatively.

Additional file 2: Table S1. Strains used throughout the study.

Additional file 3: Table S2. Oligonucleotides used in this study.

Additional file 4: Table S3. Oligonucleotides used for quantitative real-time qPCR analysis.

Abbreviations

ATCC: American type culture collection; BLAST: basic local alignment search tool; CBH1: cellobiohydrolase 1; CBH2: cellobiohydrolase 2; CRE: carbon catabolite repressor protein; DE: differentially expressed; GH: glycoside hydrolase; HPH: hygromycin B phosphotransferase; Indels: insertions and deletions; JGI: Joint Genome Institute; PDA: potato dextrose agar; QM: Quartermaster (strain collection); RUT: Rutgers (mutagenesis program); SNP: short nucleotide polymorphism; SNV: short nucleotide variants; Trire: *Trichoderma reesei*; US: United States; XYR1: xylanase regulator 1; HPH: hygromycin B phosphotransferase.

Authors' contributions

CI designed and carried out the molecular genetic studies and drafted the manuscript. JR performed the real-time qPCR analyses and helped analyze data. TA and AF carried out experiments and measurements and analyzed experimental data. BS participated in the design of the study, its coordination and helped draft the manuscript. SB, SLC, and SL sequenced and annotated the genome, and identified the mutations. AM initiated, interpreted, and discussed data and edited the manuscript. FB initiated, designed, and coordinated the study and reviewed the manuscript. All authors read and approved the final manuscript.

Author details

[1] IFP Energies Nouvelles, 1-4 Avenue de Bois-Préau, 92852 Rueil-Malmaison, France. [2] Molecular Biotechnology, Research Division Biochemical Technology, Institute of Chemical Engineering, TU-Wien, 1060 Vienna, Austria. [3] Earth and Biological Sciences Directorate, Pacific Northwest National Laboratory, Richland, WA 99354, USA. [4] Evolution Paris Seine-Institut de Biologie Paris Seine (EPS-IBPS), Sorbonne Universités, UPMC Univ Paris 06, Univ Antilles, Univ Nice Sophia Antipolis, CNRS, 75005 Paris, France. [5] École normale supérieure, PSL Research University, CNRS, Inserm, Institut de Biologie de l'École normale supérieure (IBENS), Plateforme Génomique, 75005 Paris, France. [6] Present Address: Genetics of Biofilms Unit, Department of Microbiology, Institut Pasteur, 25-28 Rue du Dr Roux, 75015 Paris, France.

Acknowledgements

The authors would like to thank Etienne Jourdier and Celine Cohen for their help during enzyme activity measurements and Professor Christian P. Kubicek for providing the cellulase-negative strain QM9978 on which this study was based.

Competing interests

The authors declare that they have no competing interests.

Funding

This work has been supported by the Austrian Science Fund (P24219-B21) to BS. JR was further supported by the TU Wien Ph.D. program molecular and elemental imaging in biosciences MEIBio.

The École normale supérieure genomic platform was supported by the France Génomique national infrastructure, funded as part of the "Investissements d'Avenir" program managed by the Agence Nationale de la Recherche (contract ANR-10-INBS-09).

This work was in part supported by the "Fondation Tuck" research fund (Grant 2009-03) to AM, SLC, and FBM.

References

1. Liu G, Qin Y, Li Z, Qu Y. Development of highly efficient, low-cost lignocellulolytic enzyme systems in the post-genomic era. Biotechnol Adv. 2013;31:962–75.
2. Bouws H, Wattenberg A, Zorn H. Fungal secretomes—nature's toolbox for white biotechnology. Appl Microbiol Biotechnol. 2008;80:381–8.
3. Druzhinina IS, Seidl-Seiboth V, Herrera-Estrella A, Horwitz BA, Kenerley CM, Monte E, Mukherjee PK, Zeilinger S, Grigoriev IV, Kubicek CP. *Trichoderma*: the genomics of opportunistic success. Nat Rev Microbiol. 2011;9:749–59.
4. Reese ET. History of the cellulase program at the U.S. army Natick Development Center. Biotechnol Bioeng Symp. 1976;6:9–20.
5. Kubicek CP, Herrera-Estrella A, Seidl-Seiboth V, Martinez DA, Druzhinina IS, Thon M, Zeilinger S, Casas-Flores S, Horwitz BA, Mukherjee PK, et al. Comparative genome sequence analysis underscores mycoparasitism as the ancestral life style of *Trichoderma*. Genome Biol. 2011;12:R40.
6. Bayer EA. Remembering Mary (1917 to 2008): editorial introduction to the thematic series on the life and lifework of Mary Mandels, first lady of cellulase research. Biotechnol Biofuels. 2009;2:23.
7. Eveleigh DE, Montenecourt BS. Increasing yields of extracellular enzymes. Adv Appl Microbiol. 1979;25:57–74.
8. Peterson R, Nevalainen H. *Trichoderma reesei* RUT-C30–thirty years of strain improvement. Microbiology. 2012;158:58–68.
9. Torigoi E, Henrique-Silva F, Escobar-Vera J, Carle-Urioste JC, Crivellaro O, El-Dorry H, El-Gogary S. Mutants of *Trichoderma reesei* are defective in cellulose induction, but not basal expression of cellulase-encoding genes. Gene. 1996;173:199–203.
10. Nitta M, Furukawa T, Shida Y, Mori K, Kuhara S, Morikawa Y, Ogasawara W. A new Zn(II)(2)Cys(6)-type transcription factor BglR regulates beta-glucosidase expression in *Trichoderma reesei*. Fungal Genet Biol. 2012;49:388–97.
11. Porciuncula Jde O, Furukawa T, Mori K, Shida Y, Hirakawa H, Tashiro K, Kuhara S, Nakagawa S, Morikawa Y, Ogasawara W. Single nucleotide polymorphism analysis of a *Trichoderma reesei* hyper-cellulolytic mutant developed in Japan. Biosci Biotechnol Biochem. 2013;77:534–43.

12. Ghassemi S, Lichius A, Bidard F, Lemoine S, Rossignol MN, Herold S, Seidl-Seiboth V, Seiboth B, Espeso EA, Margeot A, Kubicek CP. The ss-importin KAP8 (Pse1/Kap121) is required for nuclear import of the cellulase transcriptional regulator XYR1, asexual sporulation and stress resistance in *Trichoderma reesei*. Mol Microbiol. 2015;96:405–18.
13. Hakkinen M, Valkonen MJ, Westerholm-Parvinen A, Aro N, Arvas M, Vitikainen M, Penttila M, Saloheimo M, Pakula TM. Screening of candidate regulators for cellulase and hemicellulase production in *Trichoderma reesei* and identification of a factor essential for cellulase production. Biotechnol Biofuels. 2014;7:14.
14. Le Crom S, Schackwitz W, Pennacchio L, Magnuson JK, Culley DE, Collett JR, Martin J, Druzhinina IS, Mathis H, Monot F, et al. Tracking the roots of cellulase hyperproduction by the fungus *Trichoderma reesei* using massively parallel DNA sequencing. Proc Natl Acad Sci USA. 2009;106:16151–6.
15. Shida Y, Yamaguchi K, Nitta M, Nakamura A, Takahashi M, Kidokoro S, Mori K, Tashiro K, Kuhara S, Matsuzawa T, et al. The impact of a single-nucleotide mutation of bgl2 on cellulase induction in a *Trichoderma reesei* mutant. Biotechnol Biofuels. 2015;8:230.
16. Vitikainen M, Arvas M, Pakula T, Oja M, Penttila M, Saloheimo M. Array comparative genomic hybridization analysis of *Trichoderma reesei* strains with enhanced cellulase production properties. BMC Genom. 2010;11:441.
17. Porciuncula Jde O, Furukawa T, Shida Y, Mori K, Kuhara S, Morikawa Y, Ogasawara W. Identification of major facilitator transporters involved in cellulase production during lactose culture of *Trichoderma reesei* PC-3-7. Biosci Biotechnol Biochem. 2013;77:1014–22.
18. Lichius A, Bidard F, Buchholz F, Le Crom S, Martin J, Schackwitz W, Austerlitz T, Grigoriev IV, Baker SE, Margeot A, et al. Genome sequencing of the *Trichoderma reesei* QM9136 mutant identifies a truncation of the transcriptional regulator XYR1 as the cause for its cellulase-negative phenotype. BMC Genom. 2015;16:326.
19. Druzhinina IS, Schmoll M, Seiboth B, Kubicek CP. Global carbon utilization profiles of wild-type, mutant, and transformant strains of *Hypocrea jecorina*. Appl Environ Microbiol. 2006;72:2126–33.
20. Schmoll M, Zeilinger S, Mach RL, Kubicek CP. Cloning of genes expressed early during cellulase induction in *Hypocrea jecorina* by a rapid subtraction hybridization approach. Fungal Genet Biol. 2004;41:877–87.
21. Schmoll M, Franchi L, Kubicek CP. Envoy, a PAS/LOV domain protein of *Hypocrea jecorina* (Anamorph *Trichoderma reesei*), modulates cellulase gene transcription in response to light. Eukaryot Cell. 2005;4:1998–2007.
22. Koike H, Aerts A, LaButti K, Grigoriev IV, Baker SE. Comparative genomics analysis of *Trichoderma reesei* strains. Ind Biotechnol. 2013;9:352–67.
23. Castellanos F, Schmoll M, Martinez P, Tisch D, Kubicek CP, Herrera-Estrella A, Esquivel-Naranjo EU. Crucial factors of the light perception machinery and their impact on growth and cellulase gene transcription in *Trichoderma reesei*. Fungal Genet Biol. 2010;47:468–76.
24. Marie-Nelly H, Marbouty M, Cournac A, Flot JF, Liti G, Parodi DP, Syan S, Guillen N, Margeot A, Zimmer C, Koszul R. High-quality genome (re) assembly using chromosomal contact data. Nat Commun. 2014;5:5695.
25. Druzhinina IS, Kopchinskiy AG, Kubicek EM, Kubicek CP. A complete annotation of the chromosomes of the cellulase producer *Trichoderma reesei* provides insights in gene clusters, their expression and reveals genes required for fitness. Biotechnol Biofuels. 2016;9:75.
26. Jourdier E, Baudry L, Poggi-Parodi D, Vicq Y, Koszul R, Margeot A, Marbouty M, Bidard F. Proximity ligation scaffolding and comparison of two *Trichoderma reesei* strains genomes. Biotechnol Biofuels. 2017;10:151.
27. Farkas V, Sestak S, Gresíak M, Kolarova N, Labudova L, Bauer S. Induction of cellulase in *Trichoderma reesei* grown on lactose. Acta Biotechnol. 1987;5:425–9.
28. Tsitsigiannis DI, Zarnowski R, Keller NP. The lipid body protein, PpoA, coordinates sexual and asexual sporulation in *Aspergillus nidulans*. J Biol Chem. 2004;279:11344–53.
29. Xiang Q, Glass NL. Identification of vib-1, a locus involved in vegetative incompatibility mediated by het-c in *Neurospora crassa*. Genetics. 2002;162:89–101.
30. Slattery MG, Liko D, Heideman W. The function and properties of the Azf1 transcriptional regulator change with growth conditions in *Saccharomyces cerevisiae*. Eukaryot Cell. 2006;5:313–20.
31. Tian C, Beeson WT, Iavarone AT, Sun J, Marletta MA, Cate JH, Glass NL. Systems analysis of plant cell wall degradation by the model filamentous fungus *Neurospora crassa*. Proc Natl Acad Sci USA. 2009;106:22157–62.
32. Shahi S, Fokkens L, Houterman PM, Rep M. Suppressor of fusion, a *Fusarium oxysporum* homolog of Ndt80, is required for nutrient-dependent regulation of anastomosis. Fungal Genet Biol. 2016;95:49–57.
33. Li J, Wang J, Wang S, Xing M, Yu S, Liu G. Achieving efficient protein expression in *Trichoderma reesei* by using strong constitutive promoters. Microb Cell Fact. 2012;11:84.
34. Paul S, Klutts JS, Moye-Rowley WS. Analysis of promoter function in *Aspergillus fumigatus*. Eukaryot Cell. 2012;11:1167–77.
35. Sibthorp C, Wu H, Cowley G, Wong PW, Palaima P, Morozov IY, Weedall GD, Caddick MX. Transcriptome analysis of the filamentous fungus *Aspergillus nidulans* directed to the global identification of promoters. BMC Genom. 2013;14:847.
36. Durand H, Clanet M, Tiraby G. Genetic improvement of *Trichoderma reesei* for large scale cellulase production. Enzym Microb Technol. 1988;10:341–6.
37. Nevalainen KM, Palva ET. Production of extracellular enzymes in mutants isolated from *Trichoderma viride* unable to hydrolyze cellulose. Appl Environ Microbiol. 1978;35:11–6.
38. Cherry JR, Fidantsef AL. Directed evolution of industrial enzymes: an update. Curr Opin Biotechnol. 2003;14:438–43.
39. Xiong Y, Sun J, Glass NL. VIB1, a link between glucose signaling and carbon catabolite repression, is essential for plant cell wall degradation by *Neurospora crassa*. PLoS Genet. 2014;10:e1004500.
40. Cuozzo C, Porcellini A, Angrisano T, Morano A, Lee B, Di Pardo A, Messina S, Iuliano R, Fusco A, Santillo MR, et al. DNA damage, homology-directed repair, and DNA methylation. PLoS Genet. 2007;3:e110.
41. O'Hagan HM. Chromatin modifications during repair of environmental exposure-induced DNA damage: a potential mechanism for stable epigenetic alterations. Environ Mol Mutagen. 2014;55:278–91.
42. Ries L, Belshaw NJ, Ilmen M, Penttila ME, Alapuranen M, Archer DB. The role of CRE1 in nucleosome positioning within the cbh1 promoter and coding regions of *Trichoderma reesei*. Appl Microbiol Biotechnol. 2014;98:749–62.
43. Dementhon K, Iyer G, Glass NL. VIB-1 is required for expression of genes necessary for programmed cell death in *Neurospora crassa*. Eukaryot Cell. 2006;5:2161–73.
44. Katz ME, Braunberger K, Yi G, Cooper S, Nonhebel HM, Gondro C. A p53-like transcription factor similar to Ndt80 controls the response to nutrient stress in the filamentous fungus, *Aspergillus nidulans*. F1000Res. 2013;2:72.
45. Niones JT, Takemoto D. VibA, a homologue of a transcription factor for fungal heterokaryon incompatibility, is involved in antifungal compound production in the plant-symbiotic fungus *Epichloe festucae*. Eukaryot Cell. 2015;14:13–24.
46. van Munster JM, Daly P, Delmas S, Pullan ST, Blythe MJ, Malla S, Kokolski M, Noltorp EC, Wennberg K, Fetherston R, et al. The role of carbon starvation in the induction of enzymes that degrade plant-derived carbohydrates in *Aspergillus niger*. Fungal Genet Biol. 2014;72:34–47.
47. Foreman PK, Brown D, Dankmeyer L, Dean R, Diener S, Dunn-Coleman NS, Goedegebuur F, Houfek TD, England GJ, Kelley AS, et al. Transcriptional regulation of biomass-degrading enzymes in the filamentous fungus *Trichoderma reesei*. J Biol Chem. 2003;278:31988–97.
48. Derntl C, Gudynaite-Savitch L, Calixte S, White T, Mach RL, Mach-Aigner AR. Mutation of the xylanase regulator 1 causes a glucose blind hydrolase expressing phenotype in industrially used *Trichoderma* strains. Biotechnol Biofuels. 2013;6:62.
49. Mach-Aigner AR, Pucher ME, Steiger MG, Bauer GE, Preis SJ, Mach RL. Transcriptional regulation of xyr1, encoding the main regulator of the xylanolytic and cellulolytic enzyme system in *Hypocrea jecorina*. Appl Environ Microbiol. 2008;74:6554–62.
50. Ivanova C, Baath JA, Seiboth B, Kubicek CP. Systems analysis of lactose metabolism in *Trichoderma reesei* identifies a lactose permease that is essential for cellulase induction. PLoS ONE. 2013;8:e62631.
51. Portnoy T, Margeot A, Linke R, Atanasova L, Fekete E, Sandor E, Hartl L, Karaffa L, Druzhinina IS, Seiboth B, et al. The CRE1 carbon catabolite repressor of the fungus *Trichoderma reesei*: a master regulator of carbon assimilation. BMC Genom. 2011;12:269.

52. Li J, Lin L, Li H, Tian C, Ma Y. Transcriptional comparison of the filamentous fungus *Neurospora crassa* growing on three major monosaccharides D-glucose, D-xylose and L-arabinose. Biotechnol Biofuels. 2014;7:31.
53. Akel E, Metz B, Seiboth B, Kubicek CP. Molecular regulation of arabinan and L-arabinose metabolism in *Hypocrea jecorina* (*Trichoderma reesei*). Eukaryot Cell. 2009;8:1837–44.
54. Stricker AR, Mach RL, de Graaff LH. Regulation of transcription of cellulases- and hemicellulases-encoding genes in *Aspergillus niger* and *Hypocrea jecorina (Trichoderma reesei).* Appl Microbiol Biotechnol. 2008;78:211–20.
55. Coradetti ST, Craig JP, Xiong Y, Shock T, Tian C, Glass NL. Conserved and essential transcription factors for cellulase gene expression in ascomycete fungi. Proc Natl Acad Sci USA. 2012;109:7397–402.
56. Znameroski EA, Glass NL. Using a model filamentous fungus to unravel mechanisms of lignocellulose deconstruction. Biotechnol Biofuels. 2013;6:6.
57. Bischof R, Fourtis L, Limbeck A, Gamauf C, Seiboth B, Kubicek CP. Comparative analysis of the *Trichoderma reesei* transcriptome during growth on the cellulase inducing substrates wheat straw and lactose. Biotechnol Biofuels. 2013;6:127.
58. Xiong Y, Wu VW, Lubbe A, Qin L, Deng S, Kennedy M, Bauer D, Singan VR, Barry K, Northen TR, et al. A fungal transcription factor essential for starch degradation affects integration of carbon and nitrogen metabolism. PLoS Genet. 2017;13:e1006737.
59. Poggi-Parodi D, Bidard F, Pirayre A, Portnoy T, Blugeon C, Seiboth B, Kubicek CP, Le Crom S, Margeot A. Kinetic transcriptome analysis reveals an essentially intact induction system in a cellulase hyper-producer *Trichoderma reesei strain*. Biotechnol Biofuels. 2014;7:173.
60. Mandels M, Weber J, Parizek R. Enhanced cellulase production by a mutant of *Trichoderma viride*. Appl Microbiol. 1971;21:152–4.
61. Hartree EF. Determination of protein: a modification of the Lowry method that gives a linear photometric response. Anal Biochem. 1972;48:422–7.
62. Ghose TK. Measurement of cellulase activities. Pure Appl Chem. 1987;59:257–68.
63. Xiao Z, Storms R, Tsang A. Microplate-based filter paper assay to measure total cellulase activity. Biotechnol Bioeng. 2004;88:832–7.
64. Sternberg D, Mandels GR. Induction of cellulolytic enzymes in *Trichoderma reesei* by sophorose. J Bacteriol. 1979;139:761–9.
65. Hartl L, Kubicek CP, Seiboth B. Induction of the gal pathway and cellulase genes involves no transcriptional inducer function of the galactokinase in *Hypocrea jecorina*. J Biol Chem. 2007;282:18654–9.
66. Schiestl RH, Gietz RD. High efficiency transformation of intact yeast cells using single stranded nucleic acids as a carrier. Curr Genet. 1989;16:339–46.
67. Christianson TW, Sikorski RS, Dante M, Shero JH, Hieter P. Multifunctional yeast high-copy-number shuttle vectors. Gene. 1992;110:119–22.
68. Singh MV, Weil PA. A method for plasmid purification directly from yeast. Anal Biochem. 2002;307:13–7.
69. Penttila M, Nevalainen H, Ratto M, Salminen E, Knowles J. A versatile transformation system for the cellulolytic filamentous fungus *Trichoderma reesei*. Gene. 1987;61:155–64.
70. Metz B, Seidl-Seiboth V, Haarmann T, Kopchinskiy A, Lorenz P, Seiboth B, Kubicek CP. Expression of biomass-degrading enzymes is a major event during conidium development in *Trichoderma reesei*. Eukaryot Cell. 2011;10:1527–35.
71. Jourdren L, Bernard M, Dillies MA, Le Crom S. Eoulsan: a cloud computing-based framework facilitating high throughput sequencing analyses. Bioinformatics. 2012;28:1542–3.
72. Langmead B, Trapnell C, Pop M, Salzberg SL. Ultrafast and memory-efficient alignment of short DNA sequences to the human genome. Genome Biol. 2009;10:R25.
73. Li H, Handsaker B, Wysoker A, Fennell T, Ruan J, Homer N, Marth G, Abecasis G, Durbin R. The sequence alignment/map format and SAMtools. Bioinformatics. 2009;25:2078–9.
74. Anders S, Huber W. Differential expression analysis for sequence count data. Genome Biol. 2010;11:R106.
75. Li W, Wooley JC, Godzik A. Probing metagenomics by rapid cluster analysis of very large datasets. PLoS ONE. 2008;3:e3375.
76. Chen K, Wallis JW, McLellan MD, Larson DE, Kalicki JM, Pohl CS, McGrath SD, Wendl MC, Zhang Q, Locke DP, et al. BreakDancer: an algorithm for high-resolution mapping of genomic structural variation. Nat Methods. 2009;6:677–81.

9

Coexistence and competition of sulfate-reducing and methanogenic populations in an anaerobic hexadecane-degrading culture

Ting-Ting Ma[1], Lai-Yan Liu[1], Jun-Peng Rui[2,3], Quan Yuan[4], Ding-shan Feng[5], Zheng Zhou[1], Li-Rong Dai[1], Wan-Qiu Zeng[1], Hui Zhang[1] and Lei Cheng[1*]

Abstract

Background: Over three-fifths of the world's known crude oil cannot be recovered using state-of-the-art techniques, but microbial conversion of petroleum hydrocarbons trapped in oil reservoirs to methane is one promising path to increase the recovery of fossil fuels. The process requires cooperation between syntrophic bacteria and methanogenic archaea, which can be affected by sulfate-reducing prokaryotes (SRPs). However, the effects of sulfate on hydrocarbon degradation and methane production remain elusive, and the microbial communities involved are not well understood.

Results: In this study, a methanogenic hexadecane-degrading enrichment culture was treated with six different concentrations of sulfate ranging from 0.5 to 25 mM. Methane production and maximum specific methane production rate gradually decreased to 44 and 56% with sulfate concentrations up to 25 mM, respectively. There was a significant positive linear correlation between the sulfate reduction/methane production ratio and initial sulfate concentration, which remained constant during the methane production phase. The apparent methanogenesis fractionation factor (α_{app}) gradually increased during the methane production phase in each treatment, the α_{app} for the treatments with lower sulfate (0.5–4 mM) eventually plateaued at ~1.047, but that for the treatment with 10–25 mM sulfate only reached ~1.029. The relative abundance levels of *Smithella* and *Methanoculleus* increased almost in parallel with the increasing sulfate concentrations. Furthermore, the predominant sulfate reducer communities shifted from *Desulfobacteraceae* in the low-sulfate cultures to *Desulfomonile* in the high-sulfate cultures.

Conclusion: The distribution of hexadecane carbon between methane-producing and sulfate-reducing populations is dependent on the initial sulfate added, and not affected during the methane production period. There was a relative increase in hydrogenotrophic methanogenesis activity over time for all sulfate treatments, whereas the total activity was inhibited by sulfate addition. Both *Smithella* and *Methanoculleus*, the key alkane degraders and methane producers, can adapt to sulfate stress. Specifically, different SRP populations were stimulated at various sulfate concentrations. These results could help to evaluate interactions between sulfate-reducing and methanogenic populations during anaerobic hydrocarbon degradation in oil reservoirs.

Keywords: Oil reservoirs, Hydrocarbon degradation, Methane production, Sulfate reduction, Microbial interactions

*Correspondence: chenglei@caas.cn
[1] Key Laboratory of Development and Application of Rural Renewable Energy, Biogas Institute of Ministry of Agriculture, Section 4-13, Renmin South Road, Chengdu 610041, People's Republic of China
Full list of author information is available at the end of the article

Background

Crude oil is the most relied-upon source of energy supply worldwide, accounting for 32.9% of global energy consumption [1]. However, the oil recovery efficiency in most oilfields around the world is commonly less than 40% [2], indicating that a larger amount of oil trapped in subsurface reservoirs has not been efficiently explored, even with advanced techniques of enhanced oil recovery (EOR) [3]. Geological and geochemical studies suggest that the vast majority of oil reservoirs have undergone biodegradation over millennia, and oil biodegradation in most reservoirs must indeed have been anaerobic in nature [4–6]. The methanogenic degradation of crude oil to methane appears to commonly proceed in oil reservoirs over geological time-scales, and biogenic methane may account for over 20% of global conventional recoverable gas resources [5, 7, 8]. Many microbial studies in the past twenty decades have demonstrated that aliphatic and aromatic hydrocarbons can be biodegraded in anoxic environments where nitrate, sulfate, and metal ions can function as terminal electron acceptors; when those electron acceptors are limited, the hydrocarbons can be further converted to methane and carbon dioxide in a process known as methanogenic crude oil biodegradation [9]. The methanogenic process has also been observed in oil fields [10–13], which suggests the in situ bioconversion of hydrocarbons in oil reservoirs. As anaerobic microbial communities can play significant roles in increasing the amount of energy recovered and reducing the exploitation cost by converting petroleum hydrocarbons into methane and carbon dioxide [10, 14, 15], an alternative technique called microbial enhanced energy recovery (MEER) was proposed [14, 16]. However, this process takes several months and even years under laboratory incubation conditions, and may also require much longer time to complete in field-scale experiments for environmental constraints [10, 11, 17]. In addition, the microbial communities commonly utilize a small fraction of crude oil [10, 11, 17]. Hence careful exploration of the oil degradation rate and efficiency are extremely important when considering the prospects of the technique in the future [16].

Methanogenic degradation of hydrocarbons to methane requires at least two different groups of microorganisms (syntrophic bacteria and methanogenic archaea) due to thermodynamic constraints [18]. Hydrocarbon is first degraded and converted into hydrogen and/or acetate by syntrophic bacteria; the byproducts are further converted into methane by methanogens [18]. In addition to methanogenic and syntrophic consortia, sulfate-reducing prokaryotes (SRPs), which were first reported in oil reservoirs 90 years ago [19], have also been identified in diverse geographical oil reservoirs and laboratory cultures capable of methanogenic hydrocarbon degradation [12, 20–24]. SRPs not only utilize simple compounds such as H_2/CO_2 and acetate but also degrade a variety of complex organic chemicals, such as hydrocarbons and fatty acids, to produce CO_2 and/or acetate [25, 26]. In the presence of sulfate, SRPs are believed to outcompete methanogens in most cases due to their high affinities for common substrates, which divert carbon and the electron flow of organic compounds away from methanogenesis to sulfidogenesis [27–29]. When sulfate is absent or limiting, some SRPs are also able to survive by cooperating with methanogens and syntrophic bacteria [25]. Sulfate is a common constituent in oil fields, with concentrations ranging from zero to several thousand milligrams per liter [24, 30], and is a key factor affecting the activity of SRPs and the interactions between methanogenic populations and SRPs. Several studies of the methanogenic degradation of alkanes have reported that sulfate concentrations less than 2–5 mM can enhance alkane-dependent methanogenesis, while high concentrations of sulfate (10–22 mM) decreased the methanogenesis rate [11, 15, 31]. However, the interactions between methanogenic populations and SRPs during alkane degradation have not yet been fully exploited, and the contribution of them to alkane degradation has not been quantitatively described.

Previous studies have revealed that a series of novel uncultured bacteria dominated in methanogenic alkane-degrading cultures enriched from oil fields and oil-contaminated sediments [20, 21]. Among them, *Syntrophaceae*-affiliated members have always been detected [13, 15, 32–35] and have been identified as key players in alkane activation and oxidation [33, 34]. Metagenomic and single-cell genomic analysis suggested that *Syntrophaceae* activates the initial degradation of hexadecane by fumarate addition [36–38], which was further demonstrated by direct detection of methyl pentadecyl succinic acid and methyl tetradecyl succinic acid in methanogenic cultures amended with hexadecane and pentadecane, respectively [39]. The archaeal communities in the above-mentioned cultures were composed of aceticlastic and hydrogenotrophic methanogens, but the phylogenetic compositions and abundance levels of two types of methanogens varied in different cultures [11, 13, 15, 32–35, 40]. In addition, SRPs, such as *Desulfovibrionales*, *Desulfuromonadales*, and *Desulfobacterales*, also coexisted in *Syntrophaceae*-dominant, alkane-degrading cultures [15, 32, 34, 35]. However, it remains unclear how methanogenic microbial communities, including syntrophic alkane degraders and methane producers, respond to sulfate stress and which phylogenetic groups of SRPs are responsible for sulfate reduction at different sulfate concentrations.

Therefore, we incubated a highly enriched, methanogenic hexadecane-degrading culture with amendment of different sulfate concentrations (0.5, 2, 4, 10, 15, and 25 mM) to describe the carbon and electron flow between methane production and sulfate reduction during anaerobic hexadecane degradation and to evaluate the effects of sulfate on the methane production pathway. Culture-independent molecular techniques (terminal restriction fragment length polymorphism [T-RFLP], cloning and MiSeq sequencing of 16S rRNA genes) were applied to assess the shifts in microbial community compositions.

Results

Methane production and sulfate reduction

Methane production began after a lag phase of approximately 120 days and increased over a total incubation period of 421 days in the cultures in which hexadecane and variable concentrations of sulfate were present (Fig. 1a). A maximum methane production of 3.52 ± 0.02 mmol was observed in the culture with 0.5 mM sulfate. Methane production gradually decreased with increasing sulfate concentration and dropped to 1.93 ± 0.04 mmol in the 25 mM sulfate culture (Fig. 1a).

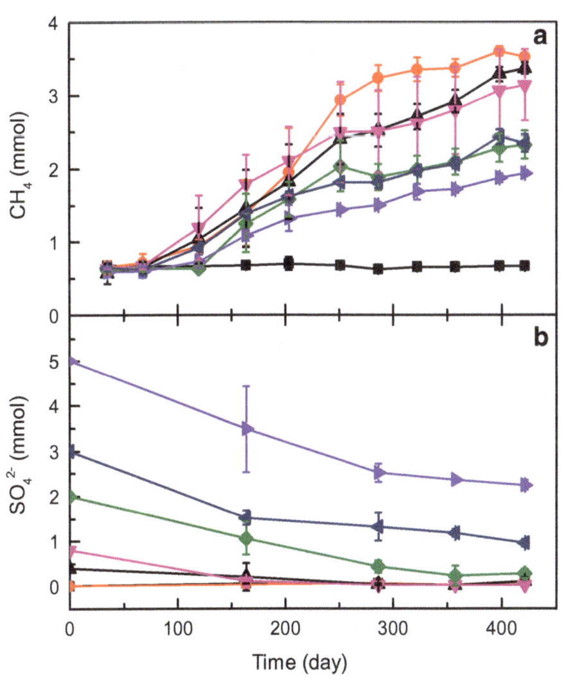

Fig. 1 Methane production (**a**) and sulfate consumption (**b**) in cultures amended with different sulfate concentrations. CK, culture without hexadecane and sulfate addition; 0.5, 2, 4, 10, 15, and 25 mM, represent cultures amended with hexadecane and 0.5, 2, 4, 10, 15, and 25 mM sulfate, respectively. Values represent means and standard errors ($n = 2$–3 replicates)

The maximum specific methane production rate (μ_{max}) declined from 0.41 ± 0.005 $month^{-1}$ in the 0.5 mM sulfate culture to 0.20 ± 0.02 $month^{-1}$ in the 25 mM sulfate culture (Table 1). In contrast, the hexadecane- and sulfate-free culture only accumulated an average of 0.66 mmol of methane during the entire incubation period (Fig. 1a). The amount of residual hexadecane decreased from 0.074 ± 0.052 mmol for the 0.5 mM sulfate culture to 0.015 ± 0.006 mmol for the 25 mM sulfate culture at the end of the incubation period (Additional file 1: Figure S1). Correspondingly, sulfate consumption increased from 0.08 ± 0.01 mmol for the 0.5 mM sulfate culture to 2.66 ± 0.11 mmol for the 25 mM sulfate culture (Fig. 1b; Table 1). A trace amount of sulfate (an average of 0.1 mM) was detected in the control culture (data not shown). No volatile fatty acid concentrations were observed above the detection limit of this method (approximately 100 μM) (data not shown).

Assuming total degradation of hexadecane via methanogenesis and/or sulfidogenesis (Eqs. 2 and 3), the percentage of hexadecane carbon recovery averaged 93.9 ± 9.0% in the 82.1–101.0% range (Table 1). The carbon balance of hexadecane consumption linked to sulfidogenic and methanogenic degradation was further evaluated. For every 1 mol of hexadecane degraded, an equivalent 12.25 mol of combined methane production and sulfate consumption occurred (Eqs. 2 and 3), consistent with the observed average values of 11.49 ± 1.11 in the 10.06–12.37 range (Table 1). However, the estimated contribution of hexadecane degradation through methanogenesis decreased from 88.6 ± 16.7% in the 0.5 mM sulfate culture to 31.6 ± 1.1% in the 25 mM sulfate culture (Fig. 2a). The decrease in electron flow to methane production was completely compensated by an increase in electron flow to sulfate reduction, which increased from 3.1 ± 0.6% in the 0.5 mM sulfate culture to 66.4 ± 2.6% in the 25 mM sulfate culture (Fig. 2a). The ratio of these two estimates, i.e., the sulfate reduction to methane production ratio, correlated positively with the initial sulfate concentrations ($R^2 = 0.94$–0.99, $p < 0.01$) (Fig. 2b). Further analysis revealed that the differences in the regression results were quite small between 164 and 421 days of incubation; the values of the slope coefficients (0.09–0.11) were also similar (Fig. 2b).

Isotopic fractionation of methane incubated at different sulfate concentrations

Isotopic analysis revealed that the $\delta^{13}C$ values of methane ($\delta^{13}CH_4$) generated were similar at the early stage of methane production (120 days), i.e., averaging −45.9 ± 2.2‰ (Fig. 3a). With increasing CH_4 accumulation, $\delta^{13}CH_4$ continued to decrease and plateaued at −61.5 ± 1.7 and −50.7 ± 1.2‰ for the 0.5–4 mM and 10–25 mM sulfate cultures, respectively, after 421 days

Table 1 Growth properties of the M82 culture incubated at different sulfate concentrations

Initial sulfate concentration (mM)	Hexadecane consumption (mmol)	Sulfate consumption (mmol)[a]	Methane production (mmol)[b]	Recovery of hexadecane carbon (%)	Maximum specific methane production rate ($month^{-1}$)	(Sulfate reduction + methane production)/hexadecane consumed
0.5	0.27 ± 0.05	0.08 ± 0.01	2.86 ± 0.02	91.7 ± 17.3	0.41 ± 0.005	11.23 ± 2.12
2	0.29 ± 0.03	0.19 ± 0.08	2.71 ± 0.01	82.1 ± 7.5	0.27 ± 0.08	10.06 ± 0.92
4	0.29 ± 0.06	0.67 ± 0.01	2.48 ± 0.49	89.7 ± 5.1	0.30 ± 0.09	10.99 ± 0.63
10	0.27 ± 0.05	1.62 ± 0.1	1.61 ± 0.24	101.0 ± 9.7	0.23 ± 0.04	12.37 ± 1.18
15	0.31 ± 0.01	1.95 ± 0.08	1.68 ± 0.13	96.9 ± 5.5	0.22 ± 0.02	11.87 ± 0.67
25	0.33 ± 0.01	2.66 ± 0.11	1.27 ± 0.04	98.1 ± 3.6	0.20 ± 0.02	12.01 ± 0.45

[a] The detected sulfate at day 421 minus the background sulfate (0.096 mmol)

[b] The detected methane at day 421 minus the background methane (0.66 mmol)

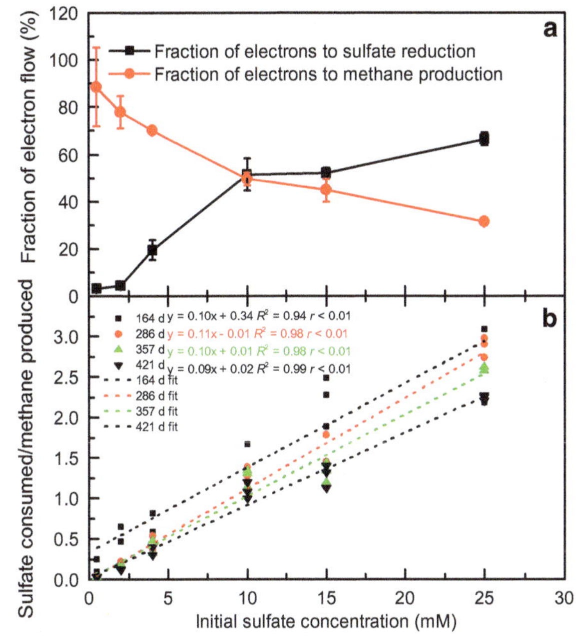

Fig. 2 Effects of sulfate on electron flow (**a**) and ratio of sulfate reduction/methane production (**b**). Values represent means and standard errors ($n = 2$–3 replicates)

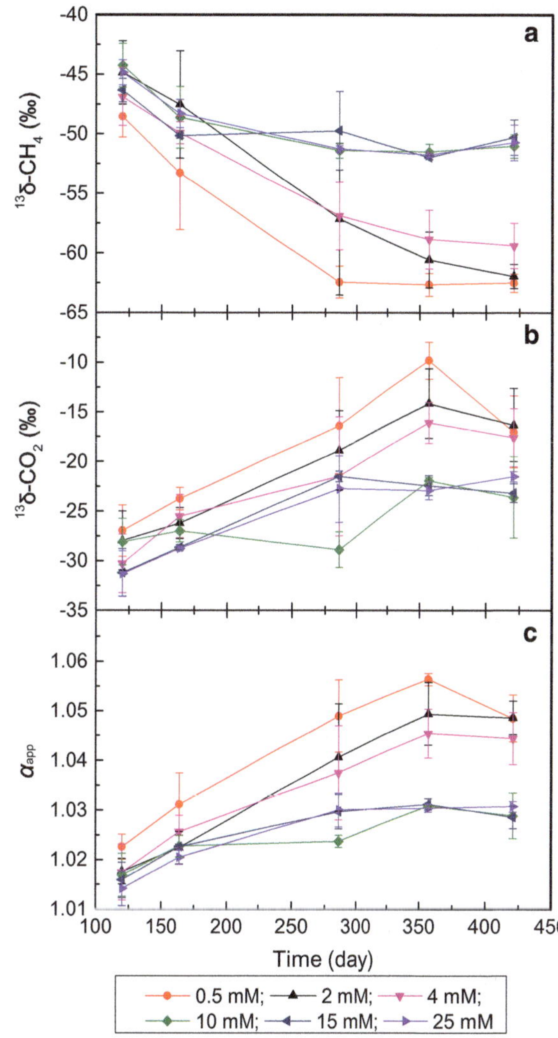

Fig. 3 Values of $\delta^{13}CH_4$ (**a**), $\delta^{13}CO_2$ (**b**) and apparent fractionation factors (α_C) (**c**). CK, culture without hexadecane and sulfate addition; 0.5, 2, 4, 10, 15, and 25 mM represent cultures amended with hexadecane and 0.5, 2, 4, 10, 15, and 25 mM sulfate, respectively. Values represent means and standard errors ($n = 2$–3 replicates)

of incubation (Fig. 3a). The CO_2 concentrations measured in the headspace were used for isotopic analysis, since CO_2 rather than bicarbonate is the active substrate for hydrogenotrophic methanogenesis [41]. The $\delta^{13}C$ value of carbon dioxide ($\delta^{13}CO_2$) was -28.7 ± 2.8‰ in all sulfate addition cultures after 120 days of incubation. The $\delta^{13}CO_2$ values continued to increase during incubation and plateaued at -17.0 ± 3.4 and -22.8 ± 1.9‰ for the 0.5–4 and 10–25 mM sulfate cultures, respectively (Fig. 3b). According to Eq. 1, reported by Whiticar et al. [42] the $\delta^{13}C$ values of CH_4 and CO_2 were used to compute the apparent fractionation factors (α_{app}). The α_{app} values initially remained at 1.017 ± 0.004 during the early stage and increased to 1.047 ± 0.004 in the 0.5–4 mM sulfate cultures. However, the α_{app} value had only increased to 1.029 ± 0.003 in the 10–25 mM sulfate cultures at the end of incubation (Fig. 3c).

$$\alpha_{app} = (\delta^{13}CO_2 + 1000)/(\delta^{13}CH_4 + 1000) \qquad (1)$$

Microbial community structures and functions

Archaeal community

The archaeal communities of the cultures with different sulfate concentrations were mainly composed of *Methanothrix* (T-RFs 228, 284 and 495 bp), *Methanolinea* (T-RFs 84 and 393 bp), *Methanobacteriaceae* (T-RF 92 bp), and *Methanoculleus* (T-RF 186 bp) (Fig. 4a; Table 2). *Methanothrix* was one of the dominant phylotypes (Fig. 4a). However, *Methanolinea*, which was represented by the 84-bp T-RF, increased substantially over time and became the predominant phylotype in the 0.5 mM sulfate culture at the end of incubation (Fig. 4a). The *Methanoculleus* populations also increased in all sulfate cultures and became the second dominant phylotype in the 2–25 mM sulfate cultures (Fig. 4a). Pearson correlation analysis revealed that the *Methanoculleus* and *Methanolinea* populations (T-RF 84 bp) were significantly and positively correlated with methane production ($r = 0.454$ and 0.352, respectively, $p < 0.01$) (Table 3). *Methanoculleus* also exhibited a positive correlation with sulfate consumption ($r = 0.351$, $p < 0.05$), whereas *Methanolinea* (T-RF 84 bp) had a significantly negative correlation ($r = -0.421$, $p < 0.01$) (Table 3). ANOSIM indicated that

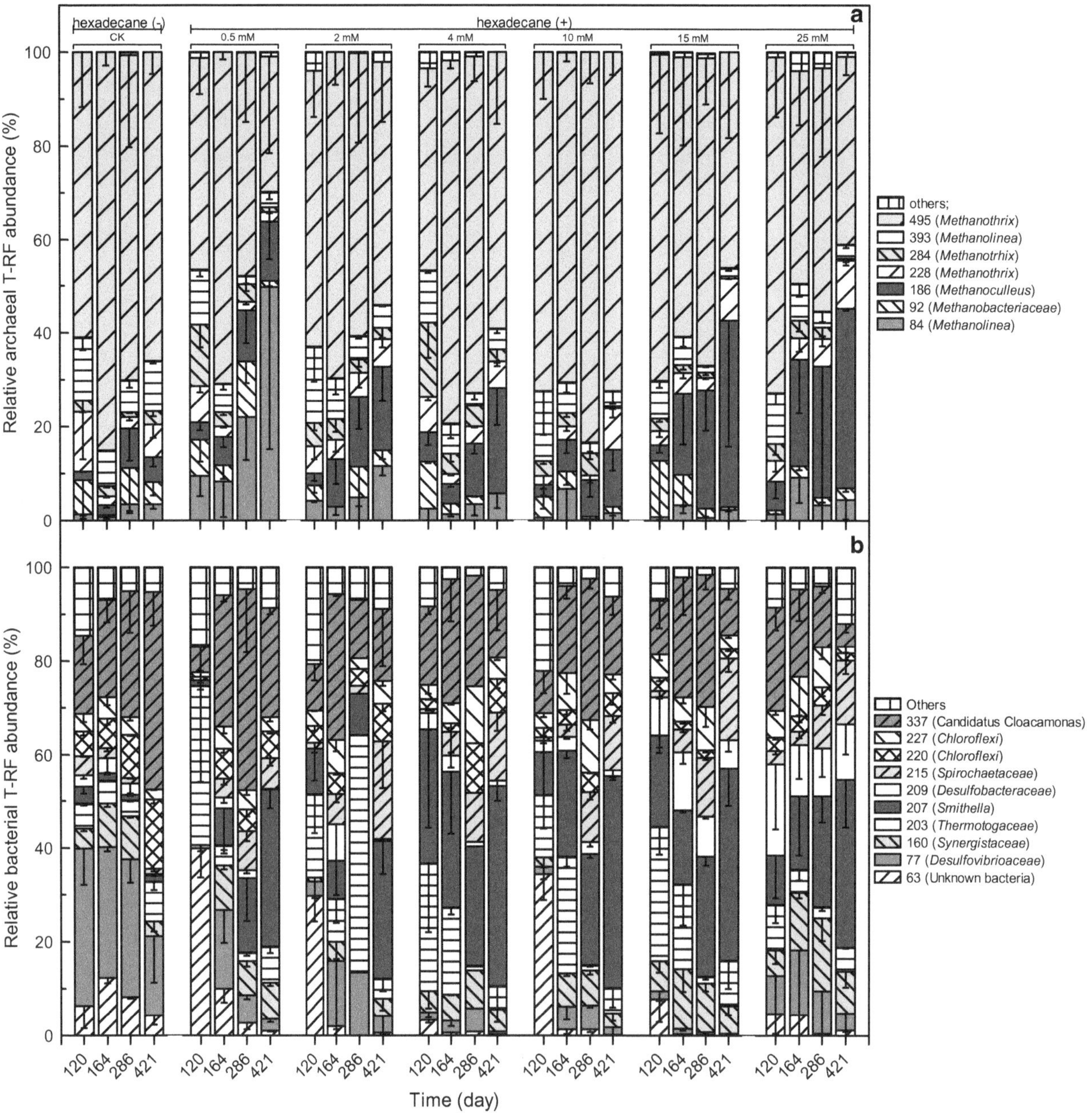

Fig. 4 Relative abundance levels of T-RFs of archaeal (**a**) and bacterial (**b**) 16S rRNA genes. CK, culture without hexadecane and sulfate addition; 0.5, 2, 4, 10, 15, and 25 mM represent cultures amended with hexadecane and 0.5, 2, 4, 10, 15, and 25 mM sulfate, respectively. Values represent means and standard errors ($n = 2$–3 replicates). The phylogenetic positions of bacterial phylotypes represented by T-RFs are as referenced in our previous report [34]

the archaeal communities from the 0.5 and 2 mM sulfate cultures were more similar to each other than cultures in which 4–25 mM sulfate was added (Table 4).

Bacterial community

The most common bacterial groups were members of *Smithella* (T-RF 207 bp), *Synergistaceae* (T-RF 160 bp), *Thermotogaceae* (T-RF 203 bp), *Spirochaetaceae* (T-RF 215 bp), *Chloroflexi* (T-RFs 220 and 227 bp), *Candidatus* Cloacamonas (T-RF 337 bp), and unknown bacteria (T-RF 63 bp) (Fig. 4b). These bacterial groups, with the exception of the 63-bp T-RF identified here, have been proposed as the core microbiome of the methanogenic hexadecane-degrading culture M82 without sulfate addition [12].

Table 2 Archaeal 16S rRNA gene clone libraries

Phylogenetic group	In silico T-RF (No. of clones)[a]			Clone type (GenBank Accession Number)	Closest species	Similarity (%)
	CK	0.5 mM	25 mM			
Methanosaetaceae	228 (2)			A4-6 (KJ735834)	*Methanosaeta harundinacea* 6Ac; NR_102896	99
	495 (12)	495 (11)	495 (10)	A18-26 (KJ735844)		96
	791 (1)			A80-6 (KR297252)		82
Methanomicrobiaceae		186 (3)	186 (12), 187 (1)	A4-7 (KJ735835)	*Methanoculleus receptaculi* ZC-2; NR_043961	99
			186 (2)	A18-28 (KJ735845)		95
Methanoregulaceae		84 (9)		A4-21 (KJ735839)	*Methanolinea tarda* NOBI-1; NR_028163	98
	84 (1), 393 (1)			A80-2 (KJ735846)		97
Methanobacteriaceae	93 (1), 95 (6)			A80-8 (KJ735840)	*Methanothermobacter thermautotrophicus* CaT2; AP011952	83
		91 (1), 92 (5)		A4-27 (KJ735841)	*Methanobacterium beijingense* 4-1; AY552778	99

[a] The 16S rRNA gene clone libraries were retrieved from the cultures after 421 days of incubation. CK, the culture without hexadecane and sulfate addition; 0.5 mM: the culture only amended with hexadecane and 0.5 mM sulfate; 25 mM: the culture amended with hexadecane and 25 mM sulfate

Table 3 Pearson correlations of T-RFs with methane production and sulfate consumption

Domain	Phylogenetic position	T-RF (bp)	Methane production	Sulfate consumption
Archaeal domain	*Methanolinea*	84	0.454[a]	−0.421[a]
	Methanolinea	393	−0.659[a]	−0.089
	Methanobacteriaceae	92	−0.073	−0.337[b]
	Methanoculleus	186	0.352[a]	0.351[b]
	Methanosaeta	228	0.082	0.528[a]
	Methanosaeta	284	−0.274[b]	−0.206
	Methanosaeta	495	−0.259	0.048
Bacterial domain	Unknown bacteria	63	−0.482[a]	−0.161
	Desulfovibrionaceae	77	−0.158	0.047
	Synergistaceae	160	−0.068	0.326[b]
	Thermotogaceae	203	−0.370[a]	−0.188
	Smithella	207	0.532[a]	0.314[b]
	Desulfobacteraceae	209	−0.333[a]	0.428[a]
	Spirochaetaceae	215	0.715[a]	0.136
	Chloroflexi	220	0.467[a]	−0.443[a]
	Chloroflexi	227	0.029	−0.061
	Candidatus Cloacamonas	337	0.262[b]	−0.499[a]

[a] Correlation is significant at the 0.01 level (2-tailed)

[b] Correlation is significant at the 0.05 level (2-tailed)

Table 4 ANOSIM pairwise comparison of similarities among archaeal communities following the addition of different sulfate concentrations

Sulfate concentration (mM)	0.5	2	4	10	15	25
0.5						
2	−0.023					
4	0.156[a]	−0.050				
10	0.319[a]	0.136	−0.015			
15	0.213[a]	0.020	−0.014	0.036		
25	0.150[a]	−0.030	0.025	0.122	−0.001	

[a] Value is significant at the 0.05 level

Table 5 ANOSIM pairwise comparison of similarities among bacterial communities following the addition of different sulfate concentrations

Sulfate concentration (mM)	0.5	2	4	10	15	25
0.5						
2	0.483[a]					
4	0.996[a]	0.056				
10	0.647[a]	0.013	−0.092			
15	0.995[a]	0.194[a]	0.045	0.090		
25	0.976[a]	0.247[a]	0.287[b]	0.217[b]	0.085	

[a] Value is significant at the 0.05 level

[b] Value is significant at the 0.01 level

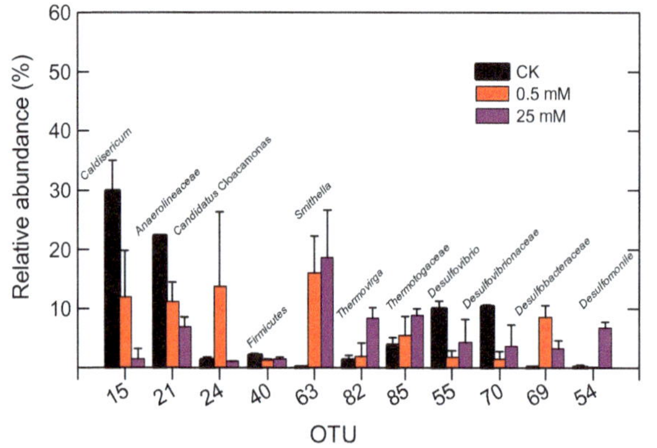

Fig. 5 Relative abundance levels of top 11 OTUs. CK, culture without hexadecane and sulfate addition; 0.5 and 25 mM represent cultures amended with hexadecane and 0.5 and 25 mM sulfate, respectively. Values represent means and standard errors ($n = 2$ replicates)

Members of the core microbiome were positively associated with methane production, except for *Thermotogaceae*, which was negatively associated with methane production ($p < 0.01$) (Table 3). Only *Smithella* exhibited positive associations with methane production and sulfate consumption (Table 3). *Desulfobacteraceae* was negatively associated with methane production but was positively correlated with sulfate consumption ($p < 0.01$) (Table 3). ANOSIM indicated that the bacterial communities could be divided into three clusters corresponding to the 0.5, 2–10, and 15–25 mM sulfate cultures, respectively (Table 5). The bacterial community structure was further evaluated through high-throughput sequencing of the 16S rRNA gene. In total, 318,839 reads with an average length of 413 bp were retrieved from the control and the 0.5 and 25 mM sulfate cultures at the end of incubation (Additional file 2: Table S1). The top 11 operational taxonomic units (OTUs) accounting for 62–75% of the sequences are shown in Fig. 5. Members of *Caldisericum* (an average of 30%), *Anaerolineaceae* (22%), and *Desulfovibrionaceae* (20%, two OTUs) were predominant in the control without sulfate and hexadecane addition (Fig. 5). The dominant phylotypes in the 0.5 mM sulfate addition cultures belonged to *Smithella* (16.1%), *Candidatus* Cloacamonas (13.8%), *Caldisericum* (12.0%), unclassified *Anaerolineaceae* (11.2%), *Desulfobacteraceae* (8.6%), and *Thermotogaceae* (5.5%), similar to the bacterial T-RFLP profile (Figs. 4b, 5). In contrast, only *Smithella* (18.7%), *Thermotogacea* (8.9%), and *Anaerolineaceae* (7.0%) were retained as the dominant members in the 25 mM sulfate culture (Fig. 5). The relative abundance of *Thermovirga* increased to above 7% in the high-sulfate culture (Fig. 5). Among the four known SRPs (*Desulfomonile*, unclassified *Desulfobacteraceae* and two OTUs in family *Desulfovibrionaceae*), members of *Desulfovibrionaceae* were dominant (19.3–21.5%, two OTUs) in the control (Fig. 5). Unclassified *Desulfobacteraceae* was predominant (7.2–10.0%) in the 0.5 mM sulfate culture, whereas *Desulfomonile* (6.1–7.5%) was predominant in the 25 mM sulfate culture (Fig. 5).

Discussion

Impacts of sulfate on hexadecane degradation and methane production

The results of this study join those of a previous body of reports demonstrating that methanogenic cultures can degrade alkanes anaerobically in the presence of sulfate [11, 15, 31]. Methane production was enhanced under low-sulfate concentrations (<2–5 mM) [15, 31], whereas significant inhibition was observed at sulfate concentrations greater than 5 mM [31]. Gieg et al. [11] reported that the extent of methanogenic degradation of crude oil alkanes was not affected by sulfate in the 10 mM range, but the methane production rate slightly decreased. The results of this study demonstrated that the rate and extent of methane production were gradually inhibited by increasing sulfate concentration (Fig. 1). However, the presence of excessive sulfate at 25 mM, which is higher than the 21 mM needed for complete conversion of hexadecane to CO_2 through sulfate reduction, did not completely inhibit methane production (Eq. 2). The concurrent methane production and sulfate reduction in the excessive presence of sulfate may be attributed to the adequate supply of noncompetitive substrates or an abundance of competitive substrates for the syntrophs and methanogens [43]. However, the possibility of cooperation between the incomplete-oxidizing sulfate reducers and methanogens cannot be ignored if incomplete oxidation of organic intermediates into H_2 or acetate through sulfate reduction presented in the cultures [44]. This study also demonstrated that sulfate reduction increased hydrocarbon degradation efficiency, although the enhancement may not be directly used to address questions of methanogenic and sulfidogenic competition [45, 46]. The increased activity of sulfate reduction after the treatment with increased sulfate concentration probably enhanced H_2 conversion rate and caused a much lower hydrogen concentration below the threshold value for hydrogenotrophic methanogens [44, 47, 48], which may facilitate the syntrophic alkane oxidation process from a thermodynamics standpoint [18].

The concept of "electron flow" was introduced by Isa et al. [49] to quantify the competition between methanogenesis and sulfidogenesis. The most common explanation for the electron distribution between sulfate reduction and methane production during organic substrate degradation is the effect of the COD/SO_4^{2-} ratio [49–51], which also includes other factors, such as H_2S, substrate concentration, and inoculum volume [49]. As expected, electron flux toward sulfidogenesis increased with increasing sulfate concentrations in this study (Fig. 2a). However, unlike most previous work showing how electron flow can be affected by the COD/SO_4^{2-} ratio [50–53], the relative distribution of electrons between the SRPs and the methanogenic populations depended on the initial sulfate addition and remained constant during methane production (Fig. 2b), which has not been reported before. The results suggested that the concentration of sulfate could be an effective indicator of hexadecane metabolism pathway under mixed electron acceptors conditions. Giving the complex composition of crude oil and diverse electron acceptors present in oil reservoirs, it is also necessary in the future to evaluate the in situ effects of sulfate on the carbon and electron flow during methanogenic crude oil degradation, which could aid in predicting the biomethane potential of crude oil in field trials.

Impact of sulfate on the methanogenic pathway

The α_{app} is a generally accepted index to coarsely estimate the dominant methane production pathway [42], with values above 1.065 and below 1.025 proposed to represent hydrogenotrophic and aceticlastic methanogenesis, respectively, according to recent reports [54, 55]. However, the α_{app} value apparently changes according to species and growth conditions, and the values measured in various cultures of hydrogenotrophic methanogens range between 1.031 and 1.077 [56, 57]. Two methanogen genera, *Methanosarcina* and *Methanothrix*, are capable of performing aceticlastic reactions [58], whereas only *Methanothrix* species were observed in this study (Fig. 4a), and the fractionation factor of this genus during aceticlastic methanogenesis is less than 1.01 [56, 59]. Moreover, the acetate concentration was always lower than the threshold value (100 μM) during incubation, indicating that carbon isotope fractionation did not occur during the conversion of acetate to CH_4 by aceticlastic methanogens [60]. Therefore, the increase in α_{app} observed during the incubation of all sulfate addition cultures likely revealed that CO_2 reduction was the predominant methanogenic pathway, although acetate disproportionation would also contribute to hexadecane degradation. Meanwhile, α_{app} decreased gradually with increasing sulfate concentration (Fig. 3c), indicating that sulfate reduction often outcompetes hydrogenotrophic methanogenesis for H_2 utilization from thermodynamic and kinetic points of view [48, 49].

Microbial community response to sulfate stress

The aceticlastic *Methanothrix* was the dominant archaea, not only in the hexadecane- and sulfate-containing cultures but also in the control culture in which hexadecane and sulfate were not added (Fig. 4a). The dominant *Methanothrix* did not show a positive relationship with either methane production or sulfate reduction (Table 3). These observations indicated that *Methanothrix* may play a minor role in methanogenic hexadecane degradation, which was not only supported by the aforementioned changes in methanogenic communities but also by increasing α_{app} values over time in all sulfate treatments. Regardless, *Methanothrix* may still play a role by leaking H_2 from acetate degradation to SRPs for sulfate reduction [61]. Both *Methanoculleus* and *Methanolinea* species, which use H_2/CO_2 for methanogenesis [62, 63], were positively associated with methane production (Fig. 4a; Table 3). The *Methanoculleus* populations steadily increased with increasing initial sulfate concentrations and became the second most abundant after *Methanothrix* in the 2–25 mM sulfate cultures. This increase could be explained by the fact that *Methanoculleus* may have a greater affinity than other hydrogenotrophic methanogens for H_2 [64] or the fact that *Methanoculleus* is involved in the syntrophic alkane oxidation process, together with SRPs [34].

Smithella is the predominant bacterial phylotype responsible for alkane degradation under methanogenic conditions [33, 34]. Detection of the *ass*A gene and methyl pentadecyl succinic acid in the methanogenic hexadecane-degrading culture has demonstrated that *Smithella* can initiate alkane degradation via fumarate addition [39]. However, no *dsr*-like gene was detected in the binned genome of strain M82_1 [39]. The observation of a positive correlation between *Smithella* abundance and sulfate addition (Fig. 4b; Table 3) may reflect cooperation rather than competition between syntrophic alkane degraders and SRPs, as supported by reported increased hexadecane degradation efficiency with increasing sulfate consumption (Additional file 1: Figure S1). The core bacterial phylotypes affiliated with *Spirochaetaceae*, *Chloroflexi*, *Candidatus* Cloacamonas, and *Thermotogaceae* were proposed to play roles as either secondary degraders during the degradation of hexadecane [12] and terephthalate [65] or contributors to scavenging anabolic products (protein and lipids), presumably derived from detrital microbial biomass [65]. Most of these bacteria were positively correlated with methane production, but their relationships with sulfate reduction were either negative or irrelevant (Table 3), which may reflect the competition of these phylotypes with SRPs.

SRPs in the methanogenic cultures have the capacity to cooperate with methanogens and syntrophic bacteria under sulfate-free conditions but also possess robustness to compete with these organisms through sulfate reduction when sulfate becomes available [66]. The abundance levels of unclassified *Desulfobacteraceae* (OTU 69) and *Desulfomonile* (OTU 54) increased, and these genera became the predominant SRPs in the low- and high-sulfate addition cultures, respectively (Fig. 5). Most members of *Desulfobacteraceae* can utilize sulfate and acetate as electron acceptors and carbon sources, respectively, and some can oxidize organic substrates completely to CO_2, whereas others perform an incomplete oxidation of organic substrates to acetate [67]. *Desulfomonile* spp., known as dehalogenators, is also capable of H_2 scavenging, and H_2 uptake can increase by more than fourfold when sulfate is used as an electron acceptor [47]. These findings suggest that the response of the sulfate reduction bacterial community to sulfate varies according to sulfate concentrations, although their ecophysiological roles in methanogenic hexadecane-degrading cultures in the presence of sulfate require further examination.

Conclusions

This study not only demonstrated the competition and coexistence of sulfate-reducing and methanogenic populations during anaerobic hexadecane degradation processes, but also revealed a linear positive relationship between the ratio of sulfate reduction/methane production and initial sulfate concentration. The removal efficiency of hexadecane could be enhanced by increased sulfate addition, although the amount of methane accumulation decreased. CO_2 reduction was the dominant methanogenic pathway, but the activity was inhibited by increasing sulfate addition. Both *Smithella* and *Methanoculleus*, the key alkane degraders and methane producers, could be adapted to the sulfate stress, and different SRP populations were stimulated at various sulfate concentrations. Taken together, our results would help to understand the interactions between methanogenic populations and SRPs during alkane degradation under mixed electron acceptors conditions.

Methods

Culture and growth conditions

A methanogenic hexadecane-degrading enrichment culture was derived from oily sludge-contaminated sediment from the Shengli oil field in eastern China [34]. The culture was inoculated (20%, vol/vol) into a modified freshwater medium and dispensed into 600-ml vials under an atmosphere of N_2 (99.999%) gas, where 100 μl of hexadecane (99%, approximately 0.34 mmol), 2 ml of 2,2,4,4,6,8,8-heptamethylnonane (HMN), and various concentrations of sodium sulfate (giving final concentrations of 0.5, 2, 4, 10, 15 and 25 mM) were added. Neither hexadecane nor sulfate was added to the control culture. The cultures were reduced with $Na_2S{\cdot}9H_2O$ (0.03%), and the pH was adjusted to 7.0 ± 0.2 by the addition of HCl or NaOH solutions prior to incubation. The vials in each set were prepared in triplicate and statically incubated at 35 °C in the dark.

Chemical methods

Methane was measured using a gas chromatograph (Agilent 7820A, USA) with a thermal conductivity detector, as previously described [12]. The maximum specific methane production rates (μ_{max}) were calculated [68] from the fitting methane production curves based on a statistical model of Slogistic 1 using ORIGINPRO 8.5 software (OriginLab, USA) The carbon isotopic compositions of CH_4 and CO_2 were determined using an IsoPrime100 mass spectrometer (United Kingdom) [34]. The liquid cultures (2–8 ml) were carefully aspirated into tubes using anoxic, sterilized syringes, and were centrifuged at 12,000×*g* for 5 min at 4 °C. The supernatants were collected and stored at −40 °C for future analysis of sulfate and volatile fatty acid (VFA) concentrations. The cell pellets were harvested and stored at −80 °C for microbial community composition analysis (see below). The sulfate concentration was measured using ion chromatography (Metrohm ICS 3000, Switzerland) with a conductivity detector. The liquid samples were diluted 10-fold with deionized water and pretreated using Cleanert IC-RP filtration columns (Agela Technologies, China) before analysis. The eluent contained 30 mM NaOH, and the flow rate was 1.3 ml/min. VFAs were analyzed by gas chromatography (GC) (Agilent 7890A, USA) with a flame ionization detector (FID) and a DB-FFAP column [69]. The residual hexadecane after 421 days of incubation was extracted using a mixture of *n*-hexane and acetone (1:1, v/v) according to a previous report [40]. A final concentration of 300 mg/l pristine was also added before extraction to evaluate the recovery efficiency of the method. The extract was analyzed by GC (Agilent 7890A, USA) using a DB-5MS capillary GC column (30 m × 0.53 mm × 1.5 μm). The GC injector and FID temperatures were set to 280 and 250 °C, respectively. The oven temperature program began at 60 °C for 2 min and increased to 300 °C at a rate of 12 °C per min, where it remained for 10 min. The carrier gas was hydrogen (99.999%) at a flow rate of 36 ml/min. The hexadecane concentration was determined from the external calibration curve generated from *n*-heptadecane, which was calibrated using the linear equation derived from the peak areas.

Electron flow distribution between sulfate reduction and methane production

The stoichiometric energy-balance equation for the complete oxidation of hexadecane via methanogenesis (Eq. 2) or coupled to sulfidogenesis (Eq. 3) can be written according to previous reports, when cell growth was not taken into consideration [15, 70].

$$C_{16}H_{34} + 7.5\ H_2O \rightarrow 12.25\ CH_4 + 3.75\ CO_2 \qquad (2)$$

$$\begin{aligned} C_{16}H_{34} &+ 12.25\ SO_4^{2-} + 24.5H^+ \\ &\rightarrow 12.25\ H_2S + 16\ CO_2 + 17\ H_2O \end{aligned} \qquad (3)$$

The distribution of electron flow between sulfidogenic and methanogenic reactions can be calculated from the amounts of sulfate reduced (*S*) and methane produced (*M*) [49]. Percent electron flow via methane production was calculated as $(M) = [M/(M + S)] \times 100$; percent electron flow via sulfate reduction was calculated as $(S) = [S/(M + S)] \times 100$, and the ratio of sulfate reduction and methane production was calculated as S/M.

Microbial analyses

Genomic DNA extraction and purification

Genomic DNA was extracted using a bead-beating method [12]. DNA products were purified with the Promega Wizard DNA Clean-up System (Promega, USA) and were stored at −80 °C.

T-RFLP analysis and clone library construction

For T-RFLP analysis, the archaeal and bacterial 16S rRNA gene fragments were amplified using the primer sets A109F/A934R and B27F/B907R, respectively, where the 5′ ends of primers B27f and A934r were labeled with 6-carboxyfluorescein (FAM) [12]. The PCR products were purified with the TIAN Universal DNA Purification Kit (Tiangen, China) and were digested at 37 °C using *Hae*III for bacterial DNA (TakaRa, Japan) and at 65 °C using *Taq*I for archaeal DNA (TakaRa, Japan) [12, 34]. Semi-quantitative DNA fragment analysis was performed as previously reported [12]. The primers and PCR used to construct the archaeal 16S rRNA gene clone library were the same as for the T-RFLP analysis, except that label-free primers were used [12].

MiSeq and bioinformatics analysis of bacterial 16S rRNA gene fragments

The bacterial primers 515F (5′-GTGYCAGCMGCCGCGGTA-3′) and 909R (5′-CCCCGYCAATTCMTTTRAGT-3′) were used for high-throughput sequencing of 16S rRNA gene PCR amplicons. The PCR mixture and program were described previously [71], and sequencing was performed on the Illumina MiSeq platform. Paired-end reads were merged using FLASh v1.2.9 [72] and were quality-filtered using Trimmomatic v0.32 [73]. Chimeras were removed using the Uchime algorithm and Usearch v 8.1.1861 [74]. The sequence reads in each sample were normalized using daisychopper.pl (http://www.festinalente.me/bioinf/downloads/daisychopper.pl). OTUs were clustered at 97% identity using Qiime v1.9.1 [75]. The phylogenetic affiliations of the representative OTU sequences were assigned using RDP Classifier at a confidence level of 80% [76]. Alpha diversity indices (Chao1, Shannon and Simpson) were calculated using Qiime [75].

Statistical analysis

Principal coordinate analysis (PCoA) of the community structures was conducted in PAST (http://folk.uio.no/ohammer/past/). The significant differences in community structures between data sets were assessed for significance with ANOSIM using PAST. Correlations between community structures and environmental factors were conducted using the R package Vegan (http://cran.r-project.org/web/packages/vegan/index.html). One-way-analysis of variance (ANOVA), and Pearson's correlations between T-RFs and environmental factors were conducted using SPSS 17.0 software (IBM, USA). The Pearson correlation coefficient (*r*) and the coefficient of determination (R^2) were used to evaluate the relationship between the sulfate reduction/methane production ratio and the initial sulfate concentrations, which were also analyzed using SPSS 17.0 software (IBM, USA).

Abbreviations

SRP: sulfate-reducing prokaryote; HMN: 2,2,4,4,6,8,8-heptamethylnonane; VFA: volatile fatty acid; T-RFLP: terminal restriction fragment length polymorphism; FAM: 6-carboxyfluorescein; OTU: operational taxonomic units; PCoA: principal coordinate analysis; ANOVA: one-way-analysis of variance; μ_{max}: maximum specific methane production rate, month^{-1}; $\delta^{13}CH_4$: $\delta^{13}C$ value of methane; $\delta^{13}CO_2$: $\delta^{13}C$ value of carbon dioxide; α_{app}: apparent fractionation factors.

Authors' contributions

LC and HZ conceived and designed this study. TTM and LRD performed experimental studies. MTT, LLY, and WQZ collected microbial diversity data (T-RFLP, cloning and sequencing of 16S rRNA genes). ZZ and DSF analyzed sulfate and hexadecane concentrations, respectively. JPR analyzed T-RFLP and MiSeq data. QY analyzed the carbon isotope data and revised the manuscript. LC wrote the manuscript. All authors read and approved the final manuscript.

Author details

[1] Key Laboratory of Development and Application of Rural Renewable Energy, Biogas Institute of Ministry of Agriculture, Section 4-13, Renmin South Road, Chengdu 610041, People's Republic of China. [2] Key Laboratory of Environmental and Applied Microbiology, Chengdu Institute of Biology of Chinese Academy of Sciences, Section 4-9, Renmin South Road, Chengdu 610041, People's Republic of China. [3] Environmental Microbiology Key Laboratory of Sichuan Province, Section 4-9, Renmin South Road, Chengdu 610041, People's Republic of China. [4] State Key Laboratory of Environmental Geochemistry, Institute of Geochemistry, Chinese Academy of Sciences, 99 Lincheng West Road, Guanshanhu District, Guiyang 550081, People's Republic of China. [5] Anhui Normal University, 1 Beijing East Road, Wuhu 241000, People's Republic of China.

Acknowledgements

We are indebted to Dr. Zhe Lyu (University of Georgia) and Dr. Peng-fei Liu (Max-Planck-Institute for Terrestrial Microbiology) for their critical comments, constructive suggestions, and extensive and vigorous edits to this work.

Competing interests

The authors declare that they have no competing interests.

Funding

This work was supported by the National High Technology Research and Development Program of China (Grant Number 2013aa064401), the National Natural Science Foundation of China (Grant Numbers 31570009, 41173088) and the Science and Technology Innovation Project of the Chinese Academy of Agriculture Science.

References

1. BP statistical review of world energy. 2016;48:48.
2. Muggeridge A, Cockin A, Webb K, Frampton H, Collins I, Moulds T, Salino P. Recovery rates, enhanced oil recovery and technological limits. Philos Trans A Math Phys Eng Sci. 2014;372:20120320.
3. Lewis RB. Microbial enhanced oil recovery (MEOR). Curr Opin Microbiol. 2010;13:316–20.
4. Head IM, Jones DM, Larter SR. Biological activity in the deep subsurface and the origin of heavy oil. Nature. 2003;426:344–52.
5. Milkov AV. Worldwide distribution and significance of secondary microbial methane formed during petroleum biodegradation in conventional reservoirs. Org Geochem. 2011;42:184–207.
6. Aitken CM, Jones DM, Larter SR. Anaerobic hydrocarbon biodegradation in deep subsurface oil reservoirs. Nature. 2004;431:291–4.
7. Katz BJ. Microbial processes and natural gas accumulations. Open Geol J. 2011;5:e83.
8. Gray ND, Sherry A, Hubert C, Dolfing J, Head IM. Methanogenic degradation of petroleum hydrocarbons in subsurface environments: remediation, heavy oil formation, and energy recovery. In: Allen I, Sariaslani S, Geoffrey MG, editors. Advances in Applied Microbiology, vol. 72. Cambridge: Academic Press; 2010. p. 137–61.
9. Widdel F, Knittel K, Galushko A. Anaerobic hydrocarbon-degrading microorganisms: an overview. In: Timmis KN, editor. Handbook of hydrocarbon and lipid microbiology. Springer: Berlin; 2010. p. 1997–2021.
10. Jones DM, Head IM, Gray ND, Adams JJ, Rowan AK, Aitken CM, Bennett B, Huang H, Brown A, Bowler BF, et al. Crude-oil biodegradation via methanogenesis in subsurface petroleum reservoirs. Nature. 2008;451:176–80.
11. Gieg LM, Duncan KE, Suflita JM. Bioenergy production via microbial conversion of residual oil to natural gas. Appl Environ Microbiol. 2008;74:3022–9.
12. Cheng L, Rui J, Li Q, Zhang H, Lu Y. Enrichment and dynamics of novel syntrophs in a methanogenic hexadecane-degrading culture from a Chinese oilfield. FEMS Microbiol Ecol. 2013;83:757–66.
13. Wang L-Y, Gao C-X, Mbadinga SM, Zhou L, Liu J-F, Gu J-D, Mu B-Z. Characterization of an alkane-degrading methanogenic enrichment culture from production water of an oil reservoir after 274 days of incubation. Int Biodeterior Biodegrad. 2011;65:444–50.
14. Parkes J. Cracking anaerobic bacteria. Nature. 1999;401:217–8.
15. Zengler K, Richnow HH, Rossello-Mora R, Michaelis W, Widdel F. Methane formation from long-chain alkanes by anaerobic microorganisms. Nature. 1999;401:266–9.
16. Suflita JM, Davidova IA, Gieg LM, Nanny M, Prince RC. Chapter 10 anaerobic hydrocarbon biodegradation and the prospects for microbial enhanced energy production. In: Rafael V-D, Rodolfo Q-R, editors. Studies in surface science and catalysis, vol. 151. New York: Elsevier; 2004. p. 283–305.
17. Cheng L, Shi S, Li Q, Chen J, Zhang H, Lu Y. Progressive degradation of crude oil n-alkanes coupled to methane production under mesophilic and thermophilic conditions. PLoS ONE. 2014;9:e113253.
18. Dolfing J, Larter SR, Head IM. Thermodynamic constraints on methanogenic crude oil biodegradation. ISME J. 2008;2:442–52.
19. Bastin ES, Greer FE, Merritt CA, Moulton G. The presence of sulphate reducing bacteria in oil field waters. Science. 1926;63:21–4.
20. Mbadinga SM, Wang L-Y, Zhou L, Liu J-F, Gu J-D, Mu B-Z. Microbial communities involved in anaerobic degradation of alkanes. Int Biodeterior Biodegrad. 2011;65:1–13.
21. Jiménez N, Richnow HH, Vogt C, Treude T, Krüger M. Methanogenic hydrocarbon degradation: evidence from field and laboratory studies. J Mol Microbiol Biotechnol. 2016;26:227–42.
22. Youssef N, Elshahed MS, McInerney MJ. Microbial processes in oil fields: culprits, problems, and opportunities. In: Allen I, Laskin SS, Geoffrey MG, editors. Advances in applied microbiology, vol. 66. Cambridge: Academic Press; 2009. p. 141–251.
23. Orphan VJ, Taylor LT, Hafenbradl D, Delong EF. Culture-dependent and culture-independent characterization of microbial assemblages associated with high-temperature petroleum reservoirs. Appl Environ Microbiol. 2000;66:700–11.
24. Guan J, Zhang BL, Mbadinga SM, Liu JF, Gu JD, Mu BZ. Functional genes (dsr) approach reveals similar sulphidogenic prokaryotes diversity but different structure in saline waters from corroding high temperature petroleum reservoirs. Appl Microbiol Biotechnol. 2014;98:1871–82.
25. Muyzer G, Stams AJM. The ecology and biotechnology of sulphate-reducing bacteria. Nat Rev Micro. 2008;6:441–54.
26. Khelifi N, Amin Ali O, Roche P, Grossi V, Brochier-Armanet C, Valette O, Ollivier B, Dolla A, Hirschler-Rea A. Anaerobic oxidation of long-chain n-alkanes by the hyperthermophilic sulfate-reducing archaeon. ISME J. 2014;8:2153–66.
27. Lovley DR, Klug MJ. Sulfate reducers can outcompete methanogens at freshwater sulfate concentrations. Appl Environ Microbiol. 1983;45:187–92.
28. Kalyuzhnyi SV, Fedorovich VV. Mathematical modelling of competition between sulphate reduction and methanogenesis in anaerobic reactors. Bioresour Technol. 1998;65:227–42.
29. Omil F, Lens P, Visser A, Hulshoff Pol LW, Lettinga G. Long-term competition between sulfate reducing and methanogenic bacteria in UASB reactors treating volatile fatty acids. Biotechnol Bioeng. 1998;57:676–85.
30. Collins AG. Chapter 8. Classification of oilfield waters. In: Collins AG, editor. Developments in petroleum science, vol. 1. New York: Elsevier; 1975. p. 253–92.
31. Siegert M, Cichocka D, Herrmann S, Gründger F, Feisthauer S, Richnow H-H, Springael D, Krüger M. Accelerated methanogenesis from aliphatic and aromatic hydrocarbons under iron- and sulfate-reducing conditions. FEMS Microbiol Lett. 2011;315:6–16.
32. Siddique T, Penner T, Klassen J, Nesbø C, Foght J. Microbial communities involved in methane production from hydrocarbons in oil sands tailings. Environ Sci Technol. 2012;46:9802–10.
33. Gray ND, Sherry A, Grant RJ, Rowan AK, Hubert CRJ, Callbeck CM, Aitken CM, Jones DM, Adams JJ, Larter SR, Head IM. The quantitative significance

of *Syntrophaceae* and syntrophic partnerships in methanogenic degradation of crude oil alkanes. Environ Microbiol. 2011;13:2957–75.

34. Cheng L, Ding C, Li Q, He Q, Dai LR, Zhang H. DNA-SIP reveals that *Syntrophaceae* play an important role in methanogenic hexadecane degradation. PLoS ONE. 2013;8:e66784.
35. Callaghan AV, Davidova IA, Savage-Ashlock K, Parisi VA, Gieg LM, Suflita JM, Kukor JJ, Wawrik B. Diversity of benyzl- and alkylsuccinate synthase genes in hydrocarbon-impacted environments and enrichment cultures. Environ Sci Technol. 2010;44:7287–94.
36. Tan B, Dong X, Sensen C, Foght J. Metagenomic analysis of an anaerobic alkane-degrading microbial culture: Potential hydrocarbon-activating pathways and inferred roles of community members. Genome. 2013;56:599–611.
37. Embree M, Nagarajan H, Movahedi N, Chitsaz H, Zengler K. Single-cell genome and metatranscriptome sequencing reveal metabolic interactions of an alkane-degrading methanogenic community. ISME J. 2013;8:757–67.
38. Tan B, Nesbo C, Foght J. Re-analysis of omics data indicates *Smithella* may degrade alkanes by addition to fumarate under methanogenic conditions. ISME J. 2014;8:2353–6.
39. Qin QS, Feng DS, Liu PF, He Q, Li X, Liu AM, Zhang H, Hu GQ, Cheng L. Metagenomic characterization of *Candidatus* Smithella cisternae strain M82_1, a syntrophic alkane-degrading bacteria, enriched from the Shengli Oil field. Microbes Environ. 2017.
40. Siddique T, Penner T, Semple K, Foght JM. Anaerobic biodegradation of longer-chain *n*-alkanes coupled to methane production in oil sands tailings. Environ Sci Technol. 2011;45:5892–9.
41. Penger J, Conrad R, Blaser M. Stable carbon isotope fractionation by methylotrophic methanogenic archaea. Appl Environ Microbiol. 2012;78:7596–602.
42. Whiticar MJ, Faber E, Schoell M. Biogenic methane formation in marine and freshwater environments: CO_2 reduction vs. acetate fermentation—isotope evidence. Geochimica et Cosmochimica Acta. 1986;50:693–709.
43. Oremland RS, Polcin S. Methanogenesis and sulfate reduction: competitive and noncompetitive substrates in estuarine sediments. Appl Environ Microbiol. 1982;44:1270–6.
44. Visser A, Beeksma I, van der Zee F, Stams AJM, Lettinga G. Anaerobic degradation of volatile fatty acids at different sulphate concentrations. Appl Microbiol Biotechnol. 1993;40:549–56.
45. Madden P, Al-Raei A, Enright AM, Chinalia FA, De Beer D, O'Flaherty V, Collins G. Effect of sulphate on low-temperature anaerobic digestion. Front Microbiol. 2014;5.
46. Vallero MVG, Treviño RHM, Paulo PL, Lettinga G, Lens PNL. Effect of sulfate on methanol degradation in thermophilic (55°C) methanogenic UASB reactors. Enzyme Microb Technol. 2003;32:676–87.
47. DeWeerd KA, Concannon F, Suflita JM. Relationship between hydrogen consumption, dehalogenation, and the reduction of sulfur oxyanions by Desulfomonile tiedjei. Appl Environ Microbiol. 1991;57:1929–34.
48. Lovley DR, Dwyer DF, Klug MJ. Kinetic analysis of competition between sulfate reducers and methanogens for hydrogen in sediments. Appl Environ Microbiol. 1982;43:1373–9.
49. Isa Z, Grusenmeyer S, Verstraete W. Sulfate reduction relative to methane production in high-rate anaerobic digestion: microbiological aspects. Appl Environ Microbiol. 1986;51:580–7.
50. Hu Y, Jing Z, Sudo Y, Niu Q, Du J, Wu J, Li Y-Y. Effect of influent COD/SO_4^{2-} ratios on UASB treatment of a synthetic sulfate-containing wastewater. Chemosphere. 2015;130:24–33.
51. Li Y-Y, Lam S, Fang HHP. Interactions between methanogenic, sulfate-reducing and syntrophic acetogenic bacteria in the anaerobic degradation of benzoate. Water Res. 1996;30:1555–62.
52. Jeong T-Y, Chung H-K, Yeom SH, Choi SS. Analysis of methane production inhibition for treatment of sewage sludge containing sulfate using an anaerobic continuous degradation process. Korean J Chem Eng. 2009;26:1319–22.
53. Mizuno O, Li YY, Noike T. Effects of sulfate concentration and sludge retention time on the interaction between methane production and sulfate reduction for butyrate. Water Sci Technol. 1994;30:45–54.
54. Conrad R. Quantification of methanogenic pathways using stable carbon isotopic signatures: a review and a proposal. Org Geochem. 2005;36:739–52.
55. Conrad R, Klose M. Stable carbon isotope discrimination in rice field soil during acetate turnover by syntrophic acetate oxidation or acetoclastic methanogenesis. Geochimica et Cosmochimica Acta. 2011;75:1531–9.
56. Valentine DL, Chidthaisong A, Rice A, Reeburgh WS, Tyler SC. Carbon and hydrogen isotope fractionation by moderately thermophilic methanogens. Geochimica et Cosmochimica Acta. 2004;68:1571–90.
57. Penning H, Plugge CM, Galand PE, Conrad R. Variation of carbon isotope fractionation in hydrogenotrophic methanogenic microbial cultures and environmental samples at different energy status. Glob Change Biol. 2005;11:2103–13.
58. Liu Y, Whitman WB. Metabolic, phylogenetic, and ecological diversity of the methanogenic Archaea. Ann NY Acad Sci. 2008;1125:171–89.
59. Penning H, Claus P, Casper P, Conrad R. Carbon isotope fractionation during acetoclastic methanogenesis by *Methanosaeta concilii* in culture and a lake sediment. Appl Environ Microbiol. 2006;72:5648–52.
60. Penning H, Conrad R. Quantification of carbon flow from stable isotope fractionation in rice field soils with different organic matter content. Org Geochem. 2007;38:2058–69.
61. Ozuolmez D, Na H, Lever MA, Kjeldsen KU, Jørgensen BB, Plugge CM. Methanogenic archaea and sulfate reducing bacteria co-cultured on acetate: teamwork or coexistence? Front Microbiol. 2015;6:492.
62. Cheng L, Qiu TL, Li X, Wang WD, Yin X, Deng Y, Zhang H. Isolation and characterization of *Methanoculleus receptaculi* sp. nov. from Shengli oil field, P. R. China. FEMS Microbiol Lett. 2008;285:65–71.
63. Imachi H, Sakai S, Sekiguchi Y, Hanada S, Kamagata Y, Ohashi A, Harada H. *Methanolinea tarda* gen. nov., sp. nov., a methane-producing archaeon isolated from a methanogenic digester sludge. Int J Syst Evol Microbiol. 2008;58:294–301.
64. Sakai S, Imachi H, Sekiguchi Y, Tseng I-C, Ohashi A, Harada H, Kamagata Y. Cultivation of methanogens under low hydrogen conditions by using the coculture method. Appl Environ Microbiol. 2009;75:4892–6.
65. Nobu MK, Narihiro T, Rinke C, Kamagata Y, Tringe SG, Woyke T, Liu W-T. Microbial dark matter ecogenomics reveals complex synergistic networks in a methanogenic bioreactor. ISME J. 2015;9:1710–22.
66. Callaghan AV, Morris BEL, Pereira IAC, McInerney MJ, Austin RN, Groves JT, Kukor JJ, Suflita JM, Young LY, Zylstra GJ, Wawrik B. The genome sequence of *Desulfatibacillum alkenivorans* AK-01: a blueprint for anaerobic alkane oxidation. Environ Microbiol. 2012;14:101–13.
67. Kuever J. The Family Desulfobacteraceae. In: Rosenberg E, DeLong EF, Lory S, Stackebrandt E, Thompson F, editors. The Prokaryotes: Deltaproteobacteria and Epsilonproteobacteria. Springer: Berlin; 2014. p. 45–73.
68. Powell GE. Interpreting gas kinetics of batch cultures. Biotechnol Lett. 1983;5:437–40.
69. Ma TT, Cheng L, Zheng ZZ, Qin Q, Dai LR, Zhang H. Effects of pH buffers on methanogenesis and methanogenic community in the cultures amended with acetate. Acta Microbiologica Sinica (In Chinese). 2014;54:1453–61.
70. So CM, Young LY. Isolation and characterization of a sulfate-reducing bacterium that anaerobically degrades alkanes. Appl Environ Microbiol. 1999;65:2969–76.
71. Li J, Rui J, Pei Z, Sun X, Zhang S, Yan Z, Wang Y, Liu X, Zheng T, Li X. Straw- and slurry-associated prokaryotic communities differ during co-fermentation of straw and swine manure. Appl Microbiol Biotechnol. 2014;98:4771–80.
72. Magoc T, Salzberg SL. FLASH: fast length adjustment of short reads to improve genome assemblies. Bioinformatics. 2011;27:2957–63.
73. Bolger AM, Lohse M, Usadel B. Trimmomatic: a flexible trimmer for Illumina sequence data. Bioinformatics. 2014;30:2114–20.
74. Edgar RC, Haas BJ, Clemente JC, Quince C, Knight R. UCHIME improves sensitivity and speed of chimera detection. Bioinformatics. 2011;27:2194–200.
75. Caporaso JG, Kuczynski J, Stombaugh J, Bittinger K, Bushman FD, Costello EK, Fierer N, Pena AG, Goodrich JK, Gordon JI. QIIME allows analysis of high-throughput community sequencing data. Nat Meth. 2010;7:335–6.
76. Wang Q, Garrity GM, Tiedje JM, Cole JR. Naïve bayesian classifier for rapid assignment of rRNA sequences into the new bacterial taxonomy. Appl Environ Microbiol. 2007;73:5261–7.

10

In vitro oxidative decarboxylation of free fatty acids to terminal alkenes by two new P450 peroxygenases

Huifang Xu[1,2], Linlin Ning[1,2,4], Wenxia Yang[1,2], Bo Fang[1,2], Cong Wang[1,2], Yun Wang[2,3], Jian Xu[2,3], Severine Collin[5], Frederic Laeuffer[5], Laurent Fourage[5] and Shengying Li[1,2*]

Abstract

Background: P450 fatty acid decarboxylases represented by the unusual CYP152 peroxygenase family member $OleT_{JE}$ have been receiving great attention recently since these P450 enzymes are able to catalyze the simple and direct production of 1-alkenes for potential applications in biofuels and biomaterials. To gain more mechanistic insights, broader substrate spectra, and improved decarboxylative activities, it is demanded to discover and investigate more P450 fatty acid decarboxylases.

Results: Here, we describe for the first time the expression, purification, and in vitro biochemical characterization of two new CYP152 peroxygenases, CYP-Aa162 and CYP-Sm46Δ29, that are capable of decarboxylating straight-chain saturated fatty acids. Both enzymes were found to catalyze the decarboxylation and hydroxylation of a broad range of free fatty acids (C_{10}–C_{20}) with overlapping substrate specificity, yet distinct chemoselectivity. CYP-Sm46Δ29 works primarily as a fatty (lauric) acid decarboxylase (66.1 ± 3.9% 1-undecene production) while CYP-Aa162 more as a fatty (lauric) acid hydroxylase (72.2 ± 0.9% hydroxy lauric acid production). Notably, the optical spectroscopic analysis of functional CYP-Sm46Δ29 revealed no characteristic P450 band, suggesting a unique heme coordination environment. Active-site mutagenesis analysis showed that substitution with the proposed key decarboxylation-modulating residues, His85 and Ile170, enhanced the decarboxylation activity of CYP-Aa162 and $P450_{BS\beta}$, emphasizing the importance of these residues in directing the decarboxylation pathway. Furthermore, the steady-state kinetic analysis of CYP-Aa162 and CYP-Sm46Δ29 revealed both cooperative and substrate inhibition behaviors which are substrate carbon chain length dependent.

Conclusions: Our data identify CYP-Sm46Δ29 as an efficient $OleT_{JE}$-like fatty acid decarboxylase. Oxidative decarboxylation chemoselectivity of the CYP152 decarboxylases is largely dependent upon the carbon chain length of fatty acid substrates and their precise positioning in the enzyme active site. Finally, the kinetic mode analysis of the enzymes could provide important guidance for future process design.

Keywords: Alkenes, Biofuels, P450 fatty acid decarboxylases, Substrate specificity, Chemoselectivity, Site-directed mutagenesis, Enzyme kinetics

Background

Cytochrome P450 (CYP) enzymes are a superfamily of important biocatalysts that perform an extraordinary breadth of biochemical reactions [1–3], and have been widely applied to oxidation of complex organic substrates and biosynthesis of chemical building block molecules [4–6]. Among diverse P450 products, terminal alkenes (i.e., α-alkenes or 1-alkenes) have significant industrial potential and outstanding economic importance, given that these hydrocarbons highly mimic the chemical composition and physical properties of fossil fuels

*Correspondence: lishengying@qibebt.ac.cn
[1] Shandong Provincial Key Laboratory of Synthetic Biology, Qingdao Institute of Bioenergy and Bioprocess Technology, Chinese Academy of Sciences, No. 189 Songling Road, Qingdao 266101, Shandong, China
Full list of author information is available at the end of the article

[7–10]. There has been widespread biotechnological interest focusing on the identification and reconstitution of enzymes and pathways capable of synthesizing biohydrocarbons in microbial hosts [7, 9, 11–17]. A sustainable biosynthetic route to 1-alkenes from biologically abundant feedstocks such as free fatty acids (FFAs) clearly represents a promising pathway. In 2011, Rude et al. [13] reported the first one-step enzymatic oxidative decarboxylation of FFAs by a P450 peroxygenase $OleT_{JE}$, isolated from *Jeotgalicoccus* sp. ATCC 8456, to yield long-chain (C_{13}–C_{19}) 1-alkenes.

In view of its tremendous potential for biological production of terminal olefins, $OleT_{JE}$ since its discovery has captured great attention within academia and industry. A fast growing number of studies have been carried out to optimize $OleT_{JE}$ catalysis and to understand its catalytic mechanism [16, 18–28]. These include the development of alternative catalytic systems (e.g., the O_2/NAD(P)H/redox partners system [16] and the light-driven H_2O_2 generation system [21]) (Scheme 1), the determination of $OleT_{JE}$ crystal structures [18, 28], the elucidation of catalytically reactive species [19, 20], and the expansion of substrate scope to structurally different aromatic carboxylic acids [29], dionic acids (to produce dienes) [30], and even some unnatural substrates including styrene, nonane, and cyclohexane [31].

Mechanistically, a majority of P450s use O_2 and NAD(P)H-dependent redox systems to catalyze diverse oxidative/monooxygenation reactions [32, 33]. However, there are a small number of P450s that evolved to naturally utilize H_2O_2 as the sole oxygen and electron donor to catalyze oxidative reactions through the so-called peroxide shunt pathway [33–35]. These enzymes are referred to as P450 peroxygenases. The best-characterized P450 peroxygenases are among the CYP152 [36] family members, including $P450_{BS\beta}$ (CYP152A1) from *Bacillus subtilis* [35] and $P450_{SP\alpha}$ (CYP152B1) from *Sphingomonas paucimobilis* [34]. $OleT_{JE}$, based on its amino acid sequence similarity, was designated as CYP152L1 [13, 18]. Within the CYP152 members that have been biochemically characterized so far, $P450_{SP\alpha}$ catalyzes exclusively the Cα hydroxylation of FFAs [34], and $P450_{BS\beta}$ generates both α- and β-hydroxy fatty acids as major products but with a small amount of 1-alkene as well [13, 35, 37]. A recently characterized CYP-MP peroxygenase [24] is able to introduce the hydroxyl group at α-, β-, γ-, δ-, and ε-positions of C_{12}–C_{18} fatty acids, but it only displayed marginal decarboxylation activity against myristic acid (C_{14}) and palmitic acid (C_{16}). Heretofore, $OleT_{JE}$ is the only one to predominantly catalyze the decarboxylation of long-chain fatty acids generating 1-alkenes as the major products and α- and β-hydroxy fatty acids as the side products (Scheme 1) [13, 16, 22]. It would be of great interest and importance to explore more P450 fatty acid decarboxylases and expand the enzyme inventory for future broader biotechnological applications.

In an attempt to screen for biocatalysts capable of converting medium-chain C_{12} fatty acid (lauric acid) to 1-undecene for potential applications in biofuels and biolubricants, two new P450 sequences, CYP-Aa162 (CYP152A8, GenBank Accession Number: WP_008340313) from *Alicyclobacillus acidocaldarius* LAA1 and CYP-Sm46 (CYP152L2, GenBank Accession Number: EKU50422) from *Staphylococcus massiliensis S46*, were identified apart from $OleT_{JE}$ and $P450_{BS\beta}$. To better harness these two uncharacterized CYP152 family members for future industrial applications, and to further our understanding of the unique decarboxylation mechanism employed by CYP152 decarboxylases, we sought to characterize the detailed biochemical properties of the two enzymes.

In this study, we cloned, expressed, and purified CYP-Aa162 and CYP-Sm46 enzymes and in vitro characterized their catalytic activities towards a panel of different carbon chain length fatty acid substrates. Analytical data demonstrated that CYP-Aa162 acts mainly as a $P450_{BS\beta}$-like fatty acid hydroxylase, whereas CYP-Sm46 acts as an efficient $OleT_{JE}$-like fatty acid decarboxylase. Their substrate conversion rates and regio- and chemoselectivity exhibited significant substrate carbon chain length dependence. In addition, our enzymatic kinetic

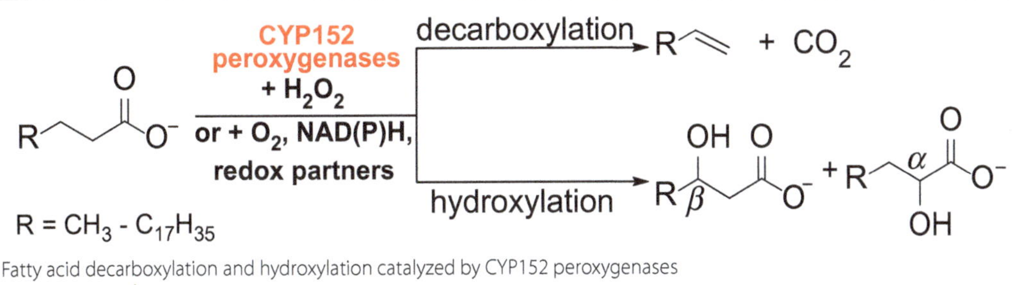

Scheme 1 Fatty acid decarboxylation and hydroxylation catalyzed by CYP152 peroxygenases

analysis suggested that the metabolism of fatty acids by the CYP152 enzymes could undergo both cooperative and substrate inhibition behaviors, providing important information and guidance for future process design.

Results

Heterologous expression and purification of CYP-Aa162 and CYP-Sm46Δ29

According to protein sequence alignment (Additional file 1: Figure S1), CYP-Aa162 has 64% (39%) sequence identity to $P450_{BS\beta}$ ($OleT_{JE}$), while CYP-Sm46 is 65% (37%) identical to $OleT_{JE}$ ($P450_{BS\beta}$). Heterologous expression of CYP-Aa162 in *E. coli* (with its encoding sequence codon-optimized) led to an appreciable level of the recombinant protein purified to near homogeneity (Additional file 1: Figure S2). However, for CYP-Sm46, the original construct of pET28b-*CYP-Sm46* (codon-optimized) showed very poor expression and solubility, which did not improve upon subcloning into an alternative expression vector pCWori [38] or by optimization of the expression conditions. This prompted us to re-examine the protein sequence of EKU50422 and its annotation. We found that this sequence has an extra long (~29 amino acids) *N* terminus with no functional annotations when compared with most other CYP152 members including those cytochrome P450 family proteins from other *Staphylococcus* species (Additional file 1: Figure S1). Analyses of the CYP-Sm46 protein sequence using several protein prediction tools such as Phobius, TMHMM 2.0, and SignalP 4.1 [39–41] also showed no positive return of either a signal peptide or transmembrane helices within the *N*-terminal region of 60 amino acids. A further BLAST search revealed a recently added (17-Jan-2015) 'truncated' entry of CYP-Sm46 (GenBank Accession Number: WP_039990689), hereafter called CYP-Sm46Δ29, which decodes from an alternative start codon (GUG) and therefore has the 29 *N*-terminal amino acids removed. Subsequent expression of this shorter version CYP-Sm46Δ29 led to significantly improved protein production and solubility, and the protein was then purified to near homogeneity (Additional file 1: Figure S2). The codon-optimized gene sequences of CYP-Aa162 and CYP-Sm46Δ29, and their corresponding amino acid sequences are shown in Additional file 2.

UV–visible spectroscopic properties of CYP-Aa162 and CYP-Sm46Δ29

The resting ferric form of CYP-Aa162 (Fig. 1a) exhibited a typical low-spin (LS) water-bound P450 heme spectrum with its Soret absorption peak at 417 nm and the smaller α- and β-bands in the visible region at ~566 and ~538 nm, respectively. These values are similar to those of $OleT_{JE}$ (418, 535, 566 nm) [18] and other LS bacterial P450s (e.g., the heme domain of CYP102A1 ($P450_{BM3}$) from *Bacillus megaterium* with maxima at 418, 534, and 568 nm) [42]. However, CYP-Aa162 displayed no obvious hyperporphyrin (split Soret) characteristic at ~360 nm, which is different from that of $OleT_{JE}$ [18]. A widely accepted criterion for the identification of cytochrome P450 is a Soret peak at around 450 nm in the reduced CO difference spectrum. Here, for CYP-Aa162, both the reduced CO difference spectrum and the absolute absorption spectrum of the dithionite-reduced Fe^{2+}–CO adducts featured a characteristic thiolate-ligated P450 Soret band at 447–448 nm (Fig. 1a and inset), which is similar to that of $P450_{BS\beta}$ with a peak at 446 nm [35].

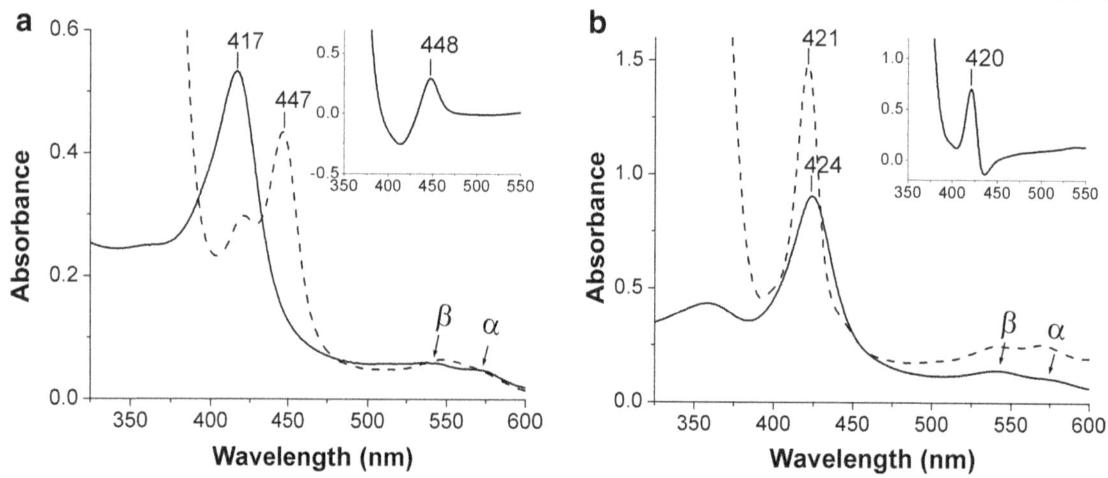

Fig. 1 UV–visible spectra of CYP-Aa162 (**a**) and CYP-Sm46Δ29 (**b**). The purified enzymes were diluted in 50 mM NaH_2PO_4 (pH 7.4) containing 300 mM NaCl and 10% glycerol, respectively. Spectra are shown for the oxidized ferric form of the CYPs (*solid lines*) and the $Na_2S_2O_4$-reduced ferrous-CO complex of the CYPs (*dashed lines*). *Insets* show the reduced CO difference spectrum of each enzyme

However, the purified ferric form of CYP-Sm46Δ29 displayed a red-shifted Soret band at 422–424 nm, followed by a small but prominent β band at 540 nm and a significantly weaker and flat α band at 571 nm (Fig. 1b). Prominent hyperporphyrin characteristics (split Soret) are visible at 360 nm. These features are similar to those reported for CYP-MP (maxima at 422, 545, and 575 nm) [24], but different from the typical 418, ~535, and ~568 nm maxima for LS water-bound P450s [18, 42]. These altered spectroscopic features may indicate an imidazole-coordinated or more likely a distal hydroxyl-coordinated ferric LS heme form as suggested in the CYP-MP [24]. Furthermore, much to our surprise, reduction of the CYP-Sm46Δ29 with sodium dithionite in the presence of CO revealed no characteristic P450 peak at ~450 nm. The maximum absorption of this Fe^{2+}–CO adduct was still at ~421 nm with enhanced α- and β-bands. Correspondingly, the reduced CO difference spectrum of CYP-Sm46Δ29 also exhibited only the P420 peak (Fig. 1b inset). Attempts to produce a detectable P450 species by altering the amount of dithionite, buffer pH, and by substrate-aided stabilization of any, if exists, P450-CO adduct as the case with P450 EpoK enzyme [43] all proved negative (Additional file 1: Figure S3). The similar 'P420' spectra were observed in all preparations of the enzyme unanimously. These data suggest that CYP-Sm46Δ29 is somehow incapable of forming a detectable cysteine thiolate-ligated Fe^{2+}–CO adduct. We speculate that this unusual spectroscopic feature may arise either from simple protonation of the proximal cysteine thiolate to a neutral thiol form [43, 44], or from dissociation of the thiolate ligand from the heme iron, followed by coordination of an alternative proximal ligand such as a histidine in the case with the pressure-induced $P420_{cam}$ [45, 46], or from other reduction-linked structural alterations that prevent the formation of thiolate-ligated ferrous-CO species. Nevertheless, these altered spectroscopic features did not affect the enzyme's peroxygenase activity as shown below.

Inability of CYP-Aa162 and CYP-Sm46Δ29 to elicit monooxygenase activity

To check if a class I electron transfer system could afford a P450 Soret feature in CYP-Aa162 and CYP-Sm46Δ29 [47], we examined the UV–visible spectra of the two enzymes in an NAD(P)H/redox proteins/CO system. As a result, while the control P450 monooxygenase $P450_{BM3}$ was successfully reduced by the addition of NADPH and exhibited an almost complete transition to P450 spectrum in the presence of lauric acid (LA) substrate (Additional file 1: Figure S4), neither CYP-Aa162 nor CYP-Sm46Δ29 showed a P450 chromophore after incubation with the redox partner proteins CamAB (i.e., putidaredoxin reductase and putidaredoxin that are able to support the in vitro activity of both P450 enzymes, see below) and their optimal cofactor NADH in the presence of LA (Fig. 2a). Further catalytic activity assay indicated that the two enzymes exhibited appreciable LA conversion activity in the H_2O_2 system. However, their activity was dramatically diminished in the O_2/CamAB/NADH reaction system when high concentrations of catalase were added to remove H_2O_2 potentially generated from the decoupling of NADH oxidation (Fig. 2b), implying that CYP-Aa162 and CYP-Sm46Δ29 have very little or

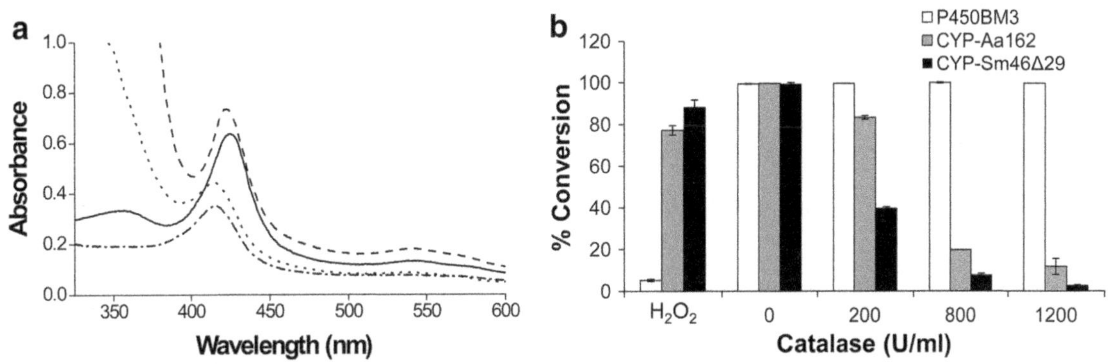

Fig. 2 UV–visible spectra and lauric acid (LA) conversion activity of CYP-Aa162 and CYP-Sm46Δ29 in the O_2/redox partners/NAD(P)H system. **a** The UV–visible spectra are, respectively, shown for the substrate LA-bound ferric form of CYP-Sm46Δ29 with a Soret band at 422 nm (*solid line*); the CamAB/NADH 'reduced' CO-bound form of CYP-Sm46Δ29 with the Soret peak still at ~422 nm (*dashed line*); the substrate LA-bound ferric form of CYP-Aa162 with a Soret band at ~418 nm (*dashed and dotted line*); the CamAB/NADH 'reduced' CO-bound form of CYP-Aa162 with an unshifted Soret peak (*dotted line*). **b** Enzymatic conversion of lauric acid by CYP-Aa162, CYP-Sm46Δ29, and the control $P450_{BM3}$ monooxygenase in the O_2/redox proteins/NAD(P)H system in the absence or presence of different concentrations of catalase. For CYP-Aa162 and CYP-Sm46Δ29, CamAB/NADH was used for the electron transfer cascade. For the self-sufficient $P450_{BM3}$ enzyme, NADPH was used as the electron donor. Conversions of the substrate in the H_2O_2 cofactor system were included as controls

no monooxygenase activity. The marginal residual activity of the two enzymes shown at 1200 U mL^{-1} catalase condition may represent some inevitable leakage of H_2O_2 to the P450 proteins, as even in the absence of putidaredoxin there was still meager activity detectable at this high catalase concentration (Additional file 1: Figure S5).

Substrate specificity and chemoselectivity of CYP-Aa162 and CYP-Sm46Δ29

Members in the CYP152 peroxygenase family are capable of utilizing H_2O_2 as the sole electron and oxygen donor to carry out catalytic reactions [33]. Among those CYP152s that have been biochemically characterized so far, $OleT_{JE}$ catalyzes predominantly the decarboxylation of long-chain fatty acids to give 1-alkenes as the main products and α- and β-hydroxy fatty acids as the side products [13]. Other CYP152 enzymes including $P450_{SP\alpha}$, $P450_{BS\beta}$, and CYP-MP catalyze primarily the hydroxylation of free fatty acids, introducing the hydroxyl group at varying carbon positions (i.e., α-, β-, γ-, δ-, or ε-position) [24, 34, 35, 37]. Here, in order to investigate the substrate specificity and chemoselectivity (decarboxylation vs hydroxylation) of the two new CYP152 enzymes, we set out to examine their catalytic reactions towards a range of different carbon chain length (C_{10}–C_{20}) FFAs using H_2O_2 as the sole cofactor.

As a result, both enzymes exhibited quite similar substrate preference profiles with lauric acid (C_{12}) being the best substrate (Fig. 3). Its conversion ratios were 77.3 ± 1.0 and 83.5 ± 3.4% by CYP-Aa162 and CYP-Sm46Δ29, respectively. In addition, CYP-Sm46Δ29 also converted myristic acid (C_{14}) efficiently (74.5 ± 3.5%). However, in terms of decarboxylation, CYP-Aa162 showed only marginal alkene production from its best substrate C_{12} (5.1 ± 0.1%), whereas CYP-Sm46Δ29 exhibited significantly higher percentage of alkene production. The optimal substrates C_{12} and C_{14} FFAs reached the alkene production efficiency of 66.1 ± 3.9 and 57.4 ± 3.6%, respectively. Unlike previously reported for $OleT_{JE}$ by others [13, 21], CYP-Sm46Δ29 favors more the medium-chain length fatty acids (C_{12} or C_{14}) rather than the long ones (C_{16}–C_{20}). The decarboxylation activity of CYP-Sm46Δ29 (Fig. 3) and the ratio of decarboxylation over hydroxylation activities (Additional file 1: Figure S6) gradually reduced after C_{12} with the increase in substrate carbon chain length, and the hydroxyl FAs became the major products in the reactions with C_{18} or C_{20} as a substrate, indicating that different substrate binding modes are likely employed by this enzyme.

In consideration of the high sequence identity (64%) and similar UV–visible absorption maxima between CYP-Aa162 and $P450_{BS\beta}$, we compared their catalytic activities in parallel. As shown in Fig. 4, the two enzymes exhibited very similar substrate preference spectra and total fatty acid conversion efficiency. However, $P450_{BS\beta}$ showed generally higher alkene production ratios compared with CYP-Aa162. The best alkene-producing substrate lauric acid (C_{12}) displayed a fivefold higher 1-undecene production (27.7 ± 0.9% versus 5.1 ± 0.1%). Nonetheless, substrate hydroxylation still represented the major chemistry of the two enzymes, which is not surprising since $P450_{BS\beta}$ was initially identified as a P450 fatty acid hydroxylase [35]. The hydroxylation versus decarboxylation ratios of the two CYPs towards their preferred C_{10}–C_{14} substrates were at least 2.5:1. Thus, we designated CYP-Aa162 as a $P450_{BS\beta}$-like fatty acid hydroxylase rather than a P450 decarboxylase.

Furthermore, to comparatively investigate the detailed product distribution, the C_{14} substrate myristic acid was reacted with the four enzymes CYP-Aa162, CYP-Sm46Δ29, $P450_{BS\beta}$, and $OleT_{JE}$, respectively. The results (Table 1) revealed that CYP-Aa162 formed

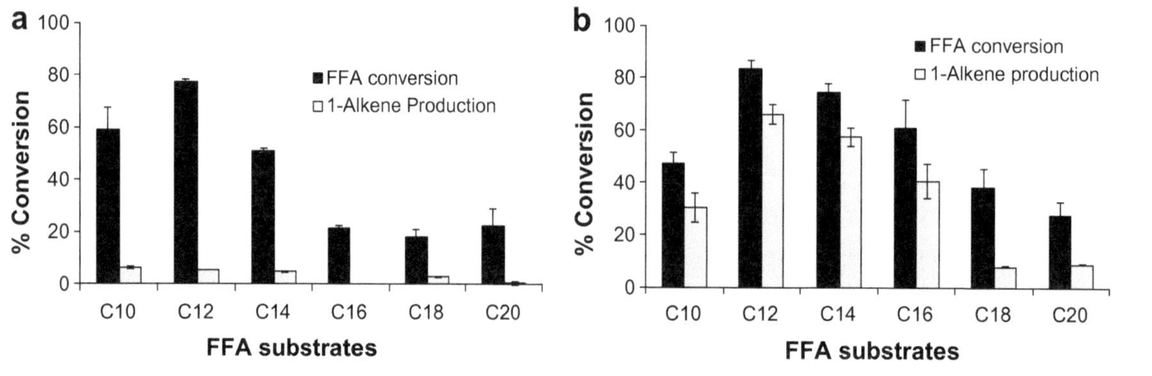

Fig. 3 Substrate preference and 1-alkene production profiles of CYP-Aa162 (**a**) and CYP-Sm46Δ29 (**b**). Substrate preference was determined by calculating the percentage conversion of each fatty acid substrate. The corresponding 1-alkene production represents the percentage of 1-alkene yield over the starting substrate. Results shown are mean ± SD of duplicated experiments

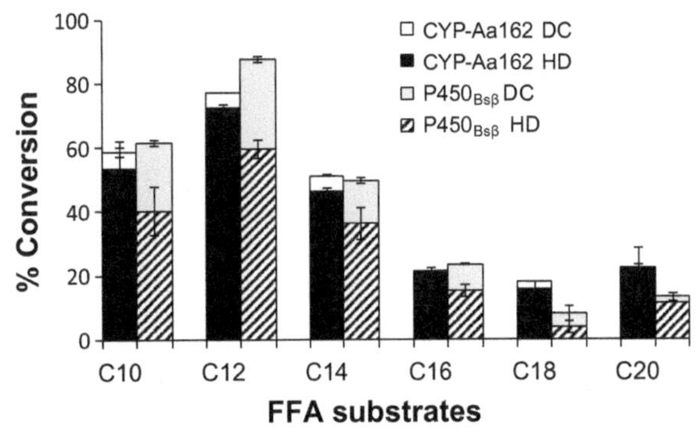

Fig. 4 Decarboxylation (DC) versus hydroxylation (HD) activities of CYP-Aa162 and $P450_{BS\beta}$. The hydroxylation activity was estimated by subtracting the alkene production from the total substrate conversion. This indirect but more convenient method was validated with C_{14} myristic acid substrate by direct measurement of its derivatized hydroxylation products. Results are shown as mean ± SD of duplicated experiments

Table 1 GC–MS analysis of catalytic activities and product distribution profiles of the selected CYP152 peroxygenases towards myristic acid (C_{14})

Enzymes	Conversion (%)	Product distribution (%)		
		1-tridecene	α-OH–C_{14}	β-OH–C_{14}
CYP-Aa162	60.1 ± 2.1	16.0 ± 1.2	78.9 ± 1.4	5.1 ± 0.6
$P450_{BS\beta}$	62.8 ± 5.0	36.8 ± 3.5	30.0 ± 4.2	33.2 ± 3.5
CYP-Sm46Δ29	73.2 ± 8.9	78.4 ± 2.9	0.4 ± 0.1	21.2 ± 0.9
$OleT_{JE}$	71.2 ± 2.9	91.8 ± 5.9	0.3 ± 0.1	7.9 ± 1.5

predominantly (78.9 ± 1.4%) the α-hydroxy myristic acid (α-OH–C_{14}), and $P450_{BS\beta}$ exhibited approximately even distribution among the three products α-OH–C_{14}, β-OH–C_{14}, and 1-tridecene. In comparison, CYP-Sm46Δ29 and $OleT_{JE}$ showed strong decarboxylation chemoselectivity. For these two enzymes, 1-tridecene was the major product, β-OH–C_{14} the minor product, while α-OH–C_{14} only accounted for a negligible proportion of the total products. In all cases, no distal hydroxylated fatty acids beyond C_β position were detected, as opposed to the mixed regioselectivity found with the CYP-MP enzyme [24]. Together, these data strongly support the notion that CYP-Aa162 functions as a $P450_{BS\beta}$-like fatty acid hydroxylase, while CYP-Sm46Δ29 acts as an efficient $OleT_{JE}$-like decarboxylase, providing another enzyme to the unusual fatty acid oxidative decarboxylation reaction (Additional file 1: Figure S7).

Site-directed mutagenesis of CYP-Aa162 and $P450_{BS\beta}$ fatty acid hydroxylases

The active-site sequence alignment of CYP-Aa162 and CYP-Sm46Δ29 with those of $P450_{BS\beta}$ and $OleT_{JE}$ revealed two major residue differences (Additional file 1: Figure S8): the histidine and isoleucine residues in the FFA decarboxylases of $OleT_{JE}$ (H85 and I170) and CYP-Sm46Δ29 (H86 and I171) are replaced by a glutamine and a valine in CYP-Aa162 (Q85 and V170) and $P450_{BS\beta}$ (Q85 and V170), respectively. According to the crystal structure of $OleT_{JE}$ (PDB: 4L40) [18], H85 and I170 are positioned at the two sides of the substrate carboxyl group with the distance of 5.1 and 3.4 Å, respectively. Previously, Rude et al. [13] reported an improved decarboxylation activity (~50%) of $P450_{BS\beta}$ Q85H variant towards palmitic acid (C_{16}), suggesting an important role of this histidine residue in regulating the decarboxylation chemoselectivity of $OleT_{JE}$. Grant et al. [20] also proposed mechanisms in which H85 of $OleT_{JE}$ may act as a proton donor to facilitate the protonation of $OleT_{JE}$-Compound II intermediate (Fe^{4+}–OH) generated from substrate C–H abstraction and thereby to initiate substrate C–C_α bond scission (i.e., decarboxylation). Moreover, our recent saturation mutagenesis analysis of $OleT_{JE}$ at H85 and I170 sites also confirmed the importance of these two amino acid residues for decarboxylation activity and substrate positioning based on the finding that all H85X (X denotes any natural amino acid) and I170X mutants exhibited abolished decarboxylation activity towards myristic acid [26].

Therefore, in view of the potential key roles of these two residues in regulating decarboxylation process and substrate positioning, we decided to substitute the corresponding amino acids in CYP-Aa162 and $P450_{Bs\beta}$ with His and Ile, and created the double mutants Aa162/Q85H/V170I and $P450_{Bs\beta}$/Q85H/V170I. Their in vitro activities towards a series of free fatty acid substrates (C_{10}–C_{20}) were evaluated along with their corresponding wild types. As a result, the $P450_{Bs\beta}$/Q85H/V170I mutant exhibited a general increase in the total substrate conversion ratios compared with its wild type, while Aa162/Q85H/V170I showed variations of increase or decrease in conversion efficiency towards different carbon chain length substrates (Fig. 5c). However, both mutants displayed enhanced decarboxylation activity to the tested substrates to different extents (Fig. 5d). This increase was more prominent in the longer carbon chain length substrates such as C_{16} palmitic acid (Fig. 5b) than in the shorter carbon chain length substrates such as C_{12} lauric acid (Fig. 5a). For C_{12} substrate, a 1.1- and a 0.9-fold increase in 1-undecene production were observed with Aa162/Q85H/V170I and $P450_{Bs\beta}$/Q85H/V170I mutant, accounting for a 1.7- and a 1.2-fold increase in the ratio of decarboxylation over hydroxylation activity, respectively (Fig. 5a). For C_{16} substrate, the production of 1-pentadecene exhibited an increase from undetectable by the wild type CYP-Aa162 to 6.9% by its double mutant Aa162/Q85H/V170I, and an increase of

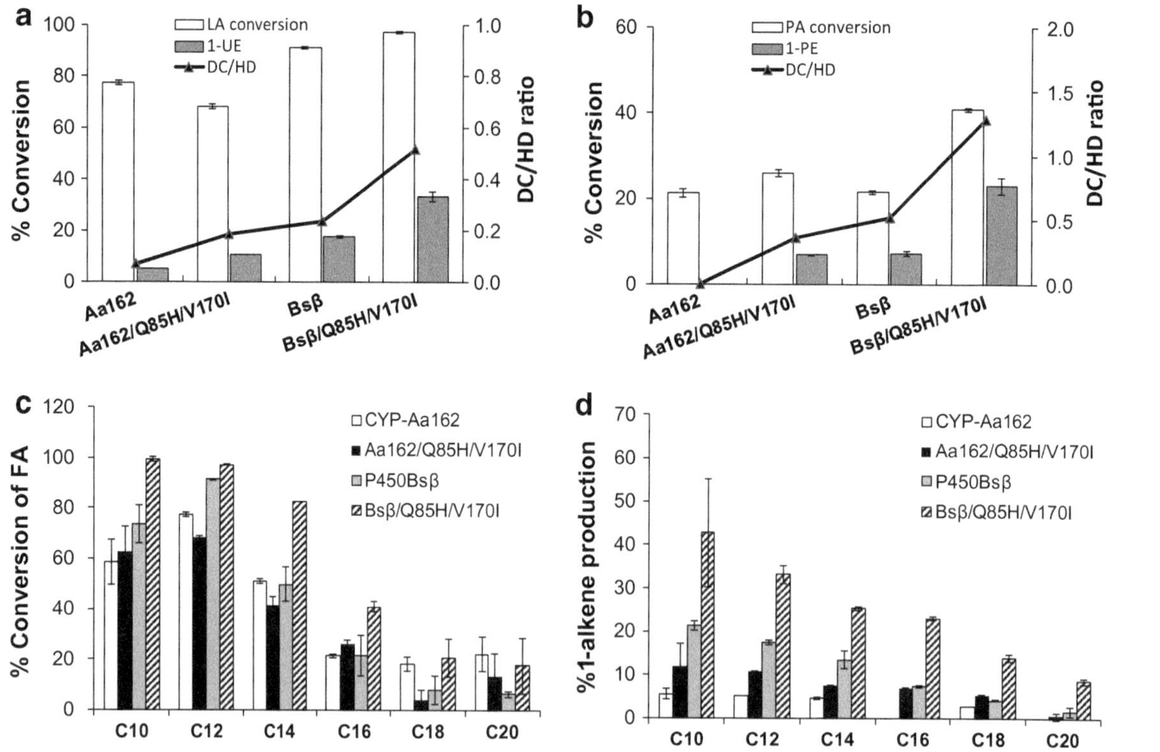

Fig. 5 Fatty acid conversion by the Aa162/Q85H/V170I and $P450_{Bs\beta}$/Q85H/V170I mutants compared to their corresponding wild-type enzymes. The hydroxylation (HD) activity was deduced by subtracting the 1-alkene production (corresponding to DC activity) from the total substrate conversion. **a** Lauric acid (LA) conversion and 1-undecene (1-UE) production; **b** palmitic acid (PA) conversion and 1-pentadecene (1-PE) production; **c** conversion ratios of different carbon chain length fatty acids (C_{10}–C_{20}) and **d** their corresponding 1-alkene production by the indicated enzymes. Results are presented as mean ± SD of duplicated experiments

2.1-fold by the $P450_{Bs\beta}$/Q85H/V170I mutant when compared with its wild type $P450_{Bs\beta}$. The latter increase led to a high decarboxylation versus hydroxylation ratio of around 1.3 in the $P450_{Bs\beta}$/Q85H/V170I mutant, making the 1-alkene the major product. Together, these results demonstrate the important role of the active-site histidine and isoleucine in regulating fatty acid decarboxylation activity of the CYP152 peroxygenases.

Steady-state kinetics of CYP-Aa162 and CYP-Sm46Δ29

To better understand the catalytic properties of CYP-Aa162 fatty acid hydroxylase and CYP-Sm46Δ29 fatty acid decarboxylase, we determined their steady-state kinetic parameters (Table 2) towards their optimal substrates C_{12} and/or C_{14} fatty acids. Since CYP-Aa162 mainly catalyzes the hydroxylation of fatty acids, trace amounts of alkene products were undetectable at the beginning of the reaction when initial rates were determined. Therefore, its kinetic constants were derived from the rates of substrate consumption. For CYP-Sm46Δ29, the kinetic constants were specifically determined by 1-alkene formation rates due to its much higher decarboxylation activity. Interestingly, it was revealed that the kinetic data of both CYP-Aa162 and CYP-Sm46Δ29 towards C_{12} substrate fit well to a sigmoidal Hill equation with a Hill coefficient value around 2 (Table 2; Additional file 1: Figure S9A, B). This may suggest that the substrate binding pocket of the two P450 enzymes could accommodate more than one C_{12} molecules and that the binding of one C_{12} molecule could facilitate the binding of the other to the same P450 enzyme, particularly at low to medium substrate concentrations (10–80 μM). Notably, the kinetic data of CYP-Sm46Δ29 towards C_{14} substrate exhibited apparent and substantial substrate inhibition (Additional file 1: Figure S9C). When the substrate concentration was in excess of 60 μM, the rate of 1-tridecene formation declined dramatically, but was not completely abolished even at a high concentration of 300 μM. Rather, it flattened off to a low activity, indicating an incomplete inhibition. Estimation of the kinetic parameters for this substrate was therefore deduced from the truncated dataset using Michaelis–Menten equation after removal of the inhibited rates at high substrate concentrations [48]. The apparent k_{cat} value of 62.5 ± 4.5 min^{-1} is comparable

Table 2 Kinetic parameters of FFA conversion catalyzed by CYP-Aa162 and CYP-Sm46Δ29

Enzymes	C_{12}^{a}				C_{14}^{b}		
	k_{cat} (min^{-1})	K_H (uM)	K_{cat}/K_H (uM^{-1} min^{-1})	Hill coefficient (n)	k_{cat} (min^{-1})	K_m (uM)	K_{cat}/K_m (uM^{-1} min^{-1})
CYP-Aa162[c]	46.1 ± 17.1	93.1 ± 37.7	0.5	2.1 ± 0.8	nd	nd	nd
CYP-Sm46Δ29[d]	24.5 ± 6.3	58.4 ± 19.0	0.4	2.2 ± 1.0	62.5 ± 4.5	41.3 ± 5.1	1.5

nd not determined

[a] Data were fitted to Hill equation ($v = V\max[S]^n / (K_H^n + [S]^n)$, *n* Hill coefficient)

[b] Substantial substrate inhibition was observed with this substrate. Data shown here were derived from Michaelis–Menten equation after omitting the inhibited rates at high substrate concentrations

[c] Initial rates were measured by the amount of substrate consumption per μM enzyme per min

[d] Initial rates were measured by the amount of 1-alkene formation per μM enzyme per min. Data (shown as mean ± SE) were calculated using OriginPro 8.0 program

to that of $OleT_{JE}$ towards C_{14} substrate (71.0 ± 8.4 min^{-1}) [26]. Collectively, these data imply that, to different carbon chain length FFA substrates, the decarboxylase CYP-Sm46Δ29 could present distinct binding modes and therefore display different kinetic behaviors.

Discussion

In this study, we successfully expressed, purified, and biochemically characterized two new CYP152 peroxygenase family members, CYP-Aa162 and CYP-Sm46Δ29, that are capable of decarboxylating straight-chain saturated fatty acids to yield 1-alkenes to different extents. Previously, Chen et al. [49] compared the in vivo alkene production by expressing a His-tagged CYP-Sm46 (there called $OleT_{SM}$, GenBank Accession Number: WP_009381667, an identical sequence to EKU50422, 453 amino acids) in *S. cerevisiae*, and only observed trace amounts of odd-numbered carbon chain alkenes (C_{13}, C_{15}, and C_{17}) from a 48-h culture of the recombinant yeast cells (total alkene titer 1.4 μg L^{-1}, 40 times lower than that of $OleT_{JE}$ 54.5 μg L^{-1}). In that study, a 'full-length' version of CYP-Sm46 was used. No protein expression data were shown, nor was enzymatic characterization performed. Here, our experiment in *E. coli* proved very poor or no soluble expression of the 'full-length' version of CYP-Sm46. Translation in bacteria is most commonly initiated at an AUG start codon, which encodes for a methionine. However, in some species, the start codon can be GUG, which within the open reading frame (ORF) would encode for a valine. *Staphylococcus aureus* has been documented to use GUG as an alternative start codon [50]. The use of alternative initiation codons can influence translation efficiency, while incorrect initiation could lead to misfolded and therefore inactive proteins. The initial amino acid sequence of CYP-Sm46 (EKU50422) was most likely mistakenly derived from the usual AUG start codon as it happens to follow a putative Shine–Dalgarno sequence (AGGATG), which may explain why we did not observe decent expression with the original EKU50422 sequence. The later version of the amino acid sequence (CYP-Sm46Δ29) adopted GUG as its start codon (WP_039990689), and a possible Shine–Dalgarno site of 5′-TGGGGG-3′ was found 7 base upstream of the putative GUG start codon in the mRNA sequence (Additional file 2). The RBS calculator 2.0 (http://www.denovodna.com/software/forward) analysis of the CYP-Sm46 mRNA sequence showed that the GUG has a fivefold higher translation rate (3.3-fold lower total initiation energy ΔG_{Total}) than that of the upstream AUG start codon. In addition, McLaughlin et al. [51] reported the identification of a very strong Shine–Dalgarno complementarity containing five G-C base pairs in the *Staphylococcus aureus* β-lactamase gene. This strong mRNA-16S rRNA binding was proposed to reduce dependence on other initiation factors in the Gram-positive bacteria such as *Bacillus* and *Staphylococcus*. Furthermore, analysis of mRNA sequences of cytochrome P450 family proteins from other *Staphylococcus* species such as *S. agnetis* (GenBank: JPRT01000001.1) and *S. pseudintermedius* (GenBank: CP002439.1) also revealed a similar pattern of the strong 5–6 G-C pairs of Shine–Dalgarno complementarity preceding a GUG start codon. Taken together, we reason that CYP-Sm46Δ29 is most likely to be the native functional form of the protein.

P420 species was generally thought to be an inactive form of cytochrome P450 enzymes [52–54]. Formation of this 'inactive' P420 was suggested to involve the protonation of the native cysteine thiolate to a neutral thiol heme ligand [44], the recruitment of a histidine to replace the native cysteinate ligand [45], or other unknown reasons. Either way leads to the failure of forming the thiolate-ligated Fe^{2+} species or to a distorted/weakened heme–thiolate bond, thus underlying the P450 → P420 transition [45]. Most P420 studies were performed with the $P450_{BM3}$ [52, 54] or $P450_{cam}$ [45, 53] monooxygenases. These enzymes employ redox partner system to transfer electrons from NAD(P)H and activate O_2 as the oxidant, whose catalytic activity depends on the formation of a specific thiolate bond between the heme iron and the absolutely conserved cysteine residue in order to afford an active conformation state [33]. P450 peroxygenases, on the other hand, unlike those monooxygenases, utilize peroxide (such as H_2O_2) to directly convert the heme iron of enzymes to a reactive ferric-hydroperoxy complex (Fe^{3+}–OOH) via the peroxide shunt pathway [33, 36], therefore bypassing the oxygen activation chemistry and the necessity of forming the thiolate-ligated ferrous-heme species. Whether the failure of observing a P450 species in CYP-Sm46Δ29 is due to the instability of the thiolate-ligated P450 species that quickly reverts to a P420 form as with EpoK [43], or an alternative proximal ligand-coordinated species as in $P420_{cam}$ [45], P450 enzymes that undergo peroxide shunt pathway are insensitive to the nature of the proximal ligand as they do not require electron transfer to the heme iron. Peroxide activation can be achieved equally well with, for example, the histidine-ligated hemoproteins [55]. Therefore, as opposed to the generally regarded 'inactive' cytochrome P420 species in the P450 monooxygenases, the peroxygenase 'P420' enzymes can be active. We reason that this difference may explain why we observed no thiolate-ligated P450 ferrous-heme species but a functional CYP-Sm46Δ29. Indeed, we observed efficient catalytic activity of CYP-Sm46Δ29 in the H_2O_2 system, but not in the H_2O_2-depleted O_2/redox proteins/NADH system (Fig. 2b). A minor increase in the activities of CYP-Aa162 and CYP-Sm46Δ29 was observed in the catalase-absent redox partner-supported reaction system (Fig. 2b). This increase may reflect the beneficial effect of slow-releasing H_2O_2 from NADH decoupling on the catalytic activity, compared with the batch-added H_2O_2 [24]. However, considering the unique spectroscopic features of the CYP-Sm46Δ29, further structural and mechanistic studies are certainly required to gain more detailed understanding of the active-site coordination environment of this enzyme and its catalytic mechanism.

The strict carbon chain length dependence of the decarboxylative activity of CYP-Sm46Δ29 is consistent with other identified decarboxylative activities in $OleT_{JE}$ [13, 16, 22], $P450_{BS\beta}$ [37], and CYP-MP [24], which further emphasizes the importance of substrate identity in controlling the hydroxylation/decarboxylation bifurcation. This substrate identity-dependent chemistry was also manifested in a recent study of $OleT_{JE}$ where only trace amounts (<1–2%) of undesired hydroxyl side products were generated when aromatic carboxylic acids were used as substrates [29]. Our finding that low alkene production in CYP-Aa162 is accompanied with a high percentage of α-hydroxyl product is also consistent with the notion that C_β–H rather than C_α–H abstraction is a prerequisite for initiation of fatty acid decarboxylation pathway. However, this could not be the sole determining factor as CYP-MP produced predominantly the β-OH fatty acid but only a meager amount of alkenes [24]. For $OleT_{JE}$, it was proposed that His85 could facilitate the decarboxylation pathway through providing proton to the Fe^{4+}–OH (compound II) intermediate to restore a water-bound Fe^{3+}–OH_2 species [18, 20]. An active-site histidine may also help with the precise substrate positioning required for the decarboxylation pathway [26]. In the absence of an active-site histidine, ·OH rebound would be the prevalent pathway, leading to accumulated hydroxyl fatty acids, such as the case with CYP-MP [24], $P450_{Bs\beta}$ [13], and CYP-Aa162 in this study. Our site-directed mutagenesis analysis of CYP-Aa162 and $P450_{Bs\beta}$ confirmed the importance of this residue in regulating decarboxylation selectivity. However, the introduction of this histidine residue did not fully reverse the chemoselectivity of either CYP-MP [24] or CYP-Aa162 (Fig. 5), albeit an increase in the decarboxylation activity, hydroxylation was still the major chemistry of CYP-Aa162/Q85H/V170I. This implies that more structural and electronic requirements need to be satisfied to be a fatty acid decarboxylase.

Makris and coworkers [19, 20] recently confirmed that $OleT_{JE}$ catalysis is initiated by the formation of an iron(IV)-oxo cation radical (compound I). The fatty acid decarboxylation bifurcation is believed to begin with a regio-specific C_β–H abstraction followed by a single-electron transfer from either a substrate carboxylate group or an additional electron from the C_β position to the incipient compound II, resulting in the generation of a substrate diradical or a substrate carbocation, respectively, to fulfill C–C_α bond scission and give rise to the terminal alkene products. Our data here together with others [13, 18–20, 23, 24] suggest that substrate C_β–H abstraction, chain length-dependent precise positioning at the active site, relative intrinsic stability of the enzyme/Fe^{4+}–OH intermediate (compound II), and the competition between ·OH rebound and abstraction of an additional substrate electron by compound II are among the decarboxylation bifurcation determinants to ensure a desaturation reaction. Given that site-directed mutagenesis did not fully boost the decarboxylation activity of CYP-Aa162, alternative approaches such as directed evolution may represent an attractive method for engineering improved alkene formation activity towards, especially, particular chain length substrates, and for thorough understanding of the enzyme structure–function relationship.

A majority of P450 catalyses follow simple Michaelis–Menten kinetics based on a single active site for the substrate–enzyme interaction [56]. However, there are growing cases that non-Michaelis–Menten kinetics were observed for some P450 reactions [48, 57–59]. In the latter cases, the non-classical kinetics apparently results from allosteric effects that commonly lead to a sigmoidal velocity curve. The Hill equation can be used to indicate cooperative kinetics. Indeed, several studies have found that some P450s are able to fit into its active site with large molecules such as cyclosporine [60] or two substrates at once [57]. The two substrates can be the same or different molecules allowing the homotropic or heterotropic cooperative kinetics, respectively. Other examples of cooperative catalysis were found with $P450_{SP\alpha}$ [61, 62] and $P450_{BS\beta}$ [62, 63] where short-chain length fatty acids (such as heptanoic acid or hexanoic acid) could act as 'decoy molecules' to facilitate the binding of non-natural substrates (such as styrene) and their oxidation by these P450 peroxygenases. In this study, cooperative binding of C_{12} lauric acid substrate to CYP-Aa162 and CYP-Sm46Δ29 enzymes was observed with a Hill coefficient of ~2 (Table 1). The smaller k_{cat} and k_{cat}/K_m values of CYP-Sm46Δ29 towards C_{12} lauric acid (24.5 ± 6.3 min^{-1} and 0.4 μM^{-1} min^{-1}) than those towards C_{14} myristic acid (62.5 ± 4.5 min^{-1} and 1.5 μM^{-1} min^{-1}) may appear to be contradictory to the qualitative substrate specificity results (Fig. 2) which demonstrated higher conversion efficiency of C_{12} over C_{14} by this enzyme. We reason that this inconsistency may stem from the substrate inhibition of this enzyme by the high substrate concentration of C_{14} (200 μM substrate was used for the substrate specificity profiling assay).

Conclusions

In this study, we have characterized two new CYP152 peroxygenases with regard to their spectroscopic characteristics, substrate specificity, decarboxylation/hydroxylation chemoselectivity, and kinetics. Our data demonstrate that CYP-Sm46Δ29 works as an efficient $OleT_{JE}$-like fatty acid decarboxylase, while CYP-Aa162 acts as a $P450_{BS\beta}$-like fatty acid hydroxylase with marginal decarboxylation chemoselectivity. Furthermore, CYP-Sm46Δ29 was found to be catalytically active without forming a stable P450 species. Its decarboxylation activity exhibited strict carbon chain length dependence, suggesting the requirement of precise active-site positioning of substrates to be decarboxylated. Moreover, towards different carbon chain length fatty acids, CYP-Sm46Δ29 undertook different kinetic behaviors (cooperative or substrate inhibition), further stressing the importance and impact of elaborate coordination of substrate within the substrate binding pocket. Taken together, our findings could provide more opportunities for further understanding the unique catalytic mechanism employed by CYP152 fatty acid decarboxylases and also serve as a guide for future bioengineering of the enzymes for improved alkene-producing activity of different carbon chain length substrates.

Methods

Materials

Fatty acid substrates, terminal alkene authentic standards, α-hydroxy myristic acid, β-hydroxy myristic acid, and *N,O*-bis(trimethylsilyl)trifluoroacetamide (BSTFA) with 1% trimethylchlorosilane were purchased from TCI (Shanghai, China). Catalase from bovine liver (# C1345) was purchased from Sigma-Aldrich (St. Louis, MO, USA). One unit is defined, as described in the manufacturer's instruction, to decompose 1.0 μmol of H_2O_2 min^{-1} at pH 7.0 at 25 °C, while the H_2O_2 concentration falls from 10.3 to 9.2 mM, measured by the rate of decrease of A_{240}. NADH and antibiotics were obtained from Solarbio (Beijing, China). Other chemicals were purchased from Amresco (Solon, OH, USA) or Sigma-Aldrich (St. Louis, MO, USA). Molecular cloning kits, such as High Purity Plasmid Miniprep Kit and Wizard SV Gel and PCR Clean-up System, were purchased from TSINGKE Biotech (Beijing, China) and Promega (Madison, WI, USA), respectively. Oligonucleotides and codon-optimized genes were synthesized by Genewiz (Suzhou, China). The *Pfu* DNA polymerases and all restriction endonucleases were obtained from Takara (Dalian, China). Ni–NTA resin used for protein purification was purchased from Qiagen (Valencia, CA, USA). The FlexiRun™ pre-mixed gel solution for SDS-PAGE was obtained from MDBio, Amicon Ultra centrifugal filters from Millipore (Billerica, MA, USA), and PD-10 desalting columns from GE Healthcare (Piscataway, NJ, USA).

Molecular cloning

The gene sequences encoding CYP-Aa162 from *Alicyclobacillus acidocaldarius* LAA1 (CYP152A8, GenBank Accession Number: WP_008340313) and CYP-Sm46 from *Staphylococcus massiliensis* strain *S46* (CYP152L2, GenBank Accession Number: EKU50422) were, respectively, codon-optimized and synthesized by Genewiz (Suzhou, China), and cloned into the expression vector pET28b between NdeI/XhoI restriction sites with a hexahistidine tag at the *N* terminus for expression in *E. coli* BL21 (DE3). The truncated *CYP-Sm46Δ29* gene (WP_039990689) was PCR-amplified from the initial construct of pET28b-*CYP-Sm46* using *Pfu* DNA polymerase and then subcloned into pET28b vector using the same restriction sites as above. The generation of double mutant gene constructs of Aa162/Q85H/V170I and $P450_{BS\beta}$/Q85H/V170I was achieved by site-directed mutagenesis via overlap extension PCR [64]. The sequences of primers used in this study are listed in Additional file 3: Table S1. All plasmid constructs were confirmed by DNA sequencing at Sangon

Biotech (Shanghai, China). Upon sequence verification, plasmids were used to transform *E. coli* BL21 (DE3) for protein expression.

Heterologous expression and purification

The *E. coli* BL21 (DE3) cells transformed with pET28b-*CYP-Sm46Δ29*, pET28b-*CYP-Aa162*, or the Q85H/V170I double mutant constructs were grown overnight at 37 °C with shaking at 220 rpm in LB medium containing 50 μg mL^{-1} kanamycin. The overnight culture was used as a seed culture to inoculate (1:100 dilution) 1–3 L of modified terrific broth containing 4% glycerol, 1 mM thiamine, trace metals [16], and the corresponding antibiotics. Cells were then grown at 37 °C for 3–4 h until the optical density at 600 nm (OD_{600}) reached ~0.6, at which point δ-aminolevulinic acid (0.5 mM final concentration) was supplemented and the expression of CYP-Sm46Δ29 was induced by the addition of 0.2 mM isopropyl-β-D-thiogalactopyranoside (IPTG). Cells were further cultured for 24 h at 18 °C before being harvested by centrifugation at 6000 rpm, 4 °C. The cell pellet was frozen at −80 °C until required.

Purification of the His-tagged protein was carried out as described by Liu et al. [16] with minor modifications. All protein purification steps were performed at 4 °C. Specifically, the cell pellets were thawed and resuspended in 40 mL lysis buffer (50 mM NaH_2PO_4, 300 mM NaCl, 10% glycerol, and 10 mM imidazole, pH 8.0) through vortexing. After cell disruption by ultrasonication, the cell lysate was centrifuged at 12,000×*g* for 30 min to remove cellular debris. To the clarified cell lysate, 1 mL of Ni–NTA resin was added and gently mixed at 4 °C for 1 h. The slurry was then loaded onto an empty column and washed with approximately 100 mL of wash buffer (50 mM NaH_2PO_4, 300 mM NaCl, 10% glycerol, and 20 mM imidazole, pH 8.0) until no proteins were detectable in flowthrough. The bound target proteins were eluted with elution buffer (50 mM NaH_2PO_4, 300 mM NaCl, 10% glycerol, and 250 mM imidazole, pH 8.0). The eluates were pooled and concentrated with an Amicon Ultra centrifugal filter (30 kDa cutoff). Imidazole contained in the protein eluates was removed by ultrafiltration and buffer exchange on a PD-10 column into storage buffer (50 mM NaH_2PO_4, 300 mM NaCl, 10% glycerol, pH 7.4). The final purified protein was flash-frozen with liquid nitrogen and stored at −80 °C for later use.

UV–visible spectroscopic characterization of CYPs

Analysis of the UV–visible spectroscopic properties of the His-tagged CYP-Aa162 and CYP-Sm46Δ29 was performed on a Cary 60 UV–visible spectrophotometer (Varian, UK). For preparation of the dithionite-reduced ferrous-CO complex of each enzyme, carbon monoxide gas was slowly bubbled into a solution of purified ferric enzyme (~4–7 μM) in 50 mM NaH_2PO_4, 300 mM NaCl, 10% glycerol, pH 7.4, immediately followed by sufficient reduction of the protein with sodium dithionite (1–3 mg) [65]. The optical absorption spectra of the ferric and ferrous-CO forms of each enzyme were recorded, respectively, before and after the addition of sodium dithionite. The CO-bound reduced difference spectrum was obtained according to the previous report [66]. In order to record the spectra in the redox partners/NAD(P)H/CO system, lauric acid (~0.7 mM) was pre-incubated with the cytochrome proteins for 5 min at 28 °C before the absorbance of ferric forms was recorded. Then, to the substrate-bound protein solution was added the CamAB/NADH (for CYP-Aa162 and CYP-Sm46Δ29) or NADPH (for the self-sufficient $P450_{BM3}$) cofactors to initiate a flavin-to-heme electron transfer. Spectra of the CO-bound NAD(P)H-reduced forms of the P450 enzymes were subsequently recorded. The protein concentration was determined by the extinction coefficient of $\varepsilon_{422\ nm} = 104\ mM^{-1}\ cm^{-1}$ as determined using the pyridine hemochromogen method [67]. The typical methodology for determination of the P450 functional concentration using $\varepsilon_{450-490\ nm}$ [66] was abandoned due to lack of the P450 content in the CYP-Sm46Δ29 spectra.

In vitro enzymatic assay

Typical assays containing 1.5 μM of each CYP enzyme (CYP-Sm46Δ29 or CYP-Aa162 or their mutant), 200 μM fatty acid substrate (one of the C_{10}–C_{20} FFAs prepared from a 20 mM stock solution in DMSO), and 220 μM H_2O_2 in 200 μL of storage buffer were carried out at 28 °C for 2 h. Reactions were quenched by the addition of 20 μL of 10 M HCl. Heptadecanoic acid was then added as the internal standard, and the mixture was extracted by 200 μL ethyl acetate. Following extraction, the organic phase was collected and analyzed by gas chromatography (GC) as described below.

For detection of the volatile C_9 nonene product generated from C_{10} fatty acid decarboxylation, 500 μL of the reaction system containing 200 μM C_{10} fatty acid substrate, 220 μM H_2O_2, 2.0 μM of enzyme, and 200 μM C_7 1-heptene as the internal standard in a 1.5 mL polytetrafluorethylene (PTFE) septum-sealed glass bottles was incubated at 28 °C for 2 h with shaking at 300 rpm. The reactions were then placed on ice and subjected to headspace sampling using a gas-tight Hamilton syringe for GC–MS analysis as described below. The standard curve of nonene was obtained by incubating different concentrations of the authentic nonene standard under the same condition as the reactions without enzyme followed by the same GC–MS analysis. After headspace sampling, the reactions were immediately mixed with ethyl acetate and the internal standard heptadecanoic acid and extracted as described above for the analysis of the remaining C_{10} substrates. With every tested substrate (C_{10}–C_{20} FFAs), a reaction without H_2O_2 was used as the control for the initial substrate concentration.

Steady-state kinetic analysis

To determine the P450 kinetic parameters, 0.1–0.5 nM of enzyme was incubated with varying concentrations of substrate (C_{12} or C_{14}) at 28 °C in a 1 mL reaction system (50 mM NaH_2PO_4, 300 mM NaCl, pH 7.4, 10% glycerol) supplemented with an excess amount of H_2O_2 as a cofactor. Aliquots (200 μL) of reactions were removed and quenched at specific time points (usually at 0, 1, 3, and 5 min) by adding 1/10 volume of 1 M HCl. Heptadecanoic acid was then added as the internal standard. Subsequent sample preparation was performed (as above) for GC analysis (as below). Initial rates were calculated from either the substrate consumption for CYP-Aa162 or the 1-alkene production for CYP-Sm46Δ29. Kinetic analyses were performed using OriginPro 8.0 program.

Gas chromatography (GC) and GC–MS

The GC analytical method for hydrocarbon and fatty acid samples was adapted from Guan et al. [68]. The analyses were performed on an Agilent 7890B gas chromatograph equipped with a capillary column HP-INNOWAX (Agilent Technologies, Santa Clara, CA, USA; cross-linked polyethylene glycerol, i.d. 0.25 μm film thickness, 30 m by 0.25 mm). The helium flow rate was set to 1 mL per min. The oven temperature was controlled initially at 40 °C for 4 min, then increased at the rate of 10 °C min^{-1} to 280 °C, and held for 5 min. The injecting temperature was set to 280 °C with the injection volume of 1 μL under splitless injection conditions. The retention times and signal intensity of fatty acids and alkenes were determined by analyzing and comparing with known authentic fatty acids (C_{10}–C_{20}), 1-alkenes (C_{11}–C_{19}), and 1-heptadecanoic acid standards [16]. For GC–MS, the gas chromatography equipment was coupled to an Agilent 5975C MSD single quadrupole mass spectrometer operated under electron ionization mode at 70 eV in the scan range of 50–500 *m/z*. For detection of the hydroxyl fatty acid products, as only authentic standards of α- and β-hydroxy myristic acids were obtained, samples extracted from the C_{14} myristic acid reactions were derivatized with an equal volume of *N,O*-bis(trimethylsilyl)trifluoroacetamide (BSTFA) with 1% trimethylchlorosilane at 72 °C for 15 min prior to GC–MS analysis. The GC–MS analysis followed the protocol developed by Rude et al. [13], except for using the Agilent J&W DB-5MS column (i.d. 0.25 μm film thickness, 50 m by 0.25 mm). Peak identity was determined by comparison of retention time and fragmentation pattern with the authentic standard compounds where available and to the National Institute of Standards and Technology, USA mass spectral database. From the reactions using myristic acid as a substrate, we found that the sum of all products (1-tridecene, α- and β-hydroxy myristic acids) almost accurately accounts for at least 99% of the substrate consumption. In addition, as only a minimal level of H_2O_2 relative to fatty acid substrate (slightly more than 1:1 molar ratio) was used, no overly oxidized products (such as ketones, di-hydroxyl fatty acids) were observed within the detection limit. Therefore, we quantified the percentage of hydroxylated products for all substrates by subtraction of the 1-alkene production from the total substrate consumption unless otherwise stated. For detection of nonene product from the C_{10} fatty acid reactions, 600 μL of the reaction headspace sample was injected into the GC–MS system using a Hamilton needle syringe. The oven temperature program was as follows: 40 °C for 2 min, then 5 °C per min to 100 °C, and held for 2 min. Quantification of the nonene product was based on the standard curve and reaction controls.

Additional files

Additional file 1: Figure S1. Protein sequence alignment of CYP-Aa162 from *A. acidocaldarius* (GenBank Accession Number: WP_008340313), $P450_{BS\beta}$ from *Bacillus subtilis str. 168* (GenBank Accession Number: NP_388092), $OleT_{JE}$ from *Jeotgalicoccus sp.* ATCC 8456 (GenBank Accession Number: ADW41779), CYP-Sm46 (labelled as Sm46 extended) from *Staphylococcus massiliensis* S46 (GenBank Accession Number: EKU50422), CYP-Sm46Δ29 (labelled as Sm46Δ29) from *S. massiliensis* (GenBank Accession Number: WP_039990689) and cytochrome P450 enzymes from other *Staphylococcus* species such as *S. agnetis* (GenBank Accession Number: KFE42911), *S. delphini* (GenBank Accession Number: WP_019165531), *S. intermedius* (GenBank Accession Number: WP_019167377) and *S. pseudintermedius* HKU10-03 (GenBank Accession Number: ADV05454). **Figure S2.** SDS-PAGE showing the purified His-tagged CYP-Aa162 (lane A) and CYP-Sm46Δ29 (lane S). Molecular sizes of the marker bands (lane M), from top to bottom, are 180, 135, 100, 75, 63, 48, 35 and 25 kDa respectively. **Figure S3.** The UV–visible spectra of CYP-Sm46Δ29 (5 μM) under different conditions. (A) The purified CYP-Sm46Δ29 was diluted in 50 mM Na_3PO_4 (pH 7.4) buffer containing 300 mM NaCl and 10% glycerol. Spectra are shown for the oxidized ferric form of the enzyme (orange line) and the ferrous-CO complex reduced by the indicated amount of $Na_2S_2O_4$. (B) The purified CYP-Sm46Δ29 was diluted in 50 mM Na_3PO_4 buffer containing 300 mM NaCl and 10% glycerol with different buffer pH as indicated. Then the absorption spectra were recorded respectively for the oxidized ferric form and the ferrous-CO adduct reduced by 10 mM $Na_2S_2O_4$. The protein precipitates at buffer pH lower than 7.0. (C) A molar excess (600 μM) of C_{12} lauric acid was pre-incubated with the enzyme at room temperature for 5 min before the absorption spectra were recorded. Binding of C_{12} FA did not seem to induce an apparent spin-state transition of the ferric heme. The Soret peak of the C_{12}-bound ferrous-CO adduct of the enzyme was still detected at 420 nm. **Figure S4.** UV–visible spectra of the self-sufficient monooxygenase $P450_{BM3}$. The substrate-bound ferric form of $P450_{BM3}$ (solid line) shows a Soret maximum at ~ 416 nm with undistinguishable β-band and a weaker α-band at 570 nm. The reduced ferrous-CO form of $P450_{BM3}$ (dashed line) generated by the subsequent NADPH-initiated electron transfer features a shifted Soret peak to 448 nm. **Figure S5.** Effect of decoupling NADH oxidation and electron transfer on the catalytic conversion of lauric acid (LA) by CYP-Aa162 and CYP-Sm46Δ29. The reactions contained 0.2 mM LA, 2.0 μM CYP-Aa162 (or

CYP-Sm46Δ29), 3.0 μM putidaredoxin reductase (PdR), 1 mM NADH in the absence and presence of 1200 U mL^{-1} catalase (Catl.). By subtracting putidaredoxin (Pdx) from the reaction system, the NADH oxidation was mandatorily decoupled from the Class I electron transfer chain to P450 enzymes. Any catalytic activity observed should be supported by the H_2O_2 generated from NADH oxidation and O_2 reduction. The percentage conversion of LA was determined by calculating the substrate consumption based on GC analysis. Results shown are mean ± SD of duplicated experiments. **Figure S6.** The ratios of free fatty acid (FFA) decarboxylation (DC) over hydroxylation (HD) by CYP-Sm46Δ29 against different FFA substrates. The decarboxylation activity was measured by detecting the 1-alkene yield using GC analytical method. The hydroxylation activity was estimated by subtracting the alkene production from the total substrate conversion. This indirect but more convenient method was validated with C_{14} myristic acid substrate by direct measurement of the BSTFA/TMCS derivatized hydroxylation products. Results are shown as mean ± SD of duplicated experiments. **Figure S7.** Phylogenetic tree for CYP-Aa162, CYP-Sm46Δ29 and other CYP152 family members. The sequences were aligned using ClustalW. The Neighbor-joining Tree was generated using MEGA 7.0 package. Bootstrap values shown next to the branches were computed from 1000 bootstrap tests. CYP-Sm46 was found most closely related to the P450 fatty acid decarboxylase $OleT_{JE}$ (CYP152L1), while CYP-Aa162 (CYP152A8) is much closer to the P450 fatty acid hydroxylase $P450_{BS\beta}$ (CYP152A1). **Figure S8.** Protein sequence alignment of $OleT_{JE}$, CYP-Sm46Δ29, CYP-Aa162 and $P450_{BS\beta}$. *: the only two residues that are distinct in the active sites of these four P450 peroxygenases, which are proposed to be important for product distribution; #: the key catalytic residue. **Figure S9.** Kinetic curves of CYP-Aa162 and CYP-Sm46Δ29 against their optimal fatty acid substrates. (A) C_{12} lauric acid substrate consumption rates by CYP-Aa162 were fitted to Hill equation; (B) 1-undecene formation rates by CYP-Sm46Δ29 were fitted to Hill equation; (C) Solid line: the plot of 1-tridecene formation rates by CYP-Sm46Δ29 as a function of increasing C_{14} myristic acid concentrations, demonstrating substantial substrate inhibition. Dotted line: a hyperbolic curve fitted with Michaelis–Menten equation after truncating the inhibited rates at high C_{14} substrate concentrations. The steady state kinetic parameters were calculated using OriginPro 8.0 and are summarized in Table 2.

Additional file 2. The original gene sequence of CYP-Sm46, and the codon-optimized gene sequences of CYP-Sm46 and CYP-Aa162, as well as their corresponding amino acid sequences.

Additional file 3: Table S1. Primers used for cloning and site-directed mutagenesis.

Abbreviations
CYP or P450: cytochrome P450 enzyme; FFA: free fatty acid; LS: low-spin heme iron; CamAB: the P450cam redox partner proteins including putidaredoxin reductase and putidaredoxin; LA: lauric acid; 1-UE: 1-undecene; TB: terrific broth; IPTG: isopropyl β-D-1-thiogalactopyranoside; Ni-NTA: nickel-nitrilotriacetic acid resin; GC–MS: gas chromatography–mass spectrometry; DC: decarboxylation; HD: hydroxylation.

Authors' contributions
SL, LF, SC, FL, HX, and JX conceived and designed the study. HX, LN, WY, BF, YW, and CW performed the experiments and analyzed the results. HX, LF, and SL wrote the manuscript. All authors read and approved the final manuscript.

Author details
[1] Shandong Provincial Key Laboratory of Synthetic Biology, Qingdao Institute of Bioenergy and Bioprocess Technology, Chinese Academy of Sciences, No. 189 Songling Road, Qingdao 266101, Shandong, China. [2] CAS Key Laboratory of Biofuels, Qingdao Institute of Bioenergy and Bioprocess Technology, Chinese Academy of Sciences, No. 189 Songling Road, Qingdao 266101, Shandong, China. [3] Single-Cell Center, Qingdao Institute of Bioenergy and Bioprocess Technology, Chinese Academy of Sciences, No. 189 Songling Road, Qingdao 266101, Shandong, China. [4] University of Chinese Academy of Sciences, Beijing 100049, China. [5] Total Refinery and Chemistry, SDR/Biofuels, Tour Coupole, 2, Pl. Jean Millier, 92400 Courbevoie, France.

Acknowledgements
We thank Prof. Zhiqi Cong at Qingdao Institute of Bioenergy and Bioprocess Technology, Chinese Academy of Sciences for his valuable discussion and suggestions. We are also grateful to Dr. Yousong Ding at University of Florida and Prof. Thomas Pochapsky at Rutgers University for their kind gifts of the plasmids for expression of $P450_{BM3}$ and CamAB, respectively.

Competing interests
The authors declare that they have no competing interests.

Funding
This work was supported by the National Science Foundation of China (NSFC 31422002 and 31770844) and the Natural Science Foundation of Shandong Province, Grant JQ201407. This study was also carried out in the framework of the Collaboration Agreement between TOTAL and QIBEBT-CAS (Qingdao Institute of Bioenergy and Bioprocess Technology, Chinese Academy of Sciences).

References

1. Denisov IG, Makris TM, Sligar SG, Schlichting I. Structure and chemistry of cytochrome P450. Chem Rev. 2005;105:2253–77.
2. Bernhardt R. Cytochromes P450 as versatile biocatalysts. J Biotechnol. 2006;124:128–45.
3. Hrycay EG, Bandiera SM. The monooxygenase, peroxidase, and peroxygenase properties of cytochrome P450. Arch Biochem Biophys. 2012;522:71–89.
4. Grogan G. Cytochromes P450: exploiting diversity and enabling application as biocatalysts. Curr Opin Chem Biol. 2011;15:241–8.
5. Guengerich FP, Munro AW. Unusual cytochrome p450 enzymes and reactions. J Biol Chem. 2013;288:17065–73.
6. Furge LL, Guengerich FP. Cytochrome P450 enzymes in drug metabolism and chemical toxicology: an introduction. Biochem Mol Biol Educ. 2006;34:66–74.
7. Steen EJ, Kang Y, Bokinsky G, Hu Z, Schirmer A, McClure A, Del Cardayre SB, Keasling JD. Microbial production of fatty-acid-derived fuels and chemicals from plant biomass. Nature. 2010;463:559–62.
8. Liao W, Heijungs R, Huppes G. Natural resource demand of global biofuels in the anthropocene: a review. Renew Sustain Energy Rev. 2012;16:996–1003.
9. Dutta K, Daverey A, Lin J-G. Evolution retrospective for alternative fuels: first to fourth generation. Renew Energy. 2014;69:114–22.
10. Timilsina GR. Biofuels in the long-run global energy supply mix for transportation. Philos Trans A Math Phys Eng Sci. 2014;372:20120323.
11. Schirmer A, Rude MA, Li X, Popova E, del Cardayre SB. Microbial biosynthesis of alkanes. Science. 2010;329:559–62.
12. Choi YJ, Lee SY. Microbial production of short-chain alkanes. Nature. 2013;502:571–4.
13. Rude MA, Baron TS, Brubaker S, Alibhai M, Del Cardayre SB, Schirmer A. Terminal olefin (1-alkene) biosynthesis by a novel p450 fatty acid decarboxylase from Jeotgalicoccus species. Appl Environ Microbiol. 2011;77:1718–27.
14. Rui Z, Li X, Zhu X, Liu J, Domigan B, Barr I, Cate JH, Zhang W. Microbial biosynthesis of medium-chain 1-alkenes by a nonheme iron oxidase. Proc Natl Acad Sci USA. 2014;111:18237–42.
15. Gu L, Wang B, Kulkarni A, Gehret JJ, Lloyd KR, Gerwick L, Gerwick WH, Wipf P, Hakansson K, Smith JL, Sherman DH. Polyketide decarboxylative chain termination preceded by o-sulfonation in curacin a biosynthesis. J Am Chem Soc. 2009;131:16033–5.
16. Liu Y, Wang C, Yan J, Zhang W, Guan W, Lu X, Li S. Hydrogen peroxide-independent production of alpha-alkenes by OleTJE P450 fatty acid decarboxylase. Biotechnol Biofuels. 2014;7:28.
17. Rui Z, Harris NC, Zhu X, Huang W, Zhang W. Discovery of a Family of Desaturase-Like Enzymes for 1-Alkene Biosynthesis. ACS Catal. 2015;5:7091–4.
18. Belcher J, McLean KJ, Matthews S, Woodward LS, Fisher K, Rigby SE, Nelson DR, Potts D, Baynham MT, Parker DA, et al. Structure and bio-

chemical properties of the alkene producing cytochrome P450 OleTJE (CYP152L1) from the *Jeotgalicoccus* sp. 8456 bacterium. J Biol Chem. 2014;289:6535–50.
19. Grant JL, Hsieh CH, Makris TM. Decarboxylation of fatty acids to terminal alkenes by cytochrome P450 compound I. J Am Chem Soc. 2015;137:4940–3.
20. Grant JL, Mitchell ME, Makris TM. Catalytic strategy for carbon-carbon bond scission by the cytochrome P450 OleT. Proc Natl Acad Sci USA. 2016;113:10049–54.
21. Zachos I, Gassmeyer SK, Bauer D, Sieber V, Hollmann F, Kourist R. Photobiocatalytic decarboxylation for olefin synthesis. Chem Commun (Camb). 2015;51:1918–21.
22. Dennig A, Kuhn M, Tassoti S, Thiessenhusen A, Gilch S, Bulter T, Haas T, Hall M, Faber K. Oxidative decarboxylation of short-chain fatty acids to 1-alkenes. Angew Chem Int Ed Engl. 2015;54:8819–22.
23. Faponle AS, Quesne MG, de Visser SP. Origin of the regioselective fatty-acid hydroxylation versus decarboxylation by a cytochrome p450 peroxygenase: what drives the reaction to biofuel production? Chemistry (Easton). 2016;22:5478–83.
24. Amaya JA, Rutland CD, Makris TM. Mixed regiospecificity compromises alkene synthesis by a cytochrome P450 peroxygenase from *Methylobacterium populi*. J Inorg Biochem. 2016;158:11–6.
25. Yan J, Liu Y, Wang C, Han B, Li S. Assembly of lipase and P450 fatty acid decarboxylase to constitute a novel biosynthetic pathway for production of 1-alkenes from renewable triacylglycerols and oils. Biotechnol Biofuels. 2015;8:34.
26. Fang B, Xu H, Liu Y, Qi F, Zhang W, Chen H, Wang C, Wang Y, Yang W, Li S. Mutagenesis and redox partners analysis of the P450 fatty acid decarboxylase $OleT_{JE}$. Sci Rep. 2017;7:44258.
27. Hsieh CH, Huang X, Amaya JA, Rutland CD, Keys CL, Groves JT, Austin RN, Makris TM. The enigmatic P450 decarboxylase OleT is capable of, but evolved to frustrate, oxygen rebound chemistry. Biochemistry. 2017;56:3347–57.
28. Matthews S, Belcher JD, Tee KL, Girvan HM, McLean KJ, Rigby SE, Levy CW, Leys D, Parker DA, Blankley RT, Munro AW. Catalytic determinants of alkene production by the cytochrome P450 peroxygenase OleTJE. J Biol Chem. 2017;292:5128–43.
29. Wang JB, Lonsdale R, Reetz MT. Exploring substrate scope and stereoselectivity of P450 peroxygenase OleTJE in olefin-forming oxidative decarboxylation. Chem Commun (Camb). 2016;52:8131–3.
30. Dennig A, Kurakin S, Kuhn M, Dordic A, Hall M, Faber K. Enzymatic oxidative tandem decarboxylation of dioic acids to terminal dienes. Eur J Org Chem. 2016;2016:3473–7.
31. Hsieh CH, Makris TM. Expanding the substrate scope and reactivity of cytochrome P450 OleT. Biochem Biophys Res Commun. 2016;476:462–6.
32. Schlichting I, Berendzen J, Chu K, Stock AM, Maves SA, Benson DE, Sweet RM, Ringe D, Petsko GA, Sligar SG. The catalytic pathway of cytochrome p450cam at atomic resolution. Science. 2000;287:1615–22.
33. Hrycay EG, Bandiera SM. Monooxygenase, peroxidase and peroxygenase properties and reaction mechanisms of cytochrome P450 enzymes. Adv Exp Med Biol. 2015;851:1–61.
34. Matsunaga I, Sumimoto T, Ueda A, Kusunose E, Ichihara K. Fatty acid-specific, regiospecific, and stereospecific hydroxylation by cytochrome P450 (CYP152B1) from *Sphingomonas paucimobilis*: substrate structure required for alpha-hydroxylation. Lipids. 2000;35:365–71.
35. Matsunaga I, Ueda A, Fujiwara N, Sumimoto T, Ichihara K. Characterization of the ybdT gene product of *Bacillus subtilis*: novel fatty acid beta-hydroxylating cytochrome P450. Lipids. 1999;34:841–6.
36. Montellano PROd. Cytochrome P450: structure, mechanism, and biochemistry. 4th ed. New York: Springer International Publishing; 2015.
37. Lee DS, Yamada A, Sugimoto H, Matsunaga I, Ogura H, Ichihara K, Adachi S, Park SY, Shiro Y. Substrate recognition and molecular mechanism of fatty acid hydroxylation by cytochrome P450 from *Bacillus subtilis*. Crystallographic, spectroscopic, and mutational studies. J Biol Chem. 2003;278:9761–7.
38. Chang MCY, Eachus RA, Trieu W, Ro D-K, Keasling JD. Engineering *Escherichia coli* for production of functionalized terpenoids using plant P450s. Nat Chem Biol. 2007;3:274–7.
39. Petersen TN, Brunak S, von Heijne G, Nielsen H. SignalP 4.0: discriminating signal peptides from transmembrane regions. Nat Methods. 2011;8:785–6.
40. Krogh A, Larsson B, von Heijne G, Sonnhammer EL. Predicting transmembrane protein topology with a hidden Markov model: application to complete genomes. J Mol Biol. 2001;305:567–80.
41. Kall L, Krogh A, Sonnhammer EL. A combined transmembrane topology and signal peptide prediction method. J Mol Biol. 2004;338:1027–36.
42. Girvan HM, Marshall KR, Lawson RJ, Leys D, Joyce MG, Clarkson J, Smith WE, Cheesman MR, Munro AW. Flavocytochrome P450 BM3 mutant A264E undergoes substrate-dependent formation of a novel heme iron ligand set. J Biol Chem. 2004;279:23274–86.
43. Ogura H, Nishida CR, Hoch UR, Perera R, Dawson JH, Ortiz de Montellano PR. EpoK, a cytochrome P450 involved in biosynthesis of the anticancer agents epothilones A and B. Substrate-mediated rescue of a P450 enzyme. Biochemistry. 2004;43:14712–21.
44. Perera R, Sono M, Sigman JA, Pfister TD, Lu Y, Dawson JH. Neutral thiol as a proximal ligand to ferrous heme iron: implications for heme proteins that lose cysteine thiolate ligation on reduction. Proc Natl Acad Sci USA. 2003;100:3641–6.
45. Sun Y, Zeng W, Benabbas A, Ye X, Denisov I, Sligar SG, Du J, Dawson JH, Champion PM. Investigations of heme ligation and ligand switching in cytochromes p450 and p420. Biochemistry. 2013;52:5941–51.
46. Austin RN, Groves JT. Alkane-oxidizing metalloenzymes in the carbon cycle. Metallomics. 2011;3:775–87.
47. Noble MA, Miles CS, Chapman SK, Lysek DA, MacKay AC, Reid GA, Hanzlik RP, Munro AW. Roles of key active-site residues in flavocytochrome P450 BM3. Biochem J. 1999;339(Pt 2):371–9.
48. Lin Y, Lu P, Tang C, Mei Q, Sandig G, Rodrigues AD, Rushmore TH, Shou M. Substrate inhibition kinetics for cytochrome P450-catalyzed reactions. Drug Metab Dispos. 2001;29:368–74.
49. Chen B, Lee DY, Chang MW. Combinatorial metabolic engineering of *Saccharomyces cerevisiae* for terminal alkene production. Metab Eng. 2015;31:53–61.
50. Yang SJ, Rice KC, Brown RJ, Patton TG, Liou LE, Park YH, Bayles KW. A LysR-type regulator, CidR, is required for induction of the *Staphylococcus aureus* cidABC operon. J Bacteriol. 2005;187:5893–900.
51. McLaughlin JR, Murray CL, Rabinowitz JC. Unique features in the ribosome binding site sequence of the gram-positive *Staphylococcus aureus* b-lactamase gene. J Biol Chem. 1981;256:11283–91.
52. Munro AW, Lindsay JG, Coggins JR, Kelly SM, Price NC. Analysis of the structural stability of the multidomain enzyme flavocytochrome P-450 BM3. Biochim Biophys Acta. 1996;1296:127–37.
53. Manna SK, Mazumdar S. Reversible inactivation of cytochrome P450 by alkaline earth metal ions: auxiliary metal ion induced conformation change and formation of inactive P420 species in CYP101. J Inorg Biochem. 2008;102:1312–21.
54. Panicco P, Astuti Y, Fantuzzi A, Durrant JR, Gilardi G. P450 versus P420: correlation between cyclic voltammetry and visible absorption spectroscopy of the immobilized heme domain of cytochrome P450 BM3. J Phys Chem B. 2008;112:14063–8.
55. Hiner AN, Raven EL, Thorneley RN, Garcia-Canovas F, Rodriguez-Lopez JN. Mechanisms of compound I formation in heme peroxidases. J Inorg Biochem. 2002;91:27–34.
56. Lin JH, Lu AY. Role of pharmacokinetics and metabolism in drug discovery and development. Pharmacol Rev. 1997;49:403–49.
57. Korzekwa KR, Krishnamachary N, Shou M, Ogai A, Parise RA, Rettie AE, Gonzalez FJ, Tracy TS. Evaluation of atypical cytochrome P450 kinetics with two-substrate models: evidence that multiple substrates can simultaneously bind to cytochrome P450 active sites. Biochemistry. 1998;37:4137–47.
58. Shou M, Mei Q, Ettore MW Jr, Dai R, Baillie TA, Rushmore TH. Sigmoidal kinetic model for two co-operative substrate-binding sites in a cytochrome P450 3A4 active site: an example of the metabolism of diazepam and its derivatives. Biochem J. 1999;340(Pt 3):845–53.
59. Houston JB, Galetin A. Modelling atypical CYP3A4 kinetics: principles and pragmatism. Arch Biochem Biophys. 2005;433:351–60.
60. Ma L, Du L, Chen H, Sun Y, Huang S, Zheng X, Kim E-S, Li S. Reconstitution of the in vitro activity of the cyclosporine-specific P450 hydroxylase

from *Sebekia benihana* and development of a heterologous whole-cell biotransformation system. Appl Environ Microbiol. 2015;81:6268–75.

61. Fujishiro T, Shoji O, Kawakami N, Watanabe T, Sugimoto H, Shiro Y, Watanabe Y. Chiral-substrate-assisted stereoselective epoxidation catalyzed by H_2O_2-dependent cytochrome $P450_{SP\alpha}$. Chem Asian J. 2012;7:2286–93.
62. Onoda H, Shoji O, Watanabe Y. Acetate anion-triggered peroxygenation of non-native substrates by wild-type cytochrome P450s. Dalton Trans. 2015;44:15316–23.
63. Shoji O, Fujishiro T, Nakajima H, Kim M, Nagano S, Shiro Y, Watanabe Y. Hydrogen peroxide dependent monooxygenations by tricking the substrate recognition of cytochrome P450BSbeta. Angew Chem Int Ed Engl. 2007;46:3656–9.
64. Ho SN, Hunt HD, Horton RM, Pullen JK, Pease LR. Site-directed mutagenesis by overlap extension using the polymerase chain reaction. Gene. 1989;77:51–9.
65. Wang ZJ, Peck NE, Renata H, Arnold FH. Cytochrome P450-catalyzed insertion of carbenoids into N–H bonds. Chem Sci. 2014;5:598–601.
66. Guengerich FP, Martin MV, Sohl CD, Cheng Q. Measurement of cytochrome P450 and NADPH-cytochrome P450 reductase. Nat Protoc. 2009;4:1245–51.
67. Berry EA, Trumpower BL. Simultaneous determination of hemes a, b, and c from pyridine hemochrome spectra. Anal Biochem. 1987;161:1–15.
68. Guan W, Zhao H, Lu X, Wang C, Yang M, Bai F. Quantitative analysis of fatty-acid-based biofuels produced by wild-type and genetically engineered cyanobacteria by gas chromatography-mass spectrometry. J Chromatogr A. 2011;1218:8289–93.

11

Bioethanol production from spent mushroom compost derived from chaff of millet and sorghum

Peter Ryden[1], Maria-Nefeli Efthymiou[1], Teddy A. M. Tindyebwa[2], Adam Elliston[1], David R. Wilson[1], Keith W. Waldron[1*] and Pradeep K. Malakar[1,3]

Abstract

Background: In Uganda, the chaff remaining from threshed panicles of millet and sorghum is a low value, lignocellulose-rich agricultural by-product. Currently, it is used as a substrate for the cultivation of edible Oyster mushrooms (*Pleurotus ostreatus*). The aim of this study was to assess the potential to exploit the residual post-harvest compost for saccharification and fermentation to produce ethanol.

Results: Sorghum and millet chaff-derived spent oyster mushroom composts minus large mycelium particles were assessed at small-scale and low substrate concentrations (5% w/v) for optimal severity hydrothermal pre-treatment, enzyme loading and fermentation with robust yeasts to produce ethanol. These conditions were then used as a basis for larger scale assessments with high substrate concentrations (30% w/v). Millet-based compost had a low cellulose content and, at a high substrate concentration, did not liquefy effectively. The ethanol yield was 63.9 g/kg dry matter (DM) of original material with a low concentration (19.6 g/L). Compost derived from sorghum chaff had a higher cellulose content and could be liquefied at high substrate concentration (30% w/v). This enabled selected furfural-resistant yeasts to produce ethanol at up to 186.9 g/kg DM of original material and a concentration of 45.8 g/L.

Conclusions: Spent mushroom compost derived from sorghum chaff has the potential to be an industrially useful substrate for producing second-generation bioethanol. This might be improved further through fractionation and exploitation of hemicellulosic moieties, and possibly the exploitation of the mycelium-containing final residue for animal feed. However, spent compost derived from millet does not provide a suitably high concentration of ethanol to make it industrially attractive. Further research on the difficulty in quantitatively saccharifying cellulose from composted millet chaff and other similar substrates such as rice husk is required.

Keywords: *Pleurotus ostreatus*, Lignocellulose, Ethanol, Pre-treatment, Cell walls, Biofuels

Background

There is increasing pressure to optimise the use of agricultural by-products for reasons of environmental and economic sustainability. In intensive cereal growing areas in Africa, residual straw is an abundant by-product and is used as a substrate for mushroom cultivation [1] and research on utilisation of sorghum wastes has demonstrated successful use of the straw as a substrate for *Pleurotus* cultivation [2]. There is also much interest in utilising it as a feedstock for second-generation (2G) biofuel production as evidenced by studies on pre-treatment and saccharification of sorghum straw and bagasse from juice-extracted sweet sorghum [3, 4], and also its pre-treatment, saccharification and fermentation [5–8].

However, 2G biofuel production from lignocellulose faces many challenges, both technical and economic. Hydrothermal/chemical pre-treatments are expensive, and saccharification and fermentation at high substrate concentration requires considerable capital investment. Also of importance is the limited availability and/or

*Correspondence: keith.waldron@quadram.ac.uk
[1] The Biorefinery Centre, Quadram Institute Bioscience, Colney, Norwich Research Park, Norwich NR4 7UA, UK
Full list of author information is available at the end of the article

prohibitive cost of the feedstock in areas where there are well-established alternative local markets.

In Uganda, millet and sorghum straw produced on marginal semi-arid land are widely used as animal fodder so are not regarded as waste. However, Oyster mushrooms (*Pleurotus ostreatus*) are cultivated widely in Uganda on substrates derived from the chaff remaining from threshed panicles of millet and sorghum. Indeed chaff, as the least valuable waste stream, has been used for commercial *Pleurotus* cultivation in south western Uganda since 1995. Since 2007, the Mushroom Training and Resource Centre (MTRC; http://www.oystermushroom.ug/) has coordinated the training and marketing of *Pleurotus* production in the region. The chaff is soaked in water for 3 days, left to ferment, sterilised in a fire-heated steel drum whilst wrapped in banana leaves, bagged and inoculated. After several flushes, the mushroom compost is discarded.

The spent mushroom compost may be considered as an interesting lignocellulosic substrate for second-generation biofuel production because it will have undergone partial degradation by the Oyster mushroom mycelium. *Pleurotus ostreatus* is a white rot fungus with peroxidase and laccase activities [9, 10] and the partial degradation of the lignocellulose might be expected to decrease the pre-treatment severity needed for saccharification. Such an effect has been observed in pre-treatments with ammonia fibre expansion of rice straw lignocellulose before and after *Pleurotus* cultivation [11].

Elliston et al. [12] developed state-of-the-art high-throughput methods for the screening of lignocellulose for saccharification to glucose and fermentation to ethanol. In this approach, the biomass is milled to a powder, hydrothermally pre-treated, and then the saccharification and fermentation is carried out with excess enzyme and at up to 10% (w/v) substrate concentration. These methods have been found to give a good indication of yields at larger scales using higher substrate concentrations of un-milled biomass steam exploded at the same severities.

The aim of this research has been (1) to employ high-throughput screening methods to study and compare the conversion of mushroom compost lignocellulose from sorghum and millet chaff, to identify the likely optimal pre-treatment and saccharification conditions (particularly enzyme loadings), and suitable inhibitor-resistant yeasts for fermentation of sugars to ethanol, and (2) to carry out simultaneous saccharification and fermentation to facilitate SSF at higher substrate concentrations, thereby maximising the concentration of ethanol produced in order to reduce downstream distillation costs and confirming optimal enzyme concentrations.

Methods

Biomass

The spent mushroom compost was supplied by the Mushroom Training and Resource Centre (MTRC), Kyanamira, Kabale, Uganda. On delivery, the bags were opened in a MSC Class 2 cabinet and the samples were transferred to 2.5, 5 or 10 L polypropylene buckets and contained in autoclave bags. Sterilisation proceeded at 127 °C for 20 min. The moisture content of the sterilised composts was measured after drying duplicate samples at 60 °C for 66 h. The stones and the larger lumps of mycelium were removed from the samples but smaller particles of mycelium were retained.

Neutral sugar analysis

Triplicate samples of 2–4 mg of milled material (<0.5 mm Retsch cyclone mill Twister, Retsch Ltd. UK) were dispersed in 65 μL 72% w/w H_2SO_4 at room temperature for 3 h then diluted to 1 M with 715 μL water and heated at 100 °C for 2.5 h then cooled on ice. 200 μL of 1.00 mg m/L 2-deoxy glucose was added. The whole sample was reduced, acetylated and quantified by gas chromatography (GC) as described [13].

Uronic acid analysis

Triplicate samples of 3–8 mg of milled material were dispersed in 200 μL 72% w/w H_2SO_4 at room temperature for 3 h then diluted to 1 M with 2.2 mL water and heated at 100 °C for 1 h then cooled on ice. 2.6 mL of water was added, and then the samples were filtered through a glass fibre filter (GF/C) into 5-mL Eppendorf tubes and frozen. Uronic acids were measured with a glucuronic acid standard using volumes of 1.8 mL $Na_2B_4O_7$/c.H_2SO_4, 0.3 mL of sample solution and 30 μL of 0.15% 3-phenyl phenol in 0.5% NaOH [14]. After storage in the dark for 30 min, absorbances were measured at 520 nm.

Small-scale hydrothermal pre-treatments

Small-scale experiments to find a suitable pre-treatment severity employed microwave irradiation (Biotage Initiator$^+$, Uppsala, Sweden) on samples Retsch-milled to <0.5 mm. The dry weights were measured after drying at 60 °C for 16 h. Samples containing 250 mg dry weight were placed in 10-mL-thick-walled glass tubes with a small Polytetrafluoroethylene (PTFE)-coated stirrer bar. Water was added to bring the total to 5 mL. The tubes were crimp-capped with silicone/PTFE septa. The combined effects of time and temperature of steam treatment are described by a severity factor [15] defined by Eq. 1:

$$\text{Severity factor} = \log_{10}(t \times \exp((T - 100)/14.75)). \quad (1)$$

Severity factors ranging from 3.00 to 4.75 are shown as hyperbolas in Additional file 1: Figure S1. A typical test for a range of severities by steam explosion on straw or woody samples might use 10-min treatments from 170 to 230 °C. In the microwave-powered method, temperatures were limited to 208 °C so as not to exceed the maximum pressure that the glass tubes can withstand but conditions of equivalent severity can be calculated (Additional file 1: Figure S1).

Large-scale steam explosion pre-treatments

One batch of each of the spent mushroom compost samples was steam exploded using a Cambi™ steam explosion pilot plant [16]. Only a small amount of KAN03 was available so 238 g was steam exploded; for all other samples 500 g was used. Warm water (50 °C) was added to the chamber (4.3 L/kg dry mass). Millet samples were steam treated for 10 min at 5.2 bar (160 °C, severity factor 2.77) and the sorghum samples were steam treated for 10 min at 14.5 bar (200 °C, severity factor 3.94). The steam-exploded slurries were centrifuged at 4200 rpm for 20 min. The dry matter in the supernatants was measured by drying samples at 40 °C for 16 h. The dry matter in the pellets was measured after drying at 40 °C for 5 days.

Saccharification

The microwave-pre-treated slurries were centrifuged in 13-mL Falcon tubes at 3150*g* for 20 min and the pellets were washed twice with water. Saccharification was performed at 50 °C for 96 h with excess enzyme at 5% (w/v) substrate concentration in a volume of 5 mL. This involved adding the following volumes to the wet pellets: 2.5 mL 0.2 M NaOAc pH 5.0, 0.1 mL (5 mg/mL) thiomersal, 1.44 mL water, 50 μL Cellic CTec2 (30 FPU/g) and 5 μL Cellic HTec2 (Novozyme).

The tubes were centrifuged at 3500 rpm for 20 min and the supernatant was poured into 5-mL Eppendorf tubes. The supernatants were diluted as follows: 0.1 mL sample + 0.9 mL water for the sorghum samples and 0.2 mL sample + 0.8 mL water for the millet samples.

Glucose monosaccharide was measured with a Megazyme kit (D-Glucose Assay Kit GOPOD Format) and a 10 mM glucose standard in 96-well plates using 3, 6 and 10 μL of sample and 7, 4 and 0 μL water. GOPOD solution (0.3 mL) was added to each well and the plate was heated at 50 °C for 20 min. Absorbance was measured at 505 nm.

Xylose monosaccharide was measured with a Megazyme kit (D-Xylose assay kit) and a 0.25 g/L xylose standard in 96-well plates using 10 μL of sample and 0.282 mL of a mixture of 12 mL water, 2.4 mL buffer, 2.4 mL NAD^+ plus ATP and 0.12 mL hexokinase. The absorbance at 340 nm was read after 4 min then 5 μL of xylose mutarotase + xylose dehydrogenase was added. The absorbance was read when the reaction had gone to completion (about 20 min).

Enzyme optimisation

The compost samples with the highest cellulose content for millet and sorghum were microwave irradiated at severity factors of 2.77 for millet (10 min at 160 °C) and 3.94 for sorghum (10 min at 200 °C). Bulk samples were prepared; 5 × 750 mg biomass + 13 mL water in 50 mL tubes. The pre-treated material was transferred to Falcon tubes, washed twice with water and centrifuged at 3150*g* for 20 min. The wet pellets were divided into 13-mL Falcon tubes to test the amount of enzyme needed with 5% substrate concentration calculated from the initial dry matter, at pH 5.0 in 0.1 M NaOAc and 0.1 g/L thiomersal in 18-h experiments at 50 °C.

Simultaneous saccharification and fermentation (SSF)

A small-scale SSF experiment was performed in a 96-well plate format in 1 mL volumes to screen the capabilities of six yeast strains against the three millet composts and four sorghum pre-treated composts at a substrate concentration of 2.5% (w/v). The millet and sorghum composts were milled to <0.5 mm and 750 mg dry mass was microwave treated for 10 min in 13 mL water at 160 and 200 °C, respectively. The pre-treated samples were centrifuged for 20 min in 15-mL Falcon tubes at 3150*g*. The pellet volumes were approximately 2 mL. The pellets were washed twice with water. The pellet moisture contents were measured in duplicate by drying small samples for 16 h at 40 °C. The supernatants were retained and the mass of solids in the supernatants was measured in triplicate on a drying balance at 105 °C to the nearest mg.

1-mL sterile 2D barcoded polypropylene tubes with screw caps in 96-well racks were obtained from Thermo Scientific (Tube TrakMates; 2D barcoded storage 1.0-mL tube screw top sterile polypropylene with caps latch rack, Thermo Scientific Matrix). Wet pellet samples (25 mg dry mass) were weighed into the tubes and two 2.5-mm glass beads were added. Water was added to bring the water content of all tubes to 100 μL after which they were autoclaved. All other additions; enzymes, yeast and yeast nitrogen base (YNB), were combined into 15 mL stock solutions.

YNB was obtained from Formedium™ (Hunstanton, CYN0201). Two concentrations were prepared, 6.9 g/L for culturing the yeasts and rinsing the pellets and 7.67 g/L for the final rinse and making up the yeast + enzyme sample to 0.9 mL to be added to the wet pellets. The solutions of yeast nitrogen base were autoclaved.

The 6 yeasts used included 5 furfural-resistant strains: *S. paradoxus*: NCYC 3277, and *S. cerevisiae*: NCYC 3312,

NCYC 3290, NCYC 3284 and NCYC 3451. In addition, NCYC 2826 which has high ethanol tolerance was also included. Yeast strains were cultured over a weekend in 10 mL YNB + sugar. These cultures were centrifuged at 3000 rpm for 5 min. The supernatants were decanted. The pellet was washed twice with YNB then with YNB at 1.11 times concentration and made up to 15 mL. 30 μL CTec2 (12 FPU/g biomass) and 3 μL HTec2 were added to each yeast suspension. The biomass samples were arrayed by rows and the yeast cultures by columns as shown (Additional file 2: Figure S2).

The rack was set on its side on a rotary plate so that the tubes were horizontal and incubated at 25 °C for 72 h at 120 rpm. Then the rack was put in a boiling water bath for 10 min. After cooling, the rack was centrifuged at 3000 rpm. The supernatants were individually filtered [4 mm syringe filters, 0.45 μm Polyvinylidene fluoride (PVDF) membrane] into GC vials for high-performance liquid chromatography (HPLC). Ethanol standards, 0.3, 0.5, 0.8, 1.0 and 1.3% v/v, were prepared.

SSF at 30% substrate concentration

Larger scale SSF experiments were carried out with two compost samples: the highest ethanol-yielding millet and sorghum composts which also had the highest glucose compositions, KAN01 and KAB08. The steam-exploded and centrifuged pellets contained too much water to be fermented at 30% substrate concentration so some water was removed through 10 μm nylon bolting cloth to make the dry/wet ratio high enough (KAN01 0.3598 g dry/g wet; KAB08 0.3816 g dry/g wet).

Wet steam-exploded biomass samples were saccharified and fermented at 30% substrate concentration in 10 mL volumes (3 g dry matter) in 49 mL plastic pots (Securitainer with tamper evident push on cap, Ampulla Limited, Cheshire, SK14 2NY, UK), and water was added. A 25.4-mm-diameter ceramic ball (2.25 L Porcelain ball charge. Capco Test Equipment, Ipswich, Suffolk, IP1 5AP, UK) was placed in the pot and the pellet of biomass + water was compressed by the ball and formed into a bowl-shaped depression. The purpose of this is so that when the yeast and enzymes are added, fermentation proceeds from the surface with an initially low substrate concentration.

The pots were warmed to 37 °C. Triplicate samples at 3 enzyme levels, 10, 15 and 20 FPU/g, and with two yeast cultures NCYC 2826 and NCYC 3312 were prepared. The yeast cultures had been grown up over 3 days from slopes, in yeast medium (YM DIFCO) at 25 °C. The cultures were centrifuged at 2000 rpm for 5 min. The pellets were washed twice with yeast nitrogen base, and then made up in 2.9 times YNB. The yeast suspensions were combined with the enzymes and added to the pots. The pots were capped. A 0.5 mm bore 25-mm-length needle was put into each pot with a cotton wool plug to let CO_2 escape. The pots were weighed, and then incubated at 37 °C and 150 rpm for 96 h. There were 6 blanks with yeasts and enzymes. The fermented material was centrifuged in 13-mL Falcon tubes at 3150*g* for 20 min. The supernatants were centrifuged again in Eppendorf tubes at 10,000*g* for 10 min. These second supernatants were filtered individually through 0.4 μm syringe filters. The filtered samples were sealed in GC vials and analysed for glucose and ethanol by HPLC using a Flexar_ FX-10 UHPLC instrument (Perkin Elmer, UK) equipped with a refractive index detector and an Aminex HPX-87H organic acid analysis column (Bio-Rad Laboratories Ltd., UK; 65 °C, mobile phase 5 mM H_2SO_4, flow rate 0.6 mL/min).

Results and discussion

Biomass sources and characterisation

Spent compost from millet and sorghum chaff were sourced from 2 districts in the Kigezi sub-region of Western Uganda; Kanungu (sampling date 09/09/2014) and Kabale (sampling date 11/09/2014), see Table 1.

All of the composts contained white lumps of mycelium, which were removed with a 4-mm sieve. One millet sample (KAN03) contained some stones >6 mm, snail shells and insect cuticles; these were also removed with the 4 mm sieve.

Polysaccharide analysis

The plant polysaccharide compositions in the spent composts, determined by acid hydrolysis of milled material, are shown in Table 2. The glucosamine content was not measured since the yeasts used cannot ferment amino sugars from the fungal mycelium. The highest cellulose content of the spent mushroom composts was 12% for compost derived from millet, and 18% for compost derived from sorghum. The cellulose levels are much lower than the 30% cellulose content of rice straw after the 3rd flush (48 days) of *Pleurotus* cultivation [17]. For comparison, the cellulose content of biomass without

Table 1 Biomass Samples, location of origin (district), quantities and moisture contents

District	Sample name	Biomass	Fresh mass (g)	Moisture (%)
Kanungu	KAN01	Millet	1356	13.1
Kanungu	KAN02	Millet	1652	10.3
Kanungu	KAN03	Millet	511	8.2
Kanungu	KAN04	Sorghum	5319	9.4
Kabale	KAB06	Sorghum	6084	9.9
Kabale	KAB07	Sorghum	5750	9.5
Kabale	KAB08	Sorghum	2068	14.7

Table 2 Monosaccharide compositions (g/kg) of the plant cell wall polysaccharides in the spent mushroom composts, hydrothermally pre-treated (microwave), and steam exploded residues ($n = 3$)

Means (sd)

Sample	Rha	Fuc	Ara	Xyl	Man	Gal	Glc	UA
Spent composts								
KAN01	4.3 (0.1)	2.9 (0.1)	44.6 (1.7)	64.6 (2.3)	11.0 (0.2)	20.1 (0.6)	108.4 (0.4)	43.7 (2.4)
KAN02	4.5 (0.1)	3.3 (0.0)	45.0 (0.2)	63.0 (0.8)	10.3 (0.2)	21.0 (0.3)	115.0 (1.5)	44.8 (4.5)
KAN03	3.0 (0.5)	1.3 (0.3)	19.6 (2.2)	41.6 (8.2)	5.5 (0.9)	9.4 (1.7)	83.8 (1.6)	29.3 (3.1)
KAN04	2.1 (0.3)	0.7 (0.1)	17.3 (3.2)	114.8 (17.0)	3.9 (0.6)	4.7 (0.8)	142.3 (18.6)	21.0 (1.6)
KAB06	2.0 (0.1)	0.6 (0.1)	17.9 (1.4)	166.0 (9.8)	3.6 (0.6)	4.7 (0.4)	181.1 (9.9)	21.7 (1.2)
KAB07	2.1 (0.2)	0.7 (0.1)	16.7 (1.2)	131.0 (8.5)	4.3 (0.5)	4.5 (0.5)	163.0 (2.2)	22.4 (1.2)
KAB08	2.1 (0.5)	0.6 (0.1)	16.9 (1.4)	139.2 (0.4)	3.8 (0.8)	5.1 (0.8)	157.8 (7.3)	25.2 (3.4)
Microwave pre-treated								
KAN01	1.8 (0.6)	1.4 (0.2)	32.8 (5.2)	58.7 (9.5)	3.3 (0.5)	10.6 (9.1)	135.0 (21.3)	nd
KAN02	1.7 (0.1)	1.4 (0.1)	21.6 (0.8)	39.8 (1.9)	3.2 (0.0)	10.7 (0.5)	123.8 (5.1)	nd
KAN03	1.8 (0.1)	0.8 (0.0)	15.9 (0.3)	33.0 (0.4)	2.9 (0.0)	7.4 (0.0)	87.3 (3.8)	nd
KAN04	1.1 (0.4)	0.0 (0.0)	5.2 (2.7)	31.1 (0.7)	0.8 (0.1)	0.6 (0.3)	146.2 (21.4)	nd
KAB06	1.0 (0.1)	0.0 (0.0)	2.4 (0.2)	42.1 (0.5)	1.1 (0.5)	0.6 (0.0)	193.0 (0.8)	nd
KAB07	0.7 (0.1)	0.1 (0.0)	2.3 (0.2)	29.6 (0.6)	0.8 (0.0)	0.6 (0.0)	128.3 (3.9)	nd
KAB08	1.4 (0.1)	0.1 (0.1)	4.7 (0.2)	64.0 (1.0)	2.0 (0.2)	1.3 (0.1)	262.4 (5.7)	nd
Steam exploded								
KAN01	2.8 (0.0)	1.7 (0.0)	30.3 (1.0)	55.9 (1.7)	6.2 (0.1)	14.2 (0.4)	146.0 (2.9)	30.0 (4.5)
KAN02	2.5 (0.2)	1.8 (0.2)	21.9 (0.6)	37.9 (1.0)	5.2 (0.1)	10.8 (0.3)	115.5 (2.9)	22.9 (5.2)
KAN03	2.5 (0.1)	1.3 (0.1)	16.7 (0.1)	36.9 (0.5)	3.8 (0.1)	8.6 (0.2)	111.4 (1.4)	21.2 (1.3)
KAN04	1.3 (0.1)	0.1 (0.0)	4.5 (0.1)	46.9 (1.4)	1.6 (0.0)	1.4 (0.0)	171.6 (4.4)	7.1 (2.0)
KAB06	1.2 (0.2)	0.1 (0.0)	4.5 (0.8)	50.3 (7.4)	1.5 (0.1)	1.4 (0.2)	159.0 (20.8)	6.4 (1.2)
KAB07	1.2 (0.0)	0.1 (0.0)	4.6 (0.2)	50.7 (0.2)	1.6 (0.0)	1.5 (0.1)	158.5 (1.8)	8.6 (1.2)
KAB08	2.2 (0.3)	0.3 (0.0)	6.5 (0.2)	78.5 (2.7)	2.5 (0.2)	2.7 (0.1)	253.7 (17.7)	7.9 (2.4)

Millet: samples KAN01-3; Sorghum: samples KAN04-8. *Rha* Rhamnose, *Fuc* Fucose, *Ara* Arabinose, *Xyl* Xylose, *Man* Mannose, *Gal* Galactose, *Glc* Glucose, *UA*, Uronic acid

mushroom cultivation is sorghum straw (32% [4]) and sweet sorghum bagasse (42 [5]; 40% [6]).

Microwave hydrothermal pre-treatment

The microwave-powered hydrothermal pre-treatment assessments were all carried out at 208 °C to ensure equal areas under the curves during the heating and cooling periods at all severities (Additional file 3: Figure S3).

The optimum pre-treatment conditions for saccharification of millet and sorghum composts were determined in triplicate (separate pre-treated samples) using excess cellulase. The saccharification results are shown for millet (Fig. 1a) and sorghum (Fig. 1b). The profiles are very different. For millet, the response to severity was essentially flat, from the lowest severity tested (1.703) to the highest (nearly 4.0). Maximum saccharification of between 60 and 70% was achieved. Nevertheless, pre-treatment enhanced saccharification potential compared with the non-pre-treated material. In contrast, sorghum showed a very different response curve. Severities >3.355 released most glucose, and for samples KAN04 and KAB08, this was close to 100%. The preferable severity factors for millet and sorghum were 2.77 and 3.94, respectively. Since sorghum composts contain appreciable amounts of xylose-containing polymers (Table 2), the release of xylose was also assessed (Fig. 1b). In keeping with other studies on lignocellulose [18], much of this degrades during pre-treatment (Fig. 1b). This degradation is likely to render the cellulose more accessible to cellulases, but it also creates fermentation inhibitors such as 2-furfural.

Enzyme optimisation

Enzyme optimisation was performed using optimally pre-treated millet (KAN01) and sorghum (KAB06), as shown in Fig. 2. The results showed that enzyme loading in the region of 12 FPU/g substrate or more was suitable.

SSF of milled and hydrothermally pre-treated spent composts

The polysaccharide compositions of the pellets from milled millet and sorghum, microwave treated for 10 min at 160 and 200 °C (severities of 2.77 and 3.94),

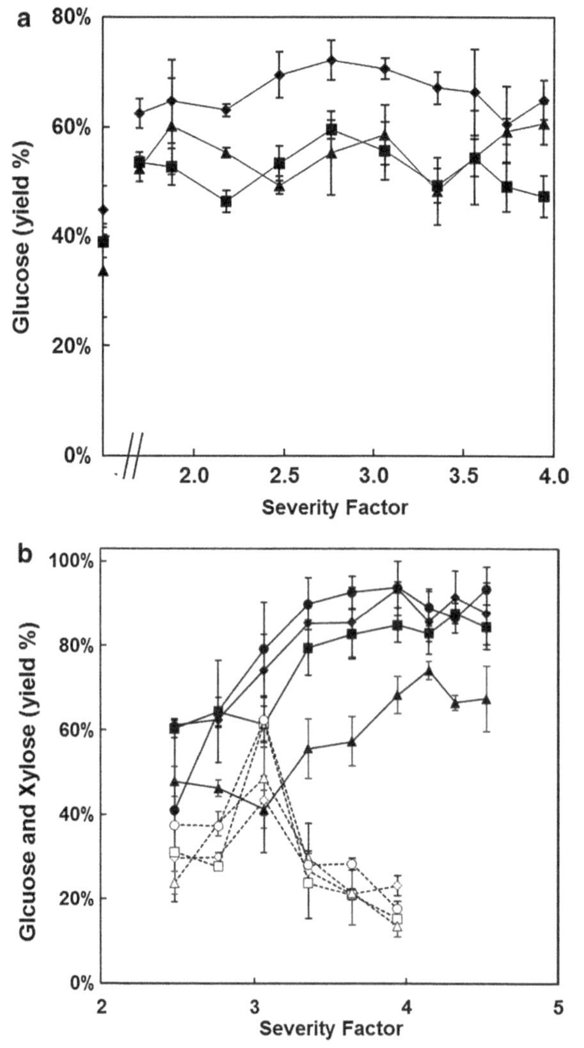

Fig. 1 **a** Mean and standard deviation (g/100 g) polymeric glucose in the original material released by cellulase and hemicellulases (30 FPU/g) from triplicate samples of spent mushroom compost derived from millet; ◆ KAN01, ■ KAN02, ▲ KAN03. The data points on the ordinate axis comprise samples without microwave pre-treatment. Xylose hydrolysis was not measured. **b** Average and standard deviation (g/100 g) polymeric glucose and xylose in the original material released by cellulase and hemicellulases from triplicate samples of spent mushroom compost derived from sorghum; filled symbols glucose, open symbols xylose: ◆◇ KAN04, ■□ KAB06, ▲△ KAB07, ●○ KAB08

respectively, were determined (Table 2). The results of the analysis show that 1) similarity in the respective compositions of composts from sorghum or millet chaffs; 2) the microwave hydrothermal pre-treatment has changed the chemistry of sorghum and millet composted lignocellulose significantly. In both cases, the level of many of the non-cellulosic sugars, particularly xylose and arabinose, were much reduced. In sorghum, xylose, for example, had been reduced by approximately 70%. In millet, which had contained relatively less xylose in the original mushroom compost, it was reduced by approximately 15%. These losses will have been due to the autohydrolysis of the non-cellulosic polysaccharides under the pre-treatment conditions.

Small-scale SSF was performed in a 96-well plate format in 1 mL volumes to screen the fermentation capabilities of six selected yeast strains against the pre-treated millet and sorghum composts. In order to optimise the fermentation, the range of yeasts included high ethanol-yielding, and furfural-resistant strains identified previously [19] with tolerance to stresses created during biorefining of lignocellulose [20]. The final concentrations of ethanol in each well are shown (Fig. 3). Interestingly, the results showed considerable variation on the basis of yeast and biomass source. The most effective yeast was NCYC 3312 for all substrates, whilst KAN01 and KAB08 were the highest yielding of the millet and sorghum composts. Broadly, the yield of ethanol (Fig. 3) generally followed the level of (cellulosic) glucose present in the pre-treated lignocellulose (Table 2).

Steam explosion

Small pilot-scale pre-treatment experiments employed steam explosion [21]. All 3 millet-derived composts were pre-treated at a severity factor that was the optimum for the sample that gave the highest glucose yield in the previous saccharification experiments (2.77, 160 °C 10 min); likewise for the 4 sorghum-derived composts (3.94, 200 °C 10 min).

The recovery of steam exploded slurry was not quantitative due to the characteristics of the apparatus, and some material (circa 10%) was lost through venting. The recovery of total mass and the distribution of that mass between supernatant and pellet is shown in Table 3.

The compositions of the steam-exploded pellets were determined (Table 2). Changes in chemical composition of the lignocellulosic residue by the pre-treatment were similar to those that occurred during the small-scale hydrothermal pre-treatment (Table 2).

SSF at 30% (w/v) substrate concentration

Using steam exploded residues, larger scale (10 mL) SSF experiments at industrially relevant substrate concentrations were carried out with the highest yielding of the millet and sorghum composts which also had the highest glucose compositions: KAN01 and KAB08. These two samples were saccharified and fermented at 30% (w/v) substrate concentration. In order to achieve the high substrate concentrations, some of the water was removed by pressing against 10 µm nylon bolting cloth (KAN01 0.3598 g dry/g wet; KAB08 0.3816 g dry/g wet). An

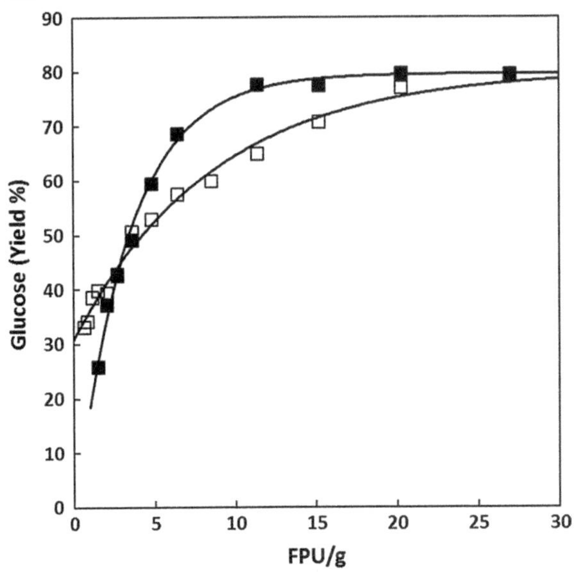

Fig. 2 Millet (□ KAN01) and sorghum (■ KAB06) saccharification with varying amounts of cellulase. $N = 1$; (g/100 g) polymeric glucose in the original material released by cellulase

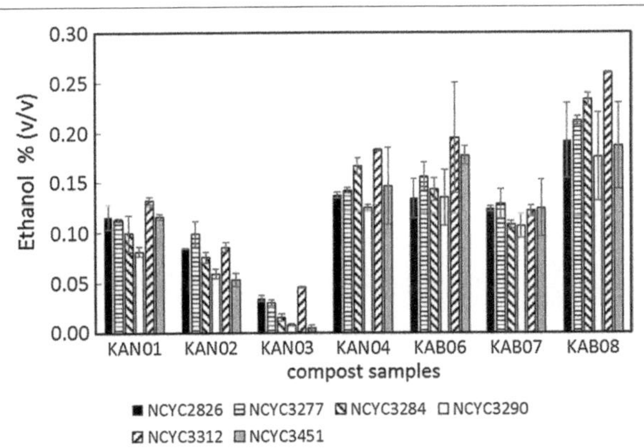

Fig. 3 A bar chart of ethanol concentration from each microwave-pre-treated compost sample in small-scale SSF experiments with six yeast strains. SSF was carried out in duplicate for 72 h at 25 °C at a substrate concentration of 2.5% (w/v) and a cellulase loading of 12 FPU/g biomass. See figure for shading regime

indication of the progress of the fermentation and when it had reached a plateau was obtained by measuring the loss in predominantly CO_2 mass (Fig. 4).

After about 30 h of SSF, a clear difference was observed between sorghum and millet fermentations in keeping with the earlier assessments (Fig. 4). After 96 h, it was observed that the sorghum samples had liquefied but the millet had not. The ethanol concentrations achieved after 96-h fermentation were much higher for sorghum than for millet (Fig. 5). Of the two yeasts assessed, strain NCYC 3312 gave only a slightly higher ethanol yield than NCYC 2826. Under these conditions, the ethanol concentration achieved from sorghum was 45.8 g/L (5.81% v/v; Fig. 5a) which is in the order of the level required for industrial distillation ($\geq$50 g/L [21, 22]). Use of high torque reactors, slightly higher substrate concentrations and optimisation of SSF conditions to increase the yield (which was in the region of 65–70% for Sorghum) might be expected to achieve this. Nevertheless, variability in the levels of cellulose in the different samples of sorghum-derived mushroom compost will need to be addressed. This and other properties of the spent compost may be related to the mushroom yield. Restricting the time of *Pleurotus* cultivation can minimise the loss of cellulose [11]. Thus, taking the compost after 2 flushes may be better than after three, although this would be likely to have a negative impact on the economics of mushroom production.

Table 3 The % recoveries of each substrate after steam explosion and its distribution between supernatant and pellet

	Starting dry matter (g)	Recovery (%)	Solids distribution	
			Supernatant (%)	Pellet (%)
KAN01	451.9	89.2	34.5	65.5
KAN02	451.7	93.8	36.2	63.8
KAN03	219.1	96.9	23.7	76.3
KAN04	466.6	91.4	22.4	77.6
KAB06	483.8	87.0	21.7	78.3
KAB07	469.6	91.7	22.1	77.9
KAB08	474.9	92.0	29.1	70.9

The visual appearance of representative samples of millet and sorghum composted material is shown in light micrographs before and after steam explosion, and after SSF (Fig. 6). The micrographs show clearly the enhanced fragmentation of the materials after steam explosion pre-treatment, reflecting the breakdown of the tissues through fracture and cell separation as found in wheat bran [23] and dissolution of some of the cell wall components. Much of the finer material appears to have been lost after SSF. However, the bulk of larger particulate material remains.

Notwithstanding the lower levels of cellulosic material in Millet chaff-derived compost, the microscopy and chemical analyses carried out on the samples fail to indicate why the millet-substrate was so much poorer than sorghum chaff-derived material in terms of ethanol yield. Both plants are monocotyledonous, and have similar classes of cell wall components. However, there may be differences in the nature of the lignin and possibly other

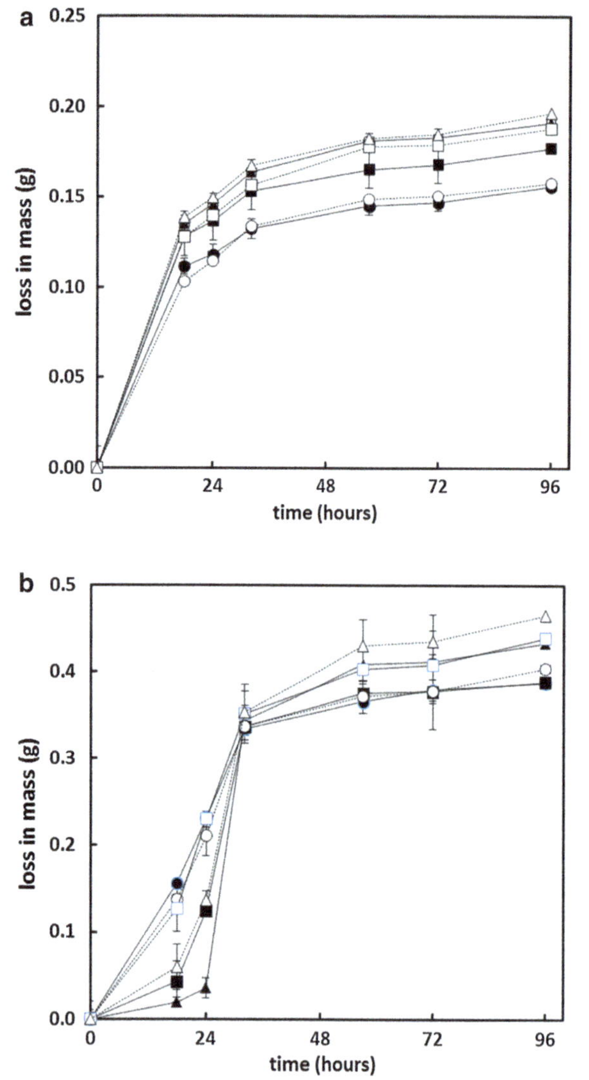

Fig. 4 **a** SSF of steam-exploded millet compost (KAN01) at 30% (w/v) substrate concentration and 3 cellulase levels: ●○ 10 FPU/g, ■□ 15 FPU/g, ▲△ 20 FPU/g substrate. *Open symbols* NCYC 2826, *closed symbols* NCYC 3312. $N = 3$. **b** SSF of steam-exploded sorghum compost (KAN08) at 30% substrate concentration and 3 cellulase levels: ●○ 10 FPU/g, ■□ 15 FPU/g, ▲△ 20 FPU/g. *Open symbols* NCYC 2826, *closed symbols* NCYC 3312. $N = 3$

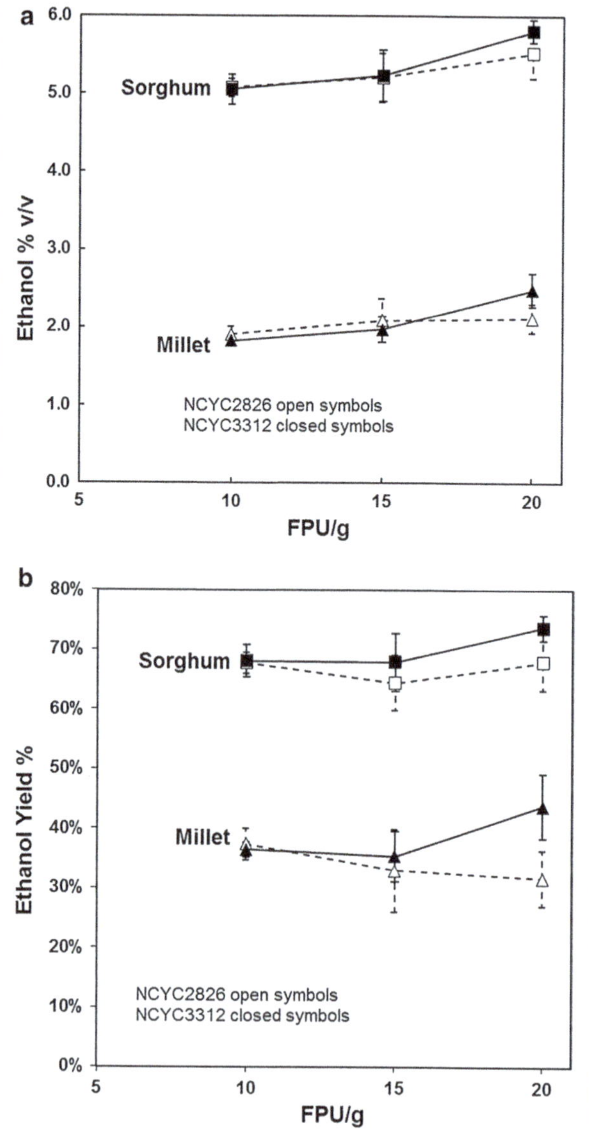

Fig. 5 Ethanol concentration after 96 h SSF at 30% substrate concentration. ■□ sorghum, ▲△ millet. *Open symbols* NCYC 2826, *closed symbols* NCYC 3312. Averages of triplicate ± standard deviation. **a** Ethanol concentration, (% v/v); (**b**) ethanol yield, % theoretical maximum assuming all the cellulose in the steam-exploded material is fermented with the stoichiometry of 1 Glucose 2 Ethanol

cell wall components such as silica. Recently, Wood et al. [18] showed that rice husk, although similar in carbohydrate composition to rice straw, was much less amenable to saccharification and fermentation after hydrothermal pre-treatment. It may be possible through breeding strategies to improve the properties of millet chaff, and this may also impact on mushroom production also.

For the future, as the Ugandan (and other African) mushroom cultivation industries develop, and move from essentially cottage industries to more sophisticated industrial scales, large localised co-product streams from e.g. spent sorghum and millet chaff mushroom compost are likely to be created. The current work has demonstrated that there is indeed potential to exploit these residues for producing bioethanol. Continued development of lignocellulose industrial biotechnology will augment such strategies. For example, the development of yeasts that can create much more valuable platform and fine chemicals would improve the economic sustainability of residue utilisation. Furthermore, the potential

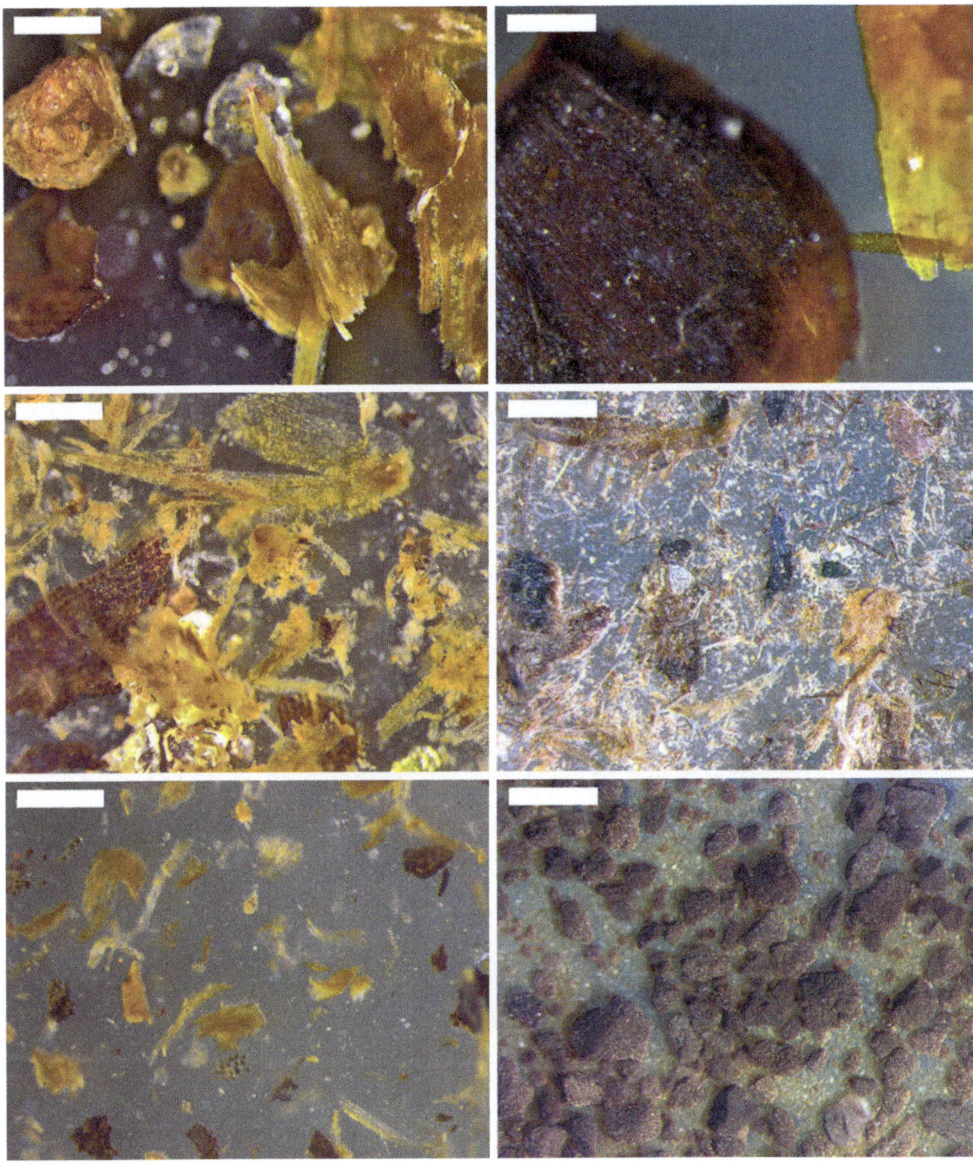

Fig. 6 Light micrographs of millet (*left*) and sorghum (*right*) composted material before pre-treatment (*top*), after steam explosion (*middle*) and after SSF (*bottom*). *Scale bars* 0.5 mm

to exploit the non-cellulosic components should not be ignored. Recent research has demonstrated the potential to recover anti-oxidants from wheat chaff [24] including the use of ultrasound-assisted extraction [25]. Such fractionation could be incorporated into the pre-treatment process step, resulting in several value-added products. Final residues, which could still include residual protein from the mycelium, might provide ingredients for animal feed. There has already been significant research to create ruminant feed ingredients from fungal-treated lignocellulosic biomass as reviewed by Kuijk et al. [26].

Although not considered in this study, *P. ostreatus* contains arabitol, pleuran (a β-glucan) and chitin in its mycelium [27]. No attempt was made to saccharify the chitin since it would not be compatible with *Saccharomyces* fermentation. Glucosamine inhibits *Saccharomyces*. Yeasts do exist which can assimilate glucosamine [28] and *Mucor circinelloides* can ferment chitin to ethanol [29] and this may provide further technological opportunities to exploit mushroom composts.

Conclusions

Three mushroom compost samples derived from millet chaff and four from sorghum chaff have been evaluated for their compositions, and propensity for saccharification and fermentation to ethanol. Millet chaff-derived compost was unsuitable and failed to produce a substantial yield even at high substrate concentration. However, sorghum chaff-derived compost, after optimisation of hydrothermal pre-treatment and saccharification, provided a good substrate for SSF. At 30% (w/v) substrate concentration, it liquefied effectively and yielded ethanol at a concentration of 5.81% (v/v) (4.8% w/v). With further improvements in % yield (circa 70%) and maybe increased substrate concentration through batch addition, it should be possible to increase the ethanol yields further.

Of the 5 furfural-resistant yeast strains tested, one strain (NCYC 3312) provided higher ethanol concentrations than a Spanish wine strain (NCYC 2826) in small-scale experiments at low substrate concentration, but showed only a slight improvement over NCYC 2826 at high substrate concentrations.

Abbreviations

2G: second generation; NCYC: National Collection of Yeast Cultures (UK); MTRC: Mushroom Training and Resource Centre; GC: gas chromatography; HPLC: high-performance liquid chromatography; PTFE: polytetrafluoroethylene; NAD: nicotinamide adenine dinucleotide; ATP: adenosine triphosphate; YNB: yeast nitrogen base; SSF: simultaneous saccharification and fermentation; PVDF: polyvinylidene fluoride.

Authors' contributions

PKM initiated the project. TAMT sourced the biomass. PR, PKM and KWW designed the experiments. PR, AE, DW and MNE performed the experiments. PR and KWW analysed the data and wrote the manuscript. All authors read and approved the final manuscript.

Author details

[1] The Biorefinery Centre, Quadram Institute Bioscience, Colney, Norwich Research Park, Norwich NR4 7UA, UK. [2] School of Biological Sciences, College of Natural Sciences, Makerere University, P.O. Box 7062, Kampala, Uganda. [3] College of Food Science and Technology, Shanghai Ocean University, 999 Hu Cheng Huan Road, Shanghai 201306, People's Republic of China.

Acknowledgements

The authors also gratefully acknowledge the provision of the yeasts by the UK National Collection of Yeast Cultures.

Competing interests

The authors declare that they have no competing interests.

Funding

QIB receives strategic funding from the Biotechnology and Biological Sciences Research Council of the UK This work was supported by (a) the AgriTT project 1579, funded by the Department for International Development (DFID) and the forum for Agricultural Research in Africa; (b) Institute Strategic Programme "Food and Health" Grant No. BB/J004545/1.

References

1. Sanchez C. Cultivation of *Pleurotus ostreatus* and other edible mushrooms. Appl Microbiol Biotechnol. 2010;85:1321–37.
2. Rani P, Kalyani N, Prathiba K. Evaluation of lignocellulosic wastes for production of edible mushrooms. Appl Biochem Biotechnol. 2008;151:151–9.
3. Choudhary R, Umagiliyage AL, Liang YN, Siddaramu T, Haddock J, Markevicius G. Microwave pretreatment for enzymatic saccharification of sweet sorghum bagasse. Biomass Bioenergy. 2012;39:218–26.
4. Vancov T, McIntosh S. Mild acid pretreatment and enzyme saccharification of *Sorghum bicolor* straw. Appl Energy. 2012;92:421–8.
5. Heredia-Olea E, Perez-Carrillo E, Serna-Saldivar SO. Production of ethanol from sweet sorghum bagasse pretreated with different chemical and physical processes and saccharified with fiber degrading enzymes. Bioresour Technol. 2013;134:386–90.
6. Matsakas L, Christakopoulos P. Fermentation of liquefacted hydrothermally pretreated sweet sorghum bagasse to ethanol at high-solids content. Bioresour Technol. 2013;127:202–8.
7. Matsakas L, Christakopoulos P. Optimization of ethanol production from high dry matter liquefied dry sweet sorghum stalks. Biomass Bioenergy. 2013;51:91–8.
8. Cardoso WS, Tardin FD, Tavares GP, Queiroz PV, Mota SS, et al. Use of sorghum straw (*Sorghum bicolor*) for second generation ethanol production: pretreatment and enzmatic hydrolysis. Quim Nova. 2013;36:623–7.
9. Stajic M, Persky L, Friesem D, Hadar Y, Wasser SP, et al. Effect of different carbon and nitrogen sources on laccase and peroxidases production by selected *Pleurotus* species. Enzyme Microb Technol. 2006;38:65–73.
10. Reddy GV, Babu PR, Komaraih P, Roy KRRM, Kothari IL. Utilization of banana waste for the production of lignolytic and cellulolytic enzymes by solid substrate fermentation using two *Pleurotus* species (*P. ostreatus* and *P. sajor-caju*). Process Biochem. 2003;38:1457–62.
11. Balan V, Sousa LD, Chundawat SPS, Vismeh R, Jones AD, Dale BE. Mushroom spent straw: a potential substrate for an ethanol-based biorefinery. J Ind Microbiol Biotechnol. 2008;35:293–301.
12. Elliston A, Wilson DR, Wellner N, Collins SRA, Roberts IN, Waldron KW. Effect of steam explosion on waste copier paper alone and in a mixed lignocellulosic substrate on saccharification and fermentation. Bioresour Technol. 2015;187:136–43.
13. Blakeney AB, Harris PJ, Henry RJ, Stone BA. A simple and rapid preparation of alditol acetates for monosaccharide analysis. Carbohyd Res. 1983;113:291–9.
14. Blumenkranz N, Asboe-Hansen G. New method for quantitative-determination of uronic acids. Anal Biochem. 1973;54:484–9.
15. Overend RP, Chornet E. Fractionation of lignocellulosics by steam-aqueous pretreatments. Philos Trans Royal Soc Lond. 1987;321:523–36.
16. Wood IP, Elliston A, Collins SRA, Wilson D, Bancroft I, Waldron KW. Steam explosion of oilseed rape straw: establishing key determinants of saccharification efficiency. Bioresour Technol. 2014;162:175–83.
17. Taniguchi M, Suzuki H, Watanabe D, Sakai K, Hoshino K, Tanaka T. Evaluation of pretreatment with *Pleurotus ostreatus* for enzymatic hydrolysis of rice straw. J Biosci Bioeng. 2005;100:637–43.
18. Wood IP, Cao HG, Tran L, Cook N, Ryden P, et al. Comparison of saccharification and fermentation of steam exploded rice straw and rice husk. Biotechnol Biofuels. 2016;5(9):193.
19. Field SJ, Ryden P, Wilson D, James SA, Roberts IN, et al. Identification of furfural resistant strains of *Saccharomyces cerevisiae* and *Saccharomyces paradoxus* from a collection of environmental and industrial isolates. Biotechnol Biofuels. 2015;26(8):33.
20. Wimalasena TT, Greetham D, Marvin ME, Liti G, Chandelia Y, et al. Phenotypic characterisation of *Saccharomyces* spp. yeast for tolerance to stresses encountered during fermentation of lignocellulosic residues to produce bioethanol. Microb Cell Fact. 2014;13(1):47.
21. Ryden P, Gautier A, Wellner N, Tapp HS, Horn SJ, et al. Changes in the composition of the main polysaccharide groups of oil seed rape straw following steam explosion and saccharification. Biomass Bioenergy. 2014;61:121–30.
22. Wingren A, Galbe M, Zacchi G. Techno-economic evaluation of producing ethanol from softwood: comparison of SSF and SHF and identification of bottlenecks. Biotechnol Progr. 2003;19:1109–17.
23. Merali Z, Collins SRA, Elliston A, Wilson DR, Kasper A, Waldron KW. Characterization of cell wall components of wheat bran following hydrothermal pretreatment and fractionation. Biotechnol Biofuels. 2015;15(8):23.

24. Antov MG, Dordevic TR. Environmental-friendly technologies for the production of antioxidant xylooligosaccharides from wheat chaff. Food Chem. 2017;235:175–80.
25. Dordevic T, Antov M. Ultrasound assisted extraction in aqueous two-phase system for the integrated extraction and separation of antioxidants from wheat chaff. Sep Purif Technol. 2017;182:52–8.
26. van Kuijk SJA, Sonnenberg ASM, Baars JJP, Hendriks WH, Cone JW. Fungal treated lignocellulosic biomass as ruminant feed ingredient: a review. Biotechnol Adv. 2015;33:191–202.
27. Rop O, Mlcek J, Jurikova T. Beta-glucans in higher fungi and their health effects. Nutr Rev. 2009;67:624–31.
28. Hui FL, Chen L, Li ZH, Niu QH, Ke T. Anton Leeuw. 2013;103:899–904.
29. Inokuma K, Takano M, Hoshino K. Direct ethanol production from N-acetylglucosamine and chitin substrates by *Mucor* species. Biochem Eng J. 2013;72:24–32.

12

Enhanced polyhydroxyalkanoate (PHA) production from the organic fraction of municipal solid waste by using mixed microbial culture

Bianca Colombo[1], Francesca Favini[1], Barbara Scaglia[1], Tommy Pepè Sciarria[1], Giuliana D'Imporzano[1], Michele Pognani[1], Anna Alekseeva[2], Giorgio Eisele[2], Cesare Cosentino[3] and Fabrizio Adani[1*]

Abstract

Background: In Europe, almost 87.6 million tonnes of food waste are produced. Despite the high biological value of food waste, traditional management solutions do not consider it as a precious resource. Many studies have reported the use of food waste for the production of high added value molecules. Polyhydroxyalkanoates (PHAs) represent a class of interesting bio-polyesters accumulated by different bacterial cells, and has been proposed for production from the organic fraction of municipal solid waste (OFMSW). Nevertheless, until now, no attention has been paid to the entire biological process leading to the transformation of food waste to organic acids (OA) and then to PHA, getting high PHA yield per food waste unit. In particular, the acid-generating process needs to be optimized, maximizing OA production from OFMSW. To do so, a pilot-scale Anaerobic Percolation Biocell Reactor (100 L in volume) was used to produce an OA-rich percolate from OFMSW which was used subsequently to produce PHA.

Results: The optimized acidogenic process resulted in an OA production of 151 g kg^{-1} from fresh OFMSW. The subsequent optimization of PHA production from OA gave a PHA production, on average, of 223 ± 28 g kg^{-1} total OA fed. Total mass balance indicated, for the best case studied, a PHA production per OFMSW weight unit of 33.22 ± 4.2 g kg^{-1} from fresh OFMSW, corresponding to 114.4 ± 14.5 g kg^{-1} of total solids from OFMSW. PHA composition revealed a hydroxybutyrate/hydroxyvalerate (%) ratio of 53/47 and Mw of $8{\cdot}10^5$ kDa with a low polydispersity index, i.e. 1.4.

Conclusions: This work showed how by optimizing acidic fermentation it could be possible to get a large amount of OA from OFMSW to be then transformed into PHA. This step is important as it greatly affects the total final PHA yield. Data obtained in this work can be useful as the starting point for considering the economic feasibility of PHA production from OFMSW by using mixed culture.

Keywords: Aerobic dynamic feeding strategy, Anaerobic percolation biocell reactor, Mixed microbial culture, Municipal solid waste, Polyhydroxyalkanoate, Sequencing batch reactor

Background

Food waste (FW) is defined as the organic material produced for human consumption and discarded, lost or degraded, primarily at manufacturing, retail and consumption stages [1]. In Europe, almost 87.6 million tonnes of FW are produced annually [2]. These wastes are characterized by both high moisture and high biodegradability, so FW creates adverse environmental impacts if landfilled (odours, fires, volatile organic compounds, groundwater contamination by leachate, global climate changes, etc.) and also, high disposal costs (88–144 $

*Correspondence: fabrizio.adani@unimi.it
[1] Gruppo Ricicla labs-DiSAA-Università degli Studi di Milano, Via Celoria 2, 20133 Milan, Italy
Full list of author information is available at the end of the article

tonne^{-1}) [1]. European and national legislation have focused on avoiding FW landfilling by treating it through a thermal process (incineration) or, more frequently, by biological processes (anaerobic digestion and composting) to be carried out on the separately collected organic fraction of municipal solid waste (OFMSW) [3]. Despite the high biological value of FW, traditional management solutions do not consider it as a precious resource, adding only a small amount of value to the final product, i.e. 60–150 $ tonne^{-1} of biomass for electricity production and 5–10 $ tonne^{-1} of biomass for compost production [4]. Therefore, supporting these activities generally requires government contributions or high tariffs for citizens, e.g. in Italy the waste tariff increased by 70% in the last 10 years.

FW is an organic matrix rich in valuable molecules such as starch, cellulose, hemicellulose, lignin, proteins, lipids and organic acids that could be managed in a more sustainably economic way by using it as a raw material for bulk chemicals production [4]. Many studies have reported the use of FW for the production of high added value molecules such as lactic acid, citric acid, succinic acid, single cell oils, enzymes and polymers [5], an approach which is more financially rewarding in comparison with compost and biomethane production (1000 $ tonne^{-1} of biomass for bulk chemicals production from FW) [6].

Among the wide varieties of bio-products obtainable from FW are the polyhydroxyalkanoates (PHAs), a class of interesting bio-polyesters accumulated by different bacterial cells under the form of granules inside the cytoplasm. PHAs are completely biodegradable and are mainly produced starting from renewable sources; their chief interesting property is their mechanical behaviour that make them comparable to common plastics [7]. Microbial PHA production can be carried out by using either pure or mixed microbial cultures (MMC). MMC with high PHA accumulation capacity have been suggested as a solution for reducing the high maintenance costs of pure cultures. Many studies related to PHA production by MMC have performed both selection of PHA-storing cultures and PHA accumulation by using just OA as substrates, since they are the direct metabolic precursors of PHA [7], while PHA production carried out by using pure cultures allows the employment of both OA and simple sugars as carbon sources. Working with mixed microbial cultures, simple sugars are not recommended as they can be used by the culture for the production of other molecules (i.e. glycogen), reducing the final PHA production yield [8, 9]. Given the high costs of synthetic OA, agro-industrial wastes, such as molasses, cheese whey, olive oil mill effluent, palm oil mill effluent, candy bar factory wastewater and many other waste streams [10–14], can be treated by acidic fermentation to produce OA which can then be transformed into PHA by MMC.

Complex wastes, such as OFMSW, have been reported in the literature to be used as substrates for PHA production by MMC [15–18]. These studies focused attention on PHA production from OA obtained by the fermentation of OFMSW, but until now, no attention has been paid to the entire process leading to the transformation of OFMSW to OA and then to PHA, i.e. to obtain overall estimates of PHA yield based on OFMSW weight unit used. These data, together with those showing the scientific feasibility in transforming organic waste into PHA, are important to estimate the effective economic sustainability of PHA production from OFMSW.

In this work, a two-stage approach, i.e. OA production from OFMSW and subsequent PHA biosynthesis from OA, is proposed.

Optimization of the acidic fermentation of OFMSW was carried out in order to exploit the biological potential of this complex matrix, obtaining a percolate rich in OA used subsequently as the substrate to produce PHA by employing MMC. Global mass balance of the process performed was achieved as well as the complete characterization of the polymer obtained.

Methods

Percolates production from OFMSW

Organic fraction of municipal solid waste collection

The source-separated OFMSW coming from street bin containers was collected at a full-scale composting plant located in northern Italy. Collection was done by following the quartering method: in brief, the OFMSW was sampled at different points of the whole mass (500 kg wet weight – w.w. – ×3 times); then the collected waste was mixed, quartered and sub-sampled until a final sample of about 300 kg w.w. material was reached. OFMSW was stored at 4 °C before the trials were set up. A representative sample of approximately 30 kg, obtained by mixing sub-samples of about 5 kg each taken randomly from sampled material, was dried and crushed to 2 mm and then used to perform analytical analyses.

The bulking agent (shredded wood) from the same plant was also collected following a similar approach.

Liquid digestate, used to irrigate OFMSW during dry anaerobic fermentation (see later), was collected directly from a continuously stirred tank reactor anaerobic digester (CSTR-AD) plant fed with OFMSW, located in northern Italy and previously described [19]. Digestate was collected from this plant in large quantities from the discharge pipe system of the post-digester after 30 days of retention time (HRT).

Experimental apparatus

The acidogenesis (SSAD) trials were carried out by developing a pilot-scale anaerobic percolation biocell reactor (APBR) that was made up of an insulated vertical cylinder (100 L of volume) of PVC material with a hermetic cap. Inside the reactor, there was a stainless steel basket with holes at the bottom, into which OFMSW was introduced from the APBR cap, as well as the digestate used to irrigate the organic wastes (Fig. 1). From the bottom of the APBR, the percolate generated daily due to digestate irrigation was extracted and weighed. Digestate irrigation was used to buffer the acidity produced during OFMSW fermentation and to remove OA produced in the APBR. This encouraged high hydrolysis performance from the organic waste, which is the limiting step in producing large amounts of OA from organic substrates [20, 21]. Moreover, large-scale OA production becomes toxic for methanogenic bacteria, limiting methane production during the trials.

Experimental setup

Three different percolation trials were carried out. For all of them, fresh OFMSW (85% w.w.) and bulking agent (chopped green waste material) (15% w.w.) were mixed just before reactor filling to avoid clogging problems in the APBR, allowing the percolation. For each trial, the APBR was filled with about 50 kg of mixture (OFMSWmix-in).

The OFMSWmix-in in the APBR was irrigated continuously by using the liquid digestate obtained from the full-scale CSTR-AD plant in order to optimize OFMSW hydrolysis and subsequent acidogenesis fermentation to produce OA. Anaerobic hydrolysis and acidic fermentation lasted for 21 days for each trial performed. During this time, irrigation with the liquid digestate (previously heated to 55 °C) was done continuously by using a peristaltic pump, adopting an OFMSWmix-in/digestate ratio of 1:0.45 kg kg^{-1} (flow rate of 0.02 L h^{-1} kg^{-1} OFMSW$_{w.w.}$) for the first trial (Trial 1), 1:0.9 kg kg^{-1} (flow rate of 0.04 L h^{-1} kg^{-1} OFMSW$_{w.w.}$) for the second trial (Trial 2), and of 1:1.8 kg kg^{-1} (flow rate of 0.08 L h^{-1} kg^{-1} OFMSW$_{w.w.}$) (day 1–8) and of 1:0.9 kg kg^{-1} (flow rate

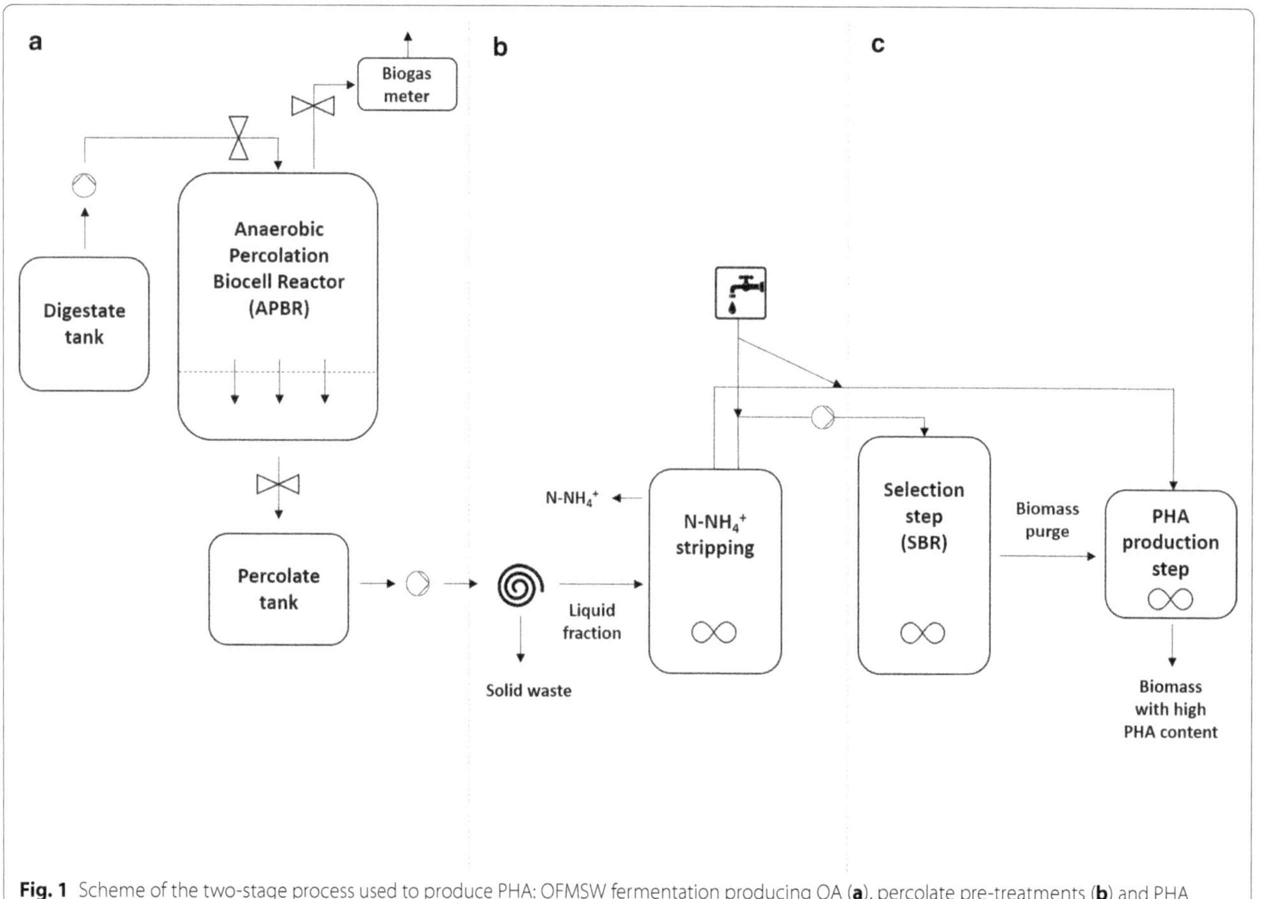

Fig. 1 Scheme of the two-stage process used to produce PHA: OFMSW fermentation producing OA (**a**), percolate pre-treatments (**b**) and PHA production (**c**)

of 0.04 L h^{-1} kg^{-1} $OFMSW_{w.w.}$) (day 9–21) for the third trial (Trial 3), respectively. The daily-generated percolates were collected from the bottom of the APBR, quantified and mixed together to get a final mixed percolate sample for each trial.

Percolates treatment before PHA production

The three percolates obtained from anaerobic acidic fermentation trials were submitted to two pre-treatments before their use for polyhydroxyalkanoate (PHA) production. The first pre-treatment was to centrifuge the percolates at 20,000*g* for 15 min, reducing suspended organic carbon from percolates. Supernatants obtained underwent a second pre-treatment adopted to remove ammonia by stripping. This step was performed by bringing supernatant-pH to 11 by adding 6-mol L^{-1} KOH and stirring it (magnetic stirring at 200 rpm) under the hood until a final C:N close to 10 was reached [11]. Then ammonia stripping was stopped and the pH was brought again to the initial value of 8 by using 3 mol L^{-1} H_2SO_4. Ammonia stripped was not trapped at lab scale and it was not considered in this work. Obviously, this step needs to be considered and further studied on a larger scale.

After pre-treatments, the three supernatants were diluted with deionized water, to produce a final COD content close to 1300 mg L^{-1} [11]. These OFMSW-derived supernatants ($OFMSW\text{-}supernatants_{in}$) were used for subsequent inoculum selection.

Substrates ($OFMSW\text{-}supernatants_{acc.}$) used for PHA accumulation tests were derived from percolates that had undergone the same treatments shown for those destined for the inoculum selection. Three $OFMSW\text{-}supernatants_{acc.}$ were obtained but, in this case, the ammonia content was lower than those reported for $OFMSW\text{-}supernatants_{in}$, since it has been reported that N starvation can determine a greater conversion of carbon into PHA because of cell growth limitation [22]. Moreover, $OFMSW\text{-}supernatants_{acc.}$ were less diluted, giving a final COD concentration of about 7500 mg L^{-1}, in order to avoid an excessive dilution of the biomass in the accumulation reactor [10].

Inoculum production and PHA-producing bacteria enrichment

The enrichment of PHA-producing bacteria was performed by using an inoculum constituted by an activated sludge (5 g total suspended solids L^{-1}) collected from the secondary sedimentation tank of a wastewater treatment plant (5.2×10^5 equivalent inhabitants) located at Peschiera Borromeo (Milan, Italy).

The enrichment in PHA-producing bacteria was carried out in a Sequencing Batch Reactor (SBR) with a working volume of 800 mL, applying an aerobic dynamic feeding (ADF) strategy [11]. In brief, the SBR cycle length was of 12 h, consisting of four discrete phases: (1) influent filling (4 min), (2) aeration (675 min), (3) settling (40 min) and (4) withdrawal of the exhausted effluent (5 min), with 1 day of hydraulic retention time (HRT) and 5 days of Sludge Retention Time (SRT), keeping the temperature at 25 °C and the pH at 8.8, this latter controlled automatically by adding 1 mol L^{-1} HCl. Aeration and agitation were provided by supplying air at 6 L min^{-1} and stirring set at 110 rpm. Pumping, aeration and stirring were automatically controlled.

The selection process lasted for 3 months; for each month, a different pre-treated supernatant was used as substrate: $OFMSW\text{-}supernatant_{in}$ 1, $OFMSW\text{-}supernatant_{in}$ 2 and $OFMSW\text{-}supernatant_{in}$ 3, respectively, for the first, second and third month of selection.

The selection trend was monitored by determining the duration of the feast phase achievable by using the dissolved oxygen (DO) concentration in the selection media [10], measured by an optical probe (FDO 925, WTW, Germany).

In particular, the feast (h)-to-famine (h) ratio (F/F ratio) was calculated as the ratio between the lengths in hours of the two phases. For a correct selection of PHA-storing bacteria, the F/F ratio had to be equal or less than 0.33 [23]. To carry out the selection of PHA-accumulating bacteria, 400 mL of activated sludge was used as the inoculum feed for each cycle, with 400 mL of $OFMSW\text{-}supernatant_{in}$. Organic Loading Rate (OLR) was kept close to 1300 mg COD L^{-1} day^{-1} and the C:N:P ratio was of about 100:13:0.4 mol C: mmole N: mmole P. Every month, two cycles were monitored in order to evaluate the performance of the selected culture. In particular, monitoring was performed between cycles 31 and cycle 45 since it was reported that the cultures reach stability after three SRTs from the beginning of the trial (cycle 31) [24].

PHA accumulation

The ability of the MMC to accumulate PHA was assessed by fed-batch assays carried out in a 300-mL working volume glass reactor, with continuous aeration and stirring. These assays consisted of feeding the substrate to 160 mL of enriched culture (at least 3 SRTs from the beginning of the selection) [24] adopting a pulse-wise feeding method. The assays were monitored continuously by measuring the concentration of the dissolved oxygen in the accumulation media [10]. In particular, substrate ($OFMSW\text{-}supernatant_{acc.}$) was fed to the reactor when DO showed a strong increase [10]. Total C dosed was calculated taking into account that the ratio of the carbon to the microorganisms had to be the same as that inside the selection reactor. The assays were stopped when no DO variation followed the substrate feeding.

For the accumulation tests, the operating conditions used were those adopted in the selection reactor, i.e. temperature of 25 °C, aeration of 6 L min^{-1} and stirring at 110 rpm.

The biomass from the selection process was submitted to accumulation tests using the same substrate as the carbon source; every month, two accumulation trials were performed in duplicate, between the cycles 31 and 45.

PHA extraction

The biomass collected after PHA accumulation tests was centrifuged at 8000*g* for 15 min, washed with 0.9% sodium chloride solution and centrifuged again at 8000*g* for 20 min. The pellet obtained was lyophilized and then suspended in chloroform (in a ratio of ca. 40 mL $CHCl_3$ g^{-1} dried cells) and left to dissolve for a period of 3 days at 37 °C [10]. The solution was then filtered to remove all undissolved material. The extracted PHA in chloroform was then precipitated by the addition of 5 volumes of methanol and allowed to settle down for 30 min [25]. The white precipitate formed was then filtered, suspended in chloroform and used to fill glass Petri dishes. Finally, chloroform was evaporated allowing polymer recovery in the form of a thin film.

Analytical procedures

OFMSW, digestate and percolates characterization

Total solids (TS), volatile solids (VS), total nitrogen (TKN), pH and ammonia ($N\text{-}NH_4^+$) (detected on fresh material) contents were determined according to the standard procedures [26]. Total OA expressed as acetic acid, were detected on fresh material and determined according to the acid titration method [27].

Substrate and biomass characterization during PHA production

The substrates (OFMSW-supernatants) fed during the selection and accumulation processes were characterized in terms of pH, TS, VS, Chemical Oxygen Demand (COD), OA content (acetate, butyrate, lactate, propionate and valerate), TKN, $N\text{-}NH_4^+$ and phosphorus (P) content.

During the selection trials, samples were taken during the cycle once in each SRT; every sample was characterized in terms of total suspended solids (TSS), volatile suspended solids (VSS), soluble COD, OA content, $N\text{-}NH_4^+$ content and PHA content. During accumulation trials, samples were taken continuously in order to measure TSS, VSS, soluble COD, OA content and PHA content. Biomass concentration was calculated as VSS according to the standard methods [10].

TSS and VSS were determined as reported by Valentino et al. in 2015 [28]. OA concentrations measured on filtered samples (filter diameter of 0.45 μm) were determined by high-performance liquid chromatography (HPLC) using a chromatograph equipped with a UV detector and Aminex HPX-87H column (column temperature 20 °C, 0.0025 M H_2SO_4 eluent, flow rate 0.6 mL min^{-1}). The OA concentrations were calculated through a standard calibration curve (20–1000 mg L^{-1} of each organic acid). The COD and the $N\text{-}NH_4^+$ content (filtered at 0.45 μm) were determined using cuvette test kits (Macherey-Nagel, Germany).

PHAs were determined by GC MS using a method adapted from Serafim et al. in 2004 [22]. Lyophilized biomass was incubated for methanolysis in a 20% v/v H_2SO_4 in MeOH solution (1 mL) and extracted with chloroform (1 mL). The mixture was digested at 100 °C for 3.5 h. After the digestion step, the organic phase (methylated monomers dissolved in chloroform) was extracted and injected (1 mL) into a gas chromatograph equipped with a detector (7980, Agilent Technologies, USA) and a ZB-Wax column (30 m, 0.25 mm internal diameter, 0.25 μm film thickness, Zebron, Phenomenex, USA), using helium as a carrier gas at 1.0 mL min^{-1}. Samples were analysed under a temperature regime starting at 40 °C, increasing to 100 °C at a rate of 20 °C min^{-1}, to 175 °C at a rate of 38 °C min^{-1} and reaching a final temperature of 220 °C at a rate of 20 °C min^{-1} for ensuring cleaning of the column after each injection. Injector and detector temperatures were at 280 and 230 °C, respectively. Hydroxybutyrate (HB) and hydroxyvalerate (HV) concentrations were determined through the use of two calibration curves, one for HB and another for HV, using standards (0.1–8 g L^{-1}) of a commercial P (HB–HV) (88/12%) (Sigma-Aldrich, Germany), and corrected using heptadecane as internal standard (concentration of approximately 1 g L^{-1}) (Sigma Aldrich, Germany).

PHA and active biomass growth yield calculation

The PHA content in cells was referred to VSS on a mass basis [PHA = (g kg^{-1} VSS)], considering VSS to be constituted by both active biomass (X) and PHA [10]. PHA was converted into COD according to the following oxidation stoichiometry: 1.67 mg COD mg^{-1} HB monomer and 1.92 mg COD mg^{-1} HV monomer [23].

Acetate, butyrate and lactate were considered as HB precursors, valerate and propionate as HV precursors [10]. X was calculated on a COD basis considering that 1 g of X contains 1.42 g of COD [23]. For the SBR, PHA storage yield was expressed in COD and referred to both COD consumed ($COD_{cons.}$) and OA consumed ($COD_{OA\text{-}cons.}$), calculated, respectively, as the ratio between the amount of PHA accumulated during the feast phase (COD_{PHA}) and the amount of COD depleted or OA depleted.

In the accumulation batches, these PHA storage yields were calculated as described before, for each pulse. In order to compare different accumulation tests, the average values of the first three pulses and for each parameter were considered [23].

In accumulation batches, PHA storage yield was also related to COD fed (COD_{in}) and OA fed (COD_{OAin}); moreover, it was also related back to total solids of OFMSW ($OFMSW_{TS}$) and to $OFMSW_{w.w.}$.

The X growth yield in the SBR was expressed in COD and referred to COD consumed ($COD_{cons.}$), calculated as the ratio between the new X produced during the feast phase (COD_X) and the amount of COD depleted [23].

PHA characterization by solid-state NMR and solution ^1H NMR, and molecular weight

Solid-state ^{13}C NMR spectra of lyophilized biomass containing PHA were recorded on a Bruker Avance 300 spectrometer operating at 75.47 MHz, using a 4 × 21 mm cylindrical zirconium rotor spun at 11,000 Hz to avoid the side bands. The ^{13}C cross-polarization magic angle spinning (CPMAS) NMR spectra were acquired using recycle delay of 8 s, ^1H 90 pulse length of 3.5 µs, 1 m contact time, acquisition time of 35 ms and from 1K to 4K scans. The ^{13}C single-pulse excitation (SPE) NMR spectra were recorded with delays of 160 s and 1–2K scans.

The chemical shifts were recorded relative to tetramethylsilane via benzene as a secondary reference.

Liquid NMR experiments on extracted PHA were performed on a Bruker 500 MHz AVANCE III NMR spectrometer (Bruker GmbH, Germany) with a 5-mm TCI cryoprobe. Deuterated chloroform (99.96%, Sigma Aldrich) was used as solvent. ^1H NMR spectra were recorded at 303 K using recycle delay of 12 s, 64K fid size and 64 scans.

PHA molecular weight was determined by HP-SEC/TDA measurement. The HPLC equipment consisted of a Viscotek system (Malvern Instrument Ltd, Malvern, UK) equipped with a Knauer HPLC pump K501, and a Biotech Degasi GPC degassing device. The detector system was a Viscotek mod. 302 Triple Detector Array (TDA), which is composed by Laser Light Scattering detector (90° and 7°; wavelength 670 nm), Refractive index (RI) detector [cell volume of 12 µL; light-emitting diode (LED) at 660 nm wavelength] and Viscosimeter detector (four capillaries with a differential Wheatstone bridge configuration). A PL GEL 20-um MIXED A column (7.5 × 300 mm) was used. Chloroform was used as the mobile phase at a flow rate of 1 mL min^{-1}. Columns, injector and detectors were maintained at 30 °C. Samples were dissolved in chloroform at concentrations of 2–8 mg mL^{-1} and filtered on a 0.2-µm membrane before injection. Injection volume was of 100 µL.

The system was calibrated with the PS narrow standard of known Mw, polydispersity and intrinsic viscosity (Malvern PolyCAL PS std 105 k). Using a standard PHA sample at different concentrations (2.3, 4.3, 6.0, 8.5 mg mL^{-1}), the differential refractive index increment (dn/dc) value was found to be equal to 0.024 and used for further calculations.

Statistical analysis

Average and standard deviation values were calculated according to standard procedures and the results were analysed by an ANOVA test. A Tukey's test was used to compare mean values and to assess the significance of the differences between mean values, adopting a fixed effects model, i.e. the digestate flow adopted modified the chemical composition of both the percolates produced and the supernatants derived.

All statistical analyses were carried out using the SPSS statistical software, version 15.0 (SPSS, Chicago, IL, USA).

Results

Percolation trials

Percolation trials carried out by adopting different irrigation regimes, i.e. varying digestate flow rate, showed differences in terms of total percolate produced, i.e. an increase of the hourly digestate flow rate determined an increase of the total percolate produced (Table 1). Again, the production of total OA increased by more than 200% when digestate flow rate doubled, when comparing Trial 1 with Trial 2 (Table 1). However, the further increase of digestate flow rate (Trial 3) did not lead to any increase of OA production and on the contrary, slightly decreased it.

Percolates obtained were similar in terms of chemical characteristics apart from the percolate coming from the third trial that appeared more diluted as regarded OA and $N\text{-}NH_4^+$ content (Table 1).

Trial 2 optimized OFMSW hydrolysis and subsequent acidogenesis fermentation, producing the highest amount of OA.

OFMSW-supernatants

Pre-treated percolates used for MMC production ($OFMSW\text{-}supernatants_{in}$) showed no statistical differences in terms of COD and $N\text{-}NH_4^+$ contents, apart from the P content and HB/HV precursors ratio which were higher for $OFMSW\text{-}supernatant_{in}$ 3 than for the other $OFMSW\text{-}supernatants_{in}$ (Table 2).

OFMSW-supernatants used to perform PHA accumulation tests ($OFMSW\text{-}supernatants_{acc.}$) were again quite similar to each other, with the exception for P content and HB/HV precursors ratio which were higher for

Table 1 Percolation trials conditions (a) and composition of OFMSWs and digestates used and of percolates produced (b)

Trial	T (°C)	Duration (day)	OFMSW (kg)	Digestate flow rate ($L\ h^{-1}\ kg^{-1}$ $OFMSW_{w.w.}$)	Percolate produced ($L\ kg^{-1}$ $OFMSW_{w.w.}$)	OA produced[a] ($g\ kg^{-1}$ $OFMSW_{w.w.}$)
a						
Trial 1	30	21	44	0.02	8.03	60.5
Trial 2	30	21	42.7	0.04	19.3	151
Trial 3	40 (day 1–8) 30 (day 9–21)	21	43.6	0.08 (1–8 day) 0.04 (9–21 day)	25.57	139

Trial	TS (%)	VS (%)	pH	OA[a] ($mg\ L^{-1}$)	$N\text{-}NH_4^+$ ($mg\ L^{-1}$)
b					
Trial 1, 2, 3					
OFMSW (average)	24.5 ± 3.9	86.8 ± 2.6	4.3 ± 0	11,663 ± 1.649	304 ± 239
Trial 1, 2, 3					
Digestate (average)	2.2 ± 0.2	70.8 ± 2.7	8.32 ± 0.1	3448 ± 1.591	2065 ± 343
Trial 1					
Percolate 1 (P1)	2.4 ± 0.49a[b]	71.2 ± 2.2b	6.91 ± 0.19a	7540 ± 1,474b	2231 ± 130b
Trial 2					
Percolate 2 (P2)	1.9 ± 1.1a	66.5 ± 2.8a	6.99 ± 0.61a	7831 ± 1,775b	2412 ± 472b
Trial 3					
Percolate 3 (P3)	2 ± 0a	73 ± 2a	6.99 ± 0.32a	5437 ± 818a	1784 ± 117a

[a] OA expressed as acetic acid, determined according to the acid titration method [27]

[b] Limited to percolates, values in the same column followed by the same letter are not statistically different (ANOVA, Tukey test, $p < 0.05$)

Table 2 Chemical composition of the substrates used in the MMC selection and PHA accumulation processes

Substrate	COD ($mg\ L^{-1}$)	$N\text{-}NH_4^+$ ($mg\ L^{-1}$)	P tot ($mg\ L^{-1}$)	C ($mmol\ L^{-1}$)	$N\text{-}NH_4^+$ ($mmol\ L^{-1}$)	P ($mmol\ L^{-1}$)	Organic acids composition (%, weight basis, HB precursors/HV precursors)
OFMSW-supernatant$_{in}$ 1	1304 ± 10a[a]	46.7 ± 4.9a	3.1 ± 0.3a	34.4 ± 0.3a	2.6 ± 0.3a	0.1 ± 0a	59.7/40.3
OFMSW-supernatant$_{in}$ 2	1322 ± 37°	56.2 ± 7.5a	3.6 ± 0.4a	34.9 ± 1a	3.1 ± 0.4a	0.12 ± 0.01a	51.2/48.8
OFMSW-supernatant$_{in}$ 3	1231 ± 63a	58.4 ± 8.1a	5.6 ± 0.6b	32.5 ± 1.7a	3.2 ± 0.5a	0.18 ± 0.02b	67.8/32.2
OFMSW-supernatant$_{acc.}$ 1	7591 ± 54a	28 ± 4a	16.9 ± 2.3a	200 ± 2a	1.6 ± 0.2a	0.55 ± 0.07a	57.2/42.8
OFMSW-supernatant$_{acc.}$ 2	7879 ± 87a	29.1 ± 2a	19.8 ± 3a	207.8 ± 2.3a	1.6 ± 0.1a	0.64 ± 0.11a	52/48
OFMSW-supernatant$_{acc.}$ 3	7253 ± 98a	26.8 ± 3a	31.1 ± 5.2b	191.3 ± 2.6a	1.5 ± 0.2a	1 ± 0b	65.1/34.9

[a] By separating OFMSW-supernatants$_{in}$ from OFMSW-supernatants$_{acc.}$, values in the same column followed by the same letter are not statistically different (ANOVA, Tukey test, $p < 0.05$)

OFMSW-supernatant$_{acc.}$ 3 than for the other OFMSW-supernatants$_{acc.}$ (Table 2).

Mixed microbial culture enrichment

The amount of PHA stored at the end of the feast phase and PHA storage yields, measured for the mixed microbial culture during the selection process, remained quite stable (Table 3) during the first 2 months of selection when OFMSW-supernatants$_{in}$ 1 and 2 were used as substrates. In particular, PHA stored at the end of the feast phase was subject to small fluctuations with an average value of 363 ± 30 g kg^{-1} VSS. PHA storage yields (expressed as COD) on COD and on OA consumed were, on average, of 0.52 ± 0.07 mg mg^{-1} $COD_{cons.}$, and of 0.86 ± 0.23 mg mg^{-1} $COD_{OA\text{-}cons.}$, respectively, indicating the main role of OA among different carbon sources in PHA production [7, 11].

Table 3 Parameters characterizing the MMC selection and PHA accumulation processes with the three OFMSW-supernatants

Substrate	PHA content[a] (g kg^{-1} VSS)	Polymer composition (ΔHB/ΔHV) (% w/w)	PHA yield[c] (mg mg^{-1} $COD_{cons.}$)	PHA yield[d] (mg mg^{-1} $COD_{OA\text{-}cons.}$)	PHA yield[e] (g kg^{-1} COD_{in})	PHA yield[f] (g kg^{-1} OA_{in})	PHA productivity (g L^{-1} day^{-1})	X growth yield[g] (mg mg^{-1} $COD_{cons.}$)
Selection								
OFMSW-supernatant$_{in}$ 1	342 ± 23	–	0.51 ± 0.02	0.85 ± 0.26	–	–	–	0.16 ± 0.01
OFMSW-supernatant$_{in}$ 2	383 ± 2	–	0.54 ± 0.07	0.87 ± 0.09	–	–	–	0.05 ± 0.01
OFMSW-supernatant$_{in}$ 3	234 ± 20	–	0.42 ± 0.04	1.15 ± 0.34	–	–	–	0.09 ± 0.01
Accumulation								
OFMSW-supernatant$_{acc.}$ 1	476 ± 41[b]	56.5 ± 3.5/43.5 ± 3.5	0.44 ± 0.17	1.13 ± 0.18	134 ± 22	230 ± 38	4.4 ± 1.3	–
OFMSW-supernatant$_{acc.}$ 2	465 ± 21[b]	54.5 ± 2.6/45.5 ± 2.6	0.52 ± 0.1	0.75 ± 0.21	128 ± 16	220 ± 28	4.3 ± 1.2	–
OFMSW-supernatant$_{acc.}$ 3	405 ± 28[b]	57 ± 2/43 ± 2	0.5 ± 0.1	0.66 ± 0.09	94 ± 5	161 ± 9	7 ± 0.9	–

[a] PHA stored at the end of the feast phase referred to VSS

[b] PHA accumulated at the end of the accumulation test referred to VSS

[c] PHA storage yield expressed as COD (COD_{PHA}) referred to $COD_{cons.}$

[d] PHA storage yield expressed as COD (COD_{PHA}) referred to $COD_{OA\text{-}cons.}$

[e] PHA produced referred to COD_{in}

[f] PHA produced referred to OA_{in}

[g] X growth yield during feast phase expressed as COD (COD_X) referred to $COD_{cons.}$

The culture fed with OFMSW-supernatant$_{in}$ 3 gave different results as the polymer content at the end of the feast phase and PHA storage yield on COD consumed were lower by 36 and 20% in comparison with the data reported for OFMSW-supernatants$_{in}$ 1 and 2. On the other hand, PHA storage yield on OA consumed increased by 34%. In any case, performance obtained with OFMSW-supernatant$_{in}$ 3 was better than that reported in the literature for similar substrates (OFMSW leachate) [17, 18] and it was comparable to those obtained with other waste substrates (e.g. fermented molasses and fermented cheese whey) [10, 29]. Probably, differences found in terms of reduction of polymer content were because OFMSW-supernatant$_{in}$ 3 was characterized by a higher presence of P that reduced PHA accumulation [30].

Feast-to-famine ratio obtained during all selection processes using the three OFMSW-supernatants$_{in}$ (F/F ratio ranged from 0.03 to 0.15) (Fig. 2), confirmed the good performance of the selection procedures, indicating successful enrichment in PHA-accumulating bacteria [23].

PHA production

Accumulation tests (Table 3) revealed a better performance for the cultures selected with OFMSW-supernatants$_{in}$ 1 and 2 than that of 3 in terms of PHA accumulation in relation to VSS. Maximum PHA content at the end of the tests fed with OFMSW-supernatants$_{acc.}$ 1 and 2 was of 467 ± 41 g kg^{-1} VSS and 465 ± 21 g kg^{-1} VSS, respectively, which were about 13% higher than that obtained in the tests fed with OFMSW-supernatant$_{acc.}$ 3 (Table 3). These results were similar to those obtained in previous work using other waste substrates (e.g. palm oil mill effluent, fermented olive oil mill pomace, crude glycerol) [31–33], indicating a reliable performance of the cultures in accumulation tests. In terms of PHA storage yields (expressed as COD) on COD consumed, PHA accumulation carried out by using the three supernatants gave substantially similar results when OFMSW-supernatants$_{acc.}$ 2 and 3 were used, i.e. 0.5 mg mg^{-1} COD$_{cons.}$. These data were higher than those detected for the accumulation trial using OFMSW-supernatant$_{acc.}$ 1.

PHA storage yields (expressed as COD) referred to as consumed-OA were very different for different supernatants. For example, PHA yield was reduced from 1.13 to 0.66 mg mg^{-1} COD$_{OA\text{-}cons.}$ when comparing the trial performed using the OFMSW-supernatant$_{acc.}$ 1 to the one using the OFMSW-supernatant$_{acc.}$ 3. Residual OA could be potentially reused to produce PHA or destined to the production of biogas and energy, supporting the entire process throughout anaerobic digestion. These solutions have not been considered in this work.

Trials performed using the OFMSW-supernatants$_{acc.}$ 1 and 2, gave the highest PHA production with respect to the total COD and OA fed, i.e., as average, 130 ± 16 g kg^{-1} COD$_{in}$ and 223 ± 28 g kg^{-1} OA$_{in}$, that

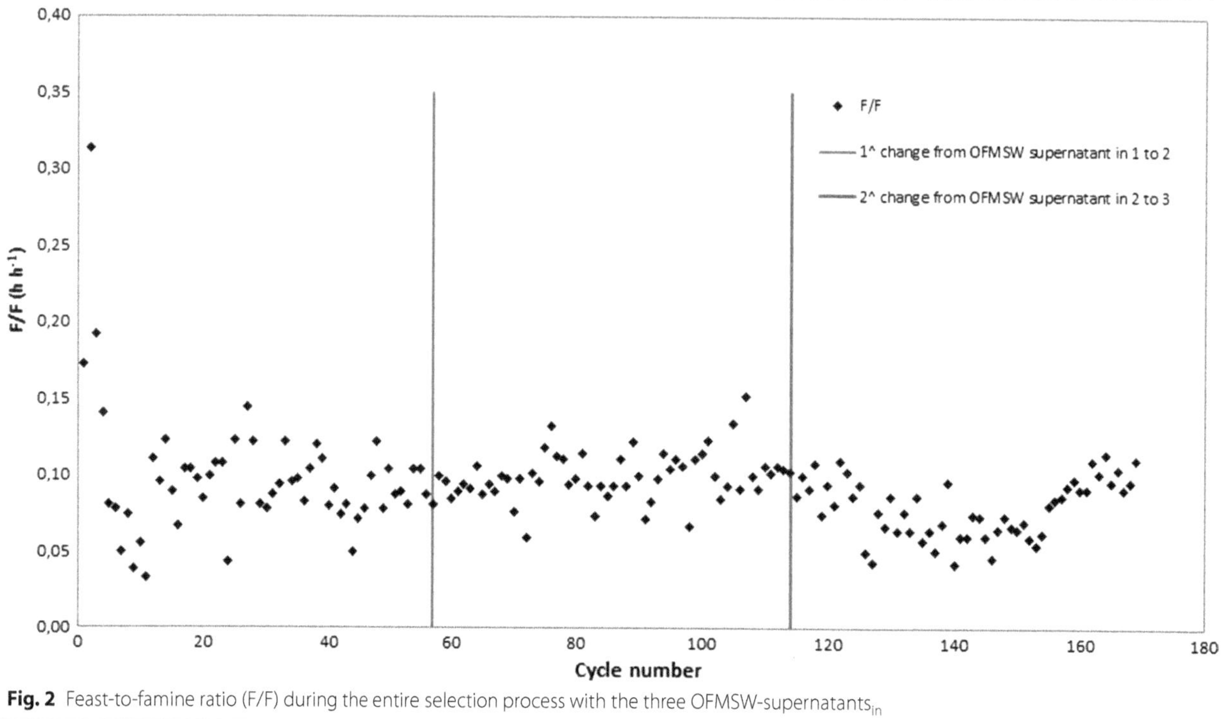

Fig. 2 Feast-to-famine ratio (F/F) during the entire selection process with the three OFMSW-supernatants$_{in}$

were 30% higher than data obtained for the culture fed with the OFMSW-supernatant$_{acc.}$ 3. Lower PHA production was probably due to the decrease of biomass which occurred in the SBR during the last month (data not reported).

PHA composition was constant among the three different accumulation tests performed, with an average value of 56 ± 3% of HB and 44 ± 3% of HV, as detected by the GC–MS approach. This composition was similar to that obtained by Amulya et al. in 2015 [17] using fermented OFMSW as substrate.

Discussion

PHA production and mass balance

Results above discussed indicated that supernatants obtained worked well as substrates to select and enrich MMC in PHA-storing bacteria. In particular, PHA content at the end of the feast phase and PHA yield on COD and on OA consumed obtained in this work were comparable or higher than those reported in the literature for selection processes of mixed microbial cultures obtained by feeding similar OFMSW leachate [17, 18] or other waste substrates (e.g. fermented molasses and fermented cheese whey) [10, 29]. Best performances obtained could be due to the fact that in this work culture enrichment was performed at 30 °C, which is an optimal temperature for microbial activity, and because a more balanced C:N ratio (close to 10) was adopted with respect to the lower values adopted by other authors [15, 18].

The good results above discussed allowed us to get PHA contents during PHA accumulation tests higher than those reported in the literature by Amulya et al. in 2015 [17] and comparable to the data reported by Zhang et al. in 2014 [16], both working on food wastes (Table 4). On the other hand, PHA contents were clearly lower than those obtained by Korkakaki et al. in 2016 [18], who obtained PHA contents of 778–784 g kg^{-1} VSS (Table 4) by feeding fermented food wastes to a MMC, achieving very similar results to those reported for pure cultures (860–870 g kg^{-1} VSS by using *Cupriavidus necator*) [34, 35] (Table 4). Nevertheless, it should be specified that those data were obtained by using a mix of synthetic volatile fatty acids (VFAs) (75–90% w/w) and OFMSW leachate (10–25% w/w) during the selection procedure. Korkakai et al. in 2016 reported that the use of pure leachate as the sole substrate to select PHA-accumulating bacteria was less successful than the use of a feed made mainly by synthetic VFAs, because the microbial culture selected with VFAs maintained a high PHA storage capacity even if fed with pure leachate.

The PHA yields obtained, i.e. 0.44–0.52 mg mg^{-1} $COD_{cons.}$ and 0.65–1.13 mg mg^{-1} $COD_{OA\text{-}cons.}$ (Table 4), were very good and comparable to those reported in the literature by using other waste substrates to produce PHA (e.g. fermented olive oil mill effluent, fermented molasses, fermented cheese whey) [10–12]. Moreover, results from this work were higher than that reported by Zhang et al. in 2014 [16] concerning the use of fermented food waste and sewage sludge as the substrate for PHA production (Table 4).

Table 5 reports the mass balance of the entire process: the percolation process producing OA (first stage) plus PHA accumulation (second stage).

Percolation trials indicated that the process parameters adopted for Trial 2 determined the highest percolate and OA production per kg of OFMSW treated, giving the best trial results among the three tested. This evidence together with the very good performance in converting OA into PHA (second stage) using the OFMSW-supernatant$_{acc.}$ 2, resulted in the highest PHA production among the three two-stage processes, i.e. 33.22 ± 4.2 g kg^{-1} of $OFMSW_{w.w.}$, corresponding to 114.4 ± 14.5 g kg^{-1} of $OFMSW_{TS}$, almost 5 times higher than that reported by Rhu et al. in 2003 [15] (25 g of PHA kg^{-1} of dry food waste).

PHA characterization

Mass balance previously discussed indicated that Trial 2 gave the best results in terms of total PHA yield. Therefore, PHA characterization was focused on the products obtained in Trial 2.

Solid-state ^{13}C NMR was used to evaluate the content of PHA in the generated biomass and to monitor the PHA extraction process. ^{13}C cross polarization NMR technique, based on the carbon magnetization transferred from ^{1}H protons, could not be used for quantification of different chemical species/functional groups. In contrast, single-pulse excitation (SPE) NMR is based on the direct ^{13}C nuclei polarization and, even if less sensitive than CP MAS NMR, can be optimized for quantitative evaluation analysis. The most critical parameter, recycle delay, was set at 160 s, in order to cover T1 values of each ^{13}C type [36]. The obtained ^{13}C SPE NMR of the reference PHA standard, biomass enriched in PHA and the biomass after PHA extraction are shown in Fig. 3.

Together with the PHA signals, the biomass spectrum contains signals compatible with proteins (large CO peak at 170–185 ppm), minor polysaccharides (large peak of anomeric O–C–O at 100–110 ppm), etc. It is worth noting that PHA-related peaks are sharper than the other signals, indicating that these polymers were present, at least partially, in their semi-crystalline form.

In spite of the optimized acquisition parameters, quantitative analysis of PHA in biomass was limited by strong signal overlapping, especially in the aliphatic region (0–80 ppm) (Fig. 3). As a tentative for PHA quantitative

Table 4 Comparison between studies related to PHA production using fermented OFMSW as substrate and this study

Culture	Substrate	Organic acids composition[a] (%, weight basis, acetate/n-butyrate/propionate/valerate/isobutyrate/lactate)	PHA content[b] (g kg^{-1} VSS)	Polymer composition PHB/PHA (w/w)	PHA yield[c] (mg mg^{-1} $COD_{cons.}$)	PHA yield[d] (mg mg^{-1} $COD_{OA\text{-}cons.}$)	PHA yield[e] (g kg^{-1} $OFMSW_{TS}$)	References
MMC	Fermented OFMSW	30/0/70/0/0/0	600	0.77	–	–	25	[15]
MMC	Fermented food waste and sewage sludge	–	477	0.38	–	0.54	–	[16]
MMC	Fermented OFMSW	51.6/22.8/21.1/1.8/2.8/0	237 ± 1	0.61	0.17	–	–	[17]
MMC	Leachate	–	784	–	–	–	–	[18][f]
MMC	Leachate	–	778	–	–	–	–	[18][g]
MMC	Percolate 1	39.8/15.3/40.3/0/0/4.6	476 ± 41	0.57 ± 0.04	0.44 ± 0.17	1.13 ± 0.18	62.9 ± 10.5	This study
MMC	Percolate 2	37.1/10.2/46.1/2.7/2.2/1.7	465 ± 21	0.54 ± 0.02	0.52 ± 0.10	0.75 ± 0.16	114 ± 14	This study
MMC	Percolate 3	45/20/29.4/2.8/1.1/1.7	405 ± 28	0.57 ± 0.01	0.50 ± 0.10	0.65 ± 0.06	100 ± 6	This study
Cupriavidus necator CCGUG 52238	Fermented food waste	6/0/0/0/0/94	860	1	–	–	–	[34]
Cupriavidus necator H16	Fermented food waste	–	870	1	–	–	–	[35]

[a] See Additional file 1

[b] PHA accumulated at the end of the accumulation test referred to VSS

[c] PHA storage yield expressed as COD (COD_{PHA}) referred to $COD_{cons.}$

[d] PHA storage yield expressed as COD (COD_{PHA}) referred to $COD_{OA\text{-}cons.}$

[e] PHA produced referred to $OFMSW_{TS}$

[f] Accumulation test with a biomass selected with a substrate made of 90% synthetic VFAs and 10% pre-treated leachate

[g] Accumulation test with a biomass selected with a substrate made of 75% synthetic VFAs and 25% pre-treated leachate

Table 5 PHA yield on OFMSW weight unit

	First stage		Second stage	Global process	
	Percolate produced ($L\ kg^{-1}$ $OFMSW_{w.w.}$)	OA produced[a] ($g\ kg^{-1}$ $OFMSW_{w.w.}$)	PHA yield[b] ($g\ kg^{-1}$ OA_{in})	PHA yield[c] ($g\ kg^{-1}$ $OFMSW_{TS}$)	PHA yield[d] ($g\ kg^{-1}$ $OFMSW_{w.w.}$)
Trial 1	8.03	60.5	230 ± 38	62.9 ± 10.5	13.9 ± 2.3
Trial 2	19.3	151	220 ± 28	114 ± 14	33.2 ± 4.2
Trial 3	25.6	139	161 ± 9	100 ± 6	22.4 ± 1.3

[a] OA expressed as acetic acid, determined according to the acid titration method [27]

[b] PHA produced referred to OA_{in}

[c] PHA produced referred to $OFMSW_{TS}$

[d] PHA produced referred to $OFMSW_{w.w}$

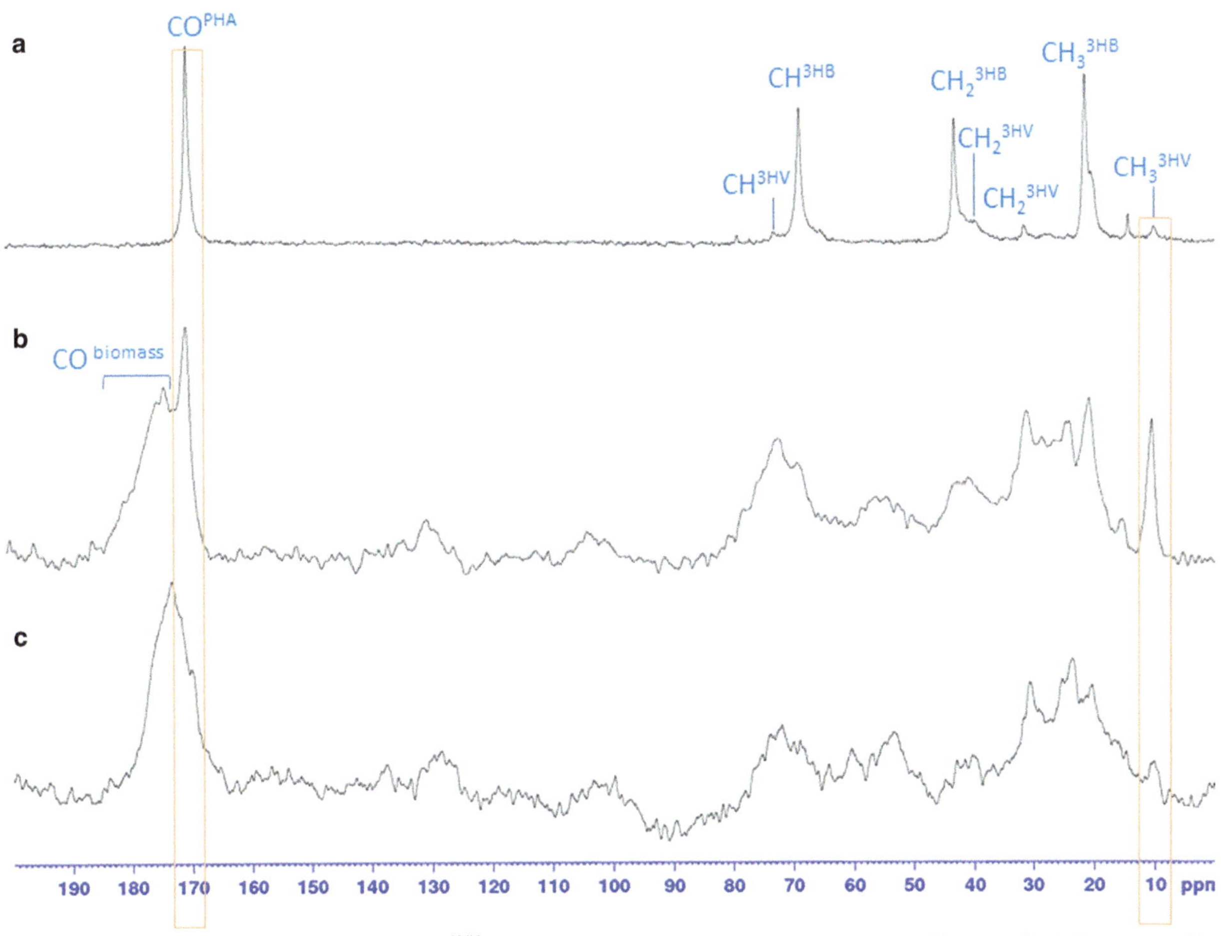

Fig. 3 ^{13}C SPE NMR spectra of the reference pure $PHA^{88/12}$ (**a**), original biomass containing PHA (**b**) and exhausted biomass after PHA extraction (**c**). *Frames* indicate CO and $CH_3$3HV signals related to PHA polymer that can be used for monitoring the synthesis and extraction process

evaluation in the original biomass, a ratio between CO^{PHA} and $CO^{biomass}$ was estimated as 30–35 molar %, using a Topspin software for peak integral simulation. Since the result significantly depends on the estimated signal width, such an approach could not be applied for the exhausted biomass, where the CO^{PHA} peak is not sufficiently high. It still can be clearly observed from the decreased signals of CO^{PHA} (~172 ppm) (Fig. 3) and well isolated CH_3^{3HV} (~10 ppm) (Fig. 3) that the biomass treated with chloroform contained PHA in lower quantity. Notably, even before extraction, it can be seen that the synthesized PHA are characterized by higher content of 3HV monomer that is in agreement with GC–MS data. Altogether, these data can be useful for monitoring the PHA content in biomass during their production and extraction.

In order to provide more detailed structural features of the extracted PHA polymers, solution ^{1}H NMR, GC–MS and GPC-TDA were applied along with solid-state NMR. By doing so, extracted PHA was characterized by ^{1}H NMR using standard PHA$^{88/12}$ for signal assignment (Fig. 4).

The PHA-related signals assigned using ^{1}H and COSY NMR (*not shown*) were in agreement with the previously published data [37]. The molar 3HB/3HV composition for commercial standard PHA$^{88/12}$ and biosynthesized PHA determined by ^{1}H NMR is reported in Table 6. Notably, quantitative data obtained for the biosynthesized PHA by ^{1}H NMR (53/47%) (Table 6) are also in agreement with GC–MS results.

Gel permeation chromatography with a TDA detection system was used to describe the molecular weight distribution of the biosynthesized PHA. Separation and detection conditions (including *dn/dc* calculation) were optimized using a commercial PHA$^{88/12}$ sample. Table 6 reports the average molecular weight Mw, polydispersity and hydrodynamic radius Rh determined for the reference PHA$^{88/12}$ and extracted PHA sample. In spite of a similar polydispersity, the biosynthesized PHA sample was characterized by higher molecular weight and

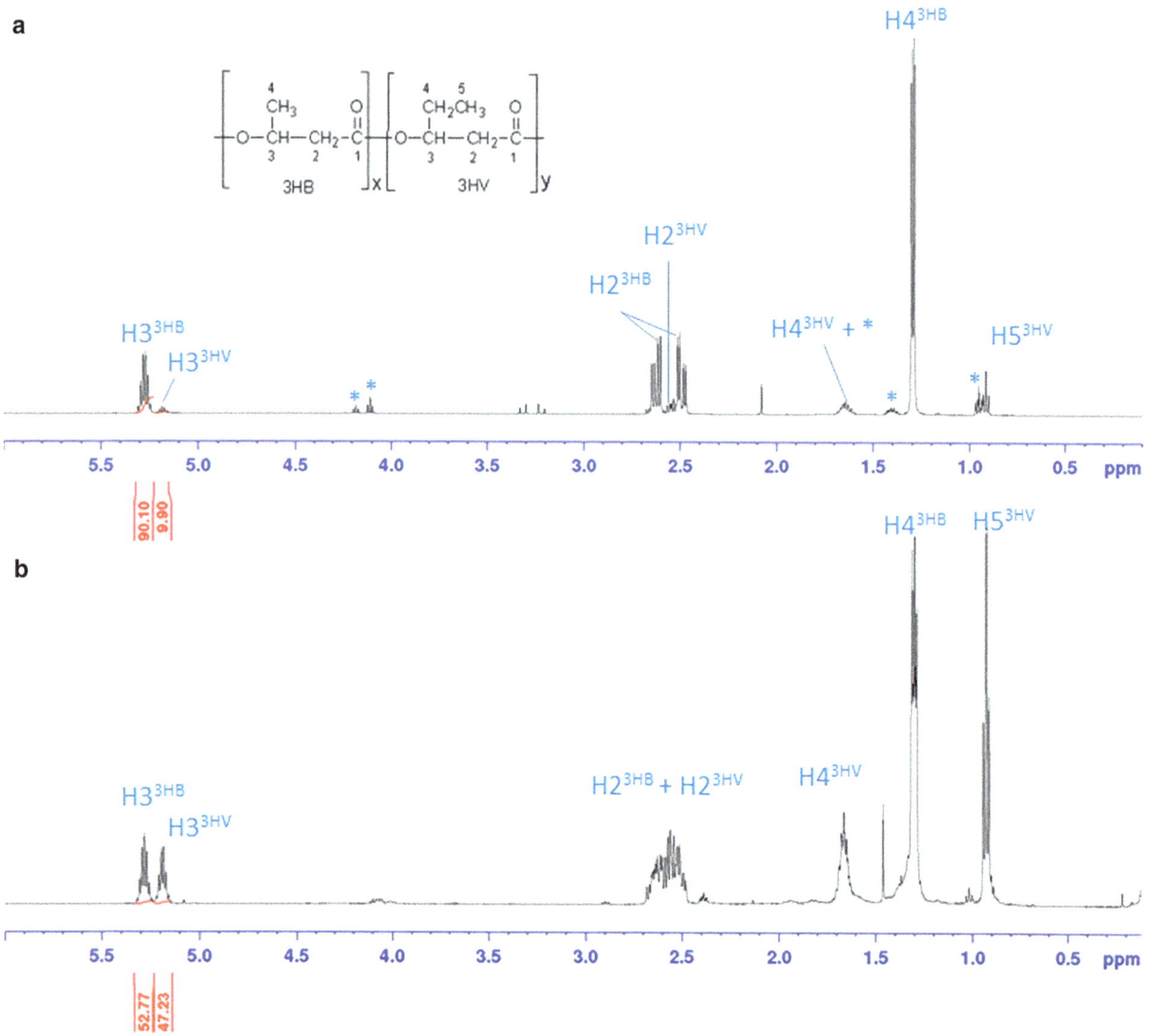

Fig. 4 ^{1}H NMR spectra of the reference PHA$^{88/12}$ (**a**) and the extracted PHA (**b**) in $CDCl_3$. *Asterisks* indicate the signals related to impurities present in commercial PHA sample

Table 6 Characterization of PHA produced in the accumulation stage of Trial 2 compared with commercial PHA$^{88/12}$

Products	HB/HV[a] (molar ratio)	Mw[b] (kDa)	Polydispersity	Rh[c] (nm)	a[d]	log$K$[d]
Commercial reference—PHA$^{88/12}$	90/10	$2{\cdot}10^5$	1.3	14	0.70	−3.8
Extracted PHA from Trial 2	53/47	$8{\cdot}10^5$	1.4	29	0.65	−3.4

[a] Molar ratio such as detected by NMR

[b] Molecular weight distribution determined by GPC-TDA

[c] Hydrodynamic Radius determined by GPC-TDA

[d] Mark–Houwink parameters a and log$K$

hydrodynamic radius. Mark–Houwink parameters a and log$K$, reflecting conformational behaviour of polymers in solution, are also reported in Table 6. The obtained a values for reference PHA$^{88/12}$ and extracted PHA are compatible with the values characteristic for flexible polymers in solution. Interestingly, these properties do not depend on the differences in molecular weight and chemical composition between the two samples.

Conclusions

Data obtained in this work can be useful as the starting point for considering both technical and economic feasibility of PHA production from OFMSW by using MMC.

The optimization of OFMSW acidic fermentation resulted a key step for the success of PHA production because it allows the production of a large amount of OA to be used to produce PHA. Moreover, the high concentrations of OA acted to prevent biogas (methane) production from consuming the substrate to be used to produce PHA. This kind of process could be rapidly implemented in a full-scale plant by adapting dry anaerobic digestion that is commonly used to produce biogas from household wastes.

This work, for the first time, reported that PHA yield referred to an OFMSW weight unit in an optimized process. Therefore, taking into consideration a PHA value of 2600–5800 \$ tonne^{-1} [38], data of total food waste produced in EU (87.6 million tonnes) and PHA yield obtained in this work (33.2 g kg^{-1} OFMSW$_{w.w.}$), in theory a total gross revenue of 7.6–16.9 billion \$ might be achieved.

Abbreviations

OA: organic acids; OFMSW: organic fraction of municipal solid waste; PHA: polyhydroxyalkanoates; FW: food waste; MMC: mixed microbial cultures; APBR: anaerobic percolation biocell reactor; SBR: sequencing batch reactor; OFMSW-supernatant$_{in}$: substrate for PHA selection phase; OFMSW-supernatants$_{acc.}$: substrate for PHA accumulation phase; W.W.: wet weight; TS: total solids; VS: volatile solids; TSS: total suspended solids; VSS: volatile suspended solids; COD$_{PHA}$: PHA expressed as COD; COD$_{in}$: COD fed; COD$_{cons.}$: COD consumed; OA$_{in}$: organic acids fed; COD$_{OA\text{-}cons.}$: organic acids consumed expressed as COD; X: active biomass; COD$_X$: active biomass expressed as COD; HB: hydroxybutyrate; HV: hydroxyvalerate.

Authors' contributions

BC designed and performed the experiments from the treatment of the percolates of OFMSW to PHA production, collected data, analysed data and drafted the manuscript. FF helped in performing and collecting data from the treatment of the percolates of OFMSW to PHA production. BS performed GC–MS analysis for PHA determination. TPS helped in designing the experiments from the treatment of the percolates of OFMSW to PHA production. GD designed the experiments related to percolates of OFMSW production. MP performed the experiments related to percolates of OFMSW production, collected data and analysed data. CC and AA designed the solid-state NMR experiments, and processed and interpreted the obtained data. CC performed the acquisition of solution and solid-state NMR spectra. GE conducted the SEC-TDA experiments and analysed the obtained data as well as interpreted the ^{1}H NMR spectra. AA gathered all the experimental data and wrote the SEC-TDA/NMR portion of the manuscript. FA designed, conceived and supervised the entire work, drafted and revised the manuscript. All authors read and approved the final manuscript.

Author details

[1] Gruppo Ricicla labs-DiSAA-Università degli Studi di Milano, Via Celoria 2, 20133 Milan, Italy. [2] Centro Alta Tecnologia Istituto di Ricerche Chimiche e Biochimiche G. Ronzoni Srl, Via Colombo 81, 20133 Milan, Italy. [3] Istituto di Ricerche Chimiche e Biochimiche G. Ronzoni, Via Colombo 81, 20133 Milan, Italy.

Acknowledgements

We acknowledge the support of Cap Holding (Peschiera Borromeo plant, Italy) who provided the sewage sludge necessary for carrying out MMC isolation for PHA production.

Competing interests

The authors declare that they have no competing interests.

Funding

This work was financed by Regione Lombardia, Italian Governement and European Community-Programma Operativo Regionale 2014–2020, Obiettivo "Investimenti in Favore della Crescita e dell'Occupazione Asse Prioritario I – Rafforzare la Ricerca, lo Sviluppo e l'Innovazione. Azione I.1.b.1.3—Sostegno alle attività collaborative di R&S per lo sviluppo di nuove tecnologie sostenibili, di nuovi prodotti e Servizi. Project ID 141082: *Renewable RAw materials valorisation for INnovative BiOplastic production from urban Waste* (RAINBOW).

References

1. Pfaltzgraff LA, De Bruyn M, Cooper EC, Budarin V, Clark JH. Food waste biomass: a resource for high-value chemicals. Green Chem. 2013;15:307–14.
2. Stenmarck Å, Jensen C, Quested T, Moates G. Estimates of European food waste levels. Stockholm: Swedish Environmental Research Institute; 2016. http://www.eu-fusions.org/phocadownload/Publications/Estimates%20of%20European%20food%20waste%20levels.pdf.
3. Girotto F, Alibardi L, Cossu R. Food waste generation and industrial uses: a review. Waste Manage. 2015;45:32–41.

4. Kiran EU, Trcinski AP, Ng WJ, Liu Y. Enzyme production from food wastes using a biorefinery concept. Waste Biomass Valorization. 2014;5:903–17.
5. Kiran EU, Trcinski AP, Liu Y. Platform chemical production from food wastes using a biorefinery concept. J Chem Technol Biotechnol. 2014;90:1364–79.
6. Tuck CO, Perez E, Horvath IT, Sheldon RA, Poliakoff M. Valorization of biomass: deriving more value from waste. Science. 2012;337:695–9.
7. Reis M, Albuquerque M, Villano M, Majone M. Mixed culture processes for polyhydroxyalkanoate production from agro-industrial surplus/wastes as feedstocks. In: Fava F, Agathos S, Young M, editors. Comprehensive biotechnology. Amsterdam: Elsevier; 2011. p. 669–83.
8. Salehizadeh H, Van Loosdrecht MCM. Production of polyhydroxyalkanoates by mixed culture: recent trends and biotechnological impor tance. Biotechnol Adv. 2014;22:261–79.
9. Ivanov V, Stabnikov V, Ahmed Z, Dobrenko S, Saliuk A. Production and applications of crude polyhydroxyalkanoate containing bioplastic from the organic fraction of municipal solid waste. Int J Environ Sci Tech. 2015;12:725–38.
10. Duque AF, Oliveira CSS, Carmo ITD, Gouveia AR, Pardelha F, Ramos AM, Reis MAM. Response of a three-stage process for PHA production by mixed microbial cultures to feedstock shift: impact on polymer composition. New Biotechnol. 2013;31:276–88.
11. Colombo B, Pepè Sciarria T, Reis M, Scaglia B, Adani F. Polyhydroxyalkanoates (PHAs) production from fermented cheese whey by using a mixed microbial culture. Bioresour Technol. 2016;218:692–9.
12. Dionisi D, Carucci G, Petrangeli Papini M, Riccardi C, Majone M, Carrasco F. Olive oil mill effluents as a feedstock for production of biodegradable polymers. Water Res. 2005;39:2076–84.
13. Lee WS, Chua ASM, Yeoh HK, Nittami T, Ngoh GC. Strategy for the biotransformation of fermented palm oil mill effluent into biodegradable polyhydroxyalkanoates by activated sludge. Chem Eng J. 2015;269:288–97.
14. Tamis J, Luzkov K, Jiang Y, Van Loosdrecht MCM, Kleerebezem R. Enrichment of *Plasticicumulans acidivorans* at pilot-scale for PHA production on industrial wastewater. J Biotechnol. 2014;192:161–9.
15. Rhu DH, Lee WH, Kim JY, Choi E. Polyhydroxyalkanoate (PHA) production from waste. Water Sci Technol. 2003;48:221–8.
16. Zhang M, Wu H, Chen H. Coupling of polyhydroxyalkanoate production with volatile fatty acid from food wastes and excess sludge. Process Saf Environ. 2014;92:171–8.
17. Amulya K, Jukuri S, Venkata Mohan S. Sustainable multistage process for enhanced productivity of bioplastics from waste remediation through aerobic dynamic feeding strategy: process integration for up-scaling. Bioresour Technol. 2015;188:231–9.
18. Korkakaki E, Mulders M, Veeken A, Rozendal R, Van Loosdrecht MC, Kleerebezem R. PHA production from the organic fraction of municipal solid waste (OFMSW): overcoming the inhibitory matrix. Water Res. 2016;96:74–83.
19. Pognani M, D'Imporzano G, Scaglia B, Adani F. Substituting energy crops with organic fraction of municipal solid waste for biogas production at farm level: a full-scale plant study. Process Biochem. 2009;44:817–21.
20. Massaccesi L, Sordi A, Micale C, Cucina M, Zadra C, Di Maria F, Gigliotti G. Chemical characterization of percolate and digestate during the hybrid solid anaerobic digestion batch process. Process Biochem. 2013;48:1361–7.
21. Di Maria F, Gigliotti G, Sordi A, Micale C, Zadra C, Massaccesi L. Hybrid solid anaerobic digestion batch: biomethane production and mass recovery from the organic fraction of solid waste. Waste Manag Res. 2013;31:869–73.
22. Serafim LS, Lemos PC, Oliveira RF, Reis MAM. Optimization of polyhydroxybutyrate production by mixed cultures submitted to aerobic dynamic feeding conditions. Biotechnol Bioeng. 2004;87:145–60.
23. Valentino F, Beccari M, Fraraccio S, Zanaroli G, Majone M. Feed frequency in a Sequencing Batch Reactor strongly affects the production of polyhydroxyalkanoates (PHAs) from volatile fatty acids. New Biotechnol. 2014;31:264–75.
24. Pardelha F, Albuquerque MGE, Reis MAM, Dias JML, Oliveira R. Flux balance analysis of mixed microbial cultures: application to the production of polyhydroxyalkanoates from complex mixtures of volatile fatty acids. J Biotechnol. 2012;162:336–45.
25. Kulkarni SO, Kanekar PP, Nilegaonkar SS, Sarnaik SS, Jog JP. Production and characterization of a biodegradable poly (hydroxybutyrate-co-hydroxyvalerate) (PHB-co-PHV) copolymer by moderately haloalkalitolerant *Halomonas campisalis* MCM B-1027 isolated from Lonar Lake, India. Bioresour Technol. 2010;101:9765–71.
26. The U.S. Department of Agriculture and The U.S. Composting Council. Test Methods for the Examination of Composting and Compost (TMECC). Houston: Edaphos Int; 2001.
27. Lahav O, Morgan B, Loewenthal RE. Rapid, simple and accurate method for measurement of VFA and carbonate alkalinity in anaerobic reactors. Environ Sci Technol. 2002;36:2736–3274.
28. Valentino F, Riccardi C, Campanari S, Pomata D, Majone M. Fate of β-hexachlorocyclohexane in the mixed microbial cultures (MMCs) three-stage polyhydroxyalkanoates (PHA) production process from cheese whey. Bioresour Technol. 2015;192:304–11.
29. Albuquerque MGE, Torres CAV, Reis MAM. Polyhydroxyalkanoate (PHA) production by a mixed microbial culture using sugar molasses: effect of the influent substrate concentration on culture selection. Water Res. 2010;44:3419–33.
30. Reddy MV, Mohan SV. Effect of substrate load and nutrients concentration on the polyhydroxyalkanoates (PHA) production using mixed consortia through wastewater treatment. Bioresour Technol. 2012;114:573–82.
31. Din MdMF, Mohanadoss P, Ujang Z, Van Loosdrecht MCM, Yunus SM, Chelliapan S, Zambare V, Olsson G. Development of Bio-PORec system for polyhydroxyalkanoates (PHA) production and its storage in mixed cultures of palm oil mill effluent (POME). Bioresour Technol. 2012;124:208–16.
32. Waller JL, Green PG, Loge FJ. Mixed-culture polyhydroxyalkanoate production from olive oil mill pomace. Bioresour Technol. 2012;120:285–9.
33. Moita R, Freches A, Lemos PC. Crude glycerol as feedstock for polyhydroxyalkanoates production by mixed microbial cultures. Water Res. 2014;58:9–20.
34. Omar FN, Rahman NA, Hafid HS, Mumtaz T, Yee PL, Hassan MA. Utilization of kitchen waste for the production of green thermoplastic polyhydroxybutyrate (PHB) by *Cupriavidus necator* CCGUG 52238. Afr J Microbiol Res. 2011;5:2873–9.
35. Hafuka A, Sakaida K, Satoh H, Takahashi M, Watanabe Y, Okabe S. Effect of feeding regimens on polyhydroxybutyrate production from food wastes by *Cupriavidus necator*. Bioresour Technol. 2011;102:3551–3.
36. Zhang L, Tang H, Hou G, Shen Y, Deng F. The domain structure and mobility of semi-crystalline poly(3-hydroxybutyrate) and poly(3-hydroxybutyrate-co-3-hydroxyvalerate: a solid-state NMR study. Polymer. 2007;48:2928–38.
37. Allen AD, Anderson WA, Ayorinde FO, Eribo BE. Biosynthesis and characterization of copolymer poly (3HB-co-3HV) from saponified *Jatropha curcas* oil by *Pseudomonas oleovorans*. J Ind Microbiol Biotechnol. 2010;37:849–56.
38. Gholami A, Mohkam M, Rasoul-Amini S, Ghasemi Y. Industrial production of polyhydroxyalkanoates by bacteria: opportunities and challenges. Minerva Biotecnol. 2016;28:59–74.

13

Metabolic engineering of *Rhodopseudomonas palustris* for the obligate reduction of *n*-butyrate to *n*-butanol

Devin F. R. Doud[1], Eric C. Holmes[1], Hanno Richter[1], Bastian Molitor[1], Georg Jander[2,3] and Largus T. Angenent[1,3,4*]

Abstract

Background: *Rhodopseudomonas palustris* is a versatile microbe that encounters an innate redox imbalance while growing photoheterotrophically with reduced substrates. The resulting excess in reducing equivalents, together with ATP from photosynthesis, could be utilized to drive a wide range of bioconversions. The objective of this study was to genetically modify *R. palustris* to provide a pathway to reduce *n*-butyrate into *n*-butanol for maintaining redox balance.

Results: Here, we constructed and expressed a plasmid-based pathway for *n*-butanol production from *Clostridium acetobutylicum* ATCC 824 in *R. palustris*. We maintained the environmental conditions in such a way that this pathway functioned as the obligate route to re-oxidize excess reducing equivalents, resulting in an innate selection pressure. The engineered strain of *R. palustris* grew under otherwise restrictive redox conditions and achieved concentrations of 1.5 mM *n*-butanol at a production rate of 0.03 g L^{-1} day^{-1} and a selectivity (i.e., products compared to the consumed substrate) of close to 40%. Since the theoretical maximum selectivity is 45%, the engineered strain converted close to its maximum selectivity.

Conclusions: The innate redox imbalance of *R. palustris* can be used to drive the reduction of *n*-butyrate into *n*-butanol after expression of a plasmid-based enzyme from a butanol-producing *Clostridium* strain.

Keywords: Metabolic engineering, Redox-driven obligate reduction, *n*-Butanol, *Rhodopseudomonas palustris*

Background

Rhodopseudomonas palustris, which is a member of the purple non-sulfur bacteria group, is often noted for its metabolic versatility. One of these metabolic activities, which is H_2 production by photofermentation, utilizes organic carbon under anaerobic conditions with light and has become an attractive application for bioenergy production [1]. The ability to produce H_2 at high concentrations is driven by an excess of reducing equivalents that accumulate during photoheterotrophic growth [2]. Here, we aim to exploit the excess reducing equivalents generated during photoheterotrophic growth to drive an engineered reduction reaction for the obligate production of the liquid fuel *n*-butanol (hereafter butanol).

When growing on acetate, McKinlay and Harwood [1] reported that approximately half of all reducing equivalents that were generated were used for biosynthesis, while the other half required re-oxidation by some alternate pathway such as CO_2 fixation, H_2 evolution, or a combination of both. These two pathways were found to be complementary in this role, and when one route was made unavailable, the other facilitated the re-oxidation

*Correspondence: l.angenent@uni-tuebingen.de
[4] Center for Applied Geosciences, University of Tübingen, Hölderlinstr. 12, 72074 Tübingen, Germany
Full list of author information is available at the end of the article

of all reducing equivalents with minimal changes occurring elsewhere in the central metabolism. However, when fed a more reduced substrate, such as *n*-butyrate (hereafter butyrate), it was observed that under non-H_2 evolving conditions, *R. palustris* could only grow if exogenous CO_2 was supplemented into the medium [2, 3]. This finding demonstrates that under these conditions, *R. palustris* generates more reducing equivalents than available electron sinks can accommodate, and ultimately suffers a lethal redox imbalance. Numerous studies have used these principles to improve H_2 yields from photofermentation using *R. palustris* [4–6].

Recently, Fixen et al. [7] engineered a strain of *R. palustris* to utilize excess reducing equivalents to reduce CO_2 into CH_4 by remodeling the nitrogenase enzyme active site and expressing the modified gene from the chromosome. A NifA* strain of *R. palustris* was chosen as the host because it expresses the nitrogenase gene constitutively even in the presence of NH_3. The result was a light-dependent *R. palustris* strain that produced CH_4 with ATP coming from cyclic photophosphorylation and electrons coming from either an organic substrate or thiosulfate. This reduction is, however, not obligatory and a large excess of H_2 was produced by the nitrogenase compared to CH_4 [7]. Another attractive route to redirect the excess reducing equivalents and to maintain redox balance when growing on butyrate is to produce the fuel butanol through an obligate reduction reaction. The advantage of the obligate reduction is the lack of side products to achieve a high selectivity (i.e., products compared to the consumed substrate).

Butanol is an attractive biofuel molecule due to its higher energy density, lower volatility, and complete intersolubility with traditional fuels compared to ethanol [8]. Butanol is traditionally produced via fermentation of monosaccharides by certain strains of Clostridia. *Clostridium acetobutylicum* ATCC 824, which has been widely studied for this activity, is generally seen as the model organism for solvent production and previously has been used as a source of genes for engineered butanol production [9–11]. In *C. acetobutylicum*, reduction of butyrate proceeds through a two-phase pathway. First, butyrate is activated to butyryl-CoA, which can occur by two different routes: (1) a two-step butyrate kinase/phosphotransbutyrylase reaction (Fig. 1); or (2) a one-step reaction with butyrate/acetoacetyl-CoA transferase transferring a CoA group from acetoacetyl-CoA to butyrate. Second, butyryl-CoA is reduced to butanol through the butyraldehyde intermediate [12, 13]. The AdhE2 enzyme from *C. acetobutylicum* ATCC 824 (hereafter $AdhE2_{824}$ and *adhE2*$_{824}$ for its gene) reduces butyryl-CoA to butyraldehyde and subsequently butanol (Fig. 1) [14]. Analysis of $AdhE2_{824}$ has revealed that the protein is

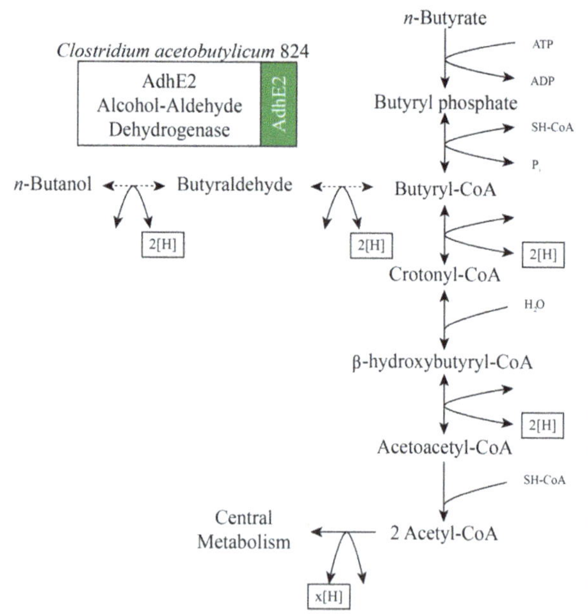

Fig. 1 Butyrate metabolism and the proposed route for butanol production in *R. palustris* by the activity of the butyryl-CoA to butanol AdhE2 enzyme. *Dotted arrows* represent the proposed pathway and *solid arrows* indicate native metabolism

a bifunctional NADH-dependent fusion protein with the N-terminal domain retaining highly conserved sequences of the aldehyde dehydrogenase family and the C-terminal domain retaining conserved sequences of the iron-containing alcohol dehydrogenase family [15]. A second alcohol dehydrogenase enzyme BdhB has also been found to aid in the reduction of butyraldehyde to butanol in *C. acetobutylicum*, though its activity is not necessary for butanol production [16].

In planning to introduce the butanol production pathway together with the existing catabolism of butyrate by *R. palustris*, we established that before butyrate can be converted to two molecules of acetyl-CoA via anaerobic β-oxidation, butyryl-CoA is first formed through native butyrate kinase/phosphotransbutyrylase enzymes (Fig. 1). Under photoheterotrophic redox imbalance, the first oxidation reaction for the metabolism of butyrate (i.e., butyryl-CoA to crotonyl-CoA in Fig. 1) would likely be inhibited by the thermodynamic limitations of the reduced cofactor pool. Because of this, we anticipated the butyryl-CoA intermediate to accumulate and provide an ideal target for shunting excess reducing equivalents for the production of butanol by introducing only *adhE2*$_{824}$ (the bifunctional aldehyde/alcohol dehydrogenase gene) into *R. palustris* (Fig. 1). After necessary optimizations, the production of butanol indicated that this assumption was correct.

A direct introduction of plasmid-based *adhE2*$_{824}$ did not result in the production of $AdhE2_{824}$ (and butanol production) due to the major incompatibilities in codon utilization preference between the *C. acetobutylicum* source and the *R. palustris* host. Therefore, we pursued two alternate attempts to introduce the activity of AdhE2 by using: (1) a gene homologous to $adhE2_{824}$ from *R. palustris* strain BisB18 (hereafter $adhE_{BisB18}$ and $AdhE_{BisB18}$ for its protein); and (2) a codon-optimized $adhE2_{824}$ (hereafter $adhE2_{opti}$ and $AdhE2_{opti}$ for its protein). In all our growth experiments, we omitted HCO_3^- (CO_2) from, and added NH_3 to, the growth medium to block the normal routes of shunting reducing equivalents, resulting in an obligate butyrate reduction. This strong selection pressure will only allow growth when butanol production occurs. In our hands, butanol production through the homologous $AdhE_{BisB18}$ did not work, but $AdhE2_{opti}$ enabled recovered growth through butanol production. Subsequently, we were able to increase the butanol selectivity and production rate in further experiments, accomplishing our aim of generating an engineered reduction reaction for the obligate production of the liquid fuel butanol in *R. palustris*.

Methods

Media and culture conditions

Rhodopseudomonas palustris CGA009 was cultured in anaerobic butyrate Rhodospirillaceae medium (RM) [17], consisting of 1.64 g $NaHCO_3$, 0.5 g KH_2PO_4, 0.5 g K_2HPO_4, 0.2 g $MgSO_4{\cdot}7\ H_2O$, 0.4 g NaCl, 0.05 g $CaCl_2{\cdot}2\ H_2O$, 0.01 g Fe-Citrate, 1.0 g $(NH_4)_2SO_4$, 1.12 mL 10 M NaOH, and 1 mL trace metals per liter. The trace-metal solution consisted of 0.25 mL concentrated HCl, 70.0 mg $ZnCl_2$, 100.0 mg $MnCl_2{\cdot}4\ H_2O$, 60.0 mg H_3BO_3, 200.0 mg $CoCl_2{\cdot}6\ H_2O$, 20.0 mg $CuCl_2{\cdot}2\ H_2O$, 20.0 mg $NiCl_2{\cdot}6\ H_2O$, and 40.0 mg $Na_2MoO_4{\cdot}2\ H_2O\ L^{-1}$. After autoclaving, 1.0 g sterile butyric acid was added as the substrate. No toxicity or inhibition of *R. palustris* growth was detected at this concentration (Additional file 1). 1 mL trace vitamins was added to the medium. The trace-vitamin solution consisted of 1.0 g *p*-aminobenzoic acid, 1.0 g thiamine, and 0.1 g biotin L^{-1}. Sterile medium (50 mL) with a final pH of 7.0 was added to autoclaved serum bottles and sparged for 20 min with sterile-filtered 80:20 N_2/CO_2 gas. For the growth of engineered strains under restrictive butyrate-to-butanol conversion conditions, $NaHCO_3$ was omitted from the medium, the addition of 10 M NaOH was increased to 2.3 mL total, and serum bottles were sparged for 20 min with N_2 instead of N_2/CO_2. In addition, 200 μg mL^{-1} of kanamycin sulfate was added for retention of the plasmid in all growth conditions with engineered strains. All serum bottle cultures were grown in an environmental growth chamber (GC8-2VH, EGC, Chagrin Falls, OH) at 30 °C with 80 μmol of photosynthetically active photons $s^{-1}\ m^{-2}$ (photons between 400 and 700 nm) illumination from both fluorescent and incandescent lamps.

DNA manipulation and cloning

We used the green fluorescence protein GFPmut2 [18] as a reporter protein to screen for the activity of different expression vectors in *R. palustris* under aerobic conditions. We tested two different plasmids as candidate expression vectors: (1) the endogenously derived pMG105 [19]; and (2) the broad-host pBBR1MCS-2 [20]. For detection by immunoreaction, we introduced a C-terminal 6X His tag *prior* to the termination site with PCR by the reverse primer (Table 1). *GFPmut2* was inserted into the pMG105 and pBBR1MCS-2 vectors under the control of a range of promoters,

Table 1 Sequences of primers used for this study

Primer	Length	Sequence (5′ → 3′)	Purpose
pBBR1MCS-2 vector			
XhoI GFPmut2 F1	60	GGCGGCCTCGAGAGGAGGATCTATTCATGAGTAAAGGAGAAGAACTTTTCACTGGAGTTG	Cloning *GFPmut2*
XbaI GFPmut2 R1	51	GCCGCCTCTAGACTATTTGTATAGTTCATCCATGCCATGTGTAATCCCAGC	
BamHI RBS $AdhE2_{824}$ F1	68	GCCGCGGATCCAGGAGGATCTATTCATGAAAGTTACAAATCAAAAAGAACTAAAACAAAA-GCTAAATG	Cloning $adhE2_{824}$
SacI 6X His $AdhE2_{824}$ R1	62	CACCCGGAGCTCTAAGTGGTGATGGTGATGATGAAATGATTTTATATAGATATCCTTAAGTT	
SacI $AdhE2_{824}$ R1	43	CACCCGGAGCTCTAAAATGATTTTATATAGATATCCTTAAGTT	
XhoI RBS $AdhE_{BisB18}$ F1	57	GACGACCTCGAGAGGAGGATCTATTCGTGACCTTATCTACCCCGTCCGACCTCGACA	Cloning $adhE_{BisB18}$
XbaI 6X His $AdhE_{BisB18}$ R1	60	CACCCGTCTAGACTAGTGGTGATGGTGATGATGTTCCGCGGCGTTCGCCGTCGCCACCGA	
XbaI $AdhE_{BisB18}$ R1	52	CAGTCACCCGTCTAGACTATTCCGCGGCGTTCGCCGTCGCCACCGACAATGT	
pMG105P vector			
XbaI GFPmut2 F1	60	GGCGGCTCTAGAAGGAGGATCTATTCATGAGTAAAGGAGAAGAACTTTTCACTGGAGTTG	Cloning *GFPmut2*
SalI GFPmut2 R1	53	GGCGGCCGGTCGACCTATTTGTATAGTTCATCCATGCCATGTGTAATCCCAGC	

Restriction sites are underlined

including the endogenous phosphoenolpyruvate carboxykinase (*ppckA*) from *R. palustris*, arabinose (*ara*), and lactose (*lac*) promoters. The $adhE_{BisB18}$ and $adhE2_{824}$ genes were amplified from *R. palustris* strain BisB18 and *C. acetobutylicum* ATCC 824, respectively, using Phusion HF polymerase (New England Biolabs, Ipswich, MA) and directionally ligated into the pBBR1MCS-2 expression vector using the XhoI/XbaI cloning sites. Restriction and ligation enzymes were purchased from Promega (Fitchburg, WI) and all constructs were verified with Sanger sequencing (Cornell University Genomics Facility, Ithaca, NY). *Escherichia coli* DH5α was used for routine cloning of constructs. Both *E. coli* and *R. palustris* were transformed by electroporation with a Bio-Rad Gene Pulser II Electroporator (Bio-Rad, Hercules, CA) with 1.8 kV, 200 Ω, 25 μF for *E. coli*, and 2.5 kV, 400 Ω, 25 μF for *R. palustris*. Electroporated cells were recovered in SOC for 1 and 3 h before plating on lysogeny broth (LB) agar plates with 50 and 200 μg mL^{-1} kanamycin sulfate for isolation of individual *E. coli* and *R. palustris* transformants, respectively.

To successfully produce $AdhE2_{824}$ in *R. palustris*, codon optimization was pursued. Optimization design was performed in-house and largely utilized the one amino acid one codon approach, except in cases of multiple codon repeats and unfavorable energetics of mRNA secondary structure (Additional file 2). $adhE2_{opti}$ with a C-terminal 6X His tag was synthesized by GenScript (Piscataway, NJ) with flanking XhoI/XbaI restriction sites that were used for directional cloning into pBBR1MCS-2.

SDS-PAGE and western blot detection

For both SDS-PAGE and subsequent western blot detection, 50 mL of late-log *E. coli* and *R. palustris* cells were harvested by centrifugation (10 min, 12,000×*g*, 4 °C) and resuspended in 1 mL 1X Laemmli sample buffer [21] with a protease inhibitor (Protease inhibitor cocktail, Promega, Fitchburg, WI). The suspension was lysed by ultrasonication on ice using three 30 s on 1 min off pulses of 20 W (Branson Sonifier 150, Emerson, Danbury, CT). After sonication, debris was removed by centrifugation and the total soluble protein concentration was determined by BCA assay (Pierce BCA protein assay kit, Thermo Scientific, Rockford, IL). Forty-five μL of 3 mg mL^{-1} soluble protein samples and 10 μL of Precision Plus Protein™ standard (Bio-Rad, Hercules, CA) were loaded into pockets of a Mini-PROTEAN® TGX gel (Bio-Rad, Hercules, CA) and run at 100 V for 1.5 h before blotting onto PVDF membranes at 100 V for 1 h. For the initial western blot, immunoreactive bands were developed using manufacturer-suggested protocols for the 1 h Western™ Standard Kit for rabbit (GenScript, Piscataway, NJ) and a 6X His primary antibody (Rabbit polyclonal, GeneTex, Irvine, CA). We used a similar protocol a second time to improve the sensitivity with these changes: we used 0.1 mL OD_{600} 1.0 cells with 0.1 mL Laemmli sample buffer. The samples were boiled for 10 min before centrifugation. 15 μL of each sample was loaded into separate gel lanes. For development, an enhanced chemiluminescence protocol was used using the manufacturer-suggested protocol for promega ECL western blotting substrate (Fitchburg, WI). The film was allowed to incubate with the PVDF membrane for 30 min *prior* to development.

Bioreactor configuration and operation

Prior to operation, sealed glass bioreactors were autoclaved and filled with 350 mL sterile RM with 1 g L^{-1} butyrate and without HCO_3^-. The bioreactors were equipped with a recirculating water jacket that was operated at 30 °C. Magnetic stir plates (IKA Works Inc., Wilmington, NC) provided stirring of the reactors at 450 rpm. The bioreactors were continuously sparged with sterile-filtered N_2 gas at a flow rate of 0.1 L min^{-1} to maintain an anaerobic environment. After sparging for 1 h to ensure anaerobic reactor conditions, 1 mL of actively growing *R. palustris* cells at an OD_{600} of 0.1 was inoculated into the bioreactor. The preculture of *R. palustris* pBBR1MCS-2 *lacp* $adhE2_{opti}$ was grown in anaerobic RM without HCO_3^-, while the preculture of *R. palustris* pBBR1MCS-2 *lacp GFPmut2* (control) was grown in aerobic RM without HCO_3^-. Gas entered the liquid phase of each bioreactor through a sparging stone to increase the gas/liquid interface for butanol stripping. Off-gas was connected to 2 Friedrich's condensers in parallel with recirculating water at 5 °C to condense the butanol that had been removed by gas stripping. Gas was released into the atmosphere after one passing through the condensers. The bioreactor volume was illuminated externally with a single 60-W incandescent lamp as used previously [22]. The bioreactors were operated in batch-mode and liquid samples were periodically taken for OD_{600} and metabolite measurements.

Metabolite detection

Butyrate, butanol, and acetate were measured via HPLC (600 HPLC, Waters, Milford, MA) with a refractive index detector and an Aminex HPX-87H column (Bio-Rad, Hercules, CA). The column temperature was set to 60 °C, and a 5-mM sulfuric acid eluent at a flow rate of 0.6 mL min^{-1} was used as the mobile phase. Butanol and butyraldehyde were measured with a gas chromatograph (HP 5890, Hewlett Packard, Palo Alto, CA) with a 7673 autoinjector and flame ionization detector. The gas chromatograph contained a custom-made packed bed glass column of 1.8 m × 2 mm id (Supelco, Sigma-Aldrich,

St. Louis, MO). The support matrix of the column was Chromosorb W/AW80 over 100 mesh; phases were preconditioned: phase A was 10% Carbowax-20M; phase B was 0.1% phosphoric acid. The inlet and detector temperatures were 220 and 240 °C, respectively. The column temperature profile was 100 °C for 2 min, a ramp of 40 °C min^{-1} to 180 °C with a 5-min hold. Butanol concentrations were verified by both HPLC and GC platforms.

Results and discussion

GFP screen of expression vectors in *R. palustris* CGA009

Different combinations of promoter and vector sequences were screened for highest levels of expression within *R. palustris*. The endogenous *ppckA* promoter of *R. palustris* is one of the few promoter sequences that had been previously validated to be active in *R. palustris* under gluconeogenic conditions [23]. However, we only observed a minimal increase in fluorescence for both the pMG105 and pBBR1MCS-2 *ppckAp GFPmut2* constructs compared to the empty vector controls (Fig. 2). While the pBBR1MCS-2 *arap GFPmut2* construct yielded slightly higher levels compared to *ppckAp*, the pBBR1MCS-2 *lacp GFPmut2* construct offered the best expression overall and was selected as the vector for all further expressions in *R. palustris* (Fig. 2). Expression from the *lac* promoter in *R. palustris* was not affected by the addition of the inducer IPTG and catabolite repression only showed a minor decrease in fluorescence (Additional file 3). This suggests that the *lac* regulatory response system utilized by *E. coli* is not functional in *R. palustris*. Although the pBBR1MCS-2 *lac* promoter system provided the best expression in *R. palustris* of all the combinations tested (Fig. 2), the fluorescent signal was still ~13 times lower than the signal from *E. coli* pBBR1MCS *lacp GFPmut2* (data not shown).

Engineering $adhE_{BisB18}$ in *R. palustris*

As determined by the lack of growth under restrictive conditions, the absence of butanol production (data not shown), and the lack of a signal in a western blot under permissive conditions (Fig. 3a, $AdhE_{BisB18}$ Aerobic), direct introduction of $adhE2_{824}$ in the pBBR1MCS-2 *lacp* construct did not result in the production of $AdhE2_{824}$ in *R. palustris*. This is likely due to the large GC discrepancy (30% vs. 65%) and codon utilization preference between *C. acetobutylicum* and *R. palustris* [24]. To address this, we looked for homologous proteins within *R. palustris*. Indeed, endogenous alcohol/aldehyde dehydrogenases with homology to $AdhE2_{824}$ have been found to function in the production of butanol from butyrate when overexpressed. This was previously demonstrated with the *fucO* gene in *E. coli*, where overexpression allowed production of butanol from an engineered reverse β-oxidation pathway where no butanol production was observed

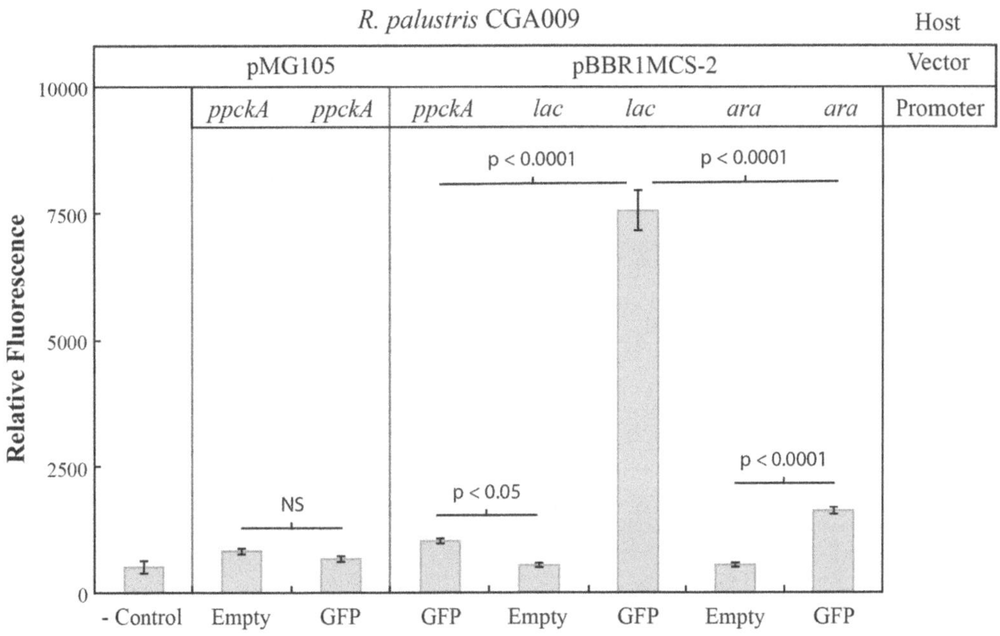

Fig. 2 Fluorescence screen of potential expression vectors in *R. palustris* CGA009. Fluorescence from each strain is expressed as relative fluorescence per 1 unit of OD_{600} in an exponentially growing culture. Technical triplicates with standard deviation and relevant *p* values are shown. Excitation/emission was measured at 480:510 nm. The control contained no cells. Different empty vectors and constructs with *GFPmut2* are designated by vector/promoter heading

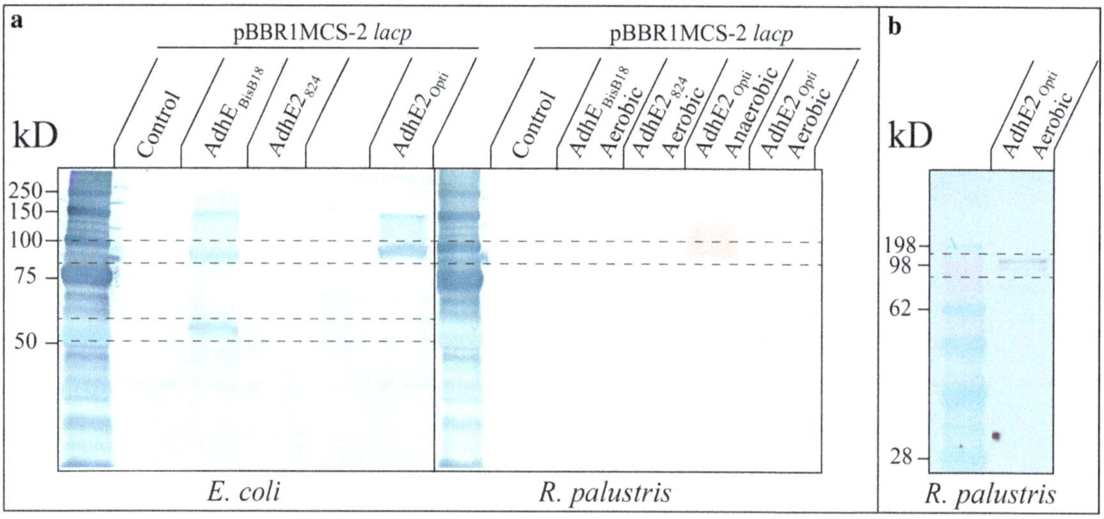

Fig. 3 Western blots for the detection of AdhE2 via the C-terminal 6X His tag: **a** in both *E. coli* and *R. palustris*. Expected $AdhE_{BisB18}$, $AdhE2_{824}$, and $AdhE2_{opti}$ sizes are 94.7, 94.3, and 94.3 kD, respectively (*outlined with dotted box*). Downfield band of ~50 kD (*outlined with dotted box*) likely corresponds to the fragmented C-terminal alcohol dehydrogenase domain where the 6X His tag is located. Protein amounts of 135 μg were loaded into all sample lanes. The *red* band in the *R. palustris* anaerobic *lane* is from photo pigments and may partially be obscuring signal from the western blot: **b** in *R. palustris* pBBR1MCS-2 *lacp* $adhE2_{opti}$ cultures. The signal size of aerobic cultures of *R. palustris* pBBR1MCS-2 *lacp* $adhE2_{opti}$ corresponded with the expected size of $AdhE2_{opti}$ (94.3 kD)

previously [25]. In an attempt to enable butanol production in *R. palustris*, a BLAST search for an endogenous enzyme yielded the identification of RPC_4481 ($AdhE_{BisB18}$) in *R. palustris* strain BisB18, which is an aldehyde/alcohol dehydrogenase with high homology to both domains of $AdhE2_{824}$.

Before expressing $adhE_{BisB18}$ as a candidate butanol-pathway gene, *R. palustris* strain BisB18 was screened for its ability to grow by producing butanol from butyrate under restrictive electron sink conditions (without HCO_3^-). Although no growth or butanol production occurred (data not shown), it is possible that wild-type production of this protein under these conditions is either inactive or insufficient for an observable effect, similar to the FucO protein in *E. coli* [25]. Because no experimental studies have previously characterized the activity of $AdhE_{BisB18}$, its highly conserved primary sequence was further investigated for its structural homology to the verified butanol-producing $AdhE2_{824}$.

Using Swiss-Pdbviewer/DeepView [26], three-dimensional structures of the individual protein domains for both $AdhE2_{824}$ and $AdhE_{BisB18}$ were generated. The predicted tertiary structure of both aldehyde/alcohol dehydrogenases appeared highly conserved with regard to overall shape, location, active site residues, channel volume, and surface charge (Additional file 4). This suggested that the two proteins likely recognized similar substrates. We, therefore, expressed $adhE_{BisB18}$ in *R. palustris* using the pBBR1MCS-2 *lacp* construct. Transformants were inoculated into RM with butyrate and with and without HCO_3^- to screen for viability and butanol production, respectively. Although transformants grew up with HCO_3^-, no growth was observed without HCO_3^- condition (data not shown), suggesting that either $AdhE_{BisB18}$ was not successfully produced by *R. palustris* CGA009 or did not have butyryl-CoA-to-butanol conversion activity.

To investigate this further, a crude cell extract of the culture with HCO_3^-, along with *E. coli* harboring the same construct, was analyzed via western blotting. Although the strong signal at the expected size of 94.5 kD for $AdhE_{BisB18}$ for *E. coli* pBBR1MCS-2 *lacp* $adhE_{BisB18}$ demonstrated that the protein was produced (Fig. 3a, $AdhE_{BisB18}$), the lack of a signal from *R. palustris* pBBR1MCS-2 *lacp* $adhE_{BisB18}$ suggested that the protein was either not produced or was produced at levels below detection (Fig. 3a, $AdhE_{BisB18}$ Aerobic). Aerobic conditions were preferred since anaerobic cultures often had photosynthetic pigments overlapping the expected size for AdhE (Fig. 3a, $AdhE2_{opti}$ Anaerobic). The lack of growth and butanol production under restrictive conditions, however, demonstrated conclusively that *R. palustris* pBBR1MCS-2 *lacp* $adhE_{BisB18}$ was not endowed with butyrate-to-butanol conversion functionality (data not shown).

Engineering $adhE2_{opti}$ in *R. palustris*

Another route to circumvent the large GC discrepancy and codon utilization preference between *C. acetobutylicum* and *R. palustris* was to perform codon optimization on $adhE2_{824}$ to enable expression. The codon utilization table for *R. palustris* CGA009 was determined by analysis of all genomic coding sequences and was retrieved from the Kazusa database [27]. The $adhE2_{opti}$ gene resulted in an increase of the average codon utilization frequency in *R. palustris* from 23% for the original $adhE2_{824}$ to 64% for the optimized $adhE2_{opti}$ (Additional file 5). Codon optimization resulted in a shift of the GC content from 32.5% for $adhE2_{824}$ to 62% for $adhE2_{opti}$. In addition, the presence of rare codons (<10% utilization frequency), which previously composed 47.4% of all codons in $adhE2_{824}$, was entirely eliminated. One disadvantage to this approach of codon optimization is that rare codons are known to induce ribosomal pausing, which has been demonstrated important for proper folding of proteins. Thus, with the elimination of all rare codons from $adhE2_{opti}$, the elimination of codon-dependent ribosomal pausing could lead to improper folding and aggregation of the target protein [28]. For comparison with $adhE2_{opti}$, the codon utilization frequency of $adhE_{BisB18}$ averaged 55% (Additional file 5), the gene possessed an overall 63% GC content, and the frequency of rare codons within the gene was 5%, demonstrating the utilization of rare codons for proper translation of proteins in endogenous alcohol/aldehyde genes (Additional file 6).

To test whether $adhE2_{opti}$ would be expressed in *R. palustris* CGA009 and enable growth by conversion of butyrate to butanol, the gene was introduced using the pBBR1MCS-2 *lacp* construct. Transformants were then inoculated into RM with butyrate and with and without HCO_3^- for growth screening and western blot analysis. This engineered strain was able to grow without HCO_3^- (restrictive conditions) and it produced a measureable amount of butanol, demonstrating the successful production of $AdhE2_{opti}$ with a functional activity. While a detectable signal was observed on a western blot for the *E. coli* cloning culture, no signal was initially detected from *R. palustris* pBBR1MCS-2 *lacp* $adhE2_{opti}$ (Fig. 3a, $AdhE2_{opti}$ Anaerobic and $AdhE2_{opti}$ Aerobic). We performed a second western blot using a more sensitive X-ray film imaging method. For aerobically grown *R. palustris* pBBR1MCS-2 *lacp* $adhE2_{opti}$, we observed a single signal at approximately 94 kD, demonstrating successful expression (Fig. 3b, $AdhE2_{opti}$ Aerobic). As stated above, GFPmut2 production from the pBBR1MCS *lac* promoter is much lower in *R. palustris* than in E. coli, so the lack of a signal in the majority of the samples is likely due to low protein production in *R. palustris*.

Butanol production rates and selectivity for *R. palustris* pBBR1MCS-2 *lacp* $adhE2_{opti}$

Because of the rescued growth and butanol production observed, *R. palustris* pBBR1MCS-2 *lacp* $adhE2_{opti}$ was further investigated for its ability to grow under both permissive and restrictive conditions and these results were compared to *R. palustris* pBBR1MCS-2 *lacp GFPmut2* as a control. *R. palustris* pBBR1MCS-2 *lacp* $adhE2_{opti}$ was capable of growth without HCO_3^- (restrictive conditions) (Fig. 4b), while this growth did not occur for the control strain. The growth for *R. palustris* pBBR1MCS-2 *lacp* $adhE2_{opti}$ occurred on a long time scale and reached a relatively low maximum cell density of ~0.06 OD_{600} (Fig. 4b) at neutral pH levels (Fig. 4d). Following the onset of the stationary phase, consumption of butyrate and production of butanol still occurred (Fig. 4f, h), but the cultures only reached a maximum butanol concentration of <0.4 mM after extended incubation (Fig. 4h).

Although only low levels of butyrate were consumed (Fig. 4f), butanol production occurred with a 39.7 ± 7.6% selectivity based on moles of carbon. The theoretical maximum for butanol production is based on all electrons derived from complete oxidation of one molecule of butyrate reducing five molecules of butyrate to butanol (6 butyrate → 5 butanol + 4 CO_2) for 83% conversion efficiency. However, based on previous empirical measurements of excess reducing equivalents of *R. palustris* growing with butyrate by McKinlay et al. [3], only an estimated 0.45 mol of butanol should be produced per mole of butyrate consumed for an effective 45% selectivity (6 butyrate → 2.7 butanol + biomass). Therefore, almost all reducing equivalents predicted to be in excess were being converted into butanol by *R. palustris* pBBR1MCS-2 *lacp* $adhE2_{opti}$ (40% vs. 45%), which we had anticipated for an obligatory reaction.

To further investigate why *R. palustris* pBBR1MCS-2 *lacp* $adhE2_{opti}$ showed low growth rates, *R. palustris* was screened for product (butanol) toxicity. However, no inhibition of growth was observed in *R. palustris* at butanol concentrations up to ~30 mM (Additional file 7), removing our concerns about a concentration of ~0.4 mM (Fig. 4h). In addition, we operated a bioreactor with product removal to investigate whether this would increase growth and butanol production rates by removing butanol from the substrate pool, thereby, improving the thermodynamics for butanol formation and eliminating any end-product inhibition. Sparging of N_2 has been previously used to remove butanol from fermentation broths [29]. Here, N_2 gas was continuously sparged into 350-mL anaerobic bioreactors to remove butanol produced by *R. palustris* pBBR1MCS-2 *lacp* $adhE2_{opti}$ in batch-mode for 6 consecutive weeks. The *R. palustris* pBBR1MCS-2 *lacp* $adhE2_{opti}$ culture grew for that entire period up to an OD_{600} of 0.12 (Additional file 8: Fig. S6A), and achieved an average maximum butanol production rate of 0.0017 g L^{-1} day^{-1} (Additional file 8: Fig. S6D) with a maximum butanol concentration of

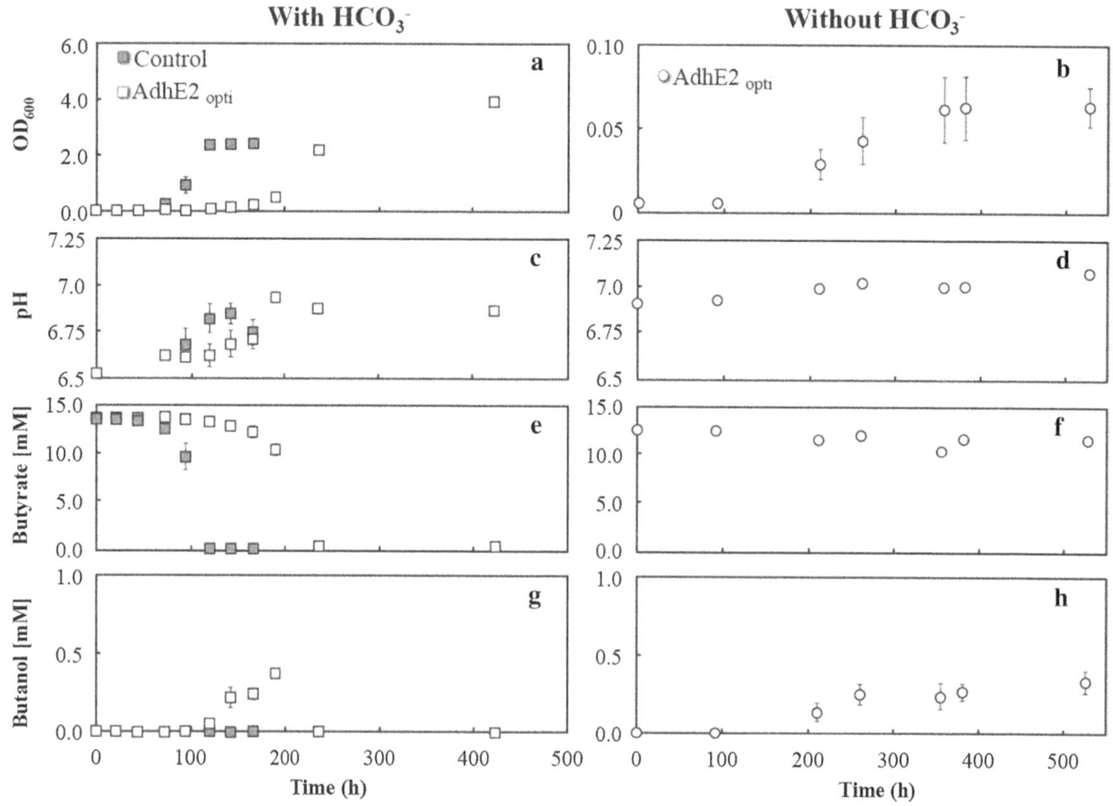

Fig. 4 Performance of *R. palustris* pBBR1MCS-2 *lacp GFPmut2* (control) and *R. palustris* pBBR1MCS-2 *lacp adhE2*$_{opti}$ in permissive conditions (with HCO_3^-, squares), and *R. palustris* pBBR1MCS-2 *lacp adhE2*$_{opti}$ in restrictive conditions (without HCO_3^-, circles): **a**, **b** growth; **c**, **d** pH; **e**, **f** concentration of butyrate in the broth; **g**, **h** butanol concentration in the broth. The results from the control and *R. palustris* pBBR1MCS-2 *lacp adhE2*$_{opti}$ are represented by *gray* and *white* symbols, respectively. Average and standard deviation of biological triplicates are shown

0.35 mM (Additional file 8: Fig. S6C) without depleting butyrate (Additional file 8: Fig. S6B). The lack of a considerable improvement compared to the earlier experiment showed that neither butanol toxicity nor thermodynamics nor end-product inhibition was a main reason for low growth rates. Finally, we found that the extraction system removed a majority of the butanol without removing any of the butyrate (Additional file 8: Fig. S6E, F).

The intermediate butyraldehyde within microbial cells is another metabolite that could have inhibited *R. palustris*, even though the maximum concentration in the supernatant was measured to be only 50 μM. However, due to the fusion nature of AdhE2$_{opti}$, we anticipate that butyraldehyde would rapidly be converted into butanol with a high flux rate, resulting in a low effective concentration [30]. Without butanol and butyraldehyde toxicity, we can only speculate about why growth and butanol production rates of our engineered strain are so low. Since AdhE2$_{opti}$ is provided as the sole route to maintaining redox balance, all excess electrons must be funneled through this pathway. When the capacity of the AdhE2$_{opti}$ pool is outweighed by the demand for regenerating oxidized reducing equivalents, this would introduce a bottleneck in the maximum rate of metabolism and growth. However, more work is required to be conclusive.

Some butanol production did also occur for *R. palustris* pBBR1MCS-2 *lacp adhE2*$_{opti}$ with HCO_3^- (permissive conditions) when exogenous CO_2 was available for maintaining redox balance (Fig. 4a, g) at a neutral pH level (Fig. 4c). Because the preculture had originally been maintained without HCO_3^-, a long lag phase and carryover of butanol producing activity was observed for *R. palustris* pBBR1MCS-2 *lacp adhE2*$_{opti}$ (Fig. 4a), resulting in a maximum concentration of 0.37 mM with 11.5 ± 2.4% selectivity (Fig. 4e, g). Though the transient production of butanol was a surprise, this lower selectivity for the permissive condition was expected because more reducing equivalents could be used in central metabolism following the introduction of exogenous HCO_3^-, resulting in much higher growth (~10× OD at 200 h), the complete depletion of butyrate (Fig. 4e), no production of acetate, and ultimate consumption of all butanol produced (Fig. 4g). This contrasts with the restrictive condition where all growth proceeds through the obligate engineered pathway, butyrate is not depleted, and all butanol produced remains in solution.

Optimizing butanol production by pregrowing *R. palustris* pBBR1MCS-2 *lacp adhE2*$_{opti}$

In an attempt to improve the maximum butanol concentrations produced by the engineered strain, 50 mL aerobic cultures was grown to mid-log phase, washed 3 times with sterile RM, resuspended in 0.5 mL RM, and used as an inoculum into 20-mL anaerobic cultures under restrictive conditions. This procedure ensured that the butyrate-to-butanol conversion was not catalyst limited as the initial OD_{600} of these cultures was >1. By using a concentrated inoculum, *R. palustris* pBBR1MCS-2 *lacp adhE2*$_{opti}$ produced butanol concentrations of greater than 1.5 mM at volumetric butanol production rates of 0.034 g L^{-1} day^{-1} (Fig. 5d, f). All butanol was produced within 100 h of inoculation (Fig. 5d, f). Following 100 h, however, the *R. palustris* pBBR1MCS-2 *lacp adhE2*$_{opti}$ culture continued to grow and consume butyrate (Fig. 5a, c) with the main carbon product switching from butanol to acetate (Fig. 5b). Acetate had not been detected before in any of the previous anaerobic experiments and in terms of maintaining redox balance, acetate is an unfavorable redox product compared to butanol. This metabolism reduced the butanol selectivity (Fig. 5e). We do not know exactly why this switch occurred, but it was likely a result from having a high biomass culture that underwent a rapid switch from aerobic to anaerobic conditions.

Conclusion

The introduction of *adhE2*$_{opti}$ into *R. palustris* via a plasmid-based construct is an example of metabolically engineered fuel production that proceeds as an obligate strategy for growth. Although growth and production rates were low for this system, our selectivity of 39.7 ± 7.6% matched well with the empirical expectation of 45.0% under these conditions. Because the low growth rates and butanol production are likely constrained by

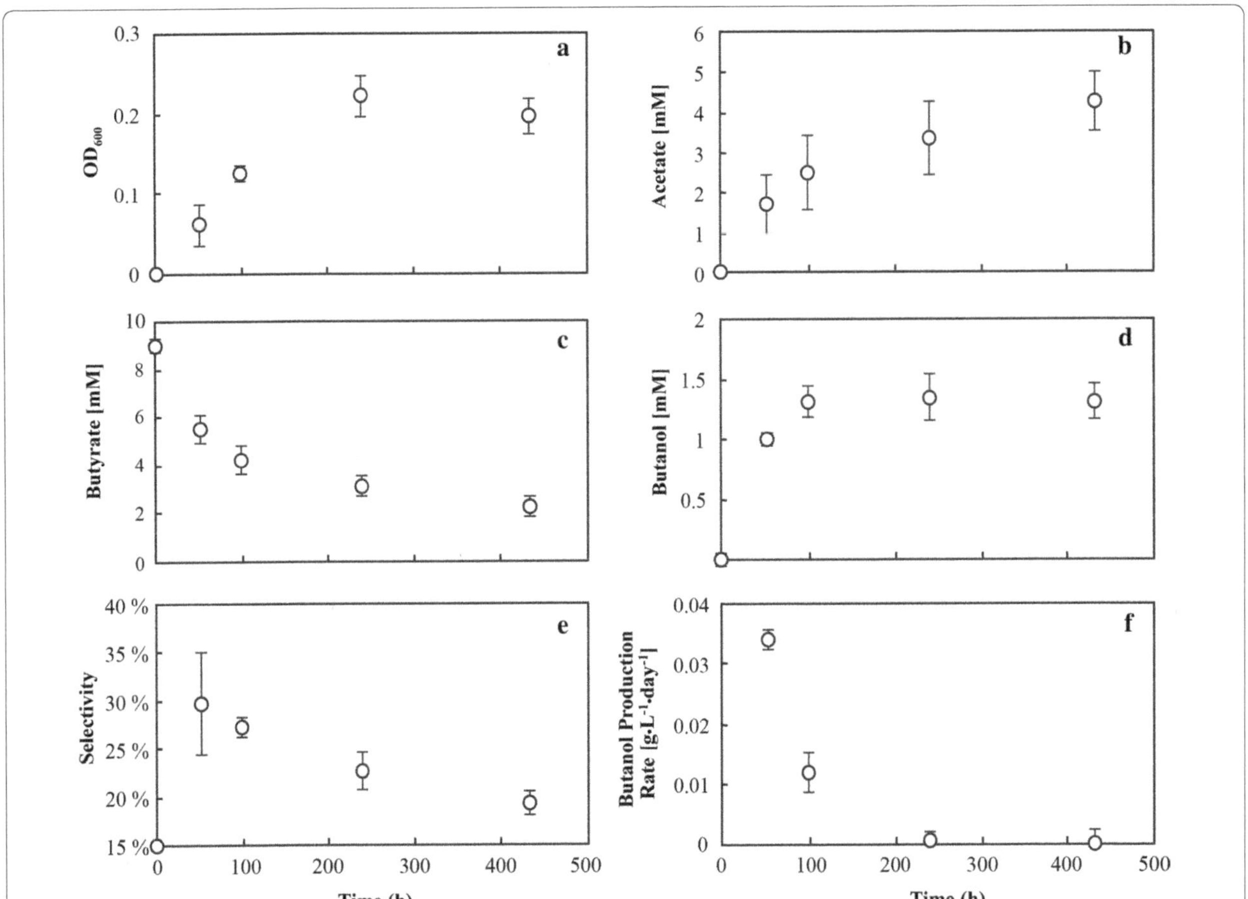

Fig. 5 Performance of *R. palustris* pBBR1MCS-2 *lacp adhE2*$_{opti}$ without HCO_3^- in bottles that were inoculated with concentrated cultures: **a** growth with corrected OD_{600} (initial OD_{600} was subtracted); **b** acetate concentration in the broth; **c** butyrate concentration in the broth; **d** butanol concentration in the broth; **e** butanol selectivity; **f** volumetric butanol production rate. *Error bars* show the standard error of three biological replicates

redox flux through the engineered pathway, rational design and directed evolution could be used to further improve the performance of this activity. If improved, the system could be easily integrated into a simple bioreactor design with highly efficient selectivity and recovery. While this proof-of-concept for harnessing an obligate reducing metabolism to drive engineered reactions used butanol as a target molecule, this redox-based driving force could be used to achieve other reactions for the production of desirable chemicals that require reduction reactions.

Additional files

Additional file 1. Growth with butyrate, containing Figure S1.

Additional file 2. Sequence.

Additional file 3. Fluorescence, containing Figure S2.

Additional file 4. Structure, containing Figure S3.

Additional file 5. Codons, containing Figure S4.

Additional file 6. Codons and GC content, containing Table S1.

Additional file 7. Growth with butanol, containing Figure S5.

Additional file 8. Growth with product removal, containing Figure S6.

Authors' contributions

DFRD designed and executed experiments, performed analysis, interpreted data, and drafted the manuscript. ECH executed experiments, performed analysis, and contributed to data interpretation. HR and BM advised on how to perform the experiments. GJ and LTA conceived the study and participated in its design. LTA coordinated the study and contributed to the drafting of the manuscript. All authors read and approved the final manuscript.

Author details

[1] Department of Biological and Environmental Engineering, Cornell University, Ithaca, NY 14853, USA. [2] Boyce Thompson Institute for Plant Research, Ithaca, NY 14853, USA. [3] Atkinson Center for a Sustainable Future, Cornell University, Ithaca, NY 14853, USA. [4] Center for Applied Geosciences, University of Tübingen, Hölderlinstr. 12, 72074 Tübingen, Germany.

Acknowledgements

We would like to thank Dr. Meena Haribal, Dr. Lubna Richter, Prof. Dr. Matt DeLisa, Dr. Jason Boock, and Jen Schmidt for their expertise and advice (all at Cornell University). We would like to thank Prof. Dr. Caroline Harwood (University of Washington) for providing the *R. palustris* strain BisB18 and Wendy Hoose (Cornell University) for providing technical assistance.

Competing interests

The authors declare that they have no competing interests.

Funding

This project and its researchers were funded in part through an Academic Venture Fund award from the Atkinson Center for a Sustainable Future to LTA and GJ, and the Advanced Research Projects Agency Energy Open 2012 program Project Number DE-AR0000312 to LTA. We acknowledge support by Deutsche Forschungsgemeinschaft and Open Access Publishing Fund of University of Tübingen.

References

1. McKinlay JB, Harwood CS. Carbon dioxide fixation as a central redox cofactor recycling mechanism in bacteria. Proc Natl Acad Sci USA. 2010;107(26):11669–75.
2. Hillmer P, Gest H. H_2 metabolism in the photosynthetic bacterium *Rhodopseudomonas capsulata*: H_2 production by growing cultures. J Bacteriol. 1977;129(2):724–31.
3. McKinlay JB, Harwood CS. Calvin cycle flux, pathway constraints, and substrate oxidation state together determine the H_2 biofuel yield in photoheterotrophic bacteria. mBio. 2011;2(2):e00323.
4. Barbosa MJ, Rocha JMS, Tramper J, Wijffels RH. Acetate as a carbon source for hydrogen production by photosynthetic bacteria. J Biotechnol. 2001;85(1):25–33.
5. Gosse JL, Engel BJ, Rey FE, Harwood CS, Scriven LE, Flickinger MC. Hydrogen production by photoreactive nanoporous latex coatings of nongrowing *Rhodopseudomonas palustris* CGA009. Biotechnol Prog. 2007;23(1):124–30.
6. Chen CY, Saratale GD, Lee CM, Chen PC, Chang JS. Phototrophic hydrogen production in photobioreactors coupled with solar-energy-excited optical fibers. Int J Hydrog Energ. 2008;33(23):6886–95.
7. Fixen KR, Zheng Y, Harris DF, Shaw S, Yang Z-Y, Dean DR, Seefeldt LC, Harwood CS. Light-driven carbon dioxide reduction to methane by nitrogenase in a photosynthetic bacterium. Proc Natl Acad Sci USA. 2016;113(36):10163–7.
8. Jin C, Yao M, Liu H, Lee C-F, Ji J. Progress in the production and application of *n*-butanol as a biofuel. Renew Sustain Energ Rev. 2011;15(8):4080–106.
9. Nolling J, Breton G, Omelchenko MV, Makarova KS, Zeng QD, Gibson R, Lee HM, Dubois J, Qiu DY, Hitti J, et al. Genome sequence and comparative analysis of the solvent-producing bacterium *Clostridium acetobutylicum*. J Bacteriol. 2001;183(16):4823–38.
10. Inui M, Suda M, Kimura S, Yasuda K, Suzuki H, Toda H, Yamamoto S, Okino S, Suzuki N, Yukawa H. Expression of *Clostridium acetobutylicum* butanol synthetic genes in *Escherichia coli*. Appl Microbiol Biotechnol. 2008;77(6):1305–16.
11. Atsumi S, Cann AF, Connor MR, Shen CR, Smith KM, Brynildsen MP, Chou KJ, Hanai T, Liao JC. Metabolic engineering of *Escherichia coli* for 1-butanol production. Metab Eng. 2008;10(6):305–11.
12. Desai RP, Harris LM, Welker NE, Papoutsakis ET. Metabolic flux analysis elucidates the importance of the acid-formation pathways in regulating solvent production by *Clostridium acetobutylicum*. Metab Eng. 1999;1(3):206–13.
13. Sillers R, Al-Hinai MA, Papoutsakis ET. Aldehyde-alcohol dehydrogenase and/or thiolase overexpression coupled with CoA transferase downregulation lead to higher alcohol titers and selectivity in *Clostridium acetobutylicum* fermentations. Biotechnol Bioeng. 2009;102(1):38–49.
14. Dürre P, Fischer RJ, Kuhn A, Lorenz K, Schreiber W, Sturzenhofecker B, Ullmann S, Winzer K, Sauer U. Solventogenic enzymes of *Clostridium acetobutylicum*: catalytic properties, genetic organization, and transcriptional regulation. FEMS Microbiol Rev. 1995;17(3):251–62.
15. Nair RV, Bennett GN, Papoutsakis ET. Molecular characterization of an aldehyde/alcohol dehydrogenase gene from *Clostridium acetobutylicum* ATCC 824. J Bacteriol. 1994;176(3):871–85.
16. Fontaine L, Meynial-Salles I, Girbal L, Yang X, Croux C, Soucaille P. Molecular characterization and transcriptional analysis of *adhE2*, the gene encoding the NADH-dependent aldehyde/alcohol dehydrogenase responsible for butanol production in alcohologenic cultures of *Clostridium acetobutylicum* ATCC 824. J Bacteriol. 2002;184(3):821–30.
17. Lee CM, Chen PC, Wang CC, Tung YC. Photohydrogen production using purple nonsulfur bacteria with hydrogen fermentation reactor effluent. Int J Hydrog Energy. 2002;27(11–12):1309–13.
18. Cormack BP, Valdivia RH, Falkow S. FACS-optimized mutants of the green fluorescent protein (GFP). Gene. 1996;173(1):33–8.
19. Inui M, Roh JH, Zahn K, Yukawa H. Sequence analysis of the cryptic plasmid pMG101 from *Rhodopseudomonas palustris* and construction of stable cloning vectors. Appl Environ Microbiol. 2000;66(1):54–63.
20. Kovach ME, Elzer PH, Hill DS, Robertson GT, Farris MA, Roop RM, Peterson KM. Four new derivatives of the broad-host-range cloning vector pBBR1MCS, carrying different antibiotic-resistance cassettes. Gene. 1995;166(1):175–6.

21. Protocols CSH. Laemmli sample buffer (2X). Cold Spring Harbor Protocols. 2006;2006(1):pdb.rec10424.
22. Doud DFR, Angenent LT. Toward electrosynthesis with uncoupled extracellular electron uptake and metabolic growth: enhancing current uptake with *Rhodopseudomonas palustris*. Environ Sci Technol Lett. 2014;1(9):351–5.
23. Inui M, Nakata K, Roh JH, Zahn K, Yukawa H. Molecular and functional characterization of the *Rhodopseudomonas palustris* no. 7 phosphoenolpyruvate carboxykinase gene. J Bacteriol. 1999;181(9):2689–96.
24. Gustafsson C, Govindarajan S, Minshull J. Codon bias and heterologous protein expression. Trends Biotechnol. 2004;22(7):346–53.
25. Dellomonaco C, Clomburg JM, Miller EN, Gonzalez R. Engineered reversal of the beta-oxidation cycle for the synthesis of fuels and chemicals. Nature. 2011;476(7360):355–9.
26. Guex N, Peitsch MC. SWISS-MODEL and the Swiss-Pdb Viewer: an environment for comparative protein modeling. Electrophoresis. 1997;18(15):2714–23.
27. Nakamura Y, Gojobori T, Ikemura T. Codon usage tabulated from international DNA sequence databases: status for the year 2000. Nucleic Acids Res. 2000;28(1):292.
28. Rosano GL, Ceccarelli EA. Rare codon content affects the solubility of recombinant proteins in a codon bias-adjusted *Escherichia coli* strain. Microb Cell Fact. 2009;8:41.
29. Qureshi N, Blaschek HP. Recovery of butanol from fermentation broth by gas stripping. Renew Energy. 2001;22(4):557–64.
30. DeLisa MP, Conrado RJ. Synthetic metabolic pipelines. Nat Biotechnol. 2009;27(8):728–9.

Limitation of cellulose accessibility and unproductive binding of cellulases by pretreated sugarcane bagasse lignin

Germano Siqueira[1,2], Valdeir Arantes[1,2], Jack N. Saddler[2], André Ferraz[1] and Adriane M. F. Milagres[1*]

Abstract

Background: The effectiveness of the enzymatic hydrolysis of cellulose in plant cell wall is strongly influenced by the access of enzymes to cellulose, which is at least in part limited by the presence of lignin. Although physicochemical treatments preceding the enzymatic catalysis significantly overcome this recalcitrance, the residual lignin can still play a role in the process. Lignin is suggested to act as a barrier, hindering cellulose and limiting the access of the enzymes. It can also unspecifically bind cellulases, reducing the amount of enzymes available to act on cellulose. However, the limiting role of the lignin present in pretreated sugarcane bagasses has not been fully understood yet.

Results: A set of sugarcane bagasses pretreated by five leading pretreatment technologies was created and used to assess their accessibility and the unproductive binding capacity of the resulting lignins. Steam explosion and alkaline sulfite pretreatments resulted in more accessible substrates, with approximately 90% of the cellulose hydrolyzed using high enzyme loadings. Enzymatic hydrolysis of alkaline-treated (NaOH) and steam-exploded sugarcane bagasses were strongly affected by unproductive binding at the lowest enzyme loading tested. Analysis of the extracted lignins confirmed the superior binding capacity of these lignins. Sulfite-based pretreatments (alkaline sulfite and acid sulfite) resulted in lignins with lower binding capacities compared to the analogue pretreatments without sulfite (alkaline and acidic). Strong acid groups present in sulfite-based pretreated substrates, attributed to sulfonated lignins, corroborated the lower binding capacities of the lignin present in these substrates. A more advanced enzyme preparation (Cellic CTec3) was shown to be less affected by unproductive binding at low enzyme loading.

Conclusions: Pretreatments that increase the accessibility and modify the lignin are necessary in order to decrease the protein binding capacity. The search for the called weak lignin-binding enzymes is of major importance if hydrolysis with low enzyme loadings is the goal for economically viable processes.

Keywords: Sugarcane bagasse, Thermochemical pretreatment, Lignin removal, Cellulose accessibility, Unproductive binding, Protein binding capacity

Background

The hierarchical organization of three main components of the composite-like structure of the plant cell walls is essential for plant development and growth. Cellulose is the major constituent and corresponds to approximately 50% of cell wall dry mass. The other components, hemicellulose, and lignin complete the resistant structure and give cell wall other vital characteristics, such as partial flexibility and hydrophobicity [1, 2]. Cellulose being the most abundant natural polymer on Earth has been the spotlight of several industrial processes either as a polymeric structure (i.e., cellulosic pulp, cellulose derivatives, and nanocellulose) or as a source of glucose for fuels and chemicals production. For cellulose depolymerization, enzymatic catalysis using cellulases and accessory enzymes has been explored over the past decades and is already available in commercial plants, but the process is still not completely understood [3].

*Correspondence: adriane@debiq.eel.usp.br
[1] Departamento de Biotecnologia, Escola de Engenharia de Lorena, Universidade de São Paulo, CP 116, Lorena, SP 12602-810, Brazil
Full list of author information is available at the end of the article

Because of the strong interaction of components, plant cell walls are highly recalcitrant and the utilization of their components in the biorefinery concept requires cell wall fractionation, at least partially. This step, known as pretreatment, is crucial to reduce cell wall recalcitrance and enhance cellulose hydrolysis [4]. Despite the many variables in a pretreatment, most of them are based on reactions either in acidic or alkaline conditions. In brief, pretreatments in acidic conditions aim hemicellulose solubilization, leaving a pretreated solid rich in cellulose and lignin. The liquid fraction containing mostly hemicellulose monomeric (or oligomeric, depending on the pretreatment conditions) sugars can be fermented to produce ethanol or other products [5–7]. The cellulose in the solid fraction can be enzymatically hydrolyzed with higher yields than untreated feedstock, mainly due to increased accessibility (hemicellulose removal and lignin relocation) [8–11]. As a matter of fact, most of the cellulosic ethanol plants around the world are based in acid pretreatment technologies [12]. Alkaline pretreatments are more efficient in lignin removal, substantially increasing cellulose digestibility, even after removing only part of the lignin [13–15]. Since most of the hemicellulose remains in the solid fraction, accessory enzymes (xylanases, esterases, and some oxidases) are necessary in the enzyme mixture [16, 17].

In both acidic and alkaline conditions, lignin solubilization can be enhanced by the presence of nucleophiles such as sulfite ions [14, 18–20]. In acid conditions, a sulfite-based pretreatment was developed in order to remove lignin and achieve better cellulose conversion yields [19, 21]. However, even under conditions in which lignin removal is favored, the residual lignin is known to play a limiting role in the enzymatic hydrolysis of the cellulose. It has been reported that the residual lignin can act as a barrier, hindering the access of cellulolytic enzymes to cellulose surface [8, 22]. Porosity of cell wall matrix, an indicative of permeability of cellulases, can be assessed by the solute exclusion [23, 24] and this technique was recently used to correlate sugarcane cell wall pore size with cellulose digestibility after lignin selective removal [25]. However, complete removal of lignin is not required to achieve considerable cellulose hydrolysis yields (i.e., higher than 70%). As a matter of fact, even pretreatments that do not result in lignin removal can generate a substrate with increased cellulose accessibility, provided that lignin relocates in the cell wall matrix [26, 27]. It has also been shown that topochemical distribution of lignin can affect the hydrolysis of sugarcane bagasse cellulose and other grasses [28–30].

Another important limitation in cellulose hydrolysis caused by lignin is unproductive binding. Differently from the barrier effect, lignin binds cellulases (in some cases irreversibly) decreasing the availability of cellulases to catalyze cellulose hydrolysis. This has become more important considering that cellulose hydrolysis at industrial scale is preferably conducted at low enzyme loading, condition at which the unproductive binding of lignin has been demonstrated to be more pronounced [31]. This behavior was observed over a wide range of substrates submitted to several different pretreatments [32–39]. Pretreatments can be designed in order to overcome, or at least partially reduce, the binding effect of lignin. For instance, the negatively sulfonated groups (in the hydrolysis pH) can cause repulsion of the also negatively charged cellulases [40], resulting in reduced unproductive binding [41, 42].

Lignin from different sources binds enzymes to a different extent. For example, wood lignin has higher binding capacity than grass lignin and lignin from the same feedstock can have different protein binding capacities, depending on the pretreatment technology and condition used [33]. Hydrophobicity of lignin and the free phenolic groups content might play a role in the protein (cellulases) binding capacity of lignins [36, 39, 43–45]. On the other hand, the enhancement of lignin hydrophilicity due to the presence of acid groups may result in reduced binding capacity, beneficial for enzymatic hydrolysis of cellulose [4, 46, 47]. In addition, the characteristics of cellulases, such as isoelectric point and hydrophobic surface, can influence the intensity of the lignin–cellulase binding [48, 49].

In order to understand the limiting role of lignin in sugarcane bagasse, an important source of sugars after enzymatic hydrolysis [50], this work first aimed to prepare a range of pretreated sugarcane bagasse (PSCB) samples with varying lignin contents and properties by pretreating bagasse with five thermochemical leading pretreatment methods. These pretreated bagasses were then used to assess their cellulose accessibility and the binding capacity of the lignins after pretreatments in alkaline and acid conditions, with and without the presence of a delignification agent (sulfite ion). Finally, to assess the improved resistance to lignin unproductive binding of a recently developed industrial cellulase preparation, sugar yields of pretreated bagasses hydrolyzed with Celluclast 1.5, a traditional cellulase mixture, and Cellic CTec3 were compared.

Methods

Preparation of the pretreated substrates

The feedstock used in this study was sugarcane bagasse obtained from a Brazilian sugarcane mill (approximately 10% moisture content) submitted to five different pretreatments, producing five pretreated sugarcane bagasses. The alkaline pretreatments (NaOH and Na_2SO_3) were conducted according to Mendes [14]. In brief, the bagasse was previously impregnated with a NaOH solution in order to achieve a solid:liquid ratio 1:10 and final alkali concentration of 5% ($w_{NaOH}/w_{bagasse}$). The bagasse was pretreated for 1 h, at 121 °C, in an autoclave. The same condition was applied for the alkaline sulfite pretreatment, except that Na_2SO_3 was added to the NaOH solution, so that the sodium sulfite concentration was 10% ($w_{Na_2SO_3}/w_{bagasse}$). The acid pretreatments were based on a previous work [47], but adapted to sugarcane bagasse. For the diluted acid pretreatment, the solid:liquid ratio was 1:10 and the H_2SO_4 concentration was 0.75% ($w_{H_2SO_4}/w_{bagasse}$). The soaked bagasse was pretreated in a rotating digester (Aurora Products Ltd, Savona, CA), at 160 °C for 30 min, with a 30-min ramp to reach the temperature. The variation of this pretreatment with sulfite was conducted by adding $NaHSO_3$ to the acid solution, resulting in a concentration of 5% ($w_{NaHSO_3}/w_{bagasse}$). Finally, for the SO_2/steam pretreatment, one of the conditions described by Ferreira-Leitão et al. [51] was used. Prior to the impregnation with SO_2, the moisture content of the bagasse was increased to 50% in order to make the absorption of the catalyst more efficient. The SO_2 concentration was 3.0% ($w_{SO_2}/w_{bagasse}$) and the impregnated bagasse was treated in a 2L Stake Tech III steam gun (Stake Technologies, Norvall, CA) at 190 °C for 5 min.

After pretreatment, all solid fractions were washed with tap water (20-fold the dry mass of pretreated bagasse), vacuum-filtered, and frozen until used.

Chemical composition

The carbohydrate and lignin contents of the raw and pretreated bagasses were determined as previously described [31]. Briefly, the carbohydrates in the substrates were acid hydrolyzed and the sugars were quantified with a Dionex (Sunnyvale, CA) HPLC (ICS-3000). Glucan content was determinate based on the released glucose, and xylan content on the released xylose, arabinose, and acetic acid. The acid insoluble lignin was gravimetrically analyzed and the acid soluble lignin was measured by reading the absorbance at 205 nm, using an extinction coefficient of 105 L g^{-1} cm^{-1} [52].

Protein concentration and enzyme activities

Protein concentration was determined by a ninhydrin-based method [53]. Filter paper activity of cellulase preparations was measured according to Ghose [54]. β-Glucosidase activity was measured hydrolyzing *p*-nitrophenyl-β-D-glucopyranoside (0.2%) and spectrophotometrically quantifying the released *p*-nitrophenol at 410 nm [55].

Cellulose digestibility

The washed substrates were enzymatically hydrolyzed using Celluclast 1.5 L (Novozymes, DK) supplemented with β-glucosidase (Novozym 188, Novozymes, DK). Enzymatic hydrolysis was conducted at 2% (w/v) solids in sodium acetate buffer (50 mM, pH 4.8), 50 °C, 120 rpm in a rotatory incubator Combi-H12 (FINEPCR, KR). The total volume was 1.5 mL, in 2.0-mL tubes. The protein loadings, based on the Celluclast protein content (130 mg/mL; 60 FPU/mL), were 2.5, 5, 10, 20, and 30 mg/g of cellulose in the PSCB. Novozym 188 (458 UI/mL) was added in order to have a β-glucosidase activity 2 times higher than the filter paper activity in the reaction mixture. After 72 h, the enzymes were inactivated by heating at 100 °C for 5 min. The glucose concentration in the hydrolysate was measured using the glucose oxidase assay [56]. Hydrolyses were analyzed in triplicates and the standard deviation was calculated.

Determination of unproductive binding capacity

Each PSCB was hydrolyzed with a combination of Celluclast and Novozym 188 or with Cellic CTec3. However, prior to the addition of the enzymes, the blocking of the lignin binding sites was done according to Kumar et al. [31]. 250 mg of bovine serum albumin (BSA) per g of cellulose was added and incubated for 1 h at 50 °C and 120 rpm. After that, 2.5 mg of protein (based on Celluclast or CTec 3 protein content) per gram of cellulose was added and the hydrolysis was conducted for 72 h. The same hydrolysis condition, but without the prior addition of BSA, was conducted and the difference in hydrolysis yields (with and without BSA) was considered an indication of unproductive binding [31]. Reaction and analysis conditions were the same as described above.

Estimation of accessible cellulose surface

Accessible cellulose surface was estimated according to Simons [57] and adapted elsewhere [58–61]. Direct Orange (Pontamine Fast Orange 6RN) and Direct Blue (Pontamine Fast Sky Blue 6BX) solutions were prepared at 10 mg/mL. The Direct Orange (DO) solution was ultra-filtered three times in Amicon apparatus with a 100-kDa membrane (filtrations stopped when approximately 80% of the initial volume was eluted). DO concentration was determined by dry weight and re-adjusted to 10 mg/mL. Molar attenuation coefficient of DO was calculated at 624 nm and DB at 455 nm (maximum absorption wavelength of each dye).

Each PSCB (10 mg, dry weight) was added to 2-mL centrifuge tubes, and 0.1 mL of phosphate buffered saline pH 6.0 (PO_4 0.3 M, NaCl 1.4 mM) and deionized water were added. The amount of water varied, since the final incubation volume (buffer, water and dye solutions) was 1.0 mL. Prior to the addition of the dye solutions (at 10 mg/mL), the mixture was incubated for 12 h, at approximately 25 °C in the bench. The volumes added were 0.025, 0.050, 0.075, 0.1, 0.15, and 0.2 mL, in each incubation condition. The tubes were vortex mixed and incubated at 70 °C in a water bath, 200 rpm, for 6 h. The mixtures were than centrifuged at 1800 rpm for 5 min and the absorbance of the supernatants was measured in spectrophotometer at 455 and 624 nm. The concentration of each dye was calculated according to the following equations:

$$[DO] = [(\varepsilon_{DB624} \times A_{455}) - (\varepsilon_{DB455} \times A_{624})] / [(\varepsilon_{DO455} \times \varepsilon_{DB624}) - (\varepsilon_{DB455} \times \varepsilon_{DO624}]$$

$$[DB] = [(\varepsilon_{DO455} \times A_{624}) - (\varepsilon_{DO624} \times A_{455})] / [(\varepsilon_{DO455} \times \varepsilon_{DB624}) - (\varepsilon_{DB455} \times \varepsilon_{DO624}],$$

where [DO] and [DB] are the concentrations of orange and blue dyes in the solutions, respectively, in mg/mL; ε_{DB455} and ε_{DB624} are the molar attenuation coefficient of the blue dye at 455 and 624 nm, respectively, in L g^{-1} cm^{-1}; ε_{DO455} and ε_{DO624} are the molar attenuation coefficient of the orange dye at 455 and 624 nm, respectively. A_{455} and A_{624} are the absorbance values of the solutions at 455 and 624 nm, respectively. The amount of adsorbed dye was calculated subtracting the amount of dye in solution before and after incubation. The amount of adsorbed dye (mg_{dye}/g_{PSCB}) was plotted against the dye concentration after incubation (mg/mL), for each PSCB. The first plateau was used for the calculation, since Direct dyes can adsorb in double layer [62]. Accessible cellulose surface was estimated as the ratio of maximum amount of adsorbed dyes (DO/DB).

Quantification of acid groups

The strong and weak acid groups in the pretreated substrates were measured by conductometric titration [63]. In 50-mL centrifuge tubes, 0.1 g of the PSCB was mixed with 1.0 M HCl and left for 12 h at 4 °C. Approximately 30 mL of deionized water was added, vortex mixed, and centrifuged at 5000 rpm for 10 min. This washing procedure was carefully repeated 6 times. After decanting the last washing water, the sample was transferred to a 100-mL beaker and 50 mL of a 0.001 M NaCL solution was added. The suspension was constantly mixed with a magnetic stirring bar and 20 μL of 0.5 M HCl solution was added. The mixture was titrated with 0.05 M NaOH, recording the conductivity of the mixture after each addition of the alkali solution, at 10 μL increments. The decrease in the conductivity after neutralization of the HCL in solution (by the addition of 200 μL of NaOH solution) was associated with the presence of strong acid groups in the substrate. The next phase of the titration curve, in which the conductivity did not change with the addition of sodium hydroxide, was associated with the presence of weak acid groups. After the neutralization of the acid groups, the conductivity increased with the addition of NaOH. To calculate the amount of acid groups in the substrates, the number of moles of NaOH required to neutralize the strong and weak acid groups in the substrate was divided by the mass of PSCB (weak and total acid groups) or lignin in the PSCB (strong acid groups).

Lignin isolation

The lignin isolation procedure was adapted from two methodologies [64, 65]. The PSCB (30 g, dry weight) was hydrolyzed with Celluclast 1.5L (65 FPU/g cellulose), β-glucosidase (120 UI/g cellulose), and Cellic HTec2 (Novozymes, DK; 20 mg protein/g cellulose), for 72 h. The mixture was centrifuged (5000 rpm, 60 min), the supernatant decanted, and the solids washed with 700 mL of distilled water. The residual solid was suspended in 400 mL of a 50 mM phosphate buffer solution (pH 7) containing 0.2 U/mL Protease type XIV (Sigma, US), 40 mg/L tetracycline, and 30 mg/L cycloheximide and the mixture was incubated for 12 h at 37 °C, 120 rpm. The residue was washed with 800 mL of 1 M NaCl, followed by 1.6 L of distilled water to remove residual carbohydrates and proteins, and then filtered. The solid was air-dried and ground in a Willey Mill to 60 mesh.

The lignin in the ground residue was solvent extracted with acidified aqueous dioxane solution (dioxane/water 85:15, v/v, containing 0.01 M HCl). In a distillation flask connected to a condenser, 200 mL of the solution was added to 10 g of ground residue. The suspension was mixed with a magnetic stirrer and heated (boiling point of the azeotrope, 86 °C) under argon atmosphere, for 2 h. The resulting suspension was filtered; the filtrated was neutralized with sodium carbonate and evaporated in a rotatory evaporator in order to reduce the volume. The solution was added dropwise to 1 L of acidified deionized water (pH 2), constantly mixing. The precipitated lignin was allowed to equilibrate with the aqueous phase for 12 h, recovered by centrifugation, washed with 800 mL of water, and freeze-dried.

Adsorption capacity of the isolated lignin

The adsorption capacity of the isolated lignins was measured by adding cellulases solution (diluted Celluclast) at different concentrations to the same amount of lignin [35]. In 2 mL centrifuge tubes, 1.0 mL of the cellulase solution (from 0.05 to 0.4 mg/mL) was added to 20 mg of lignin. The tubes were incubated at 50 °C in, 120 rpm, to simulate the hydrolysis conditions. After 3 h, the mixture was centrifuged at 10,000 rpm for 10 min and the protein concentration in the supernatant was measured by the ninhydrin assay [53]. The amount of protein adsorbed to the lignin was the difference between the protein concentration of the added cellulase solution and the supernatant after incubation. The amount of protein bound to the lignin (mg/g of lignin) was plotted against the concentration of protein in the supernatant (mg/mL). Data were fitted in a Langmuir model, using the software OriginPro 9.0.

Results and discussion

Pretreatments and chemical characteristics of the sugarcane bagasses

Sugarcane bagasse (SCB) was pretreated in alkaline and acidic conditions aiming to prepare varied substrates presenting similar glucan contents, but different levels of residual xylan and lignin. Chemical composition of PSCBs and removal of cell wall components are presented in Table 1. Cellulose content of the pretreated bagasses was similar, varying from 50.5% (NaOH PSCB) to 56.1% (SO_2/steam PSCB). As expected, cellulosic fraction was more preserved in alkaline conditions (less than 2.5% solubilization), contrasting with 13–22% solubilization in acidic pretreatments. Solubilization of the hemicellulosic fraction was also greater in acid conditions, since hydrolysis reactions of this component are favored at low pH. Xylan content of H_2SO_4 PSCB, $NaHSO_3/H_2SO_4$ PSCB, and SO_2/steam PSCB was, respectively, 13.8, 10.1, and 2.0%, whereas in NaOH PSCB and Na_2SO_3/NaOH PSCB, the xylan content was 25.6 and 26.1%, respectively.

Alkaline and acidic conditions were tested specially because of the differences in lignin solubilization. At high pH values, lignin solubilization is facilitated, and it is enhanced by the presence of sulfite ions. On the other hand, in acid conditions, lignin can polymerize, decreasing its solubility. The presence of hydrogen sulfite in acidic reactions aids lignin solubilization, preventing polymerization [41, 66]. Lignin solubilization reached 26.3 and 48.8% after NaOH and Na_2SO_3/NaOH pretreatments, respectively. Less lignin was solubilized during 3.0% SO_2/steam and H_2SO_4 pretreatments, representing 10 and 14.6%, respectively. The higher concentration of hydrogen sulfite ions present in the 9%-$NaHSO_3/H_2SO_4$ pretreatment enhanced lignin solubilization (25.4%) to a value similar to NaOH pretreatment (26.3%). As a consequence of these different delignification levels, PSCB with contrasting lignin contents was generated, ranging from 15.0% (Na_2SO_3/NaOH PSCB) to 35.6% (SO_2/steam PSCB) (Table 1). This difference was important to assess the effect of lignin in limiting the enzymatic hydrolysis of PSCB.

Pretreatments can create, expose, or even remove acid groups in lignocellulosic materials structure, which can greatly influence the subsequent enzymatic hydrolysis of polysaccharides in pretreated materials [41]. The presence of weak acids in polysaccharides (glucuronic acid for example) and also in lignin is attributed to carboxylic groups. Sulfonic acid groups are considered strong acids [63]. Total acid groups content, determined by conductometric titration (Fig. 1), varied among the PSCB. For the Na_2SO_3/NaOH and NaOH PSCB, it was found 78.5 and 35.7 mmols of acid groups per kg of PSCB, respectively. Lower values were detected under acidic conditions. The presence of strong acid groups can be considered a consequence of lignin sulfonation [67]. Therefore, the current results indicate strong sulfonation of residual lignin contained in the Na_2SO_3/NaOH PSCB (218.0 mmol/kg of lignin), whereas the hydrogen sulfite pretreatments ($NaHSO_3/H_2SO_4$ and SO_2/steam) resulted in moderate to low sulfonation degrees of PSCBs (42.8 mmol/kg of lignin and 13.9 mmol/kg of lignin, respectively) (Fig. 4b). In the Na_2SO_3/NaOH and $NaHSO_3/H_2SO_4$ pretreatments, sulfite ions promoted not only greater lignin

Table 1 Chemical composition (%) of pretreated sugarcane bagasses, solids yields, and removal of cell wall components during the pretreatments

Pretreatment	Chemical composition (g/100 g of pretreated material)			Solids yields (%)	Removal of cell wall components after pretreatment (%)		
	Glucan	Xylan	Lignin		Glucan	Xylan	Lignin
Untreated bagasse[a]	44.8 (1.0)	25.5 (0.4)	24.4 (0.7)	–	–	–	–
NaOH (5%)	50.5 (0.8)	25.6 (0.3)	20.3 (0.5)	87.3	0.0	19.4	26.3
Na_2SO_3/NaOH (10%/5%)	52.2 (0.6)	26.1 (0.4)	15.0 (0.7)	81.8	2.3	22.7	48.8
H_2SO_4 (0.75%)	53.1 (1.0)	13.8 (0.3)	28.8 (1.1)	71.1	13.5	64.1	14.6
$NaHSO_3/H_2SO_4$ (9%/0.75%)	54.0 (0.3)	10.1 (0.1)	28.3 (0.6)	63.3	21.7	76.7	25.4
Steam explosion/SO_2 (3.0%)	56.1 (0.6)	2.0 (0)	35.6 (0.8)	60.8	22.0	95.8	10.0

Values in parentheses are standard deviations

[a] Untreated bagasse contained 4.5% of water/ethanol-soluble extractives. Pretreated materials were analyzed without previous extraction

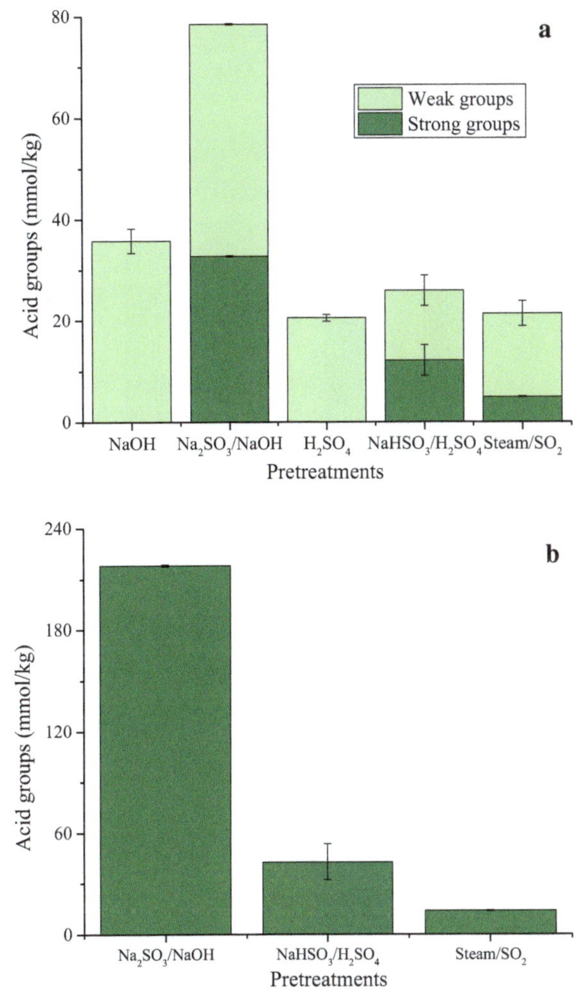

Fig. 1 Strong and weak acid groups in the PSCB (**a**) and in the lignin (**b**). Only the amount of strong acid groups was divided by the amount of lignin in each substrate

solubilization (Table 1), but also stronger sulfonation of the residual lignin (Fig. 1).

Enzymatic digestibility of the pretreated SCB

Determination of cellulose hydrolysis yields using different enzyme loadings provides information about substrate digestibility. Therefore, pretreated SCB was enzymatically hydrolyzed using Celluclast (cellulases) and Novozym 188 (source of β-glucosidase) (Fig. 2). Enzyme loading, based on protein concentration, ranged from 2.5 to 30 mg/g (protein/g of cellulose) and was calculated based only on Celluclast protein concentration. The glucan hydrolysis yields of NaOH, $NaHSO_3$/H_2SO_4, and H_2SO_4 PSCB were lower than 50%, even at the highest enzyme loading tested (30 mg/g). Moreover, the hydrolysis yields of these substrates were only slightly

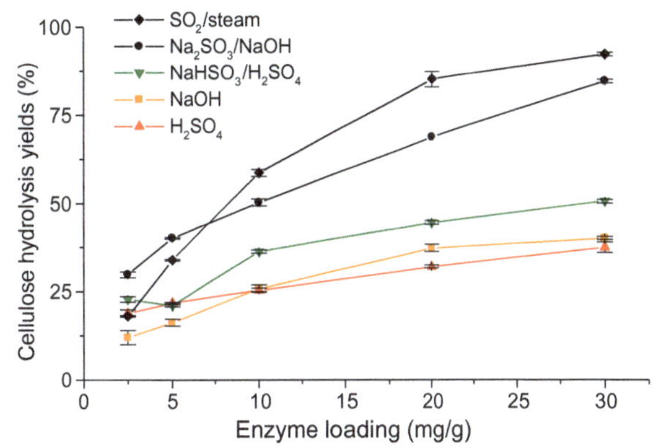

Fig. 2 PSCB cellulose hydrolysis yields after 72 h with Celluclast and Novozym 188, at varying enzyme loadings and 2% (w/w) solid loading

enhanced with increased enzyme loading (Fig. 2). On the other hand, 82–92% of the cellulose in the Na_2SO_3/NaOH PSCB and SO_2/steam PSCB was hydrolyzed, respectively, at the highest enzyme loading (30 mg/g). The hydrolysis yields of these substrates were also significantly dependent on the enzyme loading.

This experimental enzymatic hydrolysis approach enabled the comparison of the accessibility of cellulose in SCB submitted to different pretreatments. That is, if the increase in enzyme loading does not result in significant enhancement of cellulose hydrolysis yields, enzyme access to cellulose molecules are most likely limited due to the presence of lignin and/or hemicellulose (considering that the excess of enzymes could overcome any inhibitory effect). On the other hand, accessible cellulose is likely to be more intensively digested with increasing enzyme loadings [31]. Among the currently evaluated substrates, SO_2/steam, and Na_2SO_3/NaOH PSCBs were the least recalcitrant, reaching almost complete cellulose hydrolysis with the highest enzyme loading. This may indicate that the cellulose in these substrates was more accessible to the enzymes than in the other substrates. However, this can be controversial if lignin and hemicellulose contents are taken into account to explain the effect of accessibility of cellulose in the hydrolysis yields. The current data with PSCBs clearly indicated that two significantly different substrates (low xylan and high lignin contents in SO_2/steam PSCB, and high xylan and moderate lignin contents in Na_2SO_3/NaOH PSCB) presented similar digestibilities. Clearly, additional and multi-correlated characteristics of the substrates are defining their digestibilities. For example, lignin was not removed during SO_2/steam pretreatment but lignin relocation can be a key factor in increasing cellulose accessibility [66]. Associated with extensive xylan removal, this might explain the low recalcitrance of SO_2/steam PSCB, even though it had the highest lignin content. Rollin et al. [9] also observed that a pretreated switchgrass with relatively high lignin content was readily digested by cellulases. They emphasized that increasing cellulose accessibility is a key factor to reach effective enzymatic hydrolysis of lignocellulosic biomasses, even more important that lignin removal. In contrast, the Na_2SO_3/NaOH PSCB, despite presenting moderate lignin and xylan removals, contained a significantly sulfonated residual lignin, which could have contributed to a more hydrophilic and swollen substrate with high digestibility [20].

Another observation worth mentioning is the behavior of the two most digestible substrates (Na_2SO_3/NaOH and SO_2/steam PSCB) under low enzyme loadings. Even though the hydrolysis yields of SO_2/steam PSCB was higher than of Na_2SO_3/NaOH PSCB at enzyme loadings higher than 10 mg/g, at lower enzyme loadings (2.5 and 5 mg/g) that approximates desirable conditions in economically viable processes [68], the Na_2SO_3/NaOH PSCB cellulose was more hydrolyzed than the former PSCB. Different experimental approaches were then conducted in order to better understand this distinct behavior and comprehend the limiting role of lignin in the enzymatic hydrolysis of cellulose in the PSCB.

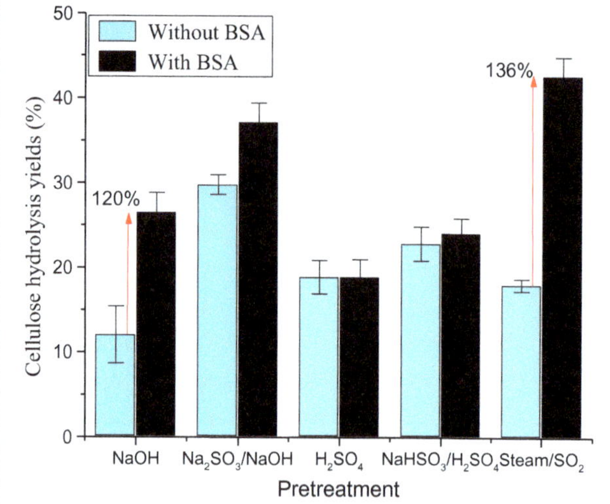

Fig. 3 PSCB cellulose hydrolysis yields after 72 h hydrolysis with Celluclast (2.5 mg/g) and Novozym 188 (2.3 UI/g), at 2% (w/w) solid loading. Experiments were run with and without 1 h-preincubation of the PSCB with BSA (250 mg/g of cellulose)

Effect of unproductive binding

An important limiting effect of lignin, especially at low enzyme loadings, is unproductive binding. Since the hydrolysis yields seemed to be dictated by cellulose accessibility at enzyme loadings higher than 5 mg/g, the effect of unproductive binding at 2.5 mg/g was investigated. To block the lignin unproductive binding sites, BSA was added to hydrolysis media 1 h before the addition of enzymes [31]. Hydrolysis yields of PSCB cellulose with and without addition of BSA are compared in Fig. 3.

Addition of BSA enhanced the cellulose hydrolysis yield of NaOH, Na_2SO_3/NaOH, and SO_2/steam PSCB. The enhancement was more pronounced in SO_2/steam PSCB (120%) and NaOH PSCB (136%). This result suggests that the low cellulose hydrolysis yields of these PSCB at low enzyme loadings could be assigned to unproductive binding of cellulolytic enzymes to the poorly charged (sulfonated) lignin (Fig. 3). Although an enhancement of hydrolysis yield was also observed in the hydrolysis of Na_2SO_3/NaOH PSCB (25%), it was much lower than in the aforementioned substrates, indicating that even though the pretreatment was conducted in alkaline conditions as in NaOH pretreatment, the effective sulfonation of Na_2SO_3/NaOH PSCB lignin may have reduced lignin negative binding influence. Sulfonated lignins bind less cellulases at hydrolysis conditions due to repulsive forces (negatively charged sulfonic groups and some of the cellulases), reducing unproductive binding [41].

Cellulose accessibility versus unproductive binding in lignin

The accessible cellulose surface was assessed through a colorimetric technique developed by Simons [54] and adapted by several authors [58–61]. The methodology is based on the combination of two dyes, direct orange (DO) and direct blue (DB), and their distinct affinity for cellulose and ability to diffuse within the substrate. DO displays higher affinity for cellulose and presents hydrodynamic radius comparable to some cellulases, which means that the orange dye may be used to estimate the accessibility of cellulose to cellulases. On the other hand, DB has lower affinity for cellulose, but is smaller than DO, allowing it to reach surfaces that DO (and consequently cellulases) cannot access [61]. By measuring the ratio of the amount of adsorbed dyes (DO/DB), it is possible to obtain some information about the accessible cellulose surface in each PSCB [69]. The amount of adsorbed dyes, as well their ratio, is shown in Table 2. The most recalcitrant substrates (NaOH, H_2SO_4, and $NaHSO_3$/H_2SO_4 PSCB) adsorbed more DB than DO, indicating that the majority of the pores that give access to cellulose fibrils in these substrates were smaller than the size of the orange dye. The less recalcitrant substrates (Na_2SO_3/NaOH and SO_2/steam PSCB) adsorbed more DO than DB, confirming that the cellulose was more exposed (accessible) in these substrates.

Table 2 Adsorbed direct orange (DO) and direct blue (DB) on PSCB after 6 h incubation at 70 °C and the ratio between them (accessible cellulose surface)

Pretreatment	Adsorbed DO (mg/g)	Adsorbed DB (mg/g)	DO/DB
NaOH	18.35	26.41	0.69
Na_2SO_3/NaOH	24.27	21.19	1.14
H_2SO_4	17.38	24.9	0.69
$NaHSO_3/H_2SO_4$	19.84	25.43	0.78
SO_2/steam	23.92	22.65	1.05

As expected, greater DO/DB values were observed in substrates with more accessible cellulose surface (higher DO adsorption), making them more easily hydrolyzed by cellulases. In fact, the more enzymatically digestible substrates showed higher DO/DB values (1.14 in Na_2SO_3/NaOH PSCB and 1.05 in SO_2/steam PSCB). In the least digestible substrates NaOH, H_2SO_4, and $NaHSO_3/H_2SO_4$ PSCB, DO/DB ratio was lower than 0.78.

In order to correlate the cellulose accessibility of the substrates with the enzymatic hydrolysis efficiency, DO/DB ratio was plotted with the cellulose hydrolysis yields after 72 h of hydrolysis at increasing enzyme loadings (Fig. 2). Except for the lowest enzyme loading (2.5 mg/g, Fig. 4), hydrolysis yields showed good correlation with accessible cellulose (DO/DB), with R^2 values higher than 0.83. Similar correlation values were also reported by Arantes and Saddler [69] for a range of substrates pretreated by different pretreatment technologies. The cellulose in Na_2SO_3/NaOH and SO_2/steam PSCB was more accessible and this is reflected in the digestibility of these substrates. Na_2SO_3/NaOH treatment resulted in the highest lignin solubilization (Table 1), which might have contributed to increase cellulose accessibility. On the other hand, even though SO_2/steam PSCB had the lowest lignin solubilization and consequently the highest lignin content of all the PSCB, cellulose accessibility was comparable to Na_2SO_3/NaOH PSCB probably because the xylan removal was almost complete and also because the steam explosion pretreatment can relocate lignin, which may have also contributed to increase cellulose accessibility [26, 70].

With the lowest enzyme loading tested, desirable condition in economically viable processes [68], R^2 value was 0.43. It has been broadly reported that, besides limiting the access of the enzymes to the cellulose surface, lignin is known to adsorb proteins. Kumar et al. [31] showed that unproductive binding of cellulases to lignin is more pronounced at low enzyme loading, and that this effect could be totally overcome by increasing the enzyme loading. Apparently the binding capacity of softwood lignin, presented in the cited work, has a higher binding capacity than sugarcane bagasse lignin, since at slightly higher enzyme loading (i.e., 5 mg/g), good correlations could be observed. To test the hypothesis that unproductive binding was limiting the hydrolysis at low enzyme loading, and consequently resulting in this poor correlation, the binding capacities of the lignins isolated from PSCB bagasse were measured. Lignin samples isolated from the PSCB were incubated in increasing concentrations of protein (Celluclast), keeping lignin concentration constant. The adsorption curves are shown in Fig. 5 and the constants extracted from Langmuir adsorption model are presented in Table 3.

Maximum protein binding capacity of lignin samples ranged from 1.6 (Na_2SO_3/NaOH PSCB) to 23.2 (SO_2/steam PSCB) mg of protein per gram of lignin. Comparing all the lignin samples, it was possible to observe that lignin sulfonation extent was a key factor in reducing the protein binding capacity of lignin. Under alkaline and acidic conditions, the presence of sulfite ions during the pretreatment results not only in enhanced lignin solubilization, but also in sulfonation of the lignin that remains in the solid fraction (Fig. 1). Sulfonic groups present in the residual lignin are strong acids and are deprotonated (negatively charged) in the experimental conditions (pH 4.8) described in this work [41]. Most of the cellulases from *T. reesei* extract (as Celluclast) have isoelectric point lower than this pH value [48] and consequently are also negatively charged in this condition. Therefore, the negative charge of sulfonated lignin can promote repulsion of cellulases, reducing the binding capacity of lignin. In fact, as a result of the addition of sulfite ions to alkaline pretreatment, the binding capacity of residual lignin reduced from 17.0 mg/g (NaOH PSCB) to 1.6 mg/g (Na_2SO_3/NaOH PSCB). Under acidic conditions, the difference was not that evident, but it was still possible to observe the reduction in the protein binding capacity of lignin, from 6.2 mg/g (H_2SO_4 PSCB lignin) to 5.8 mg/g ($NaHSO_3/H_2SO_4$ PSCB lignin). Although strong acid groups were detected in SO_2/steam PSCB (Fig. 4b), the sulfonation of this lignin was very low, resulting in a strong binding capacity.

Effect of BSA addition in the enzymatic hydrolysis of PSCB using different cocktails

As discussed earlier, unproductive binding depends not only on the lignin properties, but also on enzyme characteristics. Berlin and colleagues [32] pointed out the importance of enzyme development in the concept of weak lignin-binding enzymes. For that reason, the effect of unproductive binding was also tested in a more advanced enzyme preparation, Cellic CTec3 (Fig. 6), an industrial cellulase cocktail recently designed and

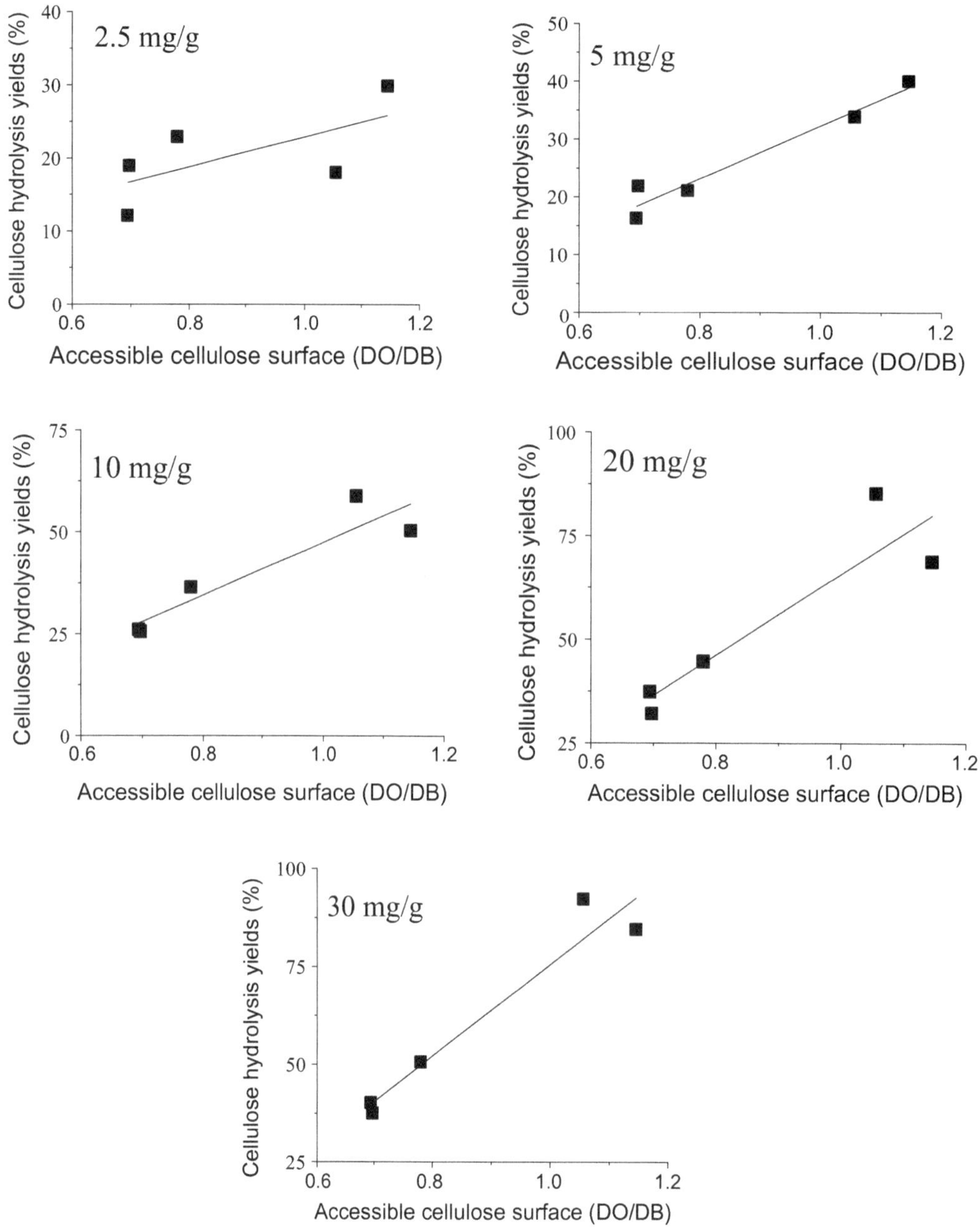

Fig. 4 Correlation between PSCB cellulose hydrolysis yields after 72 h with Celluclast and Novozym 188 and the accessible cellulose surface. NaOH PSCB (*orange*), H_2SO_4 PSCB (*red*), $NaHSO_3/H_2SO_4$ PSCB (*green*), SO_2/steam PSCB (*black*), and Na_2SO_3/NaOH PSCB (*blue*). Enzyme loadings were 2.5, 5, 10, 20, and 30 mg/g

optimized specifically for enzymatic hydrolysis of pretreated lignocellulosic substrates. With Celluclast, unproductive binding was observed for NaOH, Na_2SO_3/NaOH, and SO_2/steam PSCB (Fig. 3). With CTec3, no significant enhancement in hydrolysis yield due to BSA addition was observed in Na_2SO_3/NaOH PSCB, indicating no unproductive binding during the hydrolysis of this substrate with CTec3. Moreover, an impressive 70% cellulose hydrolysis yield was achieved with a considerable low enzyme loading (2.5 mg/g), probably as a result of more efficient enzyme combination (cellulases and accessory enzymes) that efficiently deconstructs substrates with high hemicellulose content. For SO_2/steam PSCB hydrolyzed with CTec 3, a 48% increase in hydrolysis yield was observed due to BSA addition (Fig. 6), whereas with Celluclast, the increment was 136% (Fig. 3). A lower increase

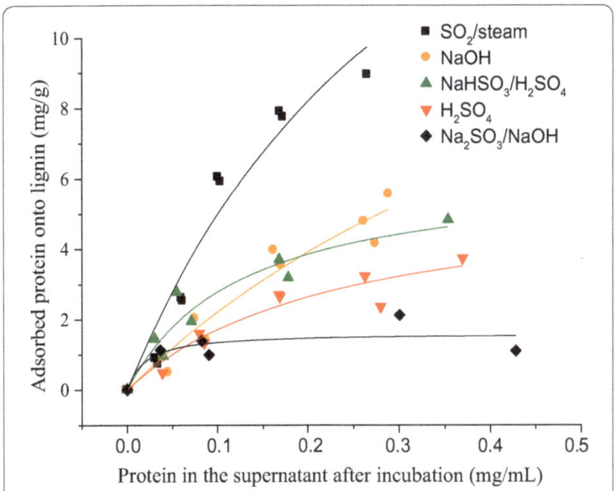

Fig. 5 Langmuir adsorption isotherms of Celluclast proteins onto isolated lignins. Lignin concentration was kept constant (20 mg/mL) and initial protein concentration varied from 0.05 to 0.45 mg/mL

Table 3 Parameters from the adsorption isotherm of the proteins from Celluclast onto lignins extracted from the PSCB

Pretreatment	P_{max} (mg/g)	K (mL/mg)	R^2
NaOH	17.0 ± 11.2	1.5 ± 0.3	0.93
Na_2SO_3/NaOH	1.6 ± 0.4	50.8 ± 70.3	0.63
H_2SO_4	6.2 ± 1.1	8.0 ± 3.2	0.90
$NaHSO_3/H_2SO_4$	5.8 ± 1.5	4.2 ± 2.1	0.91
SO_2/steam	23.2 ± 8.7	2.7 ± 1.5	0.93

Values in parentheses are standard deviations

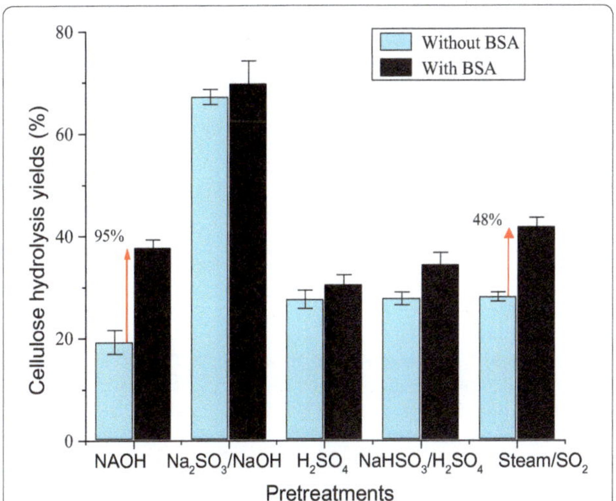

Fig. 6 PSCB cellulose hydrolysis yields after 72 h hydrolysis with Cellic CTec3 (2.5 mg/g), at 2% solids loading. Experiments were run with and without preincubation of the PSCB with BSA

was also observed with NaOH PSCB (120% with Celluclast and 95% with CTec3, Figs. 3, 6, respectively). The lower increase in hydrolysis yields in these substrates with high protein binding capacity lignins is an indication of higher tolerance of CTec3 to the presence of lignin.

Conclusions

Pretreatments that efficiently removed (Na_2SO_3/NaOH) or relocated (SO_2/steam pretreatment) lignin resulted in substrates with greater cellulose accessibility and, consequently, were more efficiently digested by cellulases. Substrates containing lignin with high protein binding capacity resulted in high unproductivity binding at low dosage of traditional cellulases as observed during hydrolysis of NaOH PSCB and SO_2/steam PSCB with Celluclast. Pretreatments that promoted sulfonation of lignin reduced its protein binding capacity as a consequence of the negative charges. In addition, a recently developed cellulose preparation, Cellic CTec3, was shown to be less affected by the unproductive binding capacity of lignin, and promoted higher hydrolysis yield, even at very low enzyme loading. These results confirm the importance of pretreatments that increase cellulose accessibility and modify the lignin in order to decrease its protein binding capacity. Also, considering that the enzymatic hydrolysis in industrial processes must be conducted at low enzyme loadings, Na_2SO_3/NaOH pretreatment might be promising, since the results show that an impressive amount of cellulose (70%) was hydrolyzed with low enzyme loadings such as 2.5 mg of protein per gram of cellulose with the modern enzyme preparation Cellic CTec3.

Authors' contributions

GS performed the experiments, data interpretation, and drafted the manuscript. VA, JS, AF and AMFM participated in the design of the study, data interpretation, and reviewed the manuscript. All authors read and approved the final manuscript.

Author details

[1] Departamento de Biotecnologia, Escola de Engenharia de Lorena, Universidade de São Paulo, CP 116, Lorena, SP 12602-810, Brazil. [2] Forest Products Biotechnology/Bioenergy Group, Faculty of Forestry, University of British Columbia, 2424 Main Mall, Vancouver, BC V6T 1Z4, Canada.

Acknowledgements

The authors thank José M. Silva and José C. Tavares for technical assistance, Richard Chandra for the valuable scientific discussions, and Sarah Teter (Novozymes) for providing Cellic CTec3. This work was supported by FAPESP (Contract Numbers 08/56256-5 and 14/06923-6), CNPq, and CAPES. Germano Siqueira thanks CAPES, FAPESP (Contract Number 12/09940-3), and CNPq for his student fellowships.

Competing interests

The authors declare that they have no competing interests.

References

1. Henriksson G, Brännvall E, Lennholm H. The trees. In: Ek M, Gellerstedt G, Henriksson G, editors. Pulp and paper chemistry and technology, vol. 1. Berlin: Walter de Gruyter; 2009. p. 14–44.
2. Aulin C, Ahola S, Josefsson P, Nishino T, Hirose Y, Österberg M, Wagberg L, et al. Nanoscale cellulose films with different crystallinities and mesostructures-their surface properties and interaction with water. Langmuir. 2009;25:7675–85.
3. dos Santos LV, de Barros Grassi MC, Gallardo JCM, Pirolla RAS, Calderón LL, de Carvalho-Netto OV, et al. Second-generation ethanol: the need is becoming a reality. Ind Biotechnol. 2016;12:40–57.
4. Galbe M, Zacchi G. Pretreatment: the key to efficient utilization of lignocellulosic materials. Biomass Bioenergy. 2012;46:70–8.
5. Olcay H, Subrahmanyam AV, Xing R, Lajoie J, Dumesic JA, Huber GW. Production of renewable petroleum refinery diesel and jet fuel feedstocks from hemicellulose sugar streams. Energy Environ Sci. 2013;6:205–16.
6. Raganati F, Olivieri G, Götz P, Marzocchella A, Salatino P. Butanol production from hexoses and pentoses by fermentation of *Clostridium acetobutylicum*. Anaerobe. 2015;34:146–55.
7. Losordo Z, McBride J, Van Rooyen J, Wenger K, Willies D, Froehlich A, et al. Cost competitive second-generation ethanol production from hemicellulose in a Brazilian sugarcane biorefinery. Biofuels, Bioprod Biorefin. 2016;10:589–602.
8. Jeoh T, Ishizawa CI, Davis MF, Himmel ME, Adney WS, Johnson DK. Cellulase digestibility of pretreated biomass is limited by cellulose accessibility. Biotechnol Bioeng. 2007;98:112–22.
9. Rollin JA, Zhu Z, Sathitsuksanoh N, Zhang YHP. Increasing cellulose accessibility is more important than removing lignin: a comparison of cellulose solvent-based lignocellulose fractionation and soaking in aqueous ammonia. Biotechnol Bioeng. 2011;108:22–30.
10. Barakat A, Chuetor S, Monlau F, Solhy A, Rouau X. Eco-friendly dry chemo-mechanical pretreatments of lignocellulosic biomass: impact on energy and yield of the enzymatic hydrolysis. Appl Energy. 2014;113:97–105.
11. Zhu M-Q, Wen J-L, Wang Z-W, Su Y-Q, Wei Q, Sun R-C. Structural changes in lignin during integrated process of steam explosion followed by alkaline hydrogen peroxide of *Eucommia ulmoides* Oliver and its effect on enzymatic hydrolysis. Appl Energy. 2015;158:233–42.
12. Silveira MHL, Morais ARC, da Costa Lopes AM, Olekszyszen DN, Bogel-Łukasik R, Andreaus J, et al. Current pretreatment technologies for the development of cellulosic ethanol and biorefineries. Chemsuschem. 2015;8:3366–90.
13. Siqueira G, Várnai A, Ferraz A, Milagres AMF. Enhancement of cellulose hydrolysis in sugarcane bagasse by the selective removal of lignin with sodium chlorite. Appl Energy. 2013;102:399–402.
14. Mendes FM, Siqueira G, Carvalho W, Ferraz A, Milagres AMF. Enzymatic hydrolysis of chemithermomechanically pretreated sugarcane bagasse and samples with reduced initial lignin content. Biotechnol Prog. 2011;27:395–401.
15. Sun S-L, Wen J-L, Ma M-G, Sun R-C. Enhanced enzymatic digestibility of bamboo by a combined system of multiple steam explosion and alkaline treatments. Appl Energy. 2014;136:519–26.
16. Hu J, Arantes V, Saddler JN. The enhancement of enzymatic hydrolysis of lignocellulosic substrates by the addition of accessory enzymes such as xylanase: is it an additive or synergistic effect? Biotechnol Biofuels. 2011;4:36.
17. Hu J, Gourlay K, Arantes V, Van Dyk JS, Pribowo A, Saddler JN. The accessible cellulose surface influences cellulase synergism during the hydrolysis of lignocellulosic substrates. Chemsuschem. 2015;8:901–7.
18. Tian S, Luo XL, Yang XS, Zhu JY. Robust cellulosic ethanol production from SPORL-pretreated lodgepole pine using an adapted strain *Saccharomyces cerevisiae* without detoxification. Bioresour Technol. 2010;101:8678–85.
19. Zhu JY, Pan XJ, Wang GS, Gleisner R. Sulfite pretreatment (SPORL) for robust enzymatic saccharification of spruce and red pine. Bioresour Technol. 2009;100:2411–8.
20. Mendes FM, Laurito DF, Bazzeggio M, Ferraz A, Milagres AMF. Enzymatic digestion of alkaline-sulfite pretreated sugar cane bagasse and its correlation with the chemical and structural changes occurring during the pretreatment step. Biotechnol Prog. 2013;29:890–5.
21. Shuai L, Yang Q, Zhu JY, Lu FC, Weimer PJ, Ralph J, et al. Comparative study of SPORL and dilute-acid pretreatments of spruce for cellulosic ethanol production. Bioresour Technol. 2010;101:3106–14.
22. Mooney CA, Mansfield SD, Tuohy MG, Saddler JN. The effect of lignin content on cellulose accessibility and enzymatic hydrolysis of softwood pulps. In: Proceedings of the 1997 Biological Science Symposium, San Francisco, CA: Tappi press; 1997, p. 259–65.
23. Stone JE, Scallan AM. A structural model for the cell wall of water-swollen wood pulp fibres based on their accessibility to macromolecules. Cellulose Chem Technol. 1968;2:343–58.
24. Grethlein HE. The effect of pore size distribution on the rate of enzymatic hydrolysis of cellulosic substrates. Bio/Technology. 1985;3:155–60.
25. Santi-Junior C, Milagres AMF, Ferraz A, Carvalho W. The effects of lignin removal and drying on the porosity and enzymatic hydrolysis of sugarcane bagasse. Cellulose. 2013;20:3165–77.
26. Várnai A, Siika-aho M, Viikari L. Restriction of the enzymatic hydrolysis of steam-pretreated spruce by lignin and hemicellulose. Enzyme Microb Technol. 2010;46:185–93.
27. Yan Z, Li J, Chang S, Cui T, Jiang Y, Yu M, et al. Lignin relocation contributed to the alkaline pretreatment efficiency of sweet sorghum bagasse. Fuel. 2015;158:152–8.
28. Siqueira G, Milagres AM, Carvalho W, Koch G, Ferraz A. Topochemical distribution of lignin and hydroxycinnamic acids in sugar-cane cell walls and its correlation with the enzymatic hydrolysis of polysaccharides. Biotechnol Biofuels. 2011;4:7.
29. Costa THF, Masarin F, Bonifácio TO, Milagres AMF, Ferraz A. The enzymatic recalcitrance of internodes of sugar cane hybrids with contrasting lignin contents. Ind Crops Prod. 2013;51:202–11.
30. Ding S-Y, Liu Y-S, Zeng Y, Himmel ME, Baker JO, Bayer EA. How does plant cell wall nanoscale architecture correlate with enzymatic digestibility? Science. 2012;338:1055–60.
31. Kumar L, Arantes V, Chandra R, Saddler J. The lignin present in steam pretreated softwood binds enzymes and limits cellulose accessibility. Bioresour Technol. 2012;103:201–8.
32. Berlin A, Gilkes N, Kurabi A, Bura R, Tu M, Kilburn D, et al. Weak lignin-binding enzymes: a novel approach to improve activity of cellulases for hydrolysis of lignocellulosics. Appl Biochem Biotechnol. 2005;121–124:163–70.
33. Nakagame S, Chandra RP, Saddler JN. The effect of isolated lignins, obtained from a range of pretreated lignocellulosic substrates, on enzymatic hydrolysis. Biotechnol Bioeng. 2010;105:871–9.
34. Chernoglazov VM, Ermolova OV, Klyosovt AA. Adsorption of high-purity endo-1,4-B-glucanases from *Trichoderma reesei* on components of lignocellulosic materials: cellulose, lignin and xylan. Enzyme Microb Technol. 1988;10:503–7.
35. Karuna N, Zhang L, Walton JH, Couturier M, Oztop MH, Master ER, et al. The impact of alkali pretreatment and post-pretreatment conditioning on the surface properties of rice straw affecting cellulose accessibility to cellulases. Bioresour Technol. 2014;167:232–40.
36. Palonen H. Role of lignin in the enzymatic hydrolysis of lignocellulose. http://lib.tkk.fi/Diss/2004/isbn9513862720/isbn9513862720.pdf (2004) Assessed 9 Dec 2016.
37. Tu M, Pan X, Saddler JN. Adsorption of cellulase on cellulolytic enzyme lignin from lodgepole pine. J Agric Food Chem. 2009;57:7771–8.
38. Rahikainen J, Mikander S, Marjamaa K, Tamminen T, Lappas A, Viikari L, et al. Inhibition of enzymatic hydrolysis by residual lignins from softwood-study of enzyme binding and inactivation on lignin-rich surface. Biotechnol Bioeng. 2011;108:2823–34.
39. Rahikainen JL, Martin-Sampedro R, Heikkinen H, Rovio S, Marjamaa K, Tamminen T, et al. Inhibitory effect of lignin during cellulose bioconversion: the effect of lignin chemistry on non-productive enzyme adsorption. Bioresour Technol. 2013;133:270–8.
40. Lu X, Zheng X, Li X, Zhao J. Adsorption and mechanism of cellulase enzymes onto lignin isolated from corn stover pretreated with liquid hot water. Biotechnol Biofuels. 2016;9:118.
41. Wang Z, Lan T, Zhu J. Lignosulfonate and elevated pH can enhance enzymatic saccharification of lignocelluloses. Biotechnol Biofuels. 2013;6:9.
42. Lou H, Zhu JY, Lan TQ, Lai H, Qiu X. pH-Induced lignin surface modification to reduce nonspecific cellulase binding and enhance enzymatic saccharification of lignocelluloses. Chemsuschem. 2013;6:919–27.
43. Sewalt VJH, Glasser WG, Beauchemin KA. Lignin impact on fiber degrada-

tion. 3. Reversal of inhibition of enzymatic hydrolysis by chemical modification of lignin and by additives. J Agric Food Chem. 1997;45:1823–8.
44. Yu Z, Gwak K-S, Treasure T, Jameel H, Chang H, Park S. Effect of lignin chemistry on the enzymatic hydrolysis of woody biomass. Chemsuschem. 2014;7:1942–50.
45. Sewalt VJH, de Oliveira W, Glasser WG, Fontenot JP. Lignin impact on fibre degradation: 2—a model study using cellulosic hydrogels. J Sci Food Agric. 1996;71:204–8.
46. Nakagame S, Chandra RP, Kadla JF, Saddler JN. Enhancing the enzymatic hydrolysis of lignocellulosic biomass by increasing the carboxylic acid content of the associated lignin. Biotechnol Bioeng. 2011;108:538–48.
47. Yang Q, Pan X. Correlation between lignin physicochemical properties and inhibition to enzymatic hydrolysis of cellulose. Biotechnol Bioeng. 2016;113:1213–24.
48. Sammond DW, Yarbrough JM, Mansfield E, Bomble YJ, Hobdey SE, Decker SR, et al. Predicting enzyme adsorption to lignin films by calculating enzyme surface hydrophobicity. J Biol Chem. 2014;289:20960–9.
49. Lai C, Tu M, Shi Z, Zheng K, Olmos LG, Yu S. Contrasting effects of hardwood and softwood organosolv lignins on enzymatic hydrolysis of lignocellulose. Bioresour Technol. 2014;163:320–7.
50. Bezerra TL, Ragauskas AJ. A review of sugarcane bagasse for second-generation bioethanol and biopower production. Biofuels, Bioprod Biorefin. 2016;10:634–47.
51. Ferreira Leitão V, Perrone CC, Rodrigues J, Franke APM, Macrelli S, Zacchi G. An approach to the utilisation of CO_2 as impregnating agent in steam pretreatment of sugar cane bagasse and leaves for ethanol production. Biotechnol Biofuels. 2010;3:7.
52. Dence C. The determination of lignin. In: Lin SY, Dence CW, editors. Methods lignin chemistry, vol. 1. Berlin: Springer; 1992. p. 33–61.
53. Mok YK, Arantes V, Saddler JN. A NaBH coupled ninhydrin-based Assay for the quantification of protein/enzymes during the enzymatic hydrolysis of pretreated lignocellulosic biomass. Appl Biochem Biotechnol. 2015;176:1564–80.
54. Ghose TK. Measurement of cellulase activities. Pure Appl Chem. 1987;59:257–68.
55. Tan LUL, Mayers P, Saddler JN. Purification and characterization of a thermostable xylanase from a thermophilic fungus *Thermoascus aurantiacus*. Can J Microbiol. 1987;33:689–92.
56. Berlin A, Maximenko V, Bura R, Kang K, Gilkes N, Saddler J. A rapid microassay to evaluate enzymatic hydrolysis of lignocellulosic substrates. Biotechnol Bioeng. 2005;93:880–6.
57. Simons FL. A stain for use in the microscopy of beaten fibers. TAPPI. 1950;33:312–4.
58. Chandra RP, Saddler JN. Use of the Simons' staining technique to assess cellulose accessibility in pretreated substrates. Ind Biotechnol. 2012;8:230–7.
59. Chandra R, Ewanick S, Hsieh C, Saddler JN. The characterization of pretreated lignocellulosic substrates prior to enzymatic hydrolysis, part 1: A modified Simons' staining technique. Biotechnol Prog. 2008;24:1178–85.
60. Chandra RP, Ewanick SM, Chung PA, Au-Yeung K, Del Rio L, Mabee W, Saddler JN. Comparison of methods to assess the enzyme accessibility and hydrolysis of pretreated lignocellulosic substrates. Biotechnol Lett. 2009;31:1217–22.
61. Yu X, Minor JL, Atalla RH. Fiber analysis mechanism of action of Simons' stain. TAPPI. 1995;78:175–80.
62. Khalfaoui M, Nakhli A, Aguir C, Omri A, M'henni MF, Ben Lamine A. Statistical thermodynamics of adsorption of dye DR75 onto natural materials and its modifications: double-layer model with two adsorption energies. Environ Sci Pollut Res. 2014;21:3134–44.
63. Katz S, Beatson RP, Scallon AM. The determination of strong and weak acidic groups in sulfite pulps. Sven Papperstidning. 1984;87:48–53.
64. Berlin A, Balakshin M, Gilkes N, Kadla J, Maximenko V, Kubo S, et al. Inhibition of cellulase, xylanase and beta-glucosidase activities by softwood lignin preparations. J Biotechnol. 2006;125:198–209.
65. Guerra A, Filpponen I, Lucia LA, Saquing C, Baumberger S, Argyropoulos DS. Toward a better understanding of the lignin isolation process from wood. J Agric Food Chem. 2006;54:5939–47.
66. Sjostrom E. Wood pulping. In: Sjos, editor. Wood chemistry: fundamentals and applications. 2nd ed. San Diego: Academic Press; 1993. p. 114–64.
67. Diniz JMBF. Determination of the sulphonic acid group content of wood pulps. Holzforschung. 1995;49:545–7.
68. Humbird D, Davis R, Tao L, Kinchin C, Hsu D, Aden A, et al. Process design and economics for biochemical conversion of lignocellulosic biomass to ethanol. Golden; 2011.
69. Arantes V, Saddler JN. Cellulose accessibility limits the effectiveness of minimum cellulase loading on the efficient hydrolysis of pretreated lignocellulosic substrates. Biotechnol Biofuels. 2011;4:3.
70. Kristensen JB, Thygesen LG, Felby C, Jørgensen H, Elder T. Cell-wall structural changes in wheat straw pretreated for bioethanol production. Biotechnol Biofuels. 2008;1:5.

15

Sustaining fermentation in high-gravity ethanol production by feeding yeast to a temperature-profiled multifeed simultaneous saccharification and co-fermentation of wheat straw

Johan O. Westman[1,2†], Ruifei Wang[1†], Vera Novy[1,3,4] and Carl Johan Franzén[1*]

Abstract

Background: Considerable progress is being made in ethanol production from lignocellulosic feedstocks by fermentation, but negative effects of inhibitors on fermenting microorganisms are still challenging. Feeding preadapted cells has shown positive effects by sustaining fermentation in high-gravity simultaneous saccharification and co-fermentation (SSCF). Loss of cell viability has been reported in several SSCF studies on different substrates and seems to be the main reason for the declining ethanol production toward the end of the process. Here, we investigate how the combination of yeast preadaptation and feeding, cell flocculation, and temperature reduction improves the cell viability in SSCF of steam pretreated wheat straw.

Results: More than 50% cell viability was lost during the first 24 h of high-gravity SSCF. No beneficial effects of adding selected nutrients were observed in shake flask SSCF. Ethanol concentrations greater than 50 g L^{-1} led to significant loss of viability and prevented further fermentation in SSCF. The benefits of feeding preadapted yeast cells were marginal at later stages of SSCF. Yeast flocculation did not improve the viability but simplified cell harvest and improved the feasibility of the cell feeding strategy in demo scale. Cultivation at 30 °C instead of 35 °C increased cell survival significantly on solid media containing ethanol and inhibitors. Similarly, in multifeed SSCF, cells maintained the viability and fermentation capacity when the temperature was reduced from 35 to 30 °C during the process, but hydrolysis yields were compromised. By combining the yeast feeding and temperature change, an ethanol concentration of 65 g L^{-1}, equivalent to 70% of the theoretical yield, was obtained in multifeed SSCF on pretreated wheat straw. In demo scale, the process with flocculating yeast and temperature profile resulted in 5% (w/w) ethanol, equivalent to 53% of the theoretical yield.

Conclusions: Multifeed SSCF was further developed by means of a flocculating yeast and a temperature-reduction profile. Ethanol toxicity is intensified in the presence of lignocellulosic inhibitors at temperatures that are beneficial to hydrolysis in high-gravity SSCF. The counteracting effects of temperature on cell viability and hydrolysis call for more tolerant microorganisms, enzyme systems with lower temperature optimum, or full optimization of the multifeed strategy with temperature profile.

*Correspondence: franzen@chalmers.se
†Johan O. Westman and Ruifei Wang contributed equally
[1] Division of Industrial Biotechnology, Department of Biology and Biological Engineering, Chalmers University of Technology, Gothenburg, Sweden
Full list of author information is available at the end of the article

Keywords: Multifeed simultaneous saccharification and co-fermentation (SSCF), High gravity, Yeast viability, Ethanol inhibition, Temperature effect, Combined stress, Flocculation, Demonstration scale, Wheat straw, Modeling

Background

Ethanol produced from lignocellulosic materials by fermentation is one of the main alternatives for replacing fossil fuels in light-duty transportation. In recent years, the commercialization of lignocellulosic ethanol has gained momentum because of two factors: the improvement of enzymes, including reduced cost and the application of lytic polysaccharide monooxygenases in enzyme cocktails [1, 2], and the construction of several commercial-scale production plants [3, 4]. However, inhibitors formed during thermochemical pretreatment of lignocelluloses, which hinder the growth and fermentation ability of the microorganisms, present a major challenge [5]. Washing the pretreated slurry removes the inhibitors and leads to rapid fermentation, but also leads to loss of fermentable sugars and increased demands for wastewater treatment [6, 7].

A number of other strategies to reduce deleterious effects of inhibitors have been studied. For example, in situ detoxification has been demonstrated using sulfur oxyanions to sulfonate inhibitory compounds and render them less reactive [8], and laccase has been used to oxidize and polymerize free phenols [9]. The yeast *Saccharomyces cerevisiae* is one of the most-used fermenting microorganisms for its efficient glucose utilization and high stress tolerance in general. The robustness of yeast cells has been improved by genetic engineering for enhanced capacities in converting inhibitors and maintaining energy/redox balances [10–13]. Knowledge on the mechanisms governing inhibitor tolerance and cellular detoxification in yeasts and other microorganisms is thus of great importance [14–17]. Desirable traits in inhibitor resistance have also been obtained through mutagenesis, genome shuffling, and evolutionary engineering [18], where a detailed understanding of the mechanisms determining microbial resistance to individual or multiple stresses is not necessary [19]. Pre-exposure of cells to inhibitors during cultivation has been shown to effectively improve fermentation performance [20, 21]. The intrinsic detoxification capability of yeast cells has been exploited by using large amounts of inocula [22, 23], increasing the local cell density by encapsulation [24] or flocculation [25], and by continuous fermentation with cell retention [26]. Furthermore, proper implementation of rapidly sedimenting flocculating cells in bioprocesses gives the possibility to omit energy-intensive centrifugation steps [27].

Fed-batch simultaneous saccharification and co-fermentation (SSCF) can minimize the effects of inhibitors on the process, as the rate at which the inhibitors are introduced can be controlled [10]. In fed-batch SSCF, pretreated raw material is fed to the fermenter where it is continuously hydrolyzed and the released sugars are simultaneously fermented into ethanol. Substrate feeding is a practical solution in processes with high solid substrate loadings, also called high-gravity fermentation [28], and promotes xylose and glucose co-consumption [29]. In contrast to other processes for handling high substrate loadings, such as separate hydrolysis and co-fermentation and combined pre-hydrolysis and SSCF, large amounts of glucose are not accumulated in fed-batch SSCF due to the continuous removal of sugars by fermentation. A high xylose-to-glucose ratio can thus be maintained throughout the process, facilitating xylose utilization by recombinant xylose-consuming yeasts by relieving the competitive inhibition by glucose on the xylose uptake via the common glucose/xylose transporters [30]. Further development of fed-batch SSCF has been reported toward multifeed SSCF, in which feedings of substrate, enzymes and cells into the SSCF reactor are coordinated in order to balance the main reactions in the process, i.e., to maintain high and balanced rates of both hydrolysis and fermentation [31, 32].

Loss of cell viability has been reported in several studies on multifeed, high-gravity SSCF of different substrates [7, 31–33]. The lack of viable cells seems to be the main reason for the declining ethanol production toward the end of the process, which may lead to incomplete utilization of the available sugars and low overall ethanol yield. Possible reasons for the reduction in viability could be lignocellulose-derived inhibitors [34], lack of nutrients/nitrogen sources [7, 35], lack of unsaturated fatty acids and ergosterol for anaerobic conditions [36], problems associated with high-gravity process or viscous media, e.g., limited mass transfer, high osmolality, and slow regulation of pH and temperature [37], and toxicity of the ethanol produced [38].

In this study, we investigate the reasons for the decline in viable cell concentrations in high-gravity SSCF of pretreated wheat straw. It has been reported that yeast flocculation leads to improved tolerance to lignocellulose-derived inhibitors [25] and ethanol [39]. The effects of yeast flocculation, to improve cell robustness in general, and also to improve the feasibility of cell feeding during multifeed SSCF, were investigated. The knowledge obtained from these developments of the multifeed SSCF process will be valuable in the further development of flexible and robust lignocellulose-based processes and cell factories, with the objectives to increase the final product concentration and promote complete conversion of the carbohydrates in the lignocellulosic raw material.

Methods

Strains

The metabolically and evolutionarily engineered xylose-fermenting *S. cerevisiae* strains KE6-12.A (originating from the hybrid diploid wine yeast USM21 via TMB3400) [Albers et al. unpublished, 40, 41, 42] and IBB10B05 (originating from CEN.PK 113-5D) [43, 44] were used in this study. Both strains were transformed with a gene cassette containing the chimeric flocculation gene *FLOw* (GenBank accession number KT264162) [45]. The cassette was amplified from the genomic DNA of CEN.PK Flow (originating from *S. cerevisiae* CEN.PK 113-7D (MATa, MAL2-8C, SUC2) [46] with the forward primer 5′-CAGAAAGGGTTCGCAAGTC-3′ and reverse primer 5′-GGCGTATTTCTACTCCAGCATTC-3′ [45]. The resulting PCR product, with flanking regions homologous to the *HO* locus, was used for homologous recombination in the two parental strains mentioned above using the lithium acetate based transformation method [47]. Transformants were selected on yeast extract, peptone, dextrose (YPD) plates containing 10 g L^{-1} yeast extract (Difco™, BD Biosciences, San Jose, CA), 20 g L^{-1} peptone (Bacto™, BD Biosciences), 20 g L^{-1} glucose (D-glucose monohydrate, Merck KGaA, Darmstadt, Germany), 20 g L^{-1} agar (Merck KGaA) and 200 μg/mL G418 (Sigma-Aldrich, Steinheim, Germany). Correct integration into the *HO* locus (which in itself should not affect cell growth [48]) and the size of the gene were confirmed by PCR using the forward primer 5′-ATGATATCCAGTTCGAGTTTATCATTATC-3′ and the reverse primer 5′-CAAATCAGTGCCGGTAACG-3′. The resulting flocculating strains were named KE-Flow and B-Flow.

Raw materials

Pretreated wheat straw and molasses used in this study were provided by SP Biorefinery Demo Plant (Örnsköldsvik, Sweden). The wheat straw was steam pretreated in a one-step vertical continuous reactor with the addition of 0.2% (w/w) H_2SO_4, at pH 2.1–2.2 and 187–188 °C for 6–7 min, by using 11.5 bar steam. After pretreatment, the biomass slurry was separated into a solid fraction and a liquid fraction (pretreatment liquor) using a filter press. The water-insoluble solids (WIS) content in the solid fraction was determined by washing a weighed amount of the moist solid fraction with excess deionized water (the wash was repeated until the glucose remaining in the wash liquid was less than 0.05 g L^{-1}), before drying in an oven at 105 °C for 24 h and weighing of the residual dried material [49]. The solid fraction, in its moist state, was used for hydrolysis and SSCF experiments. The pH of the pretreament liquor was adjusted to 5 using 50% (w/w) NaOH, and it was then filtered by vacuum filtration through sterile disposable bottle top filters with 0.2 μm PES membrane (catalog number 597–4520, ThermoFisher Scientific, Waltham, MA, USA). The filtered liquor was used in yeast propagation, spotting assays, and the SSCF experiments. The sugar composition in the solid fraction was determined using the National Renewable Energy Laboratory (NREL) analytical procedure [50].

Two batches of pretreated wheat straw, as described previously [32], were used for experiments in shake flasks and 3.6 L Labfors (INFORS HT, Switzerland) bioreactors in this study, and a third batch was used in experiments at demonstration scale and in a 30 L Techfors (INFORS HT, Switzerland) laboratory bioreactor. Material must be produced before each demonstration-scale experiment as the quantity required was too large for long-term storage. The three batches were denoted M1, M2, and M3, and their solid and liquid phases had different compositions (Table 1).

Aerobic batch and fed-batch seed cultivation

The optimal composition of growth media was evaluated separately for batch and fed-batch phases; the batch medium in shake flasks at 35 °C and 200 rpm, and the fed-batch medium in bioreactors. The batch medium contained molasses as the main carbon source supply, and pretreatment liquor to adapt the cells to the toxic environment of SSCF. The medium also contained 7.5 g L^{-1} $(NH_4)_2SO_4$, 3.5 g L^{-1} KH_2PO_4, 0.7 g L^{-1} $MgSO_4 \cdot 7H_2O$, 2 mL L^{-1} trace metal (TM) solution and 1 mL L^{-1} vitamin solution. The TM and vitamin solutions were prepared as described previously [51]. Medium compositions were investigated at combinations of 5, 7.5, and 10% (v/v) molasses, and 20, 25, and 30% pretreatment liquor. The combination that gave the best compromise between high biomass and low ethanol yield during the 24-h aerobic cultivation, showing complete utilization of the sugars, was selected for batch propagation. The fed-batch medium contained salts as in the batch medium, but no addition of vitamin or TM solutions. The fed-batch phase was started after 24 h of aerobic batch culture. Molasses contents of 0, 2.5, and 5% were investigated in the feed medium, together with 80% pretreatment liquor. The minimal molasses content that produced enough cells for SSCF was selected for feed medium.

Table 1 Compositions of solid and liquid fractions for three batches of pretreated wheat straw

Solid fraction (% WIS)				Liquid fraction (g L^{-1})			
	M1[a]	M2[a]	M3		M1[a]	M2	M3
Glucan	47.7 ± 3.5	42.4 ± 1.3	36.2 ± 5.0	Glucose	6.8	2.6 ± 0.1[b]	4.5 ± 0.2[b]
Xylan	2.3 ± 0.4	2.6 ± 0.1	4.3 ± 0.1	Xylose	12.8	22.8 ± 1.1[b]	32.7 ± 1.0[b]
Mannan	0.2 ± 0.1	0.2 ± 0.1	0.2 ± 0.08	Mannose	0.4	0.5 ± 0.1[b]	1.1 ± 0.1[b]
Galactan	0.04 ± 0.01	0	0.01 ± 0.00	Galactose	1.0	1.0 ± 0.2[b]	1.9 ± 0.1[b]
Arabinan	0.1 ± 0.01	0.08 ± 0.01	0.2 ± 0.00	Arabinose	2.0	2.8 ± 0.03[b]	4.2 ± 0.1[b]
Lignin		41.7	35.6	Acetic acid	3.8	3.2	3.7 ± 0.05
				Furfural	4.0	0.8	1.5 ± 0.04
Total		87.0	76.5	HMF	1.4	0.4	0.3 ± 0.00

All data are average of at least two separate measurements. Data shown are means ± range. M1, M2 and M3 stand for material 1, 2 and 3 received from the SP biorefinery demo plant. The ranges for acetic acid, furfural, and HMF are within 5% of the average

[a] Data from [32]

[b] Monomeric and oligomeric sugars

Cells from a glycerol stock kept at −80 °C were streaked out on a YPD plate and incubated at 30 °C for approximately 2 days. The liquid inoculum culture was inoculated with a single colony from the plate into 50–100 mL batch medium in 250–500 mL Erlenmeyer flasks and was incubated at 30 °C and a shaker speed of 200 rpm for approximately 24 h, after which the cells were harvested by centrifugation at 3000*g* for 3 min at 4 °C. Flocculating cells were deflocculated by resuspension in 50 mM EDTA followed by washing with 0.9% NaCl solution and final resuspension in 0.9% NaCl. The batch phase of the cell propagation was inoculated to an optical density (OD_{600}) of 0.5 by addition of the required amount of cell suspension.

The yeast cells used in SSCF were propagated in aerobic batch (0.5 L working volume in 3.6 L Labfors bioreactor, INFORS HT, Switzerland), followed by fed-batch culture, at a dilution rate of 0.05 h^{-1}, sparged with air, with aeration set to 1 vvm, 800 rpm stirring, a temperature of 35 °C and pH set to 5.0 by the addition of 3 M NaOH. The working volume was increased to about 2 L in the fed-batch phase and decreased to 0.5 L by harvesting part of the culture.

At every cell addition to the SSCF reactors, cells were harvested from the propagation reactor. For non-flocculating cells, OD_{600} of the culture was measured, and a separately determined correlation between the OD and the dry cell mass concentration was used to determine how much culture should be harvested. Centrifugation was performed at 3000*g* for 3 min at 4 °C, and the cell pellet was re-suspended and added to the SSCF reactors. Flocculating cells were allowed to sediment in a standing bottle. The supernatant was poured off and a concentrated cell suspension was obtained. An aliquot of the cell suspension (about 10 mL) was centrifuged at 3000*g* for 3 min at 4 °C. A separately determined correlation between the pellet volume and the dry cell weight [cell dry weight (g) = pellet volume (mL)*0.27; R^2 of the linear regression was 0.99] was used to determine the cell concentration in the suspension, and thus how much suspension should be added to the SSCF reactors.

SSCF in shake flasks

The Cellic Ctec 2 enzyme preparation (Novozymes, Denmark) was used in all SSCF experiments. The cellulase activity of the enzyme preparation was 150 ± 7 FPU mL^{-1}, measured according to the NREL protocol TP-510-42,628 [52]. Shake flask batch SSCF was carried out with working weight of 100 g in 250 mL baffled Erlenmeyer flasks as described previously [32]. Briefly, the solid fraction of pretreated wheat straw was added to the WIS content that is specified under the corresponding figures. The pH was initially adjusted to 5.0. The medium was supplemented with 0.5 g kg^{-1} $(NH_4)_2HPO_4$. When using 20% WIS, the mixture of solids and enzymes, at 10 FPU g WIS^{-1}, was preincubated at 50 °C for 2 h prior to cell inoculation, to allow partial liquefaction and mixing of the medium. SSCF was initiated by the addition of yeast cells at a cell loading of 0.02 g cell g WIS^{-1}, after reducing the temperature to 35 °C. This cell loading has previously been shown to be sufficient for efficient fermentation in shake flask SSCF on 20% WIS [32]. The flasks were shaken at 180 rpm with no pH control during fermentation. Samples were taken every 24 h for cell viability and HPLC analysis.

SSCF in bioreactors

Multifeed SSCF was performed at working weight around 1 kg in 3.6 L Labfors with one pitched blade impeller and two Rushton impellers, and in 30 L Techfors bioreactors with two Rushton impellers (INFORS HT, Switzerland). The solid fraction of pretreated wheat straw and yeast cells from a separate propagation reactor were added pulse-wise during the process. All the enzymes were added at the beginning of the process to a concentration of 10 FPU g^{-1} overall WIS, as this has previously been shown to be a suitable process option for wheat straw [32]. The medium was supplemented with 0.5 g $(NH_4)_2HPO_4$ and 125 μL Vitahop (a hop-derived fermentation enhancer, BetaTech GmbH, Schwabach, Germany) per total kg of working weight, added to reduce the risk of contamination. The reactors were operated without gas sparging.

All the experiments were started at an initial WIS of 7%. Solids were added according to predetermined hydrolysis kinetics [32]. Detailed feed profiles can be found in Additional file 1: Table S1. In the faster feeding profile, feeding was initiated after hydrolysis of 50%, rather than 60%, of the cellulose, according to model predictions, and material was fed to temporarily reach 14% (w/w) calculated apparent WIS in the reactor, rather than 13% (w/w). Samples were collected for the analysis of sugars, fermentation products, and residual inhibitors. Multifeed SSCF in bioreactors was carried out at 35 °C, however, in some experiments the temperature was reduced to 30 °C after 24 or 48 h. The pH was controlled at 5.0 by the addition of 3 M NaOH, and the agitation rate was 400 rpm.

Determination of cell concentration and viability

The cell concentration during aerobic propagation was determined by cell dry weight (CDW) and OD_{600}. Upon harvest, the flocculating yeast strains KE-Flow and B-Flow were deflocculated with 50 mM EDTA and washed with 0.9% (w/v) NaCl solution before measurement. The CDW was measured by first filtering the cell solution through a preweighed 0.45 μm filter (Sartorius, Göttingen, Germany), drying for 15 min at 150 W in a microwave oven, and cooling in a desiccator before weighing. The viable cell concentration was determined as colony forming units (CFU) during both cell propagation and SSCF. CFU were determined by plating 0.1 mL of sample, after serial dilution with 0.9% (w/v) NaCl, on YPD plates. The plates were incubated for approximately 2 days at 30 °C before colony counting. The total cell concentration was determined by counting cells under a light microscope (Leica DM 2000) in a Neubauer improved hemocytometer (Assistent, Glaswarenfabrik Karl Hecht, Germany).

Spotting assays

Concentrated (2.5×) YPD agar medium was mixed with filtered pretreatment liquor, ethanol, or both to prepare plates of 1 × YPD agar medium containing in addition to YPD agar, 50% (v/v) liquor, or 50 g L^{-1} ethanol or both. Upon harvest, the KE-Flow and B-Flow cells were treated with 50 mM EDTA, resulting in complete deflocculation, and washed with 0.9% (w/v) NaCl solution. The deflocculated cells were re-suspended to an OD_{600} of approximately 0.05. Twofold serial dilutions were made from the cell suspension, and 50 μL per dilution was spotted onto the plates. The plates were incubated at 30 or 35 °C for 2 days, after which they were examined and photographed using a Gel Doc scanner (Bio-Rad, Hercules, CA).

Analysis of sugars, fermentation products, and inhibitors

Fermentation samples were filtered through 0.2-μm nylon filters (VWR, Radnor, PA) prior to storage at −20 °C until analysis. The concentrations of the sugars (glucose, xylose, arabinose, galactose, and mannose) were analyzed using high-performance anion exchange chromatography on an ICS 3000 system (Dionex, Thermo Scientific, Sunnyvale, CA) with guard and analytical CarboPac (PA1) columns (Thermo Scientific) maintained at 30 °C, and by electrochemical detection. Milli-Q water was used for sample elution at a flow rate of 1 mL min^{-1}, and 300 mM NaOH was added post-column at a flow rate of 0.5 mL min^{-1} before the detector. The column was regenerated between sample injections using a mixed eluent consisting of (by volume) 20% Milli-Q water, 40% 300 mM NaOH, and 40% 100 mM NaOH + 85 mM sodium acetate, followed by equilibration with Milli-Q water. The fermentation products (glycerol, xylitol, and ethanol) and inhibitors [acetic acid, furfural, and 5-(hydroxymethyl)furfural] were analyzed using high-performance liquid chromatography on a Dionex Ultimate 3000 system (Dionex, Thermo Scientific) equipped with a Phenomenex Rezex ROA column (Phenomenex, Torrance, CA) and a refractive index (RI) detector (Shodex, SHOWA DENKO K.K., Tokyo, Japan). Samples were eluted at 80 °C using 5 mM H_2SO_4 at flow rate of 0.8 mL min^{-1}. The concentrations were calculated from calibration curves for standard solutions.

Results and discussion

Ethanol is a major inhibitor in high-gravity lignocellulosic ethanol production

In the fermentation of pretreated wheat straw by *S. cerevisiae* KE6-12.A in the multifeed SSCF process, the concentration of viable cells (measured by CFU) decreased to almost zero during the process, regardless of whether more or less toxic pretreated wheat straw were used, and regardless of the scales of the process (Fig. 1a, b). Similar observations have been reported with various raw materials (wheat straw, birch, and spruce) [31–33]. These results indicated that the fermentation conditions

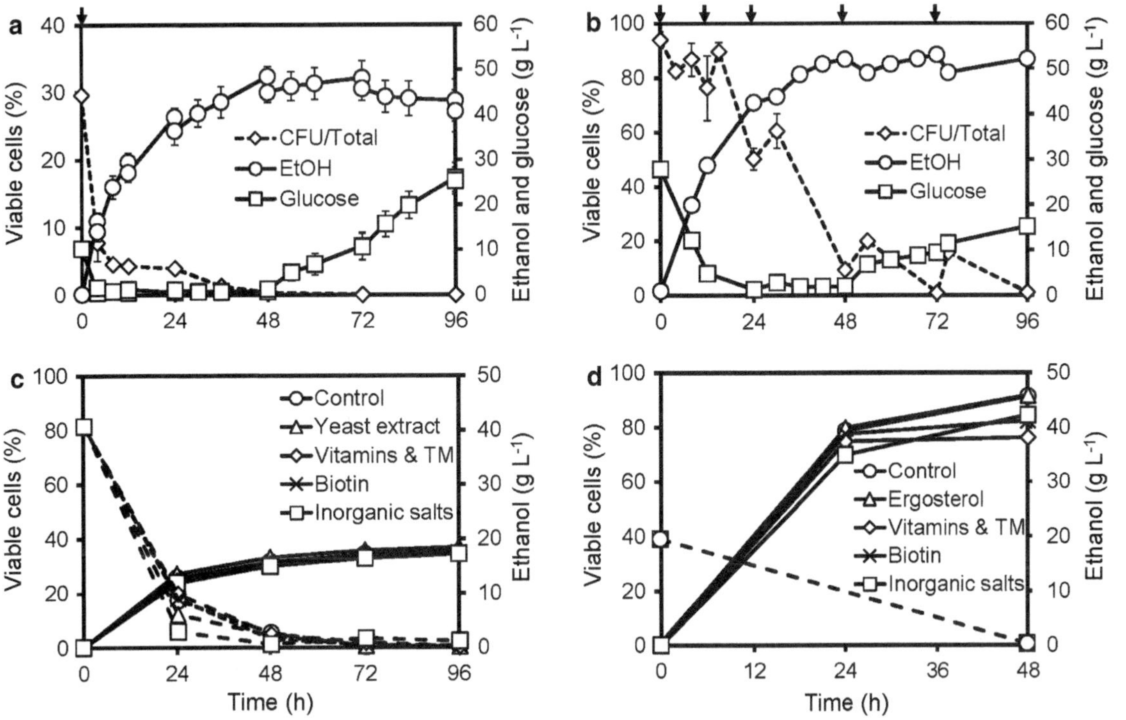

Fig. 1 Decrease of cell viability during multifeed SSCF and limited effects of nutrient supplementation. Cell viability (% of CFU/total cell counts) and concentrations of glucose and ethanol during **a** laboratory-scale SSCF of material M1 with all the yeast added initially; and **b** demonstration-scale SSCF with feeding of yeast and substrate material M2. Ethanol concentration and cell viability during **c** 7% WIS and **d** 20% WIS (with 2-h pre-hydrolysis) shake flask SSCF of material M1 with additions of indicated nutrients. *S. cerevisiae* KE6-12.A was used, and the temperature was 35 °C in all cases. *Arrows* in (**a**–**b**) indicate the loading of yeast cells. *TM* trace metals. *Error bars* show the results of duplicate experiments. *Error bars* for ethanol concentration in (**c**–**d**) are smaller than the *symbols*. The measured cell viability in (**d**) was 0 at 48 h for all cases. Panels **a** and **b** are adapted from [32]

in these lignocellulosic media were not suitable for sustaining cell viability. Cell feeding, instead of adding all the cells at the beginning of fermentation, allowed for a more effective use of the same total amount of cells, and the fermentation capacity was maintained for a longer period. However, cell feeding was not sufficient to stop the decline in the viable cell population or to ensure complete fermentation of the released sugars at the end of the process (Fig. 1b).

To investigate whether the observed decrease in viability during SSCF was due to some nutrient limitation, yeast extract (2 g kg^{-1}), vitamin solution (1 mL kg^{-1}), trace metal (TM) solution (2 mL kg^{-1}) [51], biotin (1 mg kg^{-1}), or inorganic salts [7.5 g kg^{-1} $(NH_4)_2SO_4$, 3.5 g kg^{-1} KH_2PO_4 and 0.7 g kg^{-1} $MgSO_4{\cdot}7H_2O$] were added to shake flask SSCF at a low content of water insoluble solids (WIS, 7% w/w). None of them led to significantly improved ethanol production or an increase in cell viability (Fig. 1c) The addition of nutrients was further tested in 20% WIS shake flask SSCF, in which the inhibitor concentrations were much higher. The same amounts of vitamin and TM solutions, biotin, and inorganic salts as above were used. In addition, the anaerobic growth factor ergosterol (10 mg kg^{-1}) and Tween 80 (420 mg kg^{-1}) were added, but no improvements in ethanol production were observed, and the cell viability decreased to zero within 48 h in all cases (Fig. 1d). The medium composition in control experiments was the same as that in multifeed SSCF. It was therefore concluded that the absence of viable cells at later stages of multifeed SSCF was not due to nutrient limitation.

Other factors causing stress during multifeed SSCF are the inhibitors present in the raw material and the ethanol produced. However, inhibitor concentrations do not increase to high levels in a fed-batch process due to dilution effects and continuous conversion by the yeast. In fact, a decrease in aromatic aldehyde concentrations and an almost constant acetic acid concentration were observed (data not shown). The only potentially toxic compound that increased significantly in concentration over time was the fermentation product, ethanol. Ethanol inhibition has so far not been seen as a major problem in lignocellulosic ethanol production, and has only been mentioned as a potential issue in high-gravity fermentation [37]. It is only recently that the ethanol titers reported from lignocellulosic feedstocks have exceeded the benchmark of 4% (w/w). These titers are still far below those achieved with sucrose- and starch-based substrates, of 8–12% (v/v) [53, 54], and lower than the levels generally considered to be lethal to the yeast *S. cerevisiae*. For example, the wild-type lab strain of BY4741 has been cultured in the presence of 10 and 12.5% (v/v) ethanol for 30 h, with remaining viabilities of 80 and 30%, respectively [55].

To investigate the significance of inhibition by ethanol, different levels of ethanol were added to 20% (w/w) WIS shake flask SSCF experiments. Without ethanol addition, approximately 40 g L^{-1} ethanol was produced within the first 24 h (i.e., an average fermentation rate of 1.6 g L^{-1} h^{-1}), after which the fermentation rate decreased to 0.1 g L^{-1} h^{-1} between 24 and 48 h. When approximately 50 g L^{-1} ethanol was added, some ethanol production was evident during the first 24 h (~0.5 g L^{-1} h^{-1}), after which production ceased and the ethanol concentration decreased, probably due to evaporation (−0.04 g L^{-1} h^{-1} from 24 to 48 h). The amount of ethanol produced was only 25% of that produced in the control case. When 75 g L^{-1} ethanol was added at the start of fermentation, virtually no ethanol production was seen in the shake flasks (0.08 g L^{-1} h^{-1} from 0 to 24 h) (Fig. 2). These results clearly show that at concentrations of about 50 g L^{-1} and above, ethanol is already a major inhibitor of fermentation in SSCF with 20% WIS of pretreated wheat straw. A concentration range of 40–60 g L^{-1} of ethanol has been observed to be a turning point where drastic reduction in growth rate and prolongation of the lag-phase occur for several laboratory and industrial *S. cerevisiae* strains cultivated in the medium of spruce hydrolysate [56].

To investigate whether ethanol also inhibits the hydrolytic enzymes, enzymatic hydrolysis was carried out in shake flasks on 20% WIS with the addition of 75 g L^{-1} ethanol. The results showed a 12% decrease in glucose release (Additional file 1: Figure S1). Therefore, inhibition of the enzymes by ethanol was not the main reason for the observed decreased ethanol production in the SSCF experiments with ethanol supplementations.

From these results, it is clear that at concentrations of 40–50 g L^{-1}, ethanol became inhibitory to *S. cerevisiae* KE6-12.A in SSCF at 20% WIS, with 0.02 g DW (g WIS)$^{-1}$ inoculation. Our results indicate that in high-gravity lignocellulosic ethanol production, ethanol inhibition has to be alleviated so that further increases in ethanol titer would be possible.

Flocculation simplifies cell feeding in the multifeed SSCF process

It has been shown that when cells agglomerate and form dense flocs, their tolerance to ethanol [39] and to furan

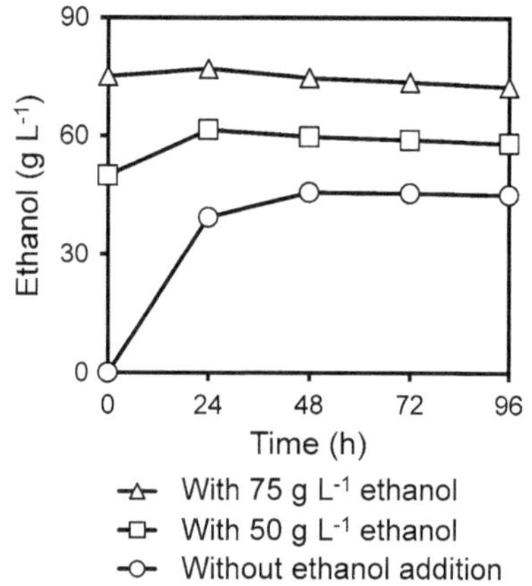

Fig. 2 Ethanol inhibition in high-gravity SSCF with *S. cerevisiae* KE6-12.A. Shake flask SSCF of 20% (w/w) WIS of steam-pretreated wheat straw (material M1) with 2-h pre-hydrolysis, 10 FPU (g WIS)$^{-1}$ enzyme dosage, at 35 °C. The addition of approximately 75 or 50 g L^{-1} ethanol severely inhibited ethanol production. Results shown are average values of duplicate experiments, and the relative difference between duplicate experiments was lower than 5%

aldehydes [25] is improved. Thus, flocculation could be a promising strategy to enhance the performance of *S. cerevisiae* in high-gravity, lignocellulosic ethanol production, especially to increase the ethanol titer. Furthermore, flocculating cells would rapidly sediment at the bottom of the reactor making cell recovery and processing easier.

The flocculating strain KE-Flow formed dense cell aggregates approximately 3–4 mm in size when cultivated in shake flasks with liquid medium. However, no improvement was observed with the KE-Flow strain compared to that with the KE6-12.A strain in shake flask SSCF (Additional file 1: Figure S2), or multifeed SSCF in bioreactors, regardless of whether all the yeast was added initially or at different times (Fig. 3). Fermentation was so slow in the later stages of SSCF that the released glucose accumulated in the reactor, and the viability decreased despite the flocculation. This suggests that dense cell flocs were not successfully formed in the SSCF reactor [25]. The SSCF medium contained a high amount of WIS, i.e., undissolved particles. These particles, together with the mixing in the reactor, probably disrupted larger flocs. These results also suggest that flocculation might not protect cells from ethanol inhibition under the conditions used in SSCF.

Although no improvement in performance was observed in SSCF when using flocculating cells, rapid sedimentation of the KE-Flow cells was observed in the propagation reactor (Fig. 4). The strong flocculation is typical for the Flo1 phenotype [57], which is illustrated by these mutants. The strain may thus be attractive from a process point of view. Using flocculating cells may

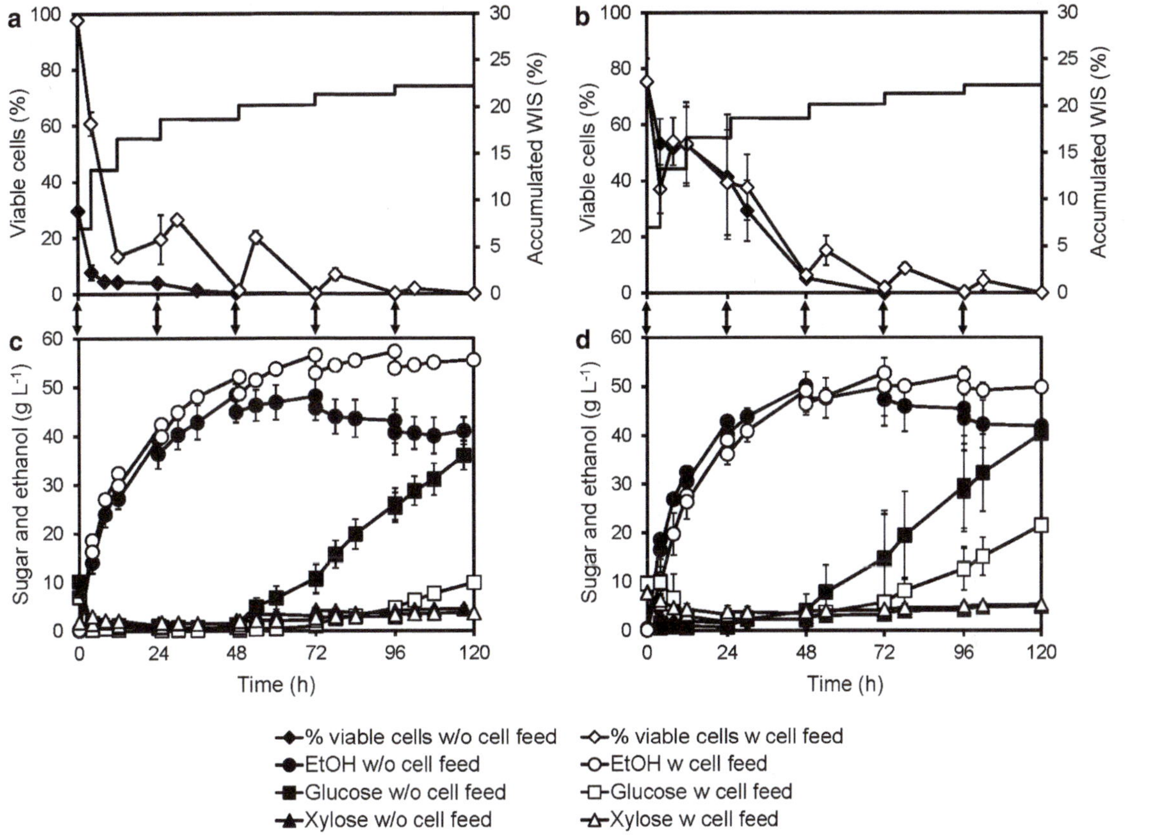

Fig. 3 Multifeed SSCF with the non-flocculating strain KE6-12.A and the flocculating strain KE-Flow. Accumulated WIS concent and percent viability (**a**, **b**) and concentrations of ethanol, glucose, and xylose (**c**, **d**) using: (**a**, **c**) KE6-12.A; and (**b**, **d**) KE-Flow in multifeed SSCF at 22% (w/w) overall WIS of material M1, at 35 °C. Cells were added initially (w/o cell feed, *filled symbols*) or fed during the process (w cell feed, *open symbols*). Feeding of cells is indicated by the *arrows*. The detailed feeding profile has been described elsewhere [32]. Values are averages from duplicate experiments, and *error bars* show the results from the individual experiments

Fig. 4 Sedimentation of KE-Flow cells in propagation reactor. At $t = 0$ s, stirring and aeration were completely stopped. Pretreatment liquor from material M2 was used in this case

simplify the harvesting and feeding of cells to the SSCF reactor.

High temperature exacerbates inhibition by ethanol and pretreatment liquor

As discussed above, inhibition by ethanol in high-gravity SSCF was greater than that expected for *S. cerevisiae*. It is unlikely that ethanol at a concentration of 40–50 g L^{-1} was solely responsible. Rather, the inhibition was due to the combined effect of ethanol, lignocellulose-derived inhibitors, and stressful SSCF conditions. In the SSCF or, generally, simultaneous saccharification and fermentation (SSF) setups, the temperature is a compromise between the optimal temperatures for the hydrolytic enzymes (45–50 °C for Cellic CTec2, Novozymes) and the fermenting microorganism (around 30 °C for *S. cerevisiae*) [58]. This may cause additional stress on the cells in the multifeed SSCF. Xiros and Olsson showed that using preadapted yeast cells and yeast extract improved the ethanol yield about 30 times at 30 °C in SSF of spruce at 20% WIS, but at higher temperatures, the effectivity of this strategy decreased [7]. These findings led us to hypothesize that the ethanol produced, in combination with the inhibitors and the relatively high temperature for the yeast (35 °C) caused the decrease in viability and fermentation capacity.

To investigate the effects of combined stresses on yeast growth, spotting assays were performed at 35 and 30 °C. The results clearly showed that incubation at the higher temperature strongly increased the inhibitory effects of both ethanol and the pretreatment liquor (Fig. 5). With the combination of ethanol and pretreatment liquor, almost no cells grew at 35 °C, showing a strong additive, possibly synergistic, effect of the three factors resembling conditions in high-gravity SSCF. This is probably the reason why no fermentation was observed when a high concentration of ethanol was added to SSCF (Fig. 2), and why further fermentation was difficult with ethanol concentrations greater than 50 g L^{-1}, in spite of adding fresh cells during the process.

Similar results were observed in spotting assays with cells obtained at different time points during the fed-batch cell propagation (data not shown). Ethanol and elevated temperature could affect both the fluidity and structure of the cell membrane [59, 60], making the cells more permeable [61] and thus vulnerable when toxic compounds are present. Similar results were also observed when the xylose-fermenting, evolutionary engineered *S. cerevisiae* strain IBB10B05 [43] was used (Additional file 1: Figure S3), suggesting that the combined inhibition due to ethanol, inhibitors, and temperature is not strain specific.

The YPD plates were incubated aerobically, indicating that the inability of the yeast to survive the combined stress was not due to a lack of sterols or oleic acid, since *S. cerevisiae* can synthesize these under aerobic conditions. This is consistent with the results of the ergosterol-supplemented batch SSCF experiments (Fig. 1d), in which the addition of sterols and oleic acid (Tween 80) did not improve the ethanol production.

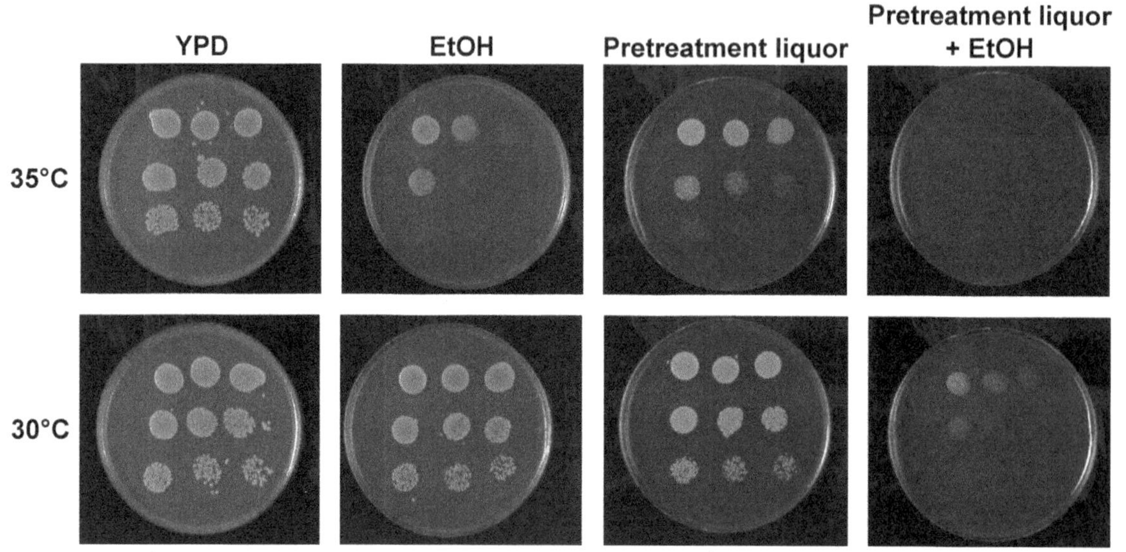

Fig. 5 Spotting assay of the KE-Flow cells. YPD agar plates were supplemented with 50 g L^{-1} ethanol, or 50% (v/v) pretreatment liquor (from material M2), or both. The cells were harvested after 24 h of fed-batch propagation. The plates were incubated at 35 or 30 °C for 48 h. Duplicate cell dilution and plating showed similar results

Reducing the temperature during SSCF improves cell viability

In order to achieve sustained fermentation and complete sugar utilization, the inhibition caused by ethanol, inhibitors, and the process temperature must be alleviated. Decreasing the temperature during SSCF could lead to improved cell viability and thus more complete fermentation. However, the risk of less complete hydrolysis has to be considered.

The temperature in the multifeed SSCF was reduced from 35 to 30 °C after 24 h. At this point, the ethanol concentration had reached approximately 40–50 g L^{-1} (cf. Fig. 3). The reduction in temperature, together with cell feeding, led to increased viability and sustained fermentation capacity until the end of the process, with no accumulation of glucose (Fig. 6). In the 30 °C case, an increase in the fermentation rate was observed when a large amount of substrate was added at 120 h. In a similar process carried out at a constant temperature of 35 °C, glucose accumulated after the same substrate addition, indicating a decrease in fermentation, and a 26% lower ethanol yield on total sugars was measured at 168 h. The measured viable cells at the later stages were significantly lower in the process carried out isothermally at 35 °C, compared to the process with temperature reduction. The decrease in fermentation and accumulation of glucose started already between 96 and 120 h, indicating that the number of viable cells might be insufficient for taking up all glucose released via enzymatic hydrolysis (Fig. 6). Reducing the temperature likely decreases the rate of cellulose hydrolysis leading to lower glucose concentrations; however, very similar ethanol profiles were obtained for both cases before 120 h. This seems due to the enhanced xylose utilization observed in the process with a decrease in the temperature (Fig. 6). The glycerol formation followed the ethanol production (see Additional file 1: Figure S4, where also acetate, furfural, and HMF concentrations are shown).

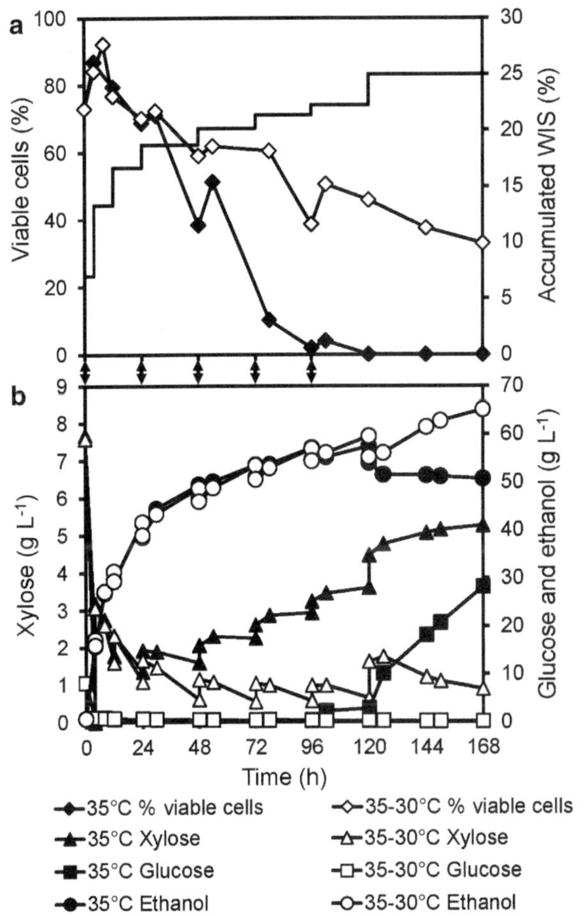

Fig. 6 Multifeed SSCF at constant temperature or with temperature reduction. **a** Cell viability (% of CFU/total cell counts) and accumulated WIS content (indicating substrate feeding); and **b** concentration of glucose, xylose, and ethanol in 25% (w/w) WIS multifeed SSCF of material M2, with the KE-Flow strain, at a constant temperature of 35 °C, or at 35 °C for 24 h, and then at 30 °C. *Arrows* between the panels indicate cell additions. Details of the substrate and cell feeding can be found in Additional file 1: Table S1

In a previous study, Mutturi and Lidén instead applied a linear increase in temperatures ranging from 32 °C to 45 °C in SSF of pretreated spruce and Arundo [62]. When using Arundo, a higher ethanol titer was obtained in SSF with the temperature profile, than those in processes operated at a constant temperature of either 32 or 39 °C. This was likely due to the fact that, first, the increasing temperature would promote enzymatic hydrolysis compared with a constant temperature at 32 °C; and second, a gradual increase in temperature may give cells better chances to adapt than when using a constant high temperature from the beginning [62]. However, it is doubtful whether this strategy would work with more toxic substrates and at higher ethanol concentrations. In the same study, when pretreated spruce which contained higher amount of inhibitors was used, no difference in ethanol titers was observed between the processes with the temperature profile and with a constant temperature 39 °C (32.2 and 33.1 g L^{-1}, respectively) [62]. These ethanol concentrations were considerably lower than the ones obtained herein, which may explain the different observations since they were below the level where we observed significant inhibition.

In conclusion, a temperature profile with a decrease in temperature after a certain period, combined with cell feeding, may offer a suitable process strategy for sustained viability and fermentation capacity at high ethanol titers.

Adapting multifeed SSCF to KE-Flow and a new material batch increased ethanol production

As shown previously, the flocculating yeast strain KE-Flow performed similar to the parental KE6-12.A strain when it was propagated and used in fermentation processes that had been designed for KE6-12.A and material M1 (Fig. 3, Additional file 1: Figure S2; Table S2) [32]. As the material M2 had different compositions from that of M1 (Table 1), the media used in the propagation step and the feeding profile in the multifeed SSCF process were adjusted for the KE-Flow strain and material M2.

In the propagation, the objective was to maximize the use of the pretreatment liquor and minimize the use of molasses, while generating enough cells with consistent capacity for fermentation in SSCF. The medium containing 5% (v/v) molasses and 25% (v/v) pretreatment liquor was selected for the initial batch propagation, and the medium with 2.5% molasses and 80% pretreatment liquor was selected for fed-batch cultivation (for details see Additional file 1: Figure S5).

The objectives of re-designing the multifeed SSCF were to speed up the substrate feeding and to maximize the use of pretreatment liquor. Faster solids feeding, while avoiding mixing problems, is preferable because it prolongs the average time that substrates are available for hydrolysis. In a process with temperature reduction, a larger fraction of substrates would be loaded at the higher temperature, which may improve the overall hydrolysis.

The faster substrate feeding was developed by using the open loop approach described previously [32], with adjustments in the threshold values used to calculate when and how much substrate should be added. In brief, the rate of hydrolysis of the substrate was predicted using a kinetic model. When the predicted substrate conversion reached a lower threshold, 50% instead of the 60% that was used in the previous experiments, a feeding event was triggered. The amount of feed was calculated from the current concentration of WIS and the maximum concentration of WIS that could be adequately mixed in the reactor. For material M2, the lab reactors could mix medium containing up to 14% (w/w) WIS, while for material M1 the upper limit was 13% (w/w). The reason for the difference in the two material batches was probably that a sieve was used to homogenize the size of the solid clumps of M2 material at the demo plant. Starting feeds earlier and adding more substrates every time resulted in a faster substrate feeding profile (Additional file 1: Table S1). The new feeding was implemented successfully in laboratory-scale multifeed SSCF of material M2 without any mixing problems.

The use of pretreatment liquor in SSCF instead of water was also investigated together with the new feeding profile. The pretreatment liquor contained sugars, thus contributing to an increase in ethanol production in the early stage of the process (Fig. 7). Ethanol titers of greater than 60 g L^{-1} were obtained at 96 h, and 65 g L^{-1} was reached after 144 h of SSCF at 22% overall WIS, equivalent to 70% of the theoretical ethanol yield from the total sugar inputs. This represents a total process ethanol yield of 186 L/ton dry straw. The calculated yield at 120 h, which can be considered a more realistic process time, is given in Additional file 1: Table S2. The relatively low overall yields reflect the facts that some of the pretreated material is used to replace other sugars during cell propagation, and that both hydrolysis and fermentation were incomplete.

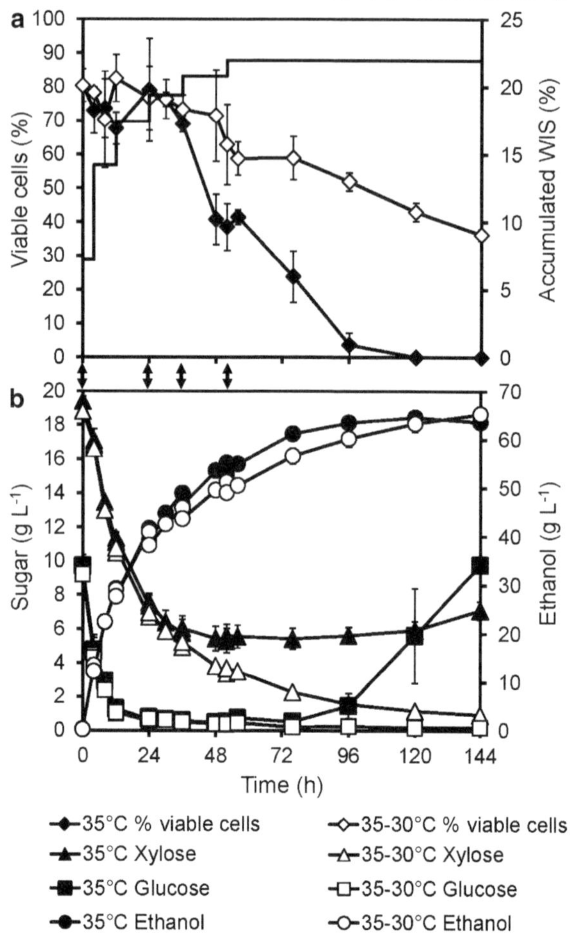

Fig. 7 Adapted multifeed SSCF with faster feeding and maximum use of pretreatment liquor. **a** Cell viability (% of CFU/total cell counts) and accumulated WIS content (indicating substrate feeding); and **b** concentration of glucose, xylose, and ethanol in 22% (w/w) WIS multifeed SSCF of material M2, with the KE-Flow strain, at a constant temperature of 35 °C, or at 35 °C for 24 h, and then at 30 °C. *Arrows* between the *panels* indicate cell additions. Values are averages from duplicate experiments, and the *error bars* show the results of the individual experiments. Details of the substrate and cell feeding can be found in Additional file 1: Table S1

Decreasing the process temperature after 24 h improved the viability of the cells and the consumption of xylose and available glucose (Fig. 7). The measured glucose concentration was close to zero at the lower temperature. Between 24 and 72 h, less ethanol was produced than at the higher temperature. Therefore, less glucose must have been available for fermentation at the lower temperature, which means the hydrolysis rate was lower. The limited hydrolysis was also illustrated by the higher residual WIS and the proportion of glucose in the WIS at the end of the process, indicating a larger residual amount of unhydrolyzed cellulose at the lower temperature (Table 2). Glycerol was formed at higher concentrations at the lower temperature, which is compatible with improved growth and/or higher xylose conversion (Additional file 1: Figure S6). The high ethanol concentrations obtained until 96 h during isothermal operation at 35 °C illustrates the benefit of feeding cells during the process (Fig. 7). Further optimization is necessary for balancing the negative effect of a lower temperature on hydrolysis vs. the beneficial effects on ethanol tolerance, viability, and fermentative activity.

Scale-up: opportunities and challenges when using flocculating yeast and a temperature profile

To investigate the feasibility of using flocculating yeast in an industrial setting, the re-designed yeast propagation and multifeed SSCF processes (Additional file 1: Table S1) were carried out in 10 m^3 reactors at the SP Biorefinery Demo Plant in Örnsköldsvik, Sweden. A new batch of material (M3) was produced. The furfural and xylose concentrations in the pretreatment liquor from this batch were higher than those in the M2 material used during the optimization of the laboratory-scale processes (the detailed compositions are given in Table 1).

When cultivated aerobically in the demonstration scale (10 m^3) on the new material, the growth of the KE-Flow cells was similar to that in the laboratory during the batch phase and the fed-batch phase until the first cell withdrawal (Additional file 1: Figure S7). At the first harvest, cells were concentrated by sedimentation inside the propagation reactor, because agitation and aeration were stopped. A high cell density slurry (OD_{600} of ~200) was thus obtained by harvesting the cells from the bottom of the reactor. The cell slurry was used directly as inoculum for SSCF.

After the first harvest, the cells did not re-suspend properly upon restarting stirring and aeration. Cell growth was therefore insufficient for the planned subsequent cell feedings to the SSCF reactor (Additional file 1: Figure S7). Hence, the cell feeding that was planned at 12 h was omitted, and the corresponding amount of cells was instead added with the last two feedings (at 30 and 52 h). Using a sedimentation tank next to the propagation reactor, as was done at the laboratory scale, would have improved the process by reducing the disturbance of the propagation process.

Since the hydrolysis rate was lower at the lower temperature of 30 °C (Table 2), further optimization of substrate feeding and temperature profile was required. The initial period of higher temperature was extended to 48 h in order to improve hydrolysis, and a final cell feeding was performed at 52 h. Actual cell dry weights added in three cell feedings were about 13, 1.3 and 3 kg, respectively. Before the last cell addition, the percentage of viable cells was 3%. After the temperature reduction, the decrease in viable cells was about 85% slower than during the 23 h (Fig. 8a) before the temperature reduction. Furthermore, the ethanol production rate appeared to be constant until the end of the process (Fig. 8c). In the end, the percentage of viable cells over total cells was approximately 8%. Of the cells added at the last feeding, approximately 45% remained viable, so the lower temperature clearly improved cell viability in the demonstration scale as well. However, the concentration of viable cells was much lower than in the laboratory-scale experiments at the corresponding time points, also after the decrease in temperature (Figs. 7a, 8a).

After 24 h of fermentation, the residual glucose concentration in the reactor leveled out at approximately 10 g L^{-1}, and the xylose concentration at approximately 25 g L^{-1} (Fig. 8c). This indicates that fermentation was significantly slower than on the laboratory scale. The difference in performance was probably due to the problems encountered in handling the flocculating yeast and process control on the larger scale. It was found, for example, that the actual WIS content in the demo reactor was higher than planned, meaning that too much solid material or too little liquid had been added to the reactor. The resulting high viscosity led to reduced mixing, hydrolysis, and fermentation. Another reason may be differences in the toxicity of the material, as M3 contained significantly higher amounts of inhibitors than M2 (Table 1).

Table 2 Residual WIS (%, w/w) and sugar composition of the WIS [%, i.e., g (100 g WIS)$^{-1}$] at 120 h of multifeed SSCF

Temperature of SSCF	WIS	Glucose	Xylose	Mannose	D-Galactose	L-Arabinose
35 °C	11.6 ± 0.8	11.9 ± 3.5	1.3 ± 0.2	1.1 ± 0.1	0.1 ± 0.0	0.2 ± 0.0
35 → 30 °C	13.0 ± 0.1	21.2 ± 0.6	1.7 ± 0.1	1.1 ± 0.0	0.0 ± 0.0	0.2 ± 0.0

The values given are the average of duplicate experiments ± the span to minimum and maximum values

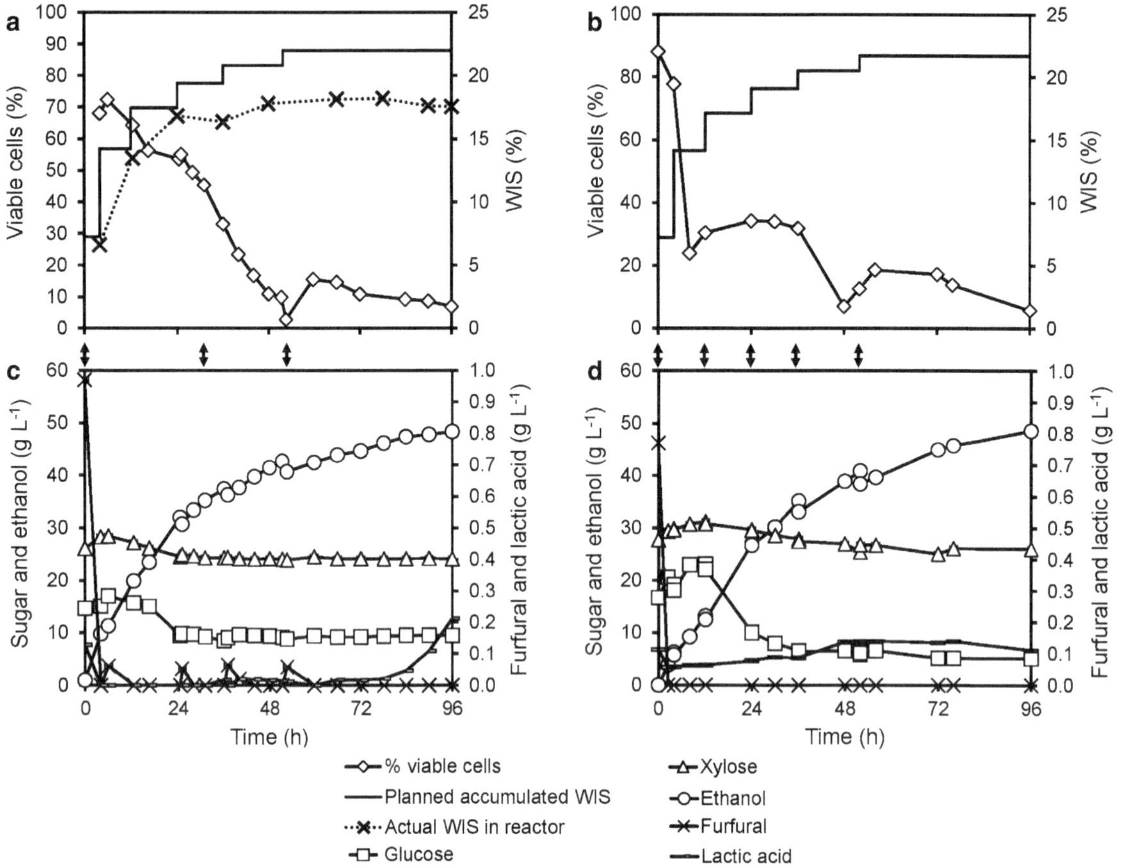

Fig. 8 Multifeed SSCF of material M3 in 10 m^3 and 30 L reactors with KE-Flow. SSCF was carried out at 35 °C for 48 h and then at 30 °C. **a** Cell viability (% of CFU/total cell counts), planned accumulation, and measured residual WIS (%, w/w) during the demonstration-scale experiment, and **b** the same experiment on an intermediate scale in a 30 L Techfors reactor; **c** concentration of sugars and fermentation products during the demo-scale experiment, and **d** during the reactor experiment. *Arrows* between the *panels* indicate cell additions

The concentration of acetate in the fermentation broth increased to 4.6 g L^{-1} at 96 h, while the glycerol concentration was lower than those in previous experiments, reflecting both poor growth and low xylose consumption (Additional file 1: Figure S8).

Although many adjustments and compromises had to be made to operate multifeed SSCF at the demonstration plant, ethanol titers above 5% (w/w) were achieved. Toward the end of the process, small amounts of lactic acid were formed, and some bacterial colonies were found on the YPD plates used for CFU determination. However, since the lactate concentration was very low (<0.3 g L^{-1}) and available glucose in the medium was high, contamination was not a major problem during these experiments.

The multifeed SSCF experiment could not be repeated on the demo scale due to limited resources, but was repeated on an intermediate scale in the laboratory using the same material. The demo process was scaled down to a 30 L Techfors reactor, i.e., roughly 10 times the scale of the 3.6 L Infors reactors used for process development. The fermentation results were consistent with those obtained from the demo plant experiment (Fig. 8; Additional file 1: Figure S8). Neither of them reached the high ethanol titers obtained during the laboratory-scale experiments using material M2 (Fig. 7). No issues with the mixing in the fermenter were observed during the experiment in the Techfors reactor. It could therefore be concluded that differences between the materials obtained after pretreatment were the main contributor to the different fermentation results.

Apart from demonstrating scalability of the process, the results from larger scales highlight the importance of consistency in the pretreatment process, in order to obtain material with similar hydrolyzability and inhibition characteristics. However, this is difficult to achieve due to variations in feedstocks and difficulties in operation on a large scale. Thus, flexibility in the SSCF process is necessary. The multifeed SSCF process is an example of an adaptable process. Adjustments can be made to the process based on the toxicity of the material by changing the amount of pretreatment liquor in the propagation step and in the SSCF, together with modifications of the feeding of cells and solids. To enable such adjustments in real time, rapid sampling methods and online measurements of sugars, inhibitors, and viable cells will be required to provide information on the status of the process, and to make feedback control of multifeed SSCF possible.

Conclusions

Results from laboratory- and demonstration-scale experiments showed the feasibility of utilizing flocculating yeast cells as an efficient way to concentrate cells for use in multifeed SSCF. Ethanol inhibits fermentation at a concentration of 50 g L^{-1} in the presence of lignocellulosic inhibitors. Flocculation did not provide a solution to the ethanol inhibition in SSCF, but is an attractive strategy for facilitating cell harvesting and processing. The fermentation capacity of the cells can be maintained, and higher ethanol titers can be achieved by reducing the temperature in the SSCF reactor when the ethanol concentration reaches an inhibitory level. These changes, together with faster substrate feeding, and replacing water with pretreatment liquor, led to an ethanol titer of 65 g L^{-1} with maintained cell viability and almost complete utilization of the fermentable sugars. Optimal performance of multifeed SSCF requires adaptation of the process to each material used.

Additional file

Additional file 1: Figure S1. Ethanol inhibition in high-gravity enzymatic hydrolysis with enzyme preparation Cellic Ctec 2. **Figure S2.** Shake flask SSCF with the non-flocculating strain KE6-12.A and the flocculating strain KE-Flow. **Figure S3.** Spotting assay of the B-Flow cells. **Figure S4.** Multi-feed SSCF at constant temperature or with temperature reduction. **Figure S5.** Investigation of medium compositions for batch and fed-batch propagation of KE-Flow with material M2. **Figure S6.** Adapted multi-feed SSCF with faster feeding and maximum use of pretreatment liquor. **Figure S7.** Comparison of cultivation of KE-Flow in 3.6 L laboratory reactors and in the demonstration plant. **Figure S8.** Multi-feed SSCF of material M3 in 10 m^3 and 30 L reactors with KE-Flow. **Table S1.** SSCF feeding schemes for laboratory scale reactors. **Table S2.** Summary of results from multi-feed SSCF experiments using steam pre-treated wheat straw.

Authors' contributions

JW designed the study, constructed the strains TE-Flow and B-Flow, performed experiments, analyzed and interpreted data, and wrote the manuscript. RW conceived and designed the study, performed the mathematical modeling, performed experiments, analyzed and interpreted data, and wrote the manuscript. VN designed the study, performed experiments, interpreted data, and participated in the writing of the manuscript. CJF conceived and designed the study, interpreted data, and wrote the manuscript. All authors read and approved the final manuscript.

Author details

[1] Division of Industrial Biotechnology, Department of Biology and Biological Engineering, Chalmers University of Technology, Gothenburg, Sweden. [2] Present Address: Chr. Hansen A/S, Bøge Allé 10-12, 2970 Hørsholm, Denmark. [3] Institute of Biotechnology and Biochemical Engineering, Graz University of Technology, Graz, Austria. [4] Present Address: Forest Products Biotechnology, Faculty of Forestry, University of British Columbia, Vancouver, BC V6T1Z4, Canada.

Acknowledgements

The authors are grateful to Prof. Bernd Nidetzky from the Graz University of Technology for providing the strain IBB10B05, and for facilitating VN's visit at the Chalmers University of Technology.

Competing interests

The authors declare they have no competing interests.

Funding

This study was funded by the Swedish Energy Agency (Grant Number P37353-1) and by the Chalmers Energy Initiative (http://www.chalmers.se/en/areas-of-advance/energy/cei/). The funding bodies had no influence on the design of the study and were not involved in the collection, analysis, and interpretation of data, nor in the writing of the manuscript.

References

1. Cannella D, Jørgensen H. Do new cellulolytic enzyme preparations affect the industrial strategies for high solids lignocellulosic ethanol production? Biotechnol Bioeng. 2014;111(1):59–68.
2. Johansen KS. Discovery and industrial applications of lytic polysaccharide mono-oxygenases. Biochem Soc Trans. 2016;44(1):143–9.
3. Bardhan S, Jose S, Godsey L. Cellulosic biofuel in the United States. In: Jose S, Bhaskar T, editors. Biomass and biofuels: advanced biorefineries for sustainable production and distribution. Boca Raton: CRC Press; 2015. p. 301.
4. Guo M, Song W, Buhain J. Bioenergy and biofuels: history, status, and perspective. Renew Sust Energ Rev. 2015;42:712–25.
5. Jönsson LJ, Alriksson B, Nilvebrant N-O. Bioconversion of lignocellulose: inhibitors and detoxification. Biotechnol Biofuels. 2013;6:16.
6. Li H, Chen H. Detoxification of steam-exploded corn straw produced by an industrial-scale reactor. Proc Biochem. 2008;43(12):1447–51.
7. Xiros C, Olsson L. Comparison of strategies to overcome the inhibitory effects in high-gravity fermentation of lignocellulosic hydrolysates. Biomass Bioenerg. 2014;65:79–90.
8. Cavka A, Alriksson B, Ahnlund M, Jönsson LJ. Effect of sulfur oxyanions on lignocellulose-derived fermentation inhibitors. Biotechnol Bioeng. 2011;108(11):2592–9.
9. Jurado M, Prieto A, Martínez-Alcalá Á, Martínez ÁT, Martínez MJ. Laccase detoxification of steam-exploded wheat straw for second generation bioethanol. Bioresour Technol. 2009;100(24):6378–84.
10. Almeida JRM, Modig T, Petersson A, Hahn-Hägerdal B, Lidén G, Gorwa-Grauslund MF. Increased tolerance and conversion of inhibitors in lignocellulosic hydrolysates by *Saccharomyces cerevisiae*. J Chem Technol Biotechnol. 2007;82(4):340–9.
11. Ask M, Mapelli V, Höck H, Olsson L, Bettiga M. Engineering glutathione biosynthesis of *Saccharomyces cerevisiae* increases robustness to inhibitors in pretreated lignocellulosic materials. Microb Cell Fact. 2013;12:87.
12. Wei N, Quarterman J, Kim SR, Cate JHD, Jin Y-S. Enhanced biofuel production through coupled acetic acid and xylose consumption by engineered yeast. Nat Commun. 2013;4:2580.
13. Liu ZL. Molecular mechanisms of yeast tolerance and in situ detoxification of lignocellulose hydrolysates. Appl Microbiol Biotechnol. 2011;90(3):809–25.
14. Wang X, Yomano LP, Lee JY, York SW, Zheng H, Mullinnix MT, Shanmugam KT, Ingram LO. Engineering furfural tolerance in *Escherichia coli* improves the fermentation of lignocellulosic sugars into renewable chemicals. Proc Nat Acad Sci USA. 2013;110(10):4021–6.
15. Koopman F, Wierckx N, de Winde JH, Ruijssenaars HJ. Identification and characterization of the furfural and 5-(hydroxymethyl)furfural degrada-

tion pathways of *Cupriavidus basilensis* HMF14. Proc Nat Acad Sci USA. 2010;107(11):4919–24.

16. Lindahl L, Genheden S, Eriksson LA, Olsson L, Bettiga M. Sphingolipids contribute to acetic acid resistance in *Zygosaccharomyces bailii*. Biotechnol Bioeng. 2016;113(4):744–53.
17. Gu H, Zhang J, Bao J. High tolerance and physiological mechanism of *Zymomonas mobilis* to phenolic inhibitors in ethanol fermentation of corncob residue. Biotechnol Bioeng. 2015;112(9):1770–82.
18. Demeke MM, Dietz H, Li Y, Foulquié-Moreno MR, Mutturi S, Deprez S, Den Abt T, Bonini BM, Liden G, Dumortier F, Verplaetse A, Boles E, Thevelein JM. Development of a D-xylose fermenting and inhibitor tolerant industrial *Saccharomyces cerevisiae* strain with high performance in lignocellulose hydrolysates using metabolic and evolutionary engineering. Biotechnol Biofuels. 2013;6:89.
19. Çakar ZP, Seker UOS, Tamerler C, Sonderegger M, Sauer U. Evolutionary engineering of multiple-stress resistant *Saccharomyces cerevisiae*. FEMS Yeast Res. 2005;5(6–7):569–78.
20. Alkasrawi M, Rudolf A, Lidén G, Zacchi G. Influence of strain and cultivation procedure on the performance of simultaneous saccharification and fermentation of steam pretreated spruce. Enz Microb Technol. 2006;38(1–2):279–86.
21. Nielsen F, Tomás-Pejó E, Olsson L, Wallberg O. Short-term adaptation during propagation improves the performance of xylose-fermenting *Saccharomyces cerevisiae* in simultaneous saccharification and co-fermentation. Biotechnol Biofuels. 2015;8:219.
22. Cantarella M, Cantarella L, Gallifuoco A, Spera A, Alfani F. Comparison of different detoxification methods for steam-exploded poplar wood as a substrate for the bioproduction of ethanol in SHF and SSF. Proc Biochem. 2004;39(11):1533–42.
23. Pienkos PT, Zhang M. Role of pretreatment and conditioning processes on toxicity of lignocellulosic biomass hydrolysates. Cellulose. 2009;16(4):743–62.
24. Westman JO, Manikondu RB, Franzén CJ, Taherzadeh MJ. Encapsulation-induced stress helps *Saccharomyces cerevisiae* resist convertible lignocellulose derived inhibitors. Int J Mol Sci. 2012;13(9):11881–94.
25. Westman JO, Mapelli V, Taherzadeh MJ, Franzen CJ. Flocculation causes inhibitor tolerance in *Saccharomyces cerevisiae* for second-generation bioethanol production. Appl Environ Microbiol. 2014;80(22):6908–18.
26. Ylitervo P, Franzén C, Taherzadeh M. Impact of furfural on rapid ethanol production using a membrane bioreactor. Energies. 2013;6(3):1604–17.
27. Westman JO, Franzén CJ. Current progress in high cell density yeast bioprocesses for bioethanol production. Biotechnol J. 2015;10(8):1185–95.
28. Hodge DB, Karim MN, Schell DJ, McMillan JD. Model-based fed-batch for high-solids enzymatic cellulose hydrolysis. Appl Biochem Biotechnol. 2009;152(1):88–107.
29. Olofsson K, Rudolf A, Lidén G. Designing simultaneous saccharification and fermentation for improved xylose conversion by a recombinant strain of *Saccharomyces cerevisiae*. J Biotechnol. 2008;134(1–2):112–20.
30. Olofsson K, Wiman M, Lidén G. Controlled feeding of cellulases improves conversion of xylose in simultaneous saccharification and co-fermentation for bioethanol production. J Biotechnol. 2010;145(2):168–75.
31. Wang R, Koppram R, Olsson L, Franzén CJ. Kinetic modeling of multi-feed simultaneous saccharification and co-fermentation of pretreated birch to ethanol. Bioresour Technol. 2014;172:303–11.
32. Wang R, Unrean P, Franzén CJ. Model-based optimization and scale-up of multi-feed simultaneous saccharification and co-fermentation of steam pre-treated lignocellulose enables high gravity ethanol production. Biotechnol Biofuels. 2016;9:88.
33. Koppram R, Olsson L. Combined substrate, enzyme and yeast feed in simultaneous saccharification and fermentation allow bioethanol production from pretreated spruce biomass at high solids loadings. Biotechnol Biofuels. 2014;7:54.
34. Palmqvist E, Hähn-Hägerdal B. Fermentation of lignocellulosic hydrolysates. II: inhibitors and mechanisms of inhibition. Bioresour Technol. 2000;74(1):25–33.
35. Jorgensen H. Effect of nutrients on fermentation of pretreated wheat straw at very high dry matter content by *Saccharomyces cerevisiae*. Appl Biochem Biotechnol. 2009;153(1–2):44–57.
36. Andreasen AA, Stier TJB. Anaerobic nutrition of *Saccharomyces cerevisiae*. I. Ergosterol requirement for growth in a defined medium. J Cell Comp Physiol. 1953;41(1):23–36.
37. Koppram R, Tomas-Pejo E, Xiros C, Olsson L. Lignocellulosic ethanol production at high-gravity: challenges and perspectives. Trends Biotechnol. 2014;32(1):46–53.
38. Stanley D, Bandara A, Fraser S, Chambers PJ, Stanley GA. The ethanol stress response and ethanol tolerance of *Saccharomyces cerevisiae*. J Appl Microbiol. 2010;109(1):13–24.
39. Smukalla S, Caldara M, Pochet N, Beauvais A, Guadagnini S, Yan C, Vinces MD, Jansen A, Prevost MC, Latgé J-P, Fink GR, Foster KR, Verstrepen KJ. *FLO1* Is a variable green beard gene that drives biofilm-like cooperation in budding yeast. Cell. 2008;135(4):726–37.
40. Tomás-Pejó E, Bonander N, Olsson L. Industrial yeasts strains for biorefinery solutions: constructing and selecting efficient barcoded xylose fermenting strains for ethanol. Biofuels Bioprod Bioref. 2014;8(5):626–34.
41. Wahlbom CF, van Zyl WH, Jönsson LJ, Hahn-Hägerdal B, Otero RRC. Generation of the improved recombinant xylose-utilizing *Saccharomyces cerevisiae* TMB 3400 by random mutagenesis and physiological comparison with *Pichia stipitis* CBS 6054. FEMS Yeast Res. 2003;3(3):319–26.
42. van der Westhuizen TJ, Pretorius IS. The value of electrophoretic fingerprinting and karyotyping in wine yeast breeding programmes. Antonie Van Leeuwenhoek. 1992;61(4):249–57.
43. Klimacek M, Kirl E, Krahulec S, Longus K, Novy V, Nidetzky B. Stepwise metabolic adaption from pure metabolization to balanced anaerobic growth on xylose explored for recombinant *Saccharomyces cerevisiae*. Microb Cell Fact. 2014;13:37.
44. Novy V, Krahulec S, Wegleiter M, Muller G, Longus K, Klimacek M, Nidetzky B. Process intensification through microbial strain evolution: mixed glucose-xylose fermentation in wheat straw hydrolyzates by three generations of recombinant *Saccharomyces cerevisiae*. Biotechnol Biofuels. 2014;7:49.
45. Westman JO, Franzén CJ. A novel chimaeric flocculation protein improves flocculation characteristics in *Saccharomyces cerevisiae*. Submitted.
46. van Dijken JP, Bauer J, Brambilla L, Duboc P, Francois JM, Gancedo C, Giuseppin MLF, Heijnen JJ, Hoare M, Lange HC, Madden EA, Niederberger P, Nielsen J, Parrou JL, Petit T, Porro D. An interlaboratory comparison of physiological and genetic properties of four *Saccharomyces cerevisiae* strains. Enzyme Microb Technol. 2000;26(9–10):706–14.
47. Gietz RD, Woods RA. Transformation of yeast by lithium acetate/single-stranded carrier DNA/polyethylene glycol method. Methods Enzymol. 2002;350:87–96.
48. Goldstein AL, McCusker JH. Three new dominant drug resistance cassettes for gene disruption in *Saccharomyces cerevisiae*. Yeast. 1999;15(14):1541–53.
49. Sluiter A, Hyman D, Payne C, Wolfe J. Determination of insoluble solids in pretreated biomass material. Laboratory analytical procedure (LAP) NREL/TP-510-42627. http://www.nrel.gov/docs/gen/fy13/42627.pdf. Issued 21 Mar 2008. National Renewable Energy Laboratory, 2008.
50. Sluiter A, Hames B, Ruiz R, Scarlata C, Sluiter J, Templeton D, Crocker D. Determination of structural carbohydrates and lignin in biomass. Laboratory analytical procedure (LAP) NREL/TP-510-42618. http://www.nrel.gov/docs/gen/fy13/42618.pdf. Issued 25 Apr 2008 (Revised 08/03/2012).
51. Verduyn C, Postma E, Scheffers WA, Van Dijken JP. Effect of benzoic acid on metabolic fluxes in yeasts: a continuous-culture study on the regulation of respiration and alcoholic fermentation. Yeast. 1992;8(7):501–17.
52. Adney B, Baker J. Measurement of cellulase activities: Laboratory Analytical Procedure (LAP). NREL/TP-510-42628 1996, http://www.nrel.gov/docs/gen/fy08/42628.pdf. Issued 12 Aug 1996.
53. Basso LC, Basso TO, Rocha SN. Ethanol production in Brazil: the industrial process and its impact on yeast fermentation. In: Bernardes MAS, editor. Biofuel production - recent developments and prospects. Rijeka: InTech; 2011. p. 85–100.
54. Bothast RJ, Schlicher MA. Biotechnological processes for conversion of corn into ethanol. Appl Microbiol Biotechnol. 2005;67(1):19–25.
55. Alper H, Moxley J, Nevoigt E, Fink GR, Stephanopoulos G. Engineering yeast transcription machinery for improved ethanol tolerance and production. Science. 2006;314(5805):1565–8.
56. Albers E, Larsson C. A comparison of stress tolerance in YPD and industrial lignocellulose-based medium among industrial and laboratory yeast strains. J Ind Microbiol Biotechnol. 2009;36(8):1085–91.
57. Stratford M, Assinder S. Yeast flocculation: Flo1 and NewFlo phenotypes and receptor structure. Yeast. 1991;7(6):559–74.

58. Olofsson K, Bertilsson M, Lidén G. A short review on SSF - an interesting process option for ethanol production from lignocellulosic feedstocks. Biotechnol Biofuels. 2008;1:7.
59. Jia K, Zhang Y, Li Y. Systematic engineering of microorganisms to improve alcohol tolerance. Eng Life Sci. 2010;10(5):422–9.
60. Lahtvee P-J, Kumar R, Hallström BM, Nielsen J. Adaptation to different types of stress converge on mitochondrial metabolism. Mol Biol Cell. 2016;27(15):2505–14.
61. Heipieper HJ, Weber FJ, Sikkema J, Keweloh H, de Bont JAM. Mechanisms of resistance of whole cells to toxic organic solvents. Trends Biotechnol. 1994;12(10):409–15.
62. Mutturi S, Lidén G. Effect of temperature on simultaneous saccharification and fermentation of pretreated spruce and arundo. Ind Eng Chem Res. 2013;52(3):1244–51.

Identification of developmental stage and anatomical fraction contributions to cell wall recalcitrance in switchgrass

Jacob D. Crowe[1], Nicholas Feringa[1], Sivakumar Pattathil[2,3], Brian Merritt[2], Cliff Foster[4], Dayna Dines[4], Rebecca G. Ong[5] and David B. Hodge[1,4,6,7*]

Abstract

Background: Heterogeneity within herbaceous biomass can present important challenges for processing feedstocks to cellulosic biofuels. Alterations to cell wall composition and organization during plant growth represent major contributions to heterogeneity within a single species or cultivar. To address this challenge, the focus of this study was to characterize the relationship between composition and properties of the plant cell wall and cell wall response to deconstruction by NaOH pretreatment and enzymatic hydrolysis for anatomical fractions (stem internodes, leaf sheaths, and leaf blades) within switchgrass at various tissue maturities as assessed by differing internode.

Results: Substantial differences in both cell wall composition and response to deconstruction were observed as a function of anatomical fraction and tissue maturity. Notably, lignin content increased with tissue maturity concurrently with decreasing ferulate content across all three anatomical fractions. Stem internodes exhibited the highest lignin content as well as the lowest hydrolysis yields, which were inversely correlated to lignin content. Confocal microscopy was used to demonstrate that removal of cell wall aromatics (i.e., lignins and hydroxycinnamates) by NaOH pretreatment was non-uniform across diverse cell types. Non-cellulosic polysaccharides were linked to differences in cell wall response to deconstruction in lower lignin fractions. Specifically, leaf sheath and leaf blade were found to have higher contents of substituted glucuronoarabinoxylans and pectic polysaccharides. Glycome profiling demonstrated that xylan and pectic polysaccharide extractability varied with stem internode maturity, with more mature internodes requiring harsher chemical extractions to remove comparable glycan abundances relative to less mature internodes. While enzymatic hydrolysis was performed on extractives-free biomass, extractible sugars (i.e., starch and sucrose) comprised a significant portion of total dry weight particularly in stem internodes, and may provide an opportunity for recovery during processing.

Conclusions: Cell wall structural differences within a single plant can play a significant role in feedstock properties and have the potential to be exploited for improving biomass processability during a biorefining process. The results from this work demonstrate that cell wall lignin content, while generally exhibiting a negative correlation with enzymatic hydrolysis yields, is not the sole contributor to cell wall recalcitrance across diverse anatomical fractions within switchgrass.

Keywords: Switchgrass, Recalcitrance, Cell wall glycans, Alkaline pretreatment

*Correspondence: hodgeda@msu.edu
[1] Department of Chemical Engineering and Materials Science, Michigan State University, East Lansing, MI, USA
Full list of author information is available at the end of the article

Background

Development of renewable sources of fuels and chemicals is necessary to improve energy security and mitigate climate change effects resulting from greenhouse gas emissions [1, 2]. Lignocellulosic biomass, which includes dedicated bioenergy crops and agricultural residues, is a widely available and largely untapped source of reduced carbon in the production of renewable fuels and chemicals [3]. A current major challenge in deconstructing plant cell walls are plant-evolved mechanisms for resisting microbial and animal degradation of plant cell walls, termed biomass recalcitrance [4]. Within a feedstock, abundance and density of cell type, structure and substitution of cell wall polymers, and organization of polymers within the cell wall can all contribute to biomass recalcitrance [4].

Plant cell walls account for the majority of lignocellulosic feedstock dry mass, and are comprised of primary and in some cases secondary cell walls depending upon the tissue type. Primary cell walls are found in growing cells, and contain cellulose microfibrils, pectins, and hemicellulose polysaccharides. Secondary cell walls are significantly thicker than primary cell walls, provide mechanical strength and structural reinforcement to the cells, are found predominantly in fiber and vascular cells post cell enlargement [5], and represent the majority of the mass of lignocellulosic feedstocks. Secondary cell walls contain cellulose microfibrils, hemicelluloses, and lignin, an aromatic heteropolymer comprising phenylpropanoid subunits. In monocot grasses, the primary hemicellulose components are glucuronoarabinoxylans ("xylans") containing arabinosyl, acetyl, and uronosyl substitutions on a β-1,4 linked xylan backbone [6]. The arabinofuranosyl residues in grass xylans can also be covalently linked via ferulic acid ester crosslinks to lignin or other xylans, and this phenomenon has been hypothesized to play an important role in promoting cell wall rigidity and decelerating expansion [7]. Mixed-linkage glucan and xyloglucans are also present as minor hemicellulose components in cell walls, and may account for a minor fraction of non-cellulosic glucan found in mature tissues [6, 8]. Lignin content and composition are often cited as one of the predominant contributors of cell wall recalcitrance [9]; however, there also exists a body of literature that has examined the role of hemicelluloses [10], structural pectins [11], and polymer interactions within the cell wall as other significant contributors to cell wall recalcitrance [12].

Switchgrass (*Panicum virgatum* L.) is a well-studied C_4 perennial grass envisioned as a promising herbaceous feedstock to supply North American cellulosic biorefineries along with other monocot grasses such as miscanthus (*Miscanthus* × *giganteus*), and corn stover (*Zea mays*) [13, 14]. Switchgrass is an attractive feedstock due to high production yields, low energy and nutrient inputs, wide adaptability, and short growing time [15]. Switchgrass grows as a clonal modular organism, with the tiller forming the main clonal module and phytomers consisting of multiple shoot meristems. Shoots comprise leaf blades, sheath leaves, stem nodes and internodes, auxiliary buds, and ligules [16]. Grasses such as switchgrass grow through elongation of stem internodes, each of which is linked at the base node to a single leaf sheath and blade. Analysis of individual internodes, as opposed to bulk plant material, is useful as internodes represent sequential stages of maturity during plant development, affecting cell wall composition and abundance of cell type within biomass [17]. Basal internodes tend to have greater lignin and cellulose content coupled with lower digestibility compared to internodes near the top of a single tiller [18]. Tissue maturity in relation to harvest times may impact switchgrass digestibility [19], including time left on field post maturation [20]. In addition, there are significant differences in compositions of stem, sheath, and leaf anatomical fractions, with the stem generally containing more lignin and being less digestible compared to the leaf [21, 22]. The distribution, structure, and extractability of non-cellulosic cell wall glycans in miscanthus stem and leaf organs sampled at distinct maturities indicated that hemicellulose structure and extractability varied with tissue maturity, and may account for significant non-lignin contributions to recalcitrance [23].

Within-plant heterogeneity for graminaceous feedstocks such as switchgrass has important impacts on several important processing variables for a biorefining process. Ranges of particle size distribution, composition, hygroscopicity and drainability, chemical pretreatment effectiveness, and extent of enzymatic hydrolysis can all result from differing responses of biomass to comminution. Physical fractionation, employed either on-field during harvest or at the biorefinery, is one potential route to addressing feedstock challenges to potentially improve agronomic, logistics, and processing–conversion outcomes. Limited work has been published on biomass fractionation in the context of pretreatment and hydrolysis. In one study, disc refining followed by air classification of corn stover yielded pith and depithed stover fractions that were subjected to dilute acid pretreatment or mild base pretreatment, quantifying mass losses and sugar release response during acid treatment [24]. Other work identified significant differences in enzymatic hydrolysis response of corn stover stems compared to other fractions [25–27].

Studies investigating tissue-specific responses have been assessed for liquid hot water pretreatment of corn stover [28] and wheat straw [29], although the only industrially relevant fractionation approach that have been studied are alteration of leaf-stem ratio in wheat straw [30]. With AFEX pretreatment, the impact of knife-milling and sieving of corn stover [31] as well manual separation of stem, sheath, leaf, and cob have been assessed [32], and found to yield fractions with differing compositions and responses to deconstruction.

The objective of this study is to characterize within-plant heterogeneity by identifying key compositional differences between switchgrass anatomical fractions at different internode maturities and relating compositional differences to enzymatic digestibility response following mild alkaline pretreatment. Structural differences are evaluated by localization of phenolics within tissue cross sections before and after alkaline pretreatment using confocal microscopy, while hemicellulose and pectin structure and extractability are evaluated using glycome profiling following chemical extraction.

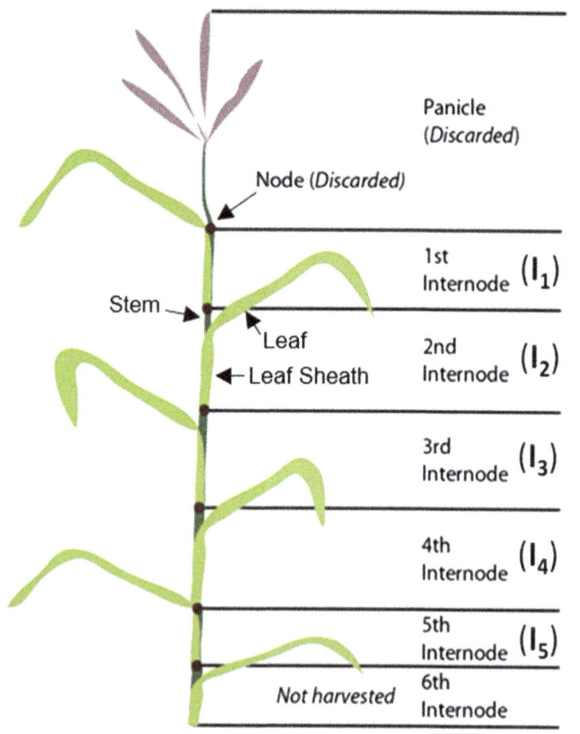

Fig. 1 Schematic of switchgrass tillers used for this study. Tillers were harvested above either the 5th or 6th internode, and then separated into internodes, discarding the nodes and panicle. Internodes were further manually subdivided into leaves, leaf sheaths, and stems. Diagram is representative and not to scale

Methods

Field conditions and biomass preparation

Switchgrass (var. "Forestburg," upland ecotype) was manually harvested post-anthesis above either the 5th or 6th internode (Fig. 1) from a field at Michigan State University (East Lansing, MI; 42°42′48.80″N by 84°28′1.41″W) between 17 and 19 September, 2014. During harvest, intact harvested tillers were dried indoors at room temperature. After harvest was completed, all tillers were dried overnight in a 45 °C oven. Internodes were then manually separated into leaves, leaf sheaths, and stems, while the nodes and panicle were discarded (Fig. 1). Whole fractions were air dried to a moisture content of ~5% (g H_2O/g total), and particle size was reduced using a Wiley Mini-Mill (Thomas Scientific, Swedesboro, NJ, USA) to pass a 30-mesh screen. Samples were stored in dry, airtight bags until further use.

Pretreatment conditions

Alkaline pretreatment was performed in 8 mL borosilicate glass tubes (Thermo Fisher Scientific, Waltham, MA, USA) with rubber lined caps. Alkali loading was performed at 10% (g NaOH/g dry biomass) loading at 10% (g biomass/g solution) loading, giving a 0.25 M NaOH concentration. Samples were vortexed and immersed in an 80 °C water bath for 1 h, with mixing every 15 min. After the elapsed pretreatment time, solids were separated from the alkaline liquor and washed with de-ionized water until pH neutral. Following washing, samples were rinsed with absolute ethanol and allowed to air dry at 45 °C to a moisture content <5%, with mass loss following pretreatment determined gravimetrically.

Composition analysis

Extractives-free cell wall material (i.e., alcohol insoluble residue or "AIR") was isolated following extraction and de-starching according to Foster et al. [33]. All composition procedures were performed on de-starched AIR, except for the determination of extractable sugars. Minor polysaccharide content (rhamnan, mannan, galactan, fucan, non-cellulosic glucan) was determined using the alditol acetate method with minor changes [33]. Cell wall acetate, glucan, xylan, arabinan, and Klason lignin content were determined for AIR using the NREL/TP 510-42618 protocol [34] with minor modifications [35]. Thioacidolysis was used to determine lignin monolignol yields [36]. Extractable free glucose, sucrose, and starch were determined on the original untreated biomass as described by Santoro et al. [37].

Alkaline saponification was used to determine the *p*-hydroxycinnamic acid content, as described previously [38], with modifications to the quantification method. Briefly, 0.5 g of biomass was treated with 25 mL of 3 M NaOH in a sealed pressure tube at 121 °C for 1 h to release esterified *p*-coumarate (*p*CA) and esterified or etherified ferulate (FA). Liquid samples were centrifuged at 13,000×*g* for 3 min. Next, 1 mL of alkaline liquor was taken, and pH adjusted to pH 1.5 using concentrated H_2SO_4 (Sigma-Aldrich Corp., St. Louis, MO, USA). Samples were centrifuged at 13,000×*g* for 3 min to remove any precipitated solids, and subsequently analyzed by HPLC (Agilent 1100 Series) with an Aminex HPX-87H column (Bio-Rad, Hercules, CA, USA) using a mobile phase of 5 mM H_2SO_4 with a 10% (v/v) ACN organic modifier. Quantification was performed using a diode-array detector with a UV wavelength of 210 nm. Standards containing FA and *p*CA ranged from 0.05 to 1 mg/mL. To account for precipitation of *p*CA and FA with pH adjustment, solubility with respect to pH curves were determined, with R^2 linear coefficients of 0.999 for both FA and *p*CA observed from the range of pH 1.4–1.8. Samples were analyzed in biological triplicate for reproducibility.

Enzymatic hydrolysis

Enzymatic hydrolysis was performed at 2.5% solids loading (g AIR biomass/g solution) in 1.5 mL microcentrifuge tubes (Posi-Click, Denville Scientific, Holliston, MA). Enzymatic hydrolysis was performed using 10 mg protein/g glucan for pretreated biomass and 30 mg protein/g glucan for untreated, de-starched, extractives-free biomass using CTec3 and HTec3 (Novozymes A/S Bagsværd, Denmark), at a CTec3:HTec3 ratio of 2:1. A buffer solution of 50 mM citric acid (pH 5.20) was used to maintain pH, and samples were incubated at 50 °C with orbital mixing at 180 rpm for either 6 or 48 h. Samples were centrifuged at 13,000×*g* for 3 min post-incubation and filtered using 22 µm mixed cellulose-ester filters (EMD Millipore, Billerica, MA). Glucose and xylose were quantified by HPLC (Agilent 1100 Series) with an Aminex HPX-87H column (Bio-Rad, Hercules, CA, USA) with a mobile phase of 5 mM H_2SO_4. Glucose yield was determined based on quantified glucose released compared to total AIR cell wall glucan available (as glucose), including non-cellulosic glucan.

Confocal microscopy

Air dried whole samples were sectioned by hand using a platinum tipped razor blade and prepared for imaging by washing with 50 mL of de-ionized water prior to immersing in water buffered at a pH of 7.0 by 20 mM Na_2PO_4 overnight. Pretreated samples were generated by NaOH pretreatment at 0.10 g NaOH/g biomass of whole cross sections, followed by washing with de-ionized water until pH neutral, and immersion in water buffered at a pH of 7.0 by 20 mM Na_2PO_4 overnight. Confocal fluorescent images were collected on an Olympus FluoView FV1000 Confocal Laser Scanning Microscope (Olympus America, Inc., Center Valley, PA, USA), configured on a fully automated Olympus IX81 inverted microscope. Images were recorded with a 10× UPlanFLN objective (NA 0.30). Blue and red autofluorescence signals were sequentially recorded, as shown in Additional file 1: Figure S1. Blue autofluorescence was excited using a 405 nm diode laser and emission was recorded using a 430–470 nm band pass emission filter. The red autofluorescence was excited using a 543 nm helium neon laser and emission was recorded using a 560 nm long pass emission filter. A series of confocal images (XYZ) were recorded through the thickness of each sample (approximately 50 µm). Images were collected in 10 µm intervals when using the 10× objective. Each confocal series was then compressed into a maximum intensity projection image and recorded in a TIF format. The color look-up table (LUT) for all images was modified with the blue display rescaled from 0–4095 to 350–800 and the red display rescaled from 0–4095 to 0–1200 to brighten images. All further processing was performed in ImageJ (http://imagej.nih.gov/ij/).

Glycome profiling

Sequential cell wall extractions and glycome profiling of switchgrass anatomical fractions were carried out as described previously [39, 40] on AIR biomass. Plant glycan-directed mAbs were from laboratory stocks (CCRC, JIM, and MAC series) available at the Complex Carbohydrate Research Center [41]. A complete description of the mABs used in this study can be found in our prior work [42].

Results

Switchgrass anatomical fractions sampled at different internodes throughout the plant representing differing levels of maturity were utilized in this work to develop an improved understanding of the impacts of within-plant heterogeneity in switchgrass on a potential cellulosic biofuels conversion process. Switchgrass samples were characterized with respect to composition and assessed for response to pretreatment and enzymatic hydrolysis with the goals of (1) identifying cell wall properties contributing to observed differences in cell wall recalcitrance, (2) understanding how the differences in composition and recalcitrance are distributed throughout the plant, and (3) potentially use this understanding to inform both plant breeding and harvesting strategies to optimally couple the plant feedstock to a conversion process.

Anatomical fraction- and internode-specific recalcitrance

Enzymatic hydrolysis before or after mild NaOH pretreatment was performed on extractives-free (AIR), de-starched cell wall material from three switchgrass anatomical fractions (leaves, leaf sheath, stem) as a function of internode (Fig. 2).

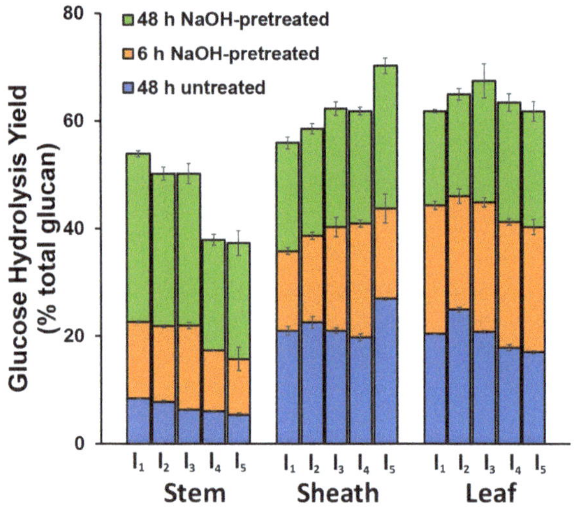

Fig. 2 Enzymatic hydrolysis glucose yields of untreated and alkaline pretreatment of anatomical fractions. Enzymatic hydrolysis of untreated biomass was performed on AIR + de-starched biomass, while alkaline pretreatment was performed using 10% w/w NaOH loading at 80 °C for 1 h. Glucose hydrolysis yields are presented as the percent of total cell wall glucan released as monomeric glucose. Enzyme loading was 30 mg protein/g glucan for untreated fractions and 10 mg protein/g glucan for alkaline pretreated fractions with a ratio of 2:1 CTec3 to HTec3. Replicates ($n = 3$) are displayed as averages with standard deviations

From the results in Fig. 2, lower glucose yields for both untreated and NaOH-pretreated cases of stem internodes indicate that stem fractions are significantly more recalcitrant compared to leaf sheath (referred to as sheath) and leaf internodes. Previous work using liquid hot water pretreatment of switchgrass also found that pretreated leaves were more digestible than stem internodes [43]. Also of note, the mass of above-ground biomass in switchgrass is known to be primarily leaves, with stems and leaf sheaths each comprising on the order of half the mass of the leaves [43]. Stem internodes also demonstrated a clear decrease in enzymatic hydrolysis yields with increasing tissue maturity (i.e., from I_1 to I_5) for both the untreated and NaOH-pretreated cases [44], although similar trends were not as clearly observed in the untreated yields of leaf sheath and leaf internodes and indicate cell wall recalcitrance increases with maturity in stem only.

NaOH-pretreated leaf sheaths again showed significantly different trends compared to stem, with a trend of increasing hydrolysis yields with increasing internode in both the 6-h and 48-h hydrolysis yield cases. Leaf hydrolysis yields became similar between internodes in the 48-h yields for the NaOH-pretreated case, contrasting the general trend of decreasing yields with increasing internode seen in the untreated and 6-h hydrolysis yield for the NaOH-pretreated cases. From these observations, the application of alkaline pretreatment resulted in a change in recalcitrance between internodes in both leaf sheath and leaf fractions, although differences in the 6- and 48-h hydrolysis yield trends in leaf internodes suggest pretreatment also had an effect on hydrolysis rate as well. The contributions of composition and cell wall biopolymer structure to these differences in recalcitrance are described in the subsequent sections.

Cell wall composition

The content of major cell wall biopolymers (cellulose, xylan, lignin), minor hemicelluloses (non-cellulosic glucan) and minor hemicellulose sugars (Rha, Ara, Man, Gal, Fuc), alkali-labile hydroxycinnamic acids (*p*-coumarate, ferulate), and lignin composition (syringyl and guaiacyl monomer yields by thioacidolysis) was quantified for stem, sheath, and leaf internodes, respectively. Key results that highlight observed trends found are presented in Fig. 3, while the complete dataset can be found in Additional file 1: Table S1.

The Klason lignin content demonstrated a clear increase with respect to internode for stem, sheath, and leaf (Fig. 3a), with results presumably corresponding to increasing maturation and development of secondary cell wall tissues following cell elongation and differentiation in older fractions. Overall, stem lignin content was higher compared to sheath and leaf, which is expected due to the abundance of sclerenchyma and xylem cells providing structural support for both upright growth and normal vascular transport [5]. It should be noted that ash was not quantified independently in this study, and may represent a portion of the quantified lignin [45]. However, ash content has been shown to be higher in abundance in leaf fractions when compared to stem and sheath fractions [43], likely limiting ash impact on trends observed in this study.

Alkali-labile hydroxycinnamic acids also varied significantly with internode, with both *p*CA and FA content decreasing between internodes I_1–I_5 for all switchgrass anatomical fractions (Fig. 3b). Decreases in *p*CA content with increasing internode maturity ranged from a 19% decrease from I_1 to I_5 in stem samples, to 53 and 45% in sheath and leaf samples, respectively. FA content also demonstrated the same trend between internodes I_1–I_5 with decreases of 47, 65, and 48% for stem, sheath, and leaf, respectively. Combined with the trend for lignin content (Fig. 3a), the inverse relationship between FA content and lignin content suggests that crosslinking of cell wall biopolymers via ferulates is an important feature of cell wall organization in tissues with less secondary cell wall lignification [18]. Prior work has concluded that total FA content decreases with increasing maturity (based on harvest date) concurrent with increasing lignin

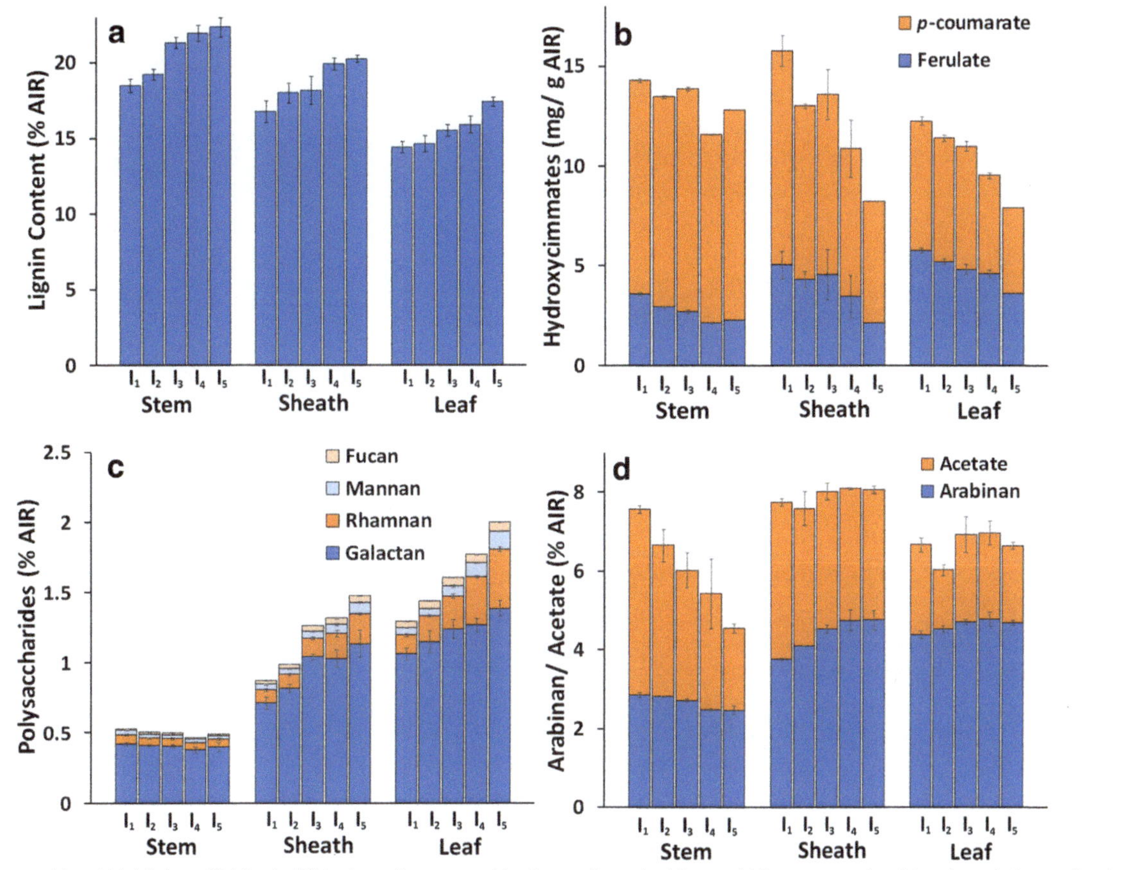

Fig. 3 Compositional highlights of (**a**) lignin (**b**) hydroxycinnamates (**c**) minor polysaccharides, and (**d**) acetate and arabinan in switchgrass fractions with respect to internode. Internodes I_1–I_5 are displayed in ascending order from *left* to *right* for each switchgrass anatomical fraction. All composition values were quantified on an extractives-free basis (AIR). Replicates ($n = 3$) displayed as average with standard deviations

content in switchgrass [45], and is in agreement with the results of the present work in which maturity is based on internode rather than harvest date.

Composition analysis of polysaccharide content revealed a similar content of xylan between internodes (average content of 24.4% for stem and sheath and 19.8% for leaf), while galactan, mannan, rhamnan, and fucan, representing minor hemicellulose and structural pectin polysaccharides were higher in sheath and leaf fractions compared to stem and showed clear trends of increased abundance with tissue maturity in sheath and leaf internodes (Fig. 3c).

Functionally, this indicates in maturing stem internodes, pectic and minor hemicellulosic polysaccharides represent a smaller fraction of total cell wall polysaccharides, and the decreases in total compositional fraction suggest minimal continued synthesis of these polysaccharides during tissue maturation. Conversely, increased overall composition content in more mature sheath and leaf internodes suggest continued synthesis and accumulation of minor polysaccharides during maturation. The exception to this trend, non-cellulosic glucan, showed an increase with respect to internode in all three anatomical fractions as well as similar overall content of non-cellulosic glucan (2.5% overall for all samples), indicating increases in either β-glucan or xyloglucan content with internode maturation (Additional file 1: Table S1).

Stem arabinan content, indicative of glucuronoarabinoxylan (GAX), arabinogalactan (AG), and rhamnogalacturonan I (RG I) pectin side chains decreased slightly in stem internodes, while increasing only in leaf sheaths (Fig. 3d). Given the small decrease in overall xylan content between internodes, it can be hypothesized that increases in arabinan content correspond to an increase in GAX or RG I side chain abundance in leaf sheath cell walls [23]. Many cell wall polysaccharides are known to be O-acetylated, including pectins, xyloglucans, and xylans. Cell wall acetyl content (Fig. 3d) decreased significantly with stem internode maturity, with only minor decreases observed with respect to sheath and leaf internode maturity, indicating similarity in acetylated cell wall components between sheath and leaf internodes compared to stem.

Non-cellulosic glucose and extractable sugar content

The accumulation of soluble sugars in the stems of grasses is of paramount importance to agriculturally important crops such as sugarcane (*Saccharum* spp.) and sweet sorghum (*Sorghum bicolor*). Soluble sugar abundance in other grasses used as feedstocks for cellulosic biofuels can be a significant fraction of total dry weight [50], with both genetic and environmental drivers responsible for starch and extractable sugar accumulation [46]. For example, soluble sugar synthesis in plants has been shown to be environment responsive, with significant accumulation of soluble sugars accompanying reduced structural carbohydrate and lignin content in switchgrass harvests subjected to drought conditions [47].

Notably, these sugars are often overlooked in many biomass-to-fuels studies or analyses and can present both processing and analytical challenges. Soluble sugars may mask structural polysaccharides in cell walls [48] resulting in false positives for carbohydrate content in screening techniques relating feedstock digestibility to biomass physical characteristics or high throughput digestibility studies that do not remove soluble sugars and starches before analysis [49]. As a processing challenge, contents of extractable sugars and starch can be degraded during many pretreatments (e.g., dilute acid or AFEX) representing a potentially significant loss of value sugars. Furthermore, sugar degradation products can present additional challenges to the biological conversion of sugars to biofuels or bioproducts. For example, in recent work using AFEX pretreatment on switchgrass subjected to drought environmental stress, imidazoles and pyrazines were formed by the reaction of ammonia with soluble sugars in high enough concentrations to significantly inhibit fermentation compared to switchgrass harvests not subjected to drought [47].

In the present work, non-cellulosic glucan (presumably as xyloglucan and β-glucan) and extractable such as sucrose, fructose, free glucose, and starches sugar contents were determined (Fig. 4) and represent a non-trivial fraction of overall feedstock mass. Stem internodes were observed to have significantly more soluble sugars compared to similar sheath and leaf internode contents, with soluble sugars combined with non-cellulosic glucan constituting between 12 and 18% of total dry weight. Starch content was dramatically different in stem internodes, accumulating to up to 8% of total biomass dry weight in internode I_5. Comparatively, sheath and leaf starch content accounted for up to 2% dry weight, and decreased with internode maturity. One likely reason for such a high stem extractible sugar content was due to sample harvesting during grain filling [50].

Non-cellulosic glucan, presumably in the form of xyloglucans, β-glucans, and glucomannans, were observed to

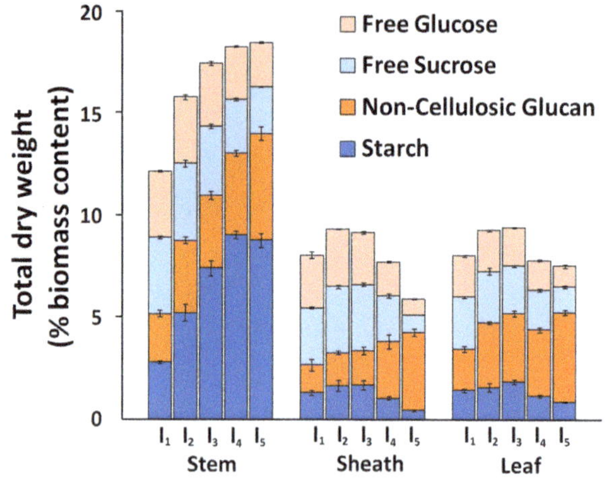

Fig. 4 Native switchgrass non-cellulosic sugar composition with respect to internode. Internodes I_1–I_5 are displayed in ascending order from *left* to *right* for each switchgrass anatomical fraction. Compositions are represented as a percent of total dry weight. Replicates ($n = 3$) displayed as average with standard deviations

accumulate from internode I_1 through I_5 for all three anatomical fractions. Although comprising a small percentage of total biomass, β-glucans are generally more easily digested compared to cellulose, while xyloglucans may be closely associated to cellulose in the primary cell wall [51]. It can also be observed that the differences in the non-cellulosic glucan can be correlated to differences in most of the minor hemicellulose and pectin sugars (i.e., Fuc, Rha, Man, Gal as presented in Fig. 3c).

Cell wall autofluorescence

At both the macroscopic and microscopic scales as shown in Fig. 5, switchgrass stem, sheath, and leaf exhibit obvious significant differences in overall morphology, tissue organization, and cell type abundance. Structural (and compositional) heterogeneity across fractions has important implications for biomass recalcitrance. As one example, sclerenchyma secondary cell wall types in stems have significantly thicker cell walls and contain a high ratio of lignified secondary cell walls compared to partially lignified parenchyma type cells [45], resulting in anatomical fractions containing high contents of sclerenchyma cells expecting to have higher recalcitrance compared to parenchyma rich fractions.

Analysis of cell wall phenolics using confocal microscopy often employs combinations of fluorescence tagging, chemical labels, and autofluorescence techniques to characterize the cell wall landscape at the micron-scale resolution [52, 53]. Using phenolic autofluorescence alone, general trends in phenolic localization and abundance can be obtained; however, the wide range of emission spectrums of phenolic compounds can limit

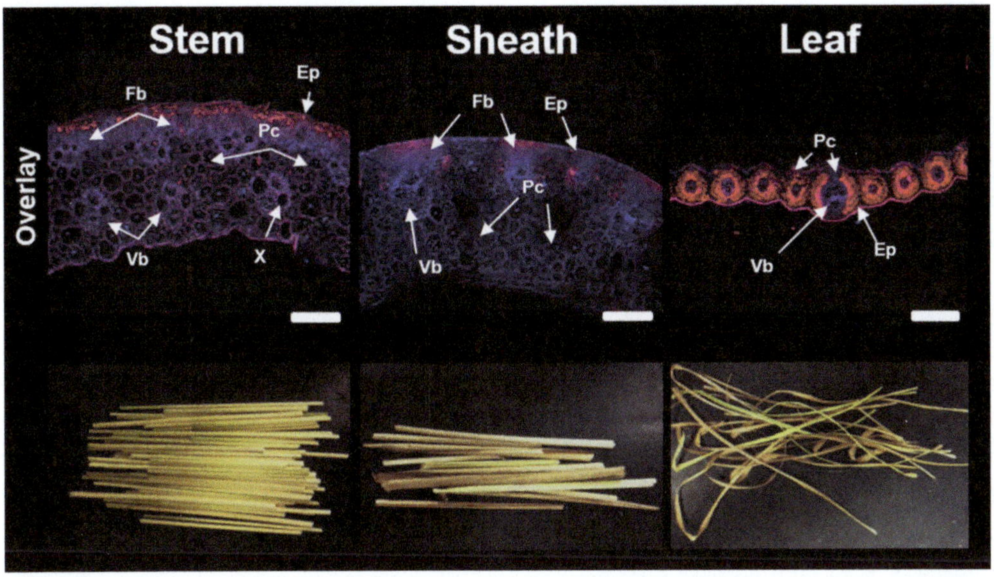

Fig. 5 Examples of differences in macroscopic morphology, cell wall organization, and aromatic abundance as shown by autofluorescence between switchgrass anatomical fractions from internode I_2. Confocal laser scanning microscopy carried out using laser emission wavelengths of 405 and 543 nm. Tissue types are denoted from the following notation: *Ep* epidermis, *Fb* fiber bundles, *Vb* vascular bundles, *Ph* phloem, *X* xylem, *Pc* parenchyma cells. *Scale bars* represent 200 μm

deconvolution of specific phenolic components [54–56]. With a low-field excitation using a diode laser at 405 nm, hydroxycinnamates have an emission spectrum of around 420–460 nm (shown and denoted as blue) along with alkaloids, flavonoids, and other phenylpropanoids [56]. Using a helium neon laser excitation at 543 nm, most phenolics fluoresce with an emission spectrum of 600–650 nm (shown and denoted as red), however hydroxycinnamates do not [55]. One convoluting factor in this technique resides in the autofluorescence of chlorophyll. With an emission spectrum in the 600–700 nm region [56], chlorophyll will certainly provide signal contribution to the red emission spectrum from chlorophyll rich tissues such as epidermal and bundle sheath regions of leaf (Figs. 5, 6).

Overlay of both autofluorescence channels of whole biomass cross sections revealed several noticeable features of each organ (Fig. 5). Stem and sheath cross sections displayed similar architecture, with atactostele patterned vascular bundles, distinct regions of fiber bundles, and similar overall thickness [57]. The only notable difference resided in distribution of fiber cells between epidermal located vascular bundles in the sheath cross section. Both displayed blue and red emission patterns in all tissue types; however, signal intensity was much stronger in secondary cell wall abundant regions such as sclerenchyma cell fiber bundles, and vascular bundles. Leaf cross-section architecture was significantly different, with vascular bundles surrounded by a bundle sheath of parenchyma cells, and a highly lignified lower midrib region. Leaf autofluorescence showed a blue emission pattern in the vascular bundles along with some epidermal tissues, while the red emission spectrum was localized to the lower midrib and bundle sheath, and was most likely due to chlorophyll autofluorescence.

The impact of NaOH pretreatment on autofluorescence signature was examined for multiple internodes to gauge both organ and internode response to pretreatment (Fig. 6). In stem and leaf sheath cross sections, the red emission spectrum was localized to secondary cell wall tissue such as vascular tissue and fiber bundles, while the blue emission spectrum was observed predominantly in parenchyma cells.

Comparing cross sections of stem internodes I_2–I_4, there was a clear increase of the blue emission spectrum in the fiber bundles, while in leaf sheaths cross sections sampled from internodes I_2–I_4, there was a clear decrease of the blue emission spectrum in the parenchyma cells. Signal intensity from the red emission spectrum decreased in fiber and vascular cells in both stem and leaf sheaths from internode I_2–I_4. In addition to changes in signal intensity, cell wall thickening was also observed in the parenchyma and interfascicular tissues of stem internode cross sections, corresponding to a maturation of tissues [18]. Similar thickening was not as apparent in sheath or leaf cross sections.

These results suggest that following NaOH pretreatment, components that fluoresced red in the stem cross

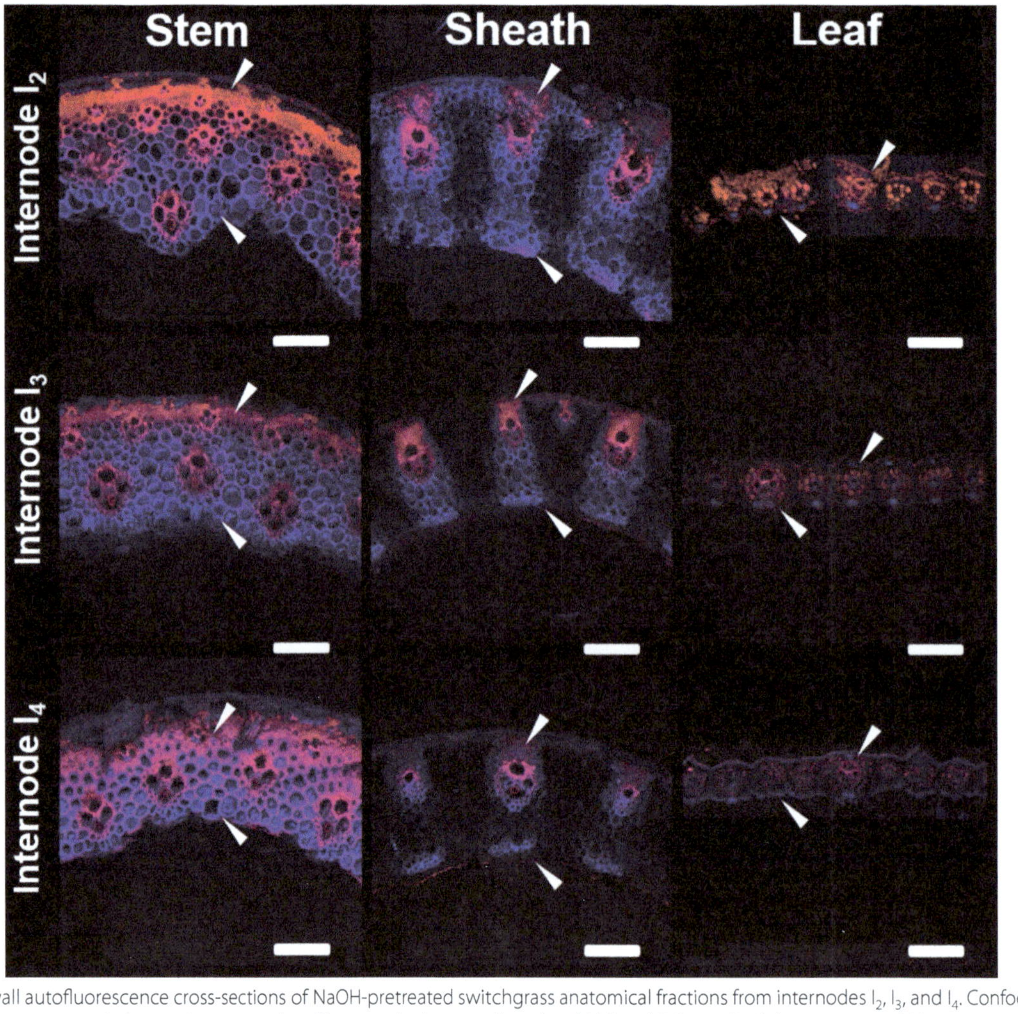

Fig. 6 Cell wall autofluorescence cross-sections of NaOH-pretreated switchgrass anatomical fractions from internodes I_2, I_3, and I_4. Confocal laser scanning microscopy carried out using an overlay of laser excitation wavelengths of 405 and 543 nm. *Scale bars* represent 200 μm

sections were either preferentially extracted due to pretreatment, or were observed in lower abundance relative to other components in more mature internodes. This may also suggest that following alkaline pretreatment, there were regions of hydroxycinnamates that remained unextracted or were more recalcitrant in stem internodes, as evidenced by the change in ratio of blue emission to red emission spectrum. The loss of both emission spectrum signals with internode maturity in sheath cross sections suggests either the removal of both components with alkaline pretreatment, or the loss of accessibility of both components for autofluorescence.

From the perspective of secondary cell wall development, more mature internode cross sections may have hydroxycinnamates in all cell types, while in younger stem internode cross sections, potential locations of hydroxycinnamates might be restricted predominantly to developing parenchyma cells. In sheath cross sections, younger internodes had potential locations of hydroxycinnamates in all cell types, with the eventual observance of the blue emission spectrum only in the vascular bundle regions. Leaf internodes following NaOH pretreatment had a noticeable decrease in both emission signal intensities, with the blue emission spectrum localized to the outer parenchyma and epidermal cells, and the red emission spectrum localized to the bundle sheath and epidermal cells. Given the autofluorescence of chlorophyll in the red emission spectrum, the decrease in red signal can be largely attributed to either chlorophyll removal from alkaline extraction, or a decrease in chlorophyll in more mature leaf internodes following senescence.

Glycome profiling

Glycome profiling was performed to assess organ-specific variations in the cell wall glycan composition and extractability across individual internodes of each anatomical fraction, I_1-stem, I_1-sheath, and I_1-leaf. In Fig. 7a, selected regions of glycome profiles highlight abundancies of xylan and pectin epitopes in cell wall extracts, while the full glycome profile results can be found in Additional file 1: Figure S2.

With the oxalate extracts, extractability of xylan epitopes varied across anatomical fractions. For instance, both I_1-stem and I_1-sheath exhibited similar xylan epitope extractability patterns with significant abundance of arabinosylated xylan epitopes recognized by mAb CCRC-M154 of xylan-4 clade [58], unsubstituted xylan epitopes recognized by several mAbs from xylan-6 clade, and unsubstituted xylan epitopes recognized by all mAbs from xylan-7 clades. The I_1-leaf oxalate extract was significantly different, showing an abundance of only arabinosylated xylan epitopes recognized by mAb CCRC-M154 of xylan-4 clade along with significantly reduced abundance of unsubstituted xylan epitopes recognized by mAbs from xylan-7 clades. Differences across anatomical fractions were also observed in the carbonate extract, with all three carbonate extracts containing most classes of xylan epitopes from unsubstituted and substituted regions. The I_1-leaf sample again exhibited differences in the pattern of extractability with a marginal reduction in the overall abundancies of these epitopes, namely the Me-GlcA substituted xylan epitopes (xylan-5) and complete absence of xylan-6 epitopes. Xylan-3 epitopes were observed only in the I_1-sheath, while all xylan clades were observed in the I_1-stem extract.

Likely due to grass susceptibility to alkaline pretreatment [38], the 1 M KOH extract showed the largest portion of total carbohydrates released, with significant xylan epitope abundance in all three sample extracts. I_1-leaf exhibited the strongest binding intensity for xylan epitopes 4 through 7 in the 1 M KOH extract, and this trend was mirrored in the 4 M KOH fraction. Similar

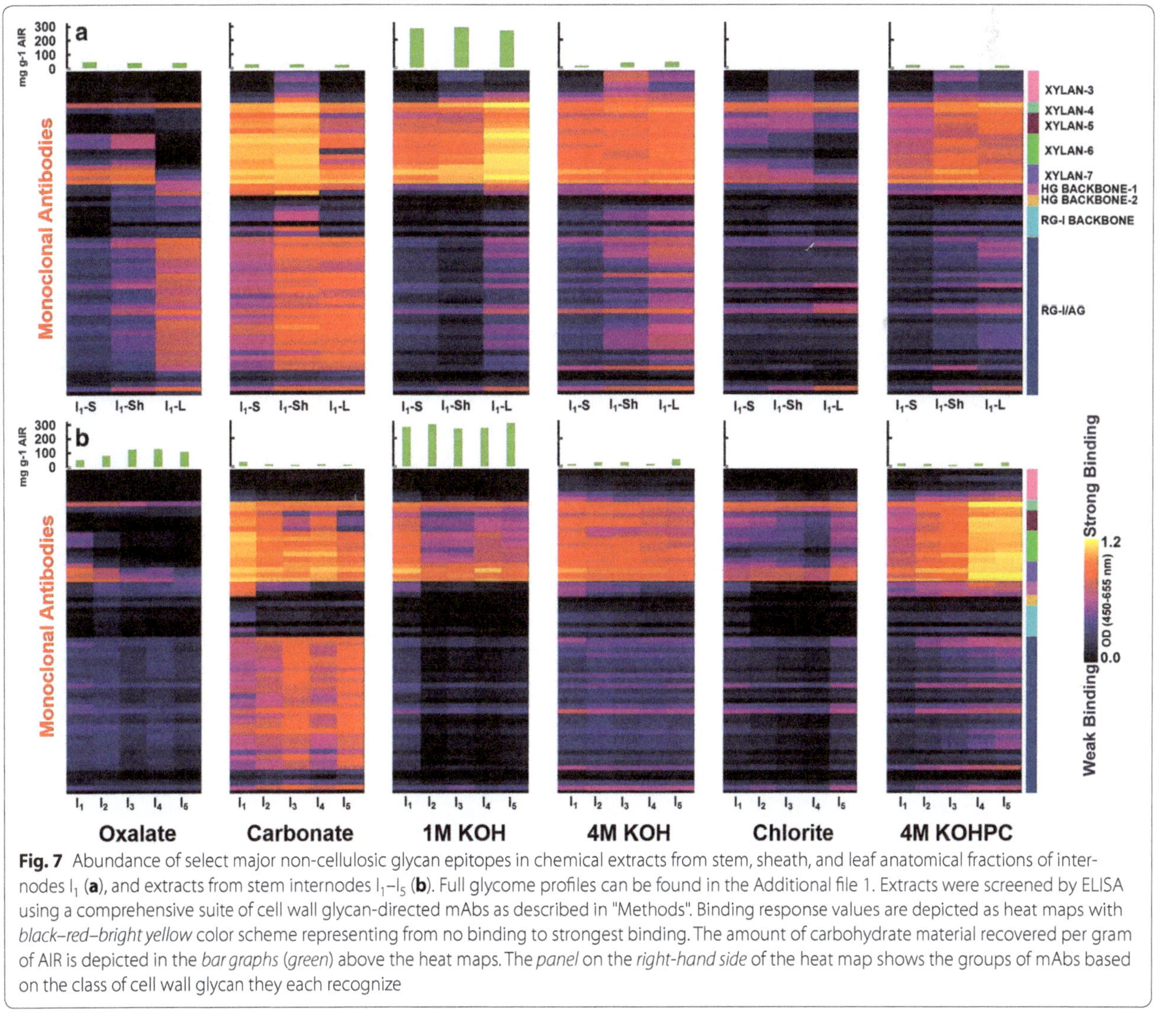

Fig. 7 Abundance of select major non-cellulosic glycan epitopes in chemical extracts from stem, sheath, and leaf anatomical fractions of internodes I_1 (**a**), and extracts from stem internodes I_1–I_5 (**b**). Full glycome profiles can be found in the Additional file 1. Extracts were screened by ELISA using a comprehensive suite of cell wall glycan-directed mAbs as described in "Methods". Binding response values are depicted as heat maps with *black–red–bright yellow* color scheme representing from no binding to strongest binding. The amount of carbohydrate material recovered per gram of AIR is depicted in the *bar graphs* (*green*) above the heat maps. The *panel* on the *right-hand side* of the heat map shows the groups of mAbs based on the class of cell wall glycan they each recognize

patterns of xylan epitopes composition and extractability were recognized in the 4 M KOH extract; however, at this point, binding intensities across samples were largely similar and were probably the residual alkaline susceptible xylans that remained after the 1 M KOH extract [42]. The chlorite and subsequent 4 M KOH PC extracts were used to determine carbohydrates more tightly bound with lignin. Interestingly, the chlorite extract showed overall similar epitope recognition patterns for all three extracts as the oxalate extract, with the I_1-leaf containing only xylan-4 and xylan-7 epitopes. Subsequent 4 M KOH PC extraction showed similar overall epitope extractability patterns and binding intensity as the 1 M KOH extract, with I_1-sheath and I_1-leaf extracts having the most intense binding for xylan-4 to xylan-7 epitopes.

From these results, it can be hypothesized that xylan integration within the cell walls of each anatomical fraction are not equivalent. Specifically, xylan association in the leaf fraction appeared to be more tightly integrated within the cell wall, as leaf xylan required a more severe basic 1 M KOH extract to observe significant xylan epitopes. Recognition of xylan-5 epitopes in the carbonate and 1 M KOH fraction was unexpected, as more substituted xylans such as arabinoxylan are thought to exist in primary cell walls, while less substituted and branched xylans are associated with the secondary cell wall [59]. This result may indicate an abundance in arabinosylation of leaf xylans compared to stem xylan in these switchgrass fractions [60]. Furthermore, higher xylan epitope recognition in the 4 M KOH PC suggests that xylan is more integrated with lignin in the leaf fraction, which may be a result of hemicelluloses in the leaf contributing more towards leaf internode recalcitrance compared to higher lignin containing fractions such as stem and sheath.

In terms of pectin epitope trends, while I_1-stem and I_1-sheath displayed marginal pectin epitope abundancies in the oxalate extract, the majority of recognized epitopes and highest abundancies was in the I_1-leaf extract. Specifically, epitopes of the homogalacturonan backbone (HG backbone 1 and 2) and pectic arabinogalactan epitopes (RG-I/AG) were broadly recognized in the I_1-leaf extract, with weak to moderate binding also occurring in the I_1-sheath extract. More intense binding and recognition for the pectic arabinogalactan epitopes were observed in the carbonate extract for I_1-sheath and I_1-leaf, with a moderate amount of recognition in the I_1-stem extract. In addition, HG 1 epitopes were observed in high intensity for all three extracts, and rhamnogalacturonan-I backbone (RG-I backbone) was recognized highest in the I_1-sheath extract. Overall, a higher abundance of pectin backbone and pectic arabinogalactan (AG) epitopes was noted in I_1-leaf samples in 1 M, 4 M KOH, and 4 M KOH PC extracts, while patterns in chlorite extracts were largely similar across samples.

From the pectin epitope recognition patterns, it is clear the stem anatomical fraction generally had lower overall pectin content that was also less recalcitrant. Sheath and leaf anatomical fractions displayed similar epitope recognition, suggesting similar pectin structures; however, differential epitope binding intensity suggests that each anatomical fraction has a unique organization of pectin in the cell wall. Pectic AGs are the major galactose containing pectin found in grasses [61], with the higher content of galactan found in the cell walls of leaf extracts suggesting higher overall pectic AGs content compared to stem samples [23].

Glycome profiling for internode-specific extractability

Stem internodes I_1–I_5 were subjected to glycome profiling to examine trends in the non-cellulosic polysaccharide extractability and structure with respect to tissue maturity. As earlier, selected epitope profiles for xylan and pectins are shown in Fig. 7b, while the complete glycome profile can be found in Additional file 1: Figure S3.

Oxalate and carbonate extracts showed binding of xylan-4 and xylan-7 epitopes in all internodes, with the significantly enhanced appearance of xylan-6 epitopes in the carbonate fraction. There was a trend of reduced xylan-5 epitope abundance in the more mature internodes, corresponding to a decrease in extractable Me-GlcA substituted xylans in oxalate, carbonate, and 1 M KOH extracts. The 1 M KOH extract showed significant epitope extractability of xylan-4 and xylan-7, indicating that a large fraction of the total polysaccharides released in the 1 M KOH extracts were unsubstituted xylans or arabinoxylated xylans. The 4 M KOH fraction showed similar abundance of xylan-4, xylan-5, xylan-6, and xylan-7 epitopes among all developmental stages of internodes, while the 4 M KOH PC extracts showed an increasing trend of xylan mAb extractability in more mature internodes.

Taken together, these results indicate that internode maturity impacts xylan deconstruction and extractability. Specifically, higher abundance of xylans were recognized in less severe chemical extracts for younger internodes, while harsh extractions after chlorite delignification showed the opposite trend, indicating lignification plays a greater role in xylan recalcitrance in more mature internodes [62]. Pectin epitope abundance was largely observed in the carbonate and 4 M KOH PC extracts, with an increase in binding intensity of pectic AG epitopes with maturing internodes. In addition, a clear trend in decreasing binding strength was observed with the epitopes for the HG backbone 1, which is an indication of a transition from an abundance of primary cell walls to the development of secondary cell wall thickening [63]. In the 4 M KOH PC extracts, a similar trend was observed for the pectic AG as in the carbonate fraction; however, the HG backbone 1 epitope abundance increased with internode maturity. Pectin trends indicate that less severe extracts result in more extractable pectins from more mature internodes, which would indicate that pectins more closely associate within the cell wall in less mature internodes or are more easily liberated compared to xylans. At the same time, however, the trend observed post chlorite delignification suggests that there is a second fraction of pectins that are closely associated with lignin, as indicated by the increased pectin and pectin backbone abundance in more mature internodes. Pectin-lignin associations have been proposed in grasses in prior studies [23, 64].

Discussion

Results for plant cell wall composition together with cell wall susceptibility to deconstruction by a coupled alkaline pretreatment and enzymatic hydrolysis demonstrate different switchgrass anatomical fractions possess significant differences both within fractions of varying maturity, and between different anatomical fractions. These trends are shown as a summary of Pearson's correlation coefficients between cell wall properties and hydrolysis yields for either individual anatomical fractions or for all pooled fractions are presented in Fig. 8, while data from a few illustrative identified correlations are plotted in Fig. 9.

First, lignin content was found to exhibit a negative correlation with untreated and pretreated in stems and leaves (Fig. 8), while demonstrating a strong, significant negative correlation when all anatomical fractions are pooled for 48 h hydrolysis yields following pretreatment ($R = -0.68$; p value $= 0.005$; Figs. 8 and 9a). As similar trends are observed for untreated hydrolysis yields and pretreated hydrolysis yields, indicating the trend is related to anatomical fraction structure and maturation rather than pretreatment response. Stems contain the highest content of lignin of any of the anatomical fractions (Fig. 3a), with tissue maturity-dependent lignification and cell wall thickening [18] observed in confocal microscopy cross sections (Fig. 6). These results together suggest lignin is a defining feature of recalcitrance in the cell walls of developing stems. Notably, lignin is a well-known contributor to cell wall recalcitrance, and as one example, our previous work in diverse maize lines demonstrated that initial lignin content was a strong predictor of glucose hydrolysis yields for no pretreatment, while lignin content after mild NaOH pretreatment was a strong predictor of glucose hydrolysis yields comparable to what was performed in the present work [38]. Glycome profiling results do suggest that hemicelluloses may play a parallel role in inhibiting cell wall deconstruction, as a noted decrease in xylan epitope extractability was observed in maturing stem internodes until alkaline extraction post chlorite delignification (Fig. 7); however, higher lignification reducing xylan accessibility to extraction may have contributed to the observed trends.

Variability in lignin composition may also play a significant role in anatomical fraction-specific differences of recalcitrance. A strong negative correlation was observed between lignin and ferulate content for all fractions, shown in Fig. 9b ($R^2 = 0.923$; p-value $= 10^{-8}$). Coupled

		Xyl	Ara	Man	Gal	Rha	Fuc	Ac	Lignin	*p* CA	FA
Stem Internode	Unpretreated 48 h	0.79	0.92	0.90	0.72	0.66	0.87	0.97	-1.00	0.27	0.92
	NaOH pretreated 6 h	0.96	0.95	0.75	0.74	0.46	0.96	0.88	-0.82	0.59	0.84
	NaOH pretreated 48 h	0.97	0.98	0.85	0.84	0.55	0.99	0.89	-0.86	0.67	0.91
Leaf Sheath	Unpretreated 48 h	-0.76	0.27	0.60	0.40	0.52	0.41	-0.35	0.44	-0.61	-0.72
	NaOH pretreated 6 h	-0.59	0.94	0.91	0.96	0.93	0.96	-0.90	0.93	-0.96	-0.91
	NaOH pretreated 48 h	-0.76	0.83	0.93	0.91	0.92	0.92	-0.75	0.85	-0.90	-0.92
Leaf Blade	Unpretreated 48 h	0.74	-0.51	-0.83	-0.65	-0.76	-0.60	-0.62	-0.76	0.76	0.64
	NaOH pretreated 6 h	0.79	-0.56	-0.93	-0.76	-0.90	-0.68	-0.31	-0.85	0.93	0.78
	NaOH pretreated 48 h	0.83	0.33	-0.35	-0.04	-0.26	0.15	-0.12	-0.24	0.42	0.14
All Fractions	Unpretreated 48 h	-0.33	0.90	0.52	0.81	0.48	0.77	-0.21	-0.63	-0.63	0.59
	NaOH pretreated 6 h	-0.51	0.97	0.67	0.93	0.65	0.91	-0.35	-0.75	-0.78	0.68
	NaOH pretreated 48 h	-0.37	0.92	0.67	0.86	0.63	0.85	-0.15	-0.68	-0.71	0.56

Positive correlation
Pearson's Correlation Coefficient: 1.0, 0.8, 0.6, 0.4, 0.2, 0, -0.2, -0.4, -0.6, -0.8, -1.0
Negative correlation

Fig. 8 Summary of Pearson's correlation coefficients between plant cell wall properties and glucose hydrolysis yields for each anatomical fraction and for pooled samples. Data demonstrate strong correlations between many properties and yields

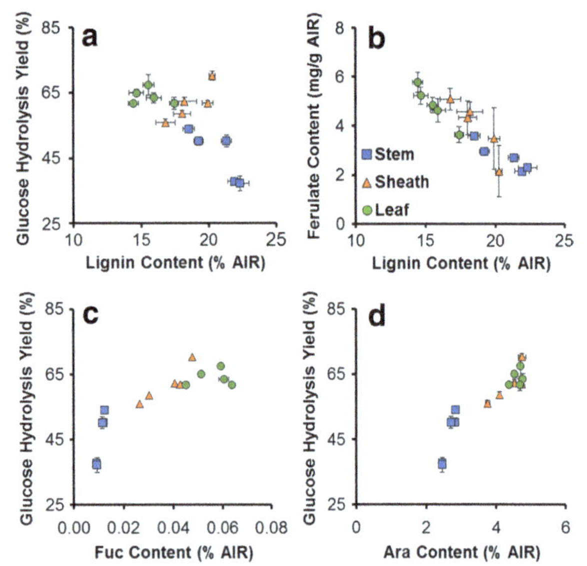

Fig. 9 Select correlations between data relationships between either lignin content or ferulic acid against glucose hydrolysis yield represented as scatter plots. **a** 48-h glucose hydrolysis yields for NaOH-pretreat biomass versus lignin content, **b** ferulate content in relation to total lignin, and glucose hydrolysis yield with respect to **c** fucose and **d** arabinan content

with the observed decrease in ferulate content with increasing internodes shown in Fig. 3b and as reported previously in switchgrass stem internodes [18], suggesting that ferulate content within the cell wall is dependent upon both tissue maturity and lignin content. Ester crosslinking of cell wall biopolymers via ferulates is considered to be one of the major contributing features cell wall recalcitrance in low-lignin grass cell walls [7, 65]. Notably, ester crosslinks are susceptible to saponification during mild alkaline pretreatment relative to ether linkages between other monolignols, and may potentially contribute significantly to enzymatic hydrolysis yields following alkaline pretreatment.

Similar to stem internode composition, lignification increased in maturing leaf sheath cell walls; however, the positive correlation between 48-h glucose hydrolysis yields following NaOH pretreatment and lignin content in leaf sheath samples (Fig. 9a) suggests that other cell wall components, such as hemicellulose substitution, pectin abundance, and polysaccharide–lignin associations may be a more significant contributor to cell wall recalcitrance. In the leaf sheath, significant positive correlations were observed between minor polysaccharide sugars (Rha, Ara, Fuc, Gal, Man) and hydrolysis yields following pretreatment (Figs. 8, 9c, d), and considering the leaf sheath cell wall composition of each of these minor sugars can be more than double that in stem internodes (Fig. 3c), indicating these polymers may play a more important role in cell wall recalcitrance. Of note, arabinan, which is present as substitutions on the xylan backbone in GAX and to a substantially lesser extent as arabinogalactan (AG), is substantially higher in leaf sheath cell walls than in stem internodes and exhibits a strong positive correlation to glucose hydrolysis yields in pretreated biomass (Fig. 9d). The implications of this are that increased substitution of xylan with arabinosyl substitutions suggests a less tightly crosslinked cell wall matrix that may be more accessible to enzymatic deconstruction [21, 66]. This finding is in agreement with the work of Costa et al. [62], who found that the Ara:Xyl ratio in different anatomical fractions isolated from sugarcane stems was a strong predictor of enzymatic hydrolysis yields in un-pretreated biomass.

Although possessing similar arabinan content, the same trend was not observed in leaf internodes; lower cellulose, xylan, and lignin content in leaf internodes suggest that due to limited lignification in leaf internodes compared to stem internodes (Fig. 2), structural pectins and minor hemicellulose components with lower overall compositions playing a more significant role in cell wall recalcitrance. Results suggest that a relatively internode-independent, uniform response was observed in leaf cross-section phenolic autofluorescence to alkaline pretreatment (Fig. 6), corresponding to similar response to lignin removal. The effectiveness of enzymatic hydrolysis of leaf internodes following NaOH pretreatment further suggests that alkaline susceptible cell wall components did not have a differential effect upon leaf cell wall digestibility. Although displaying comparable yields to alkaline pretreated leaf sheaths, leaf cell walls may be more susceptible to a pretreatment that predominantly removes more polysaccharide components, such as liquid hot water pretreatment, which has shown to dramatically improve leaf digestibility [67].

Technologies for biomass fractionation or extraction employed either on-field or at a centralized processing facility offer the potential to simultaneously improve agronomic, logistics, and processing outcomes. As an example, on-site field fractionation of switchgrass could capitalize on more recalcitrant stem internodes residing at the bottom of the tiller, harvesting upper less mature internodes while leaving mature internodes on-field for soil coverage and nutrient retention. Given stem and sheath leaf fractions constitute half of leaf dry weight in prior reporting [43], loss of biomass yield would have to be weighed against increases in downstream sugar yields due to less recalcitrant materials representing a larger total fraction of biomass.

Additionally, grass stems can accumulate a substantial amount of soluble sugars in storage parenchyma cells surrounding the vascular tissues [68]. Given that soluble sugars can represent more than 10% of the switchgrass total dry weight in this study, water extraction may be proposed as a route to yield a stream rich in fermentable sugars. Sugarcane undergoes mechanical fractionation coupled with counter-current hot water extraction to remove soluble, fermentable sugars. Flow-through liquid hot water pretreatments are well-established pretreatment technologies [69]; counter-current processing configurations can be envisioned that integrate soluble sugar extraction with liquid hot water pretreatment, each taking place at their respective optimal temperatures.

Conclusions

In this work, we were able to demonstrate that significant heterogeneity in cell wall structure, composition, properties within switchgrass results in substantially different feedstock response to processing in a cellulosic biorefinery. Feedstock heterogeneity may offer the potential to either tailor harvest approach, harvest time, or employ physical fractionation during processing to optimize processing outcomes such as sugar yield. Using switchgrass physically fractionated by anatomical fraction and internode, we identified compositional differences that impact recalcitrance and related results to macroscopic observations. Specifically, we showed that while all anatomical fractions experienced different extents of lignification between internodes, stem internode, leaf sheath, and leaf blade fractions were shown to have different structural features that dominate cell wall recalcitrance. Lignin content was identified as exhibiting a strong correlation to cell wall recalcitrance in stem internodes and initial recalcitrance in leaves. Leaf sheath and leaf cell wall recalcitrance were also shown to be impacted significantly by hemicellulose content and substitution, as well as structural pectin content. Ferulate and *p*-coumarate content were inversely correlated to lignin content in all fractions and decreased significantly with increasing internode maturity. Additionally, soluble sugars can account for a non-trivial fraction of total switchgrass dry weight, and represent a potential avenue to generate an enriched stream of easily extracted sugars to improve process economics.

Additional file

Additional file 1: Table S1. Compiled AIR + destarched cell wall polysaccharides, acetate, and lignin content for five internodes of three switchgrass anatomical fractions. Samples are presented as an average (n = 3) with standard deviations following in additional Table S2. **Table S2.** Standard deviations for compiled AIR + destarched cell wall polysaccharides, acetate, and lignin content for five internodes of three switchgrass anatomical fractions. Samples are presented as an average (n = 3) with standard deviations following. **Figure S1.** Cell wall autofluorescence of switchgrass organ cross-sections from internode I_2. Confocal laser scanning microscopy carried out using laser excitation wavelengths of 405 nm and 543 nm. Tissue types are denoted from the following notation: Ep – epidermis, Fb – fiber bundles, Vb – vascular bundles, Ph – phloem, X – xylem, Pc – Parenchyma cells. Scale bars represent 200 μm. **Figure S2.** Heat map abundance of major non-cellulosic glycan epitopes in oxalate, carbonate, 1 M KOH, 4 M KOH, chlorite, and 4 M KOH PC extracts from cell walls of stem, sheath, and leaf samples of internodes I_1. Extracts were screened by ELISA using a comprehensive suite of cell wall glycan-directed mAbs as described in materials and methods. Binding response values are depicted as heat maps with black-red-bright yellow color scheme representing from no binding to strongest binding. The amount of carbohydrate material recovered per gram of AIR is depicted in the bar graphs (purple) above the heat maps. The panel on the right-hand side of the heat map shows the groups of mAbs based on the class of cell wall glycan they each recognize. **Figure S3.** Heat map abundance of major non-cellulosic glycan epitopes in oxalate, carbonate, 1 M KOH, 4 M KOH, chlorite, and 4 M KOHPC extracts from stem internodes I_1-I_5. Extracts were screened by ELISA using a comprehensive suite of cell wall glycan-directed mAbs as described in materials and methods. Binding response values are depicted as heat maps with black-red-bright yellow color scheme representing from no binding to strongest binding. The amount of carbohydrate material recovered per gram of AIR is depicted in the bar graphs (purple) above the heat maps. The panel on the right-hand side of the heat map shows the groups of mAbs based on the class of cell wall glycan they each recognize.

Abbreviations

AG: arabinogalactan; AIR: alcohol insoluble residue; FA: ferulate; GAX: glucuronoarabinoxylan; HG: homogalacturonan; *p*CA: *p*-coumarate; RG: rhamnogalacturonan; XG: xyloglucan.

Authors' contributions

DBH and RGO conceived the work. JDC and DBH wrote the manuscript. NF and JDC performed the pretreatment and enzymatic hydrolysis. CF and DD performed the composition analysis. JDC performed the confocal microscopy. SP and BM performed the glycome profiling. All authors read and approved the final manuscript.

Author details

[1] Department of Chemical Engineering and Materials Science, Michigan State University, East Lansing, MI, USA. [2] Complex Carbohydrate Research Center, University of Georgia, Athens, GA, USA. [3] Bioenergy Science Center, Oak Ridge National Laboratory, Oak Ridge, TN, USA. [4] DOE-Great Lakes Bioenergy Research Center, Michigan State University, East Lansing, MI, USA. [5] Department of Chemical Engineering, Michigan Technological University, Houghton, MI, USA. [6] Department of Biosystems & Agricultural Engineering, Michigan State University, East Lansing, MI, USA. [7] Department of Civil, Environmental and Natural Resources Engineering, Luleå University of Technology, Luleå, Sweden.

Acknowledgements

The authors would like to thank Dr. Melinda Frame (Michigan State University) for assistance with confocal microscopy sample preparation and image collection.

Competing interests

The authors declare that they have no competing interests.

Funding

Jacob Crowe was supported in part by a grant from the U.S. National Science Foundation (NSF CBET 1336622). RGO was supported by the U.S. Department of Energy Great Lakes Bioenergy Research Center (DOE BER Office of Science DE-FC02-07ER64494). Glycome profiling studies were supported by the BioEnergy Science Center administered by Oak Ridge National Laboratory and funded by Grant DE-AC05-00OR22725 from the Office of Biological and

Environmental Research, Office of Science, U.S. Department of Energy. Generation of the CCRC series of cell wall glycan-directed monoclonal antibodies was supported by the National Science Foundation Plant Genome Program (DBI-0421683 and IOS-0923992).

References

1. Melillo JM, Richmond TC, Yohe GW. Climate change impacts in the United States: The third national climate assessment. Washington: US Global Change Research Program; 2014. p. 841.
2. Tilman D, Socolow R, Foley JA, Hill J, Larson E, Lynd L, Pacala S, Reilly J, Searchinger T, Somerville C, et al. Beneficial biofuels-the food, energy, and environment trilemma. Science. 2009;325:270–1.
3. U.S. Department of Energy. U.S. Billion-ton update: biomass supply for a bioenergy and bioproducts industry. In: Perlack D, Stokes BJ, editors. ORNL/TM-2011/224. Oak Ridge: Oak Ridge National Laboratory; 2011.
4. Himmel ME, Ding SY, Johnson DK, Adney WS, Nimlos MR, Brady JW, Foust TD. Biomass recalcitrance: engineering plants and enzymes for biofuels production. Science. 2007;315:804–7.
5. Albersheim PD, Roberts K, Sederoff R, Staehelin A. Plant cell walls: Garland Science. 1st ed. New York: Taylor & Francis Group; 2011.
6. Pauly M, Gille S, Liu L, Mansoori N, Souza A, Schultink A, Xiong G. Hemicellulose biosynthesis. Planta. 2013;238:627–42.
7. Grabber JH, Ralph J, Lapierre C, Barrière Y. Genetic and molecular basis of grass cell-wall degradability. I. Lignin–cell wall matrix interactions. C R Biol. 2004;327:455–65.
8. Scheller HV, Ulvskov P. Hemicelluloses. Annu Rev Plant Biol. 2010;61:263–89.
9. Wilkerson CG, Mansfield SD, Lu F, Withers S, Park JY, Karlen SD, Gonzales-Vigil E, Padmakshan D, Unda F, Rencoret J, et al. Monolignol ferulate transferase introduces chemically labile linkages into the lignin backbone. Science. 2014;344:90–3.
10. DeMartini JD, Pattathil S, Miller JS, Li HJ, Hahn MG, Wyman CE. Investigating plant cell wall components that affect biomass recalcitrance in poplar and switchgrass. Energy Environ Sci. 2013;6:898–909.
11. Lionetti V, Francocci F, Ferrari S, Volpi C, Bellincampi D, Galletti R, D'Ovidio R, De Lorenzo G, Cervone F. Engineering the cell wall by reducing de-methyl-esterified homogalacturonan improves saccharification of plant tissues for bioconversion. Proc Natl Acad Sci USA. 2010;107:616–21.
12. Cass CL, Lavell AA, Santoro N, Foster CE, Karlen SD, Smith RA, Ralph J, Garvin DF, Sedbrook JC. Cell wall composition and biomass recalcitrance differences within a genotypically diverse set of *Brachypodium distachyon* inbred lines. Front Plant Sci. 2016;7:16.
13. McLaughlin SB, Kszos LA. Development of switchgrass (*Panicum virgatum*) as a bioenergy feedstock in the United States. Biomass Bioenerg. 2005;28:515–35.
14. Somerville C, Youngs H, Taylor C, Davis SC, Long SP. Feedstocks for lignocellulosic biofuels. Science. 2010;329:790–2.
15. Parrish DJ, Casler MD, Monties A. The evolution of switchgrass as an energy crop. In: Monties A, editor. Switchgrass: a valuable biomass crop for energy. London: Springer; 2012. p. 1–28.
16. Bouton JH. Molecular breeding of switchgrass for use as a biofuel crop. Curr Opin Genet Dev. 2007;17:553–8.
17. Jung HG, Casler MD. Maize stem tissues: cell wall concentration and composition during development. Crop Sci. 2006;46:1793–800.
18. Sarath G, Baird LM, Vogel KP, Mitchell RB. Internode structure and cell wall composition in maturing tillers of switchgrass (*Panicum virgatum* L.). Biores Technol. 2007;98:2985–92.
19. Dien BS, Jung HJG, Vogel KP, Casler MD, Lamb JFS, Iten L, Mitchell RB, Sarath G. Chemical composition and response to dilute-acid pretreatment and enzymatic saccharification of alfalfa, reed canarygrass, and switchgrass. Biomass Bioenerg. 2006;30:880–91.
20. Sarath G, Mitchell RB, Sattler SE, Funnell D, Pedersen JF, Graybosch RA, Vogel KP. Opportunities and roadblocks in utilizing forages and small grains for liquid fuels. J Ind Microbiol Biotechnol. 2008;35:343–54.
21. Vogel KP, Casler MD, Dien BS. Switchgrass biomass composition traits and their effects on its digestion by ruminants and bioconversion to ethanol. Crop Sci. 2016;57:275–81.
22. Sarath G, Dien B, Saathoff AJ, Vogel KP, Mitchell RB, Chen H. Ethanol yields and cell wall properties in divergently bred switchgrass genotypes. Biores Technol. 2011;102:9579–85.
23. da Costa RMF, Pattathil S, Avci U, Lee SJ, Hazen SP, Winters A, Hahn MG, Bosch M. A cell wall reference profile for Miscanthus bioenergy crops highlights compositional and structural variations associated with development and organ origin. New Phytol. 2016;213:1710–25.
24. Bootsma JA, Shanks BH. Hydrolysis characteristics of tissue fractions resulting from mechanical separation of corn stover. Appl Biochem Biotechnol. 2005;125:27–39.
25. Crofcheck CL, Montross MD. Effect of stover fraction on glucose production using enzymatic hydrolysis. Trans Am Soc Agric Eng. 2004;47:841–4.
26. Montross MD, Crofcheck CL. Effect of stover fraction and storage method on glucose production during enzymatic hydrolysis. Biores Technol. 2004;92:269–74.
27. Hansey CN, Lorenz AJ, de Leon N. Cell wall composition and ruminant digestibility of various maize tissues across development. Bioenerg Res. 2010;3:28–37.
28. Ding SY, Liu YS, Zeng Y, Himmel ME, Baker JO, Bayer EA. How does plant cell wall nanoscale architecture correlate with enzymatic digestibility? Science. 2012;338:1055–60.
29. Hansen MA, Hidayat BJ, Mogensen KK, Jeppesen MD, Jørgensen B, Johansen KS, Thygesen LG. Enzyme affinity to cell types in wheat straw (*Triticum aestivum* L.) before and after hydrothermal pretreatment. Biotechnol Biofuels. 2013;6:54.
30. Zhang H, Fangel J, Willats WGT, Selig MJ, Lindedam J, Jorgensen H, Felby C. Assessment of leaf/stem ratio in wheat straw feedstock and impact on enzymatic conversion. GCB Bioenerg. 2014;6:90–6.
31. Chundawat SPS, Balan V, Dale BE. Effect of particle size based separation of milled corn stover on AFEX pretreatment and enzymatic digestibility. Biotechnol Bioeng. 2007;96:219–31.
32. Garlock RJ, Chundawat SPS, Balan V, Dale BE. Optimizing harvest of corn stover fractions based on overall sugar yields following ammonia fiber expansion pretreatment and enzymatic hydrolysis. Biotechnol Biofuels. 2009;2:29.
33. Foster CE, Martin TM, Pauly M. Comprehensive compositional analysis of plant cell walls (lignocellulosic biomass) part II: carbohydrates. J Vis Exp. 2010;37:e1837.
34. Sluiter JB, Ruiz RO, Scarlata CJ, Sluiter AD, Templeton DW. Compositional analysis of lignocellulosic feedstocks. 1. Review and description of methods. J Agric Food Chem. 2010;58:9043–53.
35. Li M, Foster C, Kelkar S, Pu Y, Holmes D, Ragauskas A, Saffron C, Hodge D. Structural characterization of alkaline hydrogen peroxide pretreated grasses exhibiting diverse lignin phenotypes. Biotechnol Biofuels. 2012;5:38.
36. Foster CE, Martin TM, Pauly M. Comprehensive compositional analysis of plant cell walls (lignocellulosic biomass) part I: lignin. J Vis Exp. 2010;37:e1745.
37. Santoro N, Cantu SL, Tornqvist C-E, Falbel TG, Bolivar JL, Patterson SE, Pauly M, Walton JD. A high-throughput platform for screening milligram quantities of plant biomass for lignocellulose digestibility. Bioenerg Res. 2010;3:93–102.
38. Li M, Heckwolf M, Crowe JD, Williams DL, Magee TD, Kaeppler SM, de Leon N, Hodge DB. Cell-wall properties contributing to improved deconstruction by alkaline pre-treatment and enzymatic hydrolysis in diverse maize (*Zea mays* L.) lines. J Exp Bot. 2015;66:4305–15.
39. DeMartini JD, Pattathil S, Avci U, Szekalski K, Mazumder K, Hahn MG, Wyman CE. Application of monoclonal antibodies to investigate plant cell wall deconstruction for biofuels production. Energy Environ Sci. 2011;4:4332–9.
40. Pattathil S, Avci U, Miller JS, Hahn MG. Immunological approaches to plant cell wall and biomass characterization: glycome profiling. In: Biomass conversion. Berlin: Springer; 2012. p. 61–72.
41. Pattathil S, Avci U, Baldwin D, Swennes AG, McGill JA, Popper Z, Bootten T, Albert A, Davis RH, Chennareddy C, et al. A comprehensive toolkit of plant cell wall glycan-directed monoclonal antibodies. Plant Physiol. 2010;153:514–25.
42. Pattathil S, Hahn MG, Dale BE, Chundawat SP. Insights into plant cell wall structure, architecture, and integrity using glycome profiling of native and AFEX™-pre-treated biomass. J Exp Bot. 2015;66:4279–94.
43. Hu ZJ, Foston M, Ragauskas AJ. Comparative studies on hydrothermal pretreatment and enzymatic saccharification of leaves and internodes of alamo switchgrass. Biores Technol. 2011;102:7224–8.
44. Shen H, Fu CX, Xiao XR, Ray T, Tang YH, Wang ZY, Chen F. Developmental

control of lignification in stems of lowland switchgrass variety Alamo and the effects on saccharification efficiency. Bioenerg Res. 2009;2:233–45.
45. Grabber J, Jung G, Hill R. Chemical composition of parenchyma and sclerenchyma cell walls isolated from orchardgrass and switchgrass. Crop Sci. 1991;31:1058–65.
46. Chuck GS, Tobias C, Sun L, Kraemer F, Li C, Dibble D, Arora R, Bragg JN, Vogel JP, Singh S, et al. Overexpression of the maize Corngrass1 microRNA prevents flowering, improves digestibility, and increases starch content of switchgrass. Proc Natl Acad Sci USA. 2011;108:17550–5.
47. Ong RG, Higbee A, Bottoms S, Dickinson Q, Xie D, Smith SA, Serate J, Pohlmann E, Jones AD, Coon JJ, et al. Inhibition of microbial biofuel production in drought-stressed switchgrass hydrolysate. Biotechnol Biofuels. 2016;9:237.
48. Garlock RJ, Balan V, Dale BE, Ramesh Pallapolu V, Lee YY, Kim Y, Mosier NS, Ladisch MR, Holtzapple MT, Falls M, et al. Comparative material balances around pretreatment technologies for the conversion of switchgrass to soluble sugars. Biores Technol. 2011;102:11063–71.
49. Decker SR, Carlile M, Selig MJ, Doeppke C, Davis M, Sykes R, Turner G, Ziebell A. Reducing the effect of variable starch levels in biomass recalcitrance screening. In: Himmel ME, editor. Biomass conversion: methods and protocols, methods in molecular biology. Berlin: Springer; 2012.
50. Slewinski TL. Non-structural carbohydrate partitioning in grass stems: a target to increase yield stability, stress tolerance, and biofuel production. J Exp Bot. 2012;63:4647–70.
51. Park YB, Cosgrove DJ. A revised architecture of primary cell walls based on biomechanical changes induced by substrate-specific endoglucanases. Plant Physiol. 2012;158:1933–43.
52. Tobimatsu Y, Wagner A, Donaldson L, Mitra P, Niculaes C, Dima O, Kim JI, Anderson N, Loque D, Boerjan W, et al. Visualization of plant cell wall lignification using fluorescence-tagged monolignols. Plant J. 2013;76:357–66.
53. O'brien T, Feder N, McCully ME. Polychromatic staining of plant cell walls by toluidine blue. Protoplasma. 1964;59:368–73.
54. Donaldson L. Softwood and hardwood lignin fluorescence spectra of wood cell walls in different mounting media. IAWA J. 2013;34:3–19.
55. García-Plazaola JI, Fernández-Marín B, Duke SO, Hernández A, López-Arbeloa F, Becerril JM. Autofluorescence: biological functions and technical applications. Plant Sci. 2015;236:136–45.
56. Hutzler P, Fischbach R, Heller W, Jungblut TP, Reuber S, Schmitz R, Veit M, Weissenböck G, Schnitzler J P. Tissue localization of phenolic compounds in plants by confocal laser scanning microscopy. J Exp Bot. 1998;49:953–65.
57. Matos DA, Whitney IP, Harrington MJ, Hazen SP. Cell walls and the developmental anatomy of the *Brachypodium distachyon* stem internode. PLoS ONE. 2013;8:9.
58. Schmidt D, Schuhmacher F, Geissner A, Seeberger PH, Pfrengle F. Automated synthesis of arabinoxylan-oligosaccharides enables characterization of antibodies that recognize plant cell wall glycans. Chem Eur J. 2015;21:5709–13.
59. Suzuki K, Kitamura S, Kato Y, Itoh T. Highly substituted glucuronoarabinoxylans (hsGAXs) and low-branched xylans show a distinct localization pattern in the tissues of *Zea mays* L. Plant Cell Physiol. 2000;41:948–59.
60. de Souza AP, Leite DCC, Pattathil S, Hahn MG, Buckeridge MS. Composition and structure of sugarcane cell wall polysaccharides: implications for second-generation bioethanol production. Bioenerg Res. 2013;6:564–79.
61. Mohnen D. Pectin structure and biosynthesis. Curr Opin Plant Biol. 2008;11:266–77.
62. Costa THF, Vega-Sanchez ME, Milagres AMF, Scheller HV, Ferraz A. Tissue-specific distribution of hemicelluloses in six different sugarcane hybrids as related to cell wall recalcitrance. Biotechnol Biofuels. 2016;9:99.
63. Cosgrove DJ. Growth of the plant cell wall. Nat Rev Mol Cell Biol. 2005;6:850–61.
64. Shen H, Poovaiah CR, Ziebell A, Tschaplinski TJ, Pattathil S, Gjersing E, Engle NL, Katahira R, Pu Y, Sykes R, et al. Enhanced characteristics of genetically modified switchgrass (*Panicum virgatum* L.) for high biofuel production. Biotechnol Biofuels. 2013;6:71.
65. Ralph J. Hydroxycinnamates in lignification. Phytochem Rev. 2010;9:65–83.
66. Busse-Wicher M, Grantham NJ, Lyczakowski JJ, Nikolovski N, Dupree P. Xylan decoration patterns and the plant secondary cell wall molecular architecture. Biochem Soc Trans. 2016;44:74–8.
67. Zeng M, Ximenes E, Ladisch MR, Mosier NS, Vermerris W, Huang C-P, Sherman DM. Tissue-specific biomass recalcitrance in corn stover pretreated with liquid hot-water: enzymatic hydrolysis (part 1). Biotechnol Bioeng. 2012;109:390–7.
68. Rae AL, Grof CP, Casu RE, Bonnett GD. Sucrose accumulation in the sugarcane stem: pathways and control points for transport and compartmentation. Field Crops Res. 2005;92:159–68.
69. Mosier N, Wyman C, Dale B, Elander R, Lee YY, Holtzapple M, Ladisch M. Features of promising technologies for pretreatment of lignocellulosic biomass. Biores Technol. 2005;96:673–86.

Permissions

All chapters in this book were first published in BFB, by BioMed Central; hereby published with permission under the Creative Commons Attribution License or equivalent. Every chapter published in this book has been scrutinized by our experts. Their significance has been extensively debated. The topics covered herein carry significant findings which will fuel the growth of the discipline. They may even be implemented as practical applications or may be referred to as a beginning point for another development.

The contributors of this book come from diverse backgrounds, making this book a truly international effort. This book will bring forth new frontiers with its revolutionizing research information and detailed analysis of the nascent developments around the world.

We would like to thank all the contributing authors for lending their expertise to make the book truly unique. They have played a crucial role in the development of this book. Without their invaluable contributions this book wouldn't have been possible. They have made vital efforts to compile up to date information on the varied aspects of this subject to make this book a valuable addition to the collection of many professionals and students.

This book was conceptualized with the vision of imparting up-to-date information and advanced data in this field. To ensure the same, a matchless editorial board was set up. Every individual on the board went through rigorous rounds of assessment to prove their worth. After which they invested a large part of their time researching and compiling the most relevant data for our readers.

The editorial board has been involved in producing this book since its inception. They have spent rigorous hours researching and exploring the diverse topics which have resulted in the successful publishing of this book. They have passed on their knowledge of decades through this book. To expedite this challenging task, the publisher supported the team at every step. A small team of assistant editors was also appointed to further simplify the editing procedure and attain best results for the readers.

Apart from the editorial board, the designing team has also invested a significant amount of their time in understanding the subject and creating the most relevant covers. They scrutinized every image to scout for the most suitable representation of the subject and create an appropriate cover for the book.

The publishing team has been an ardent support to the editorial, designing and production team. Their endless efforts to recruit the best for this project, has resulted in the accomplishment of this book. They are a veteran in the field of academics and their pool of knowledge is as vast as their experience in printing. Their expertise and guidance has proved useful at every step. Their uncompromising quality standards have made this book an exceptional effort. Their encouragement from time to time has been an inspiration for everyone.

The publisher and the editorial board hope that this book will prove to be a valuable piece of knowledge for researchers, students, practitioners and scholars across the globe.

List of Contributors

Hehong Wei, Ying Shi, Xiaonian Ma and Jin Liu
Institute for Food and Bioresource Engineering, Department of Energy and Resources Engineering and BIC-ESAT, College of Engineering, Peking University, Beijing 100871, China

Yufang Pan and Hanhua Hu
Key Laboratory of Algal Biology, Institute of Hydrobiology, Chinese Academy of Sciences, Wuhan 430072, China

Yantao Li
Institute of Marine and Environmental Technology, University of Maryland Center for Environmental Science and University of Maryland Baltimore County, Baltimore, MA 21202, USA

Ming Luo
Guangdong Provincial Key Laboratory of Applied Botany, Key Laboratory of South China Agricultural Plant Molecular Analysis and Genetic Improvement, South China Botanical Garden, Chinese Academy of Sciences, Guangzhou 510650, China

Henri Gerken
School of Sustainable Engineering and the Built Environment, Arizona State University Polytechnic campus, Mesa, AZ 85212, USA

Way Cern Khor, Stephen Andersen, Han Vervaeren and Korneel Rabaey
Department of Biochemical and Microbial Technology, Centre for Microbial Ecology and Technology (CMET), Ghent University, Coupure Links 653, 9000 Ghent, Belgium

Piotr Chylenski, Dejan M. Petrović, Gerdt Müller, Svein Jarle Horn and Vincent G. H. Eijsink
Faculty of Chemistry, Biotechnology and Food Science, Norwegian University of Life Sciences, P.O. Box 5003, 1432 Ås, Norway

Marie Dahlström, Oskar Bengtsson and Martin Lersch
Borregaard AS, P.O.Box 162, 1701 Sarpsborg, Norway

Matti Siika-aho
VTT Technical Research Centre of Finland, P.O. Box 1000, 02044 VTT Espoo, Finland

Sriram Satagopan and F. Robert Tabita
Department of Microbiology, The Ohio State University, 484 West 12th Avenue, Columbus, OH 43210-1292, USA

Yuan Sun and Jon R. Parquette
Department of Chemistry and Biochemistry, The Ohio State University, 100 West 18th Avenue, Columbus, OH 43210-1185, USA

Viktor Klassen, Olga Blifernez-Klassen, Daniel Wibberg, Anika Winkler, Jörn Kalinowski and Olaf Kruse
Department of Biology/Center for Biotechnology (CeBiTec), Bielefeld University, Universitätsstrasse 27, 33615 Bielefeld, Germany

Clemens Posten
Institute of Life Science Engineering (KIT), Bioprocess Engineering, University of Karlsruhe, Fritz-Haber-Weg 2, 76131 Karlsruhe, Germany

Meng Li, Jun Wang, Fu Du, Boubacar Diallo and Guang Hui Xie
College of Agronomy and Biotechnology, China Agricultural University, Beijing 100193, China
National Energy R&D Center for Non-food Biomass, China Agricultural University, Beijing 100193, China

Satoshi Katahira, Nobuhiko Muramoto, Kenro Tokuhiro and Risa Nagura
Bioinspired Systems Research-Domain, Toyota Central R&D Labs., Inc., 41-1, Yokomichi, Nagakute, Aichi 480-1192, Japan

Shigeharu Moriya
RIKEN Center for Sustainable Resource Science, 1-7-22 Suehiro-cho, Tsurumi-ku, Yokohama, Kanagawa 230-0045, Japan

Nobuki Tada, Noriko Yasutani and Toru Onishi
Biotechnology and Afforestation Laboratory, New Business Planning Div, Toyota Motor Corporation, 1099 Marune, Kurozasa-cho, Miyoshi, Aichi 470-0201, Japan

Moriya Ohkuma
RIKEN Center for Sustainable Resource Science, 3-1-1 Koyadai, Tsukuba, Ibaraki 305-0074, Japan
Japan Collection of Microorganisms,RIKEN BioResource Center, 3-1-1 Koyadai, Tsukuba, Ibaraki 305-0074, Japan

Thiziri Aouam, Antoine Margeot and Frédérique Bidard
IFP Energies Nouvelles, 1-4 Avenue de Bois-Préau, 92852 Rueil-Malmaison, France

Christa Ivanova
IFP Energies Nouvelles, 1-4 Avenue de Bois-Préau, 92852 Rueil-Malmaison, France
Genetics of Biofilms Unit, Department of Microbiology, Institut Pasteur, 25-28 Rue du Dr Roux, 75015 Paris, France.

Alexa Frischmann, Bernhard Seiboth and Jonas Ramoni
Molecular Biotechnology, Research Division Biochemical Technology, Institute of Chemical Engineering, TU-Wien, 1060 Vienna, Austria

Scott E. Baker
Earth and Biological Sciences Directorate, Pacific Northwest National Laboratory, Richland, WA 99354, USA

Stéphane Le Crom
Evolution Paris Seine-Institut de Biologie Paris Seine (EPS-IBPS), Sorbonne Universités, UPMC Univ Paris 06, Univ Antilles, Univ Nice Sophia Antipolis, CNRS, 75005 Paris, France

Sophie Lemoine
École normale supérieure, PSL Research University, CNRS, Inserm, Institut de Biologie de l'École normale supérieure (IBENS), Plateforme Génomique, 75005 Paris, France

Ting-Ting Ma, Lai-Yan Liu, Zheng Zhou, Li-Rong Dai, Wan-Qiu Zeng, Hui Zhang and Lei Cheng
Key Laboratory of Development and Application of Rural Renewable Energy, Biogas Institute of Ministry of Agriculture, Section 4-13, Renmin South Road, Chengdu 610041, People's Republic of China

Jun-Peng Rui
Key Laboratory of Environmental and Applied Microbiology, Chengdu Institute of Biology of Chinese
Academy of Sciences, Section 4-9, Renmin South Road, Chengdu 610041, People's Republic of China
Environmental Microbiology Key Laboratory of Sichuan Province, Section 4-9, Renmin South Road, Chengdu 610041, People's Republic of China

Quan Yuan
State Key Laboratory of Environmental Geochemistry, Institute of Geochemistry, Chinese Academy of Sciences, 99 Lincheng West Road, Guanshanhu District, Guiyang 550081, People's Republic of China

Ding-shan Feng
Anhui Normal University, 1 Beijing East Road, Wuhu 241000, People's Republic of China

Huifang Xu, Shengying Li, Wenxia Yang, Bo Fang and Cong Wang
Shandong Provincial Key Laboratory of Synthetic Biology, Qingdao Institute of Bioenergy and Bioprocess Technology, Chinese Academy of Sciences, No. 189 Songling Road, Qingdao 266101, Shandong, China
CAS Key Laboratory of Biofuels, Qingdao Institute of Bioenergy and Bioprocess Technology,
Chinese Academy of Sciences, No. 189 Songling Road, Qingdao 266101, Shandong, China

Linlin Ning
Shandong Provincial Key Laboratory of Synthetic Biology, Qingdao Institute of Bioenergy and Bioprocess Technology, Chinese Academy of Sciences, No. 189 Songling Road, Qingdao 266101, Shandong, China
CAS Key Laboratory of Biofuels, Qingdao Institute of Bioenergy and Bioprocess Technology,
Chinese Academy of Sciences, No. 189 Songling Road, Qingdao 266101, Shandong, China
University of Chinese Academy of Sciences, Beijing 100049, China

Yun Wang and Jian Xu
CAS Key Laboratory of Biofuels, Qingdao Institute of Bioenergy and Bioprocess Technology,
Chinese Academy of Sciences, No. 189 Songling Road, Qingdao 266101, Shandong, China
Single-Cell Center, Qingdao Institute of Bioenergy and Bioprocess Technology, Chinese Academy of Sciences, No. 189 Songling Road, Qingdao 266101, Shandong, China

Severine Collin, Frederic Laeuffer and Laurent Fourage
Total Refinery and Chemistry, SDR/Biofuels, Tour Coupole, 2, PI. Jean Millier, 92400 Courbevoie, France

Peter Ryden, Maria-Nefeli Efthymiou, Adam Elliston, David R. Wilson and Keith W. Waldron
The Biorefinery Centre, Quadram Institute Bioscience, Colney, Norwich Research Park, Norwich NR4 7UA, UK

Teddy A. M. Tindyebwa
School of Biological Sciences, College of Natural Sciences, Makerere University, P.O. Box 7062, Kampala, Uganda

Pradeep K. Malakar
The Biorefinery Centre, Quadram Institute Bioscience, Colney, Norwich Research Park, Norwich NR4 7UA, UK
College of Food Science and Technology, Shanghai Ocean University, 999 Hu Cheng Huan Road, Shanghai 201306, People's Republic of China

Bianca Colombo, Francesca Favini, Barbara Scaglia, Tommy Pepè Sciarria, Giuliana D'Imporzano, Michele Pognani and Fabrizio Adani
Gruppo Ricicla labs-DiSAA-Università degli Studi di Milano, Via Celoria 2, 20133 Milan, Italy

Anna Alekseeva and Giorgio Eisele
Centro Alta Tecnologia Istituto di Ricerche Chimiche e Biochimiche G. Ronzoni Srl, Via Colombo 81, 20133 Milan, Italy

Cesare Cosentino
Istituto di Ricerche Chimiche e Biochimiche G. Ronzoni, Via Colombo 81, 20133 Milan, Italy

Devin F. R. Doud, Eric C. Holmes, Hanno Richter and Bastian Molitor
Department of Biological and Environmental Engineering, Cornell University, Ithaca, NY 14853, USA

Georg Jander
Boyce Thompson Institute for Plant Research, Ithaca, NY 14853, USA
Atkinson Center for a Sustainable Future, Cornell University, Ithaca, NY 14853, USA

Largus T. Angenent
Department of Biological and Environmental Engineering, Cornell University, Ithaca, NY 14853, USA
Boyce Thompson Institute for Plant Research, Ithaca, NY 14853, USA.
Atkinson Center for a Sustainable Future, Cornell University, Ithaca, NY 14853, USA
Center for Applied Geosciences, University of Tübingen, Hölderlinstr. 12, 72074 Tübingen, Germany

André Ferraz and Adriane M. F. Milagres
Departamento de Biotecnologia, Escola de Engenharia de Lorena, Universidade de São Paulo, CP 116, Lorena, SP 12602-810, Brazil

Germano Siqueira and Valdeir Arantes
Departamento de Biotecnologia, Escola de Engenharia de Lorena, Universidade de São Paulo, CP 116, Lorena, SP 12602-810, Brazil
Forest Products Biotechnology/Bioenergy Group, Faculty of Forestry, University of British Columbia, 2424 Main Mall, Vancouver, BC V6T 1Z4, Canada

Jack N. Saddler
Forest Products Biotechnology/Bioenergy Group, Faculty of Forestry, University of British Columbia, 2424 Main Mall, Vancouver, BC V6T 1Z4, Canada

Johan O. Westman
Division of Industrial Biotechnology, Department of Biology and Biological Engineering, Chalmers University of Technology, Gothenburg, Sweden
Chr. Hansen A/S, Bøge Allé 10-12, 2970 Hørsholm, Denmark

Ruifei Wang and Carl Johan Franzén
Division of Industrial Biotechnology, Department of Biology and Biological Engineering, Chalmers University of Technology, Gothenburg, Sweden

Vera Novy
Division of Industrial Biotechnology, Department of Biology and Biological Engineering, Chalmers University of Technology, Gothenburg, Sweden
Institute of Biotechnology and Biochemical Engineering, Graz University of Technology, Graz, Austria
Forest Products Biotechnology,Faculty of Forestry, University of British Columbia, Vancouver, BC V6T1Z4, Canada

Jacob D. Crowe and Nicholas Feringa
Department of Chemical Engineering and Materials Science, Michigan State University, East Lansing, MI, USA

Sivakumar Pattathil
Complex Carbohydrate Research Center, University of Georgia, Athens, GA, USA
Bioenergy Science Center, Oak Ridge National Laboratory, Oak Ridge, TN, USA

Brian Merritt
Complex Carbohydrate Research Center, University of Georgia, Athens, GA, USA

Cliff Foster and Dayna Dines
DOE-Great Lakes Bioenergy Research Center, Michigan State University, East Lansing, MI, USA

Rebecca G. Ong
Department of Chemical Engineering, Michigan Technological University, Houghton, MI, USA

David B. Hodge
Department of Chemical Engineering and Materials Science, Michigan State University, East Lansing, MI, USA
DOE-Great Lakes Bioenergy Research Center, Michigan State University, East Lansing, MI, USA
Department of Biosystems & Agricultural Engineering, Michigan State University, East Lansing, MI, USA
Department of Civil, Environmental and Natural Resources Engineering, Luleå University of Technology, Luleå, Sweden

Index

A

Alkaline Pretreatment, 67, 189, 192, 210, 212, 214, 218, 222

Alkenes, 130-131, 134, 137-138, 141-143

Ammonia Inhibition, 67

Aromatics, 210

B

Bioconversion, 43, 67, 71, 101, 207, 224

Bioenergy Sorghum, 69-76, 78, 81-82

Bioethanol Production, 69-70, 85-86, 145, 193, 208, 225

Biofuel, 1-2, 14, 16-18, 30, 41, 57-58, 65, 67-68, 70-71, 83-84, 143, 145-146, 172, 180, 207-208, 224-225

Biogas, 29, 57-60, 62-68, 117, 128, 164, 169-170

C

Caproic Acid, 19-29

Cell Wall Glycans, 210, 225

Cell Walls, 70, 145, 182-183, 192, 211, 216, 220-225

Cell-free Systems, 43

Cellulase, 30-31, 33, 37-39, 42, 102-104, 107-111, 114-116, 149-152, 183, 185-186, 189, 192-193, 208

Cellulase Production, 102-104, 107-111, 115-116

Cellulose Accessibility, 182-183, 188-189, 191-193

Chain Elongation, 19, 21-23, 28-29

Chemoselectivity, 130, 134-135, 138-139

CO_2 Fixation, 43, 49, 171

Combined Stress, 195

Continuous Anaerobic Fermentation, 57

D

Decane, 19-20, 23-25, 27-28, 118

Diacylglycerol Acyltransferase, 1, 14, 16-18

E

Electrolysis, 19-20, 23-27, 29

Enzymatic Degradation, 30-31, 38

Enzymatic Hydrolysis, 30-32, 34-42, 107, 154, 182-184, 186-193, 200, 203, 210-211, 213-214, 221-225

Enzyme Kinetics, 130

Ethanol, 20-23, 25, 27-28, 41, 56, 69-70, 72, 75-79, 81-83, 85-86, 89-94, 96, 98, 100-101, 113, 145-146, 148, 150-155, 183, 186, 192-196, 198-209, 212, 224

Ethanol Inhibition, 195, 200, 207

F

Fermentation, 19-29, 31, 57-68, 83, 85-86, 89-95, 98, 100-101, 129, 145-154, 156-158, 161, 169, 180-181, 192, 194-195, 197-200, 202-209, 216

Firmicutes, 22, 28, 57, 63-64, 85, 87-88, 94

Flocculation, 194-196, 200, 207-208

Fossil Fuel, 43, 58

Functional Characterization, 1, 4

G

Genetic Engineering, 1, 12-13, 86, 195

Genome Analysis, 103-104, 109

H

Herbaceous Biomass, 210

Heterogeneity, 210-213, 216, 223

High Gravity, 195, 208

Hydrocarbon Degradation, 117, 125, 128

L

Lactic Acid, 19-25, 27-28, 157

Lignin Removal, 37-38, 182-183, 188, 222

Lignocellulose, 30-31, 42, 101, 116, 145-146, 149-150, 152, 192-193, 195-196, 202, 207-208, 224

Linear Discriminant Analysis Model, 70

M

Metabolic Engineering, 18, 55, 100-101, 143, 171, 180

Methane, 28-29, 57-62, 64-68, 117-121, 123-129, 158, 169, 180

Methane Production, 28, 67, 117, 119-121, 123-129, 158

Microalga, 1, 14, 16-18, 67-68

Microalgae Mono-substrate, 64

Microbial Interactions, 117

Mixed Microbial Culture, 156, 162, 170

Municipal Solid Waste, 156-157, 169-170

Mutagenesis, 17, 52, 85-86, 90-94, 98-99, 102-104, 108-109, 114, 130, 135, 138-139, 142-144, 208

N

Nannochloropsis Oceanica, 1, 14, 17

Near Infrared Spectroscopy, 70, 82-83

Nitrogen Limitation, 57, 64

O

Obligate Reduction, 171-172
Oil Reservoirs, 117-118, 128
Optimal Variable Selection, 70, 78
Oxidative Decarboxylation, 130, 135, 143

P

Pathway Engineering, 43
Pleurotus Ostreatus, 145-146, 154
Polyhydroxyalkanoate (PHA), 156, 159, 170
Pre-treatment, 103, 145-146, 149-154, 159, 224
Promoter, 13, 15, 42, 97, 99, 102-104, 107-108, 110-112, 114-115, 175, 177
Protein Binding Capacity, 182, 189, 191
Pulping, 30, 32, 37, 39, 41, 193

R

Re-oxidation, 171
Recalcitrance, 30, 41-42, 57-58, 64, 70-71, 182-183, 188, 192, 210-211, 213-214, 216, 220-225
Reticulitermes Speratus, 85-86, 100
Rhodopseudomonas Palustris, 56, 171, 173, 180-181
Rubisco, 43-56

S

Saccharification, 30, 32-42, 83, 145-147, 149-152, 154, 192, 194-195, 202, 208, 224-225
Saccharomyces Cerevisiae, 4, 18, 85-86, 100-101, 143, 154, 207-208
Site-directed Mutagenesis, 17, 91, 99, 130, 138-139, 144
Substrate Specificity, 5, 13, 17, 86, 130, 134, 139
Sugarcane Bagasse, 38, 182-183, 186, 189, 192-193
Sulfate Reduction, 117-121, 124-127, 129
Switchgrass, 188, 210-218, 221-225

T

Temperature Effect, 195
Theoretical Ethanol Yield, 69-70, 72, 75-79, 81-82, 204
Thermochemical Pretreatment, 182, 195
Transcriptome, 102-103, 109, 115-116
Translocation, 102-110, 112, 114
Triacylglycerol, 1, 16-18
Trichoderma Reesei, 31-32, 40, 42, 102-104, 111-112, 114-116, 192

U

Unproductive Binding, 182-184, 188-191

V

Vib1, 102-105, 107-112, 114-115

X

Xylose Fermentation, 85-86, 89-95, 98,
100-101, 208

Y

Yeast Viability, 195